An Introduction to Radiation Chemistry

An Introduction to

Radiation Chemistry

Second Edition

J. W. T. SPINKS

President Emeritus
University of Saskatchewan
Saskatoon, Saskatchewan, Canada

Formerly, Head of the Chemistry Department,
Dean of Graduate Studies and President
University of Saskatchewan

R. J. WOODS

Associate Professor of Chemistry
University of Saskatchewan
Saskatoon, Saskatchewan, Canada

A Wiley-Interscience Publication
John Wiley & Sons
New York · London · Sydney · Toronto

Library of Congress Cataloging in Publication Data:

Spinks, John William Tranter, 1908–
    An introduction to radiation chemistry.

    "A Wiley-Interscience publication."
    Includes bibliographical references and index.
    1. Radiation chemistry.   I. Woods, Robert James,
joint author.   II. Title.
QD636.S64 1976   541'.38   75–46589
ISBN 0–471–81670–1

Printed in the United States of America

10 9 8 7 6 5 4 3 2 1

To Mary and Peggy

PREFACE TO THE SECOND EDITION

It is now ten years since the first edition of this book appeared, and the time seems ripe for a second edition. We have been encouraged to proceed with a second edition by the kind remarks about the first edition, received from all over the world, and by the instant success of a Russian translation. Additionally, enough changes have taken place in the subject of radiation chemistry in the last decade to warrant making some changes in content and emphasis; pulse radiolysis is now a well developed subject and forms part of an exciting new chapter on radiolysis kinetics, and radiation synthesis is attracting increasing attention as are also industrial applications. Nuclear power can also be said to have come of age, and an increasing number of people, connected in one way or another with the nuclear industry, need to have more than a passing knowledge of radiation chemistry. In this number we include nuclear engineers, nuclear physicists, reactor operators, and even members of Boards of Directors of power corporations!

This edition then is meant to be recent and thorough enough for the graduate student to read with profit and yet general enough to be read by those interested overall in the effects of radiation on materials. Acutally, the number of references now available is so large that an up-to-date coverage of the subject of radiation chemistry within the covers of one book is quite impossible. Thus no attempt has been made to make the coverage exhaustive. However, enough recent references have been given to provide points of entry into the literature.

Once more we are grateful to a number of our colleagues, in particular Drs. McCallum, Eager, Cormack and Walz, who read parts of the revised chapters, and Miss Muriel Stein, who took care of secretarial matters.

J. W. T. Spinks
R. J. Woods

Saskatoon, Saskatchewan
Canada
December 1975

PREFACE

This book arose from a course of lectures on radiation chemistry given to graduate students at the University of Saskatchewan over a number of years, initially by the senior author, Dr. Spinks, and then by Dr. Woods. At first the number of books and review articles bearing on the subject was extremely limited, and it is only within the last year or so that the log-jam has broken and a number of books, on specific aspects of the subject, have appeared, such as Allen's *Radiation Chemistry of Water and Aqueous Solutions*, and Charlesby's *Atomic Radiation and Polymers*. There is also the excellent series of publications, *Actions Chimiques et Biologiques des Radiations*, edited by Haissinsky. However, there is as yet no unified textbook on the subject, and the time seemed ripe to produce such a book, if only to spur others to write a better one.

A more serious reason is that the interest in the field and the number of publications related to it have increased enormously during the last decade. This is due in part to the intrinsic interest of the subject and in part to the imperative necessity of our understanding the fundamental principles underlying the action of radiation on matter in this atomic age. It is also due to the relative ease with which large sources of radiation can now be obtained and to an awakening interest of industry in the possibilities of launching a radiation-chemical industry.

Important, too, is the fact that one can now see the broad outlines of a definite "philosophy" of radiation chemical research. With the help of the physicist we can measure the amount of radiation energy absorbed by a given system. Thanks also to the physicist, we now have a good general idea of the series of physical events following the absorption of radiation. We can measure the final products of the reaction, and we can also detect with certainty and measure many of the reactive species involved as intermediates in the "chemical" part of the process, thus making it possible to give a reasonably accurate explanation of the courses of a given radiation chemical reaction.

Radiation chemistry has benefited enormously from a number of powerful new experimental techniques, such as e.p.r., mass spectrometry, gas chromatography, and isotope dilution, and has benefited from theoretical developments in chemical kinetics, such as the theory of chain reactions. In a sense radiation chemistry stands on the shoulders of a number of earlier disciplines such as chemical kinetics, photochemistry, spectroscopy, radiochemistry, and radiology, and even an initial publication in radiation chemistry can be expected to adopt a more definitive and unified approach than was possible for the early textbooks on chemical kinetics and photochemistry. The authors are strengthened in this view by a rather personal experience, extending over thirty-five years, in the field of photochemistry, spectroscopy, chemical kinetics, radiochemistry, and radiation chemistry, *as they were developing.*

It may be argued that such a broad experience must inevitably result in thinness of coverage, but it did, by good chance, happen to cover those subjects helpful to a thorough understanding of radiation chemistry.

The approach to the subject is, of necessity, academic since both authors are academics. However, the approach is at the same time practical, and reasonably down to earth in that the authors have had first-hand experience of most of the matters dealt with. The methodology of the subject is also clearly stated, and the industrial possibilities are clearly indicated. Indeed, the greatest possible gratification to the authors would be that this book should contribute, even in a modest way, to the widespread use of radiation in industry.

We wish to acknowledge our special indebtedness to C. J. Mackenzie, past President of the National Research Council of Canada, and to Sir John Cockcroft, one-time Director of the joint Canadian-United Kingdom Atomic Energy Project, who were responsible for our early entrance into atomic energy work, and to the late Dr. E. W. R. Steacie who, while President of the National Research Council, gave strong financial support to university radiation research programs. We are indebted in a personal way to a number of our colleagues at the University of Saskatchewan, in particular to Drs. Johns and Cormack of the Department of Physics, who introduced us to the mysteries of radiation energy absorption and measurement, and to Dr. K. J. McCallum, Head of the Department of Chemistry, who kindly read the greater part of the manuscript. The senior author (J.W.T.S.) is particularly grateful to his co-author for cheerfully accepting a much greater part of the writing of this book than had been initially agreed upon.

Finally we must record our particular appreciation to our wives, who showed great understanding and forbearance while the book was in the course of preparation.

J. W. T. Spinks
R. J. Woods

Saskatoon, Saskatchewan,
Canada
December 1963

CONTENTS

An Introduction to Radiation Chemistry

CHAPTER 1

Introduction

Development of Radiation Chemistry

Radiation chemistry may be defined as the study of the chemical effects produced in a system by the absorption of ionizing radiation. Included in this definition are the chemical effects produced by radiation from radioactive nuclei (α-, β-, and γ-rays),[1] by high-energy charged particles (electrons, protons, deuterons, etc.), and by electromagnetic radiation of short wavelength (x-rays with a wavelength less than about 250 Å, i.e., with an energy greater than about 50 eV).

Electromagnetic radiation of rather longer wavelength, in the ultraviolet and visible regions of the spectrum, may also initiate chemical reactions, though in this case ionization does not occur and reaction is brought about via electronically excited species. The reactions of these excited species, unaccompanied by ionization, make up the subject of photochemistry. The chief difference between radiation chemistry and photochemistry lies in the energy of the radiation that initiates the reaction, the energy of

[1] For clarity, the terms α-, β-, and γ-ray are confined to the radiations emitted by radioactive nuclei while similar radiations produced by other means are referred to as accelerated helium nuclei, accelerated electrons, and x-rays, respectively. The separation of the radioactive and artificial radiations in this way has some further justification in the case of β- and γ-rays, since their energy distributions are generally different from those of the corresponding artificial radiations (see Chapter 2).

1

the particles and photons concerned in radiation chemistry being very much greater than the energy of the photons causing photochemical reactions. Thus in photochemistry each photon excites only one molecule and, by the use of monochromatic light, it is often possible to produce a single, well-defined, excited state in a particular component in the system. The excited species are distributed essentially uniformly in any plane at right angles to the direction of the beam of light. In radiation chemistry each photon or particle can (via secondary electrons in the case of photons) ionize or excite a large number of molecules, which are distributed along its track (Fig. 1.1). The high-energy

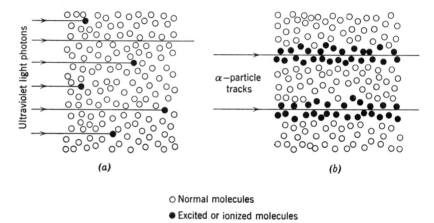

○ Normal molecules

● Excited or ionized molecules

FIGURE 1.1   *Absorption by matter of (a) ultraviolet light photons; (b) α-particles.*

photons and particles are not selective and may react with any molecule lying in their path, raising it to any one of its possible ionized or excited states. Subsequently the different energy-rich species react to give a complex mixture of products, in contrast to the relatively small number of products from a photochemical reaction. The radiation-chemical reaction may be complicated further by effects due to the high initial concentration of ionized and excited species in the tracks, particularly in the liquid and solid phases.

Ions and excited molecules are also formed by electric discharges in gases and give rise to chemical effects similar to those produced by ionizing radiation. However it is not a simple matter to measure the energy transferred to the active species and the experimental results are qualitative rather than quantitative; they are generally treated separately from radiation chemistry.

Closely related to radiation chemistry is the study of the chemical effects that accompany nuclear reactions, brought about, for example, by slow neutrons or by radioactive decay. In so far as the products are charged particles with excess kinetic energy, the ionization and excitation that they bring about in slowing down to thermal energies will initiate typical radiation-chemical changes. However, the nuclear reaction may make it possible to identify the nuclei involved and to trace the products in which they appear if, for example, the product nuclei are radioactive. The study of the chemical changes involving such nuclei is termed hot-atom chemistry and is a branch of nuclear chemistry.

In the past, radiation chemistry was sometimes called radiochemistry. This is generally avoided at present, and the term radiochemistry is confined to chemical studies with radioactive isotopes where the isotope is not used as a source of radiation.

DEVELOPMENT OF RADIATION CHEMISTRY

Radiation chemistry can be said to have had its origin with the discovery of x-rays by Röentgen in 1895 and of radioactivity by Becquerel in the following year. The discoveries followed the observations that discharge tubes and uranium salts respectively give off penetrating rays that can pass through opaque materials and activate photographic emulsions. Both x-rays and the rays from uranium were found to render air electrically conductive, and by measuring the rate of discharge of a charged electroscope, it was possible to estimate the intensity of the radiation. By comparing the amount of ionization produced by various uranium minerals and salts, Mme. Curie eventually discovered polonium and radium in 1898. The discovery of these elements and the isolation of radium in appreciable amounts was important to the study of radiation chemistry, since it made available a relatively powerful source of radiation, and it was not long before the chemical effects induced by the α-rays from radium and radium emanation (radon) were being examined.

One of the first reactions to be observed and studied was the action of the radiation from radium on water. Curie and Debierne (1901) found that a hydrated radium salt produced gas continuously, and Giesel (1902) observed the evolution of gas from aqueous solutions of radium bromide. Ramsay and Soddy (1903) showed that the gas evolved was a mixture of hydrogen and oxygen, which led Cameron and Ramsay (1907) to suggest that the action of the radiation might be analogous to the electrolytic decomposition of water, though the analogy did not hold in other cases, and an attempt to deposit copper from a solution of copper sulfate with the α-particles from radon was unsuccessful. The quantitative data on the decomposition of water reported by Ramsay and Soddy were used by Bragg (1907) in making the first comparison between the chemical and ionizing effects of α-particles. Bragg calculated that the number of molecules of water decomposed was approximately equal to the number of ions that would have been produced in air by the radiation. Three years later Mme. Curie proposed that the primary effect of high-energy radiation on any substance is the formation of ions, which are precursors of the chemical action.

The relationship between ionization and chemical action was put on a firmer basis by Lind (1) who studied the formation of ozone in oxygen under the influence of α-radiation, and was able to measure for the first time both the ionization produced and the amount of chemical change in the same medium. Yields of about 0.5 molecule of ozone per ion pair were found at this time. Lind (2) also collected the available data on the chemical action of α-rays and from them calculated values of the *ion-pair*, or *ionic*, *yield* (M/N), the ratio of the number of molecules undergoing change (M) to the number of ion-pairs formed (N), which was regarded as the radiation-chemical equivalent of the photochemical quantum yield. The results showed that ionization and chemical action were closely related and were proportional to each other.

It was later realized that the ion-pair yields were very often greater than the corresponding quantum yields and an explanation for this was found in the "ion-cluster" theory associated with the names of Lind (3) and Mund (4). The essential concept of this theory was that the ions could act as nuclei to which neutral molecules were drawn and held by polarization forces. When a cluster was neutralized by an ion, or cluster, of opposite charge, the heat of neutralization was believed to be shared between the molecules in the cluster, which might then all undergo chemical reaction. Since the number of molecules reacting depended on the size of the clusters rather than on the number of ions formed, ion-pair yields greater than unity could readily be explained. For example, acetylene is polymerized by α-particles to give a polymer, cuprene, with an ion-pair yield of about 20, and this could be visualized as taking place as follows:

$$C_2H_2 \xrightarrow{\alpha} C_2H_2^+ + e^-$$

$$C_2H_2^+ + 19C_2H_2 \rightarrow (19C_2H_2, C_2H_2^+)$$

$$(19C_2H_2, C_2H_2^+) + e^- \rightarrow C_{40}H_{40} \qquad (\text{cuprene})$$

(The symbol $\xrightarrow{}$ is used to distinguish reactions brought about by the absorption of ionizing radiation.)

While a great many experimental observations were explained satisfactorily on the basis of the ion-cluster theory, more recent work has raised several major objections both to the adoption of the theory as a general mechanism for radiation-chemical reactions and to the supposition that ions are the sole precursors of the chemical effects. In the first place, Eyring, Hirschfelder, and Taylor (5) calculated that large clusters were theoretically improbable and drew attention to the fact that the average energy lost in forming an ion pair in a gas (denoted by the symbol W) is appreciably greater than the first ionization potential (I) for the gas, determined by means of the mass-spectrograph; W is often about twice as large as I. They concluded that the excess energy ($W - I$) could be used to form electronically excited molecules, such as are produced in photochemical reactions. Both the ions and excited molecules were considered capable of giving rise to free radicals. Free-radical mechanisms based on these assumptions explained satisfactorily results obtained in earlier work on the α-particle-induced *ortho-para* hydrogen conversion and on the synthesis and decomposition of hydrogen bromide. Second, Essex and his co-workers (6) studied gaseous reactions induced by α-particles both in the presence and absence of an electric field. Their results showed that even in the presence of the field, when ion neutralization should be eliminated or very much reduced, the chemical yield is not greatly affected. Finally, both chemical and physical evidence has shown that atoms and free radicals, which are not accounted for on the basis of the ion-cluster theory, are present in a number of reaction systems. In consequence it is now accepted that both ions and excited molecules are formed initially during the absorption of the ionizing radiation and that free radicals are generally produced subsequently to play an important part in the chemical reactions that follow. Chemical reactions involving ions are not excluded, however, and ion-molecule reactions are believed to be important steps in many radiolyses.

During the early years of radiation chemistry, gas-phase studies proved most profitable but some work was also carried out on condensed systems. The chief source of radiation, radium, was generally used as a source of α-particles, though a few experiments were carried out with β- and γ-rays by using filters to remove the less-penetrating radiation. The sources available were not large and work was retarded by the long irradiation times needed to produce sufficient products for analysis (Kailan in Vienna irradiated samples for as long as 3 years); Kailan found the ion-pair yields in liquids to be much smaller than the corresponding quantum yields, in contrast to reactions in gases where the ion-pair yield was often greater than the quantum yield.

The development of powerful x-ray machines for industrial and medical purposes introduced an intense source of radiation that was more penetrating than α-particles and better suited to the irradiation of bulk liquid and solid samples. At the same time, interest in the biological effects of x-rays focused attention upon the chemical effects produced in water and aqueous solutions. For example, Fricke, in the United States, used x-rays to irradiate a variety of aqueous solutions, including solutions of simple organic compounds, and found that both oxidation and reduction might occur. In solutions, in contrast to reactions in gaseous systems, there was found to be a marked difference between radiation-chemical and photochemical reactions. Whereas in the latter energy is absorbed only by the solute, while the solvent itself is inert, in the former energy is absorbed by both solvent and solute and, at least in dilute solutions where the energy is absorbed almost entirely by the solvent, the direct action of the radiation on the solute molecules is relatively unimportant. This introduced the concept of *indirect action* in which, in a dilute solution, energy is absorbed from ionizing radiation by the solvent molecules and reaction between the "activated" solvent and the solute follows (7, 8), though the nature of the activation was not immediately known. Fricke (9) found that irradiation of aqueous solutions of ferrous sulfate, formic acid, and methanol with light of wavelengths between 2000 and 1850 Å produced oxidation similar to that produced by x-rays, suggesting that "activated water" was water in an excited state. Later work has led to the conclusion that irradiation of water gives hydrogen atoms (or some equivalent reducing species) and hydroxyl radicals, as suggested by Debierne in 1914 (10), and that the reactions of these radicals are responsible for most of the chemical results observed. However, highly excited water molecules may be intermediates in the formation of these radicals.

A good historical survey of the radiation chemistry of water appears in Chapter 1 of *The Radiation Chemistry of Water* by I. G. Draganic and Z. D. Draganić (11). A review of the development of the Fricke dosimeter is given by Allen (12). Allen has also written a history of the radiolysis of water (13). Hart and Platzman have written an historical development of radiation chemistry and the radiolysis of water (14).

The ion-pair yield, M/N, used as a measure of radiation-chemical yields in the gas-phase was also used when referring to yields in liquids, though in this case the value of N, the number of ions formed, was not known and had to be calculated assuming a value of W, the mean energy loss in forming an ion pair in the liquid. This itself was not known and the value of W for air (32.5 eV) was usually taken, without any particular justification. An alternative way of representing the yield is to relate it to the energy absorbed, which in principle can be measured directly, and the term G *value* was introduced to denote the number of molecules changed for each 100 eV of energy absorbed. Thus $G(X)$ refers to the number of molecules of a product X formed on irradiation per 100 eV of energy absorbed and $G(-Y)$ refers in the same way to the loss of a material Y that is destroyed on irradiation. Subscripts are sometimes added to further identify the G value, e.g., $G(X)_\alpha$ to represent the yield of X formed on irradiation with α-particles. The use of G values has the advantage that it does not imply, as does the ion-pair yield, that the chemical action is controlled by the number of ions formed, and G values have now become the customary means of expressing radiation-chemical yields. The G value is related to the ion-pair yield by the following expression,

$$G = \frac{M}{N} \times \frac{100}{W}$$

where W (eV) is the mean energy required to form an ion pair in the material being irradiated. Thus for ion-pair yields calculated assuming W to be 32.5 eV,

$$G \sim \frac{3M}{N}$$

Until about 1942 the number of workers studying radiation chemistry was limited; but in that year and succeeding years large groups were set up to meet the needs in this field of atomic energy programs. Indeed it was at this time that the subject received the name radiation chemistry, to distinguish it from the chemistry of the radioactive nuclides that retained the name radiochemistry.

With the development of atomic energy programs, a variety of particle accelerators became available and have been used to provide high-energy radiation for various specific radiation-chemical problems. Some machines, such as the Van de Graaf accelerator and the linear accelerator, have proved particularly adaptable to the needs of the radiation chemist. They have been especially useful in the study of the solvated electron and in the study of very fast reactions (see Chapters 5 and 7). The development of nano- and picosecond techniques has been a most exciting advance in the last decade and has had a major impact on other related areas of science.

The radiation chemist has also benefitted from the increasing availability of isotopes such as Co-60 and Sr-90 which provide intense, adaptable and relatively cheap sources of radiation. They have led to numerous studies in chemical radiation synthesis and to a few uses of radiation in chemical industry (Chapter 10).

REFERENCES

1. S. C. Lind, *Monatsh. Chem.*, **33**, 295 (1912); *Am. Chem. J.*, **47**, 397 (1912).
2. S. C. Lind, *J. Phys. Chem.*, **16**, 564 (1912).
3. S. C. Lind, *The Chemical Effects of Alpha-particles and Electrons*, Chemical Catalog Co., Inc., New York, 1928; G. Glockler and S. C. Lind, *The Electrochemistry of Gases and Other Dielectrics*, Wiley, New York, 1939; R. S. Livingstone and S. C. Lind, *J. Am. Chem. Soc.*, **58**, 612 (1936); S. C. Lind, *J. Chem. Phys.*, **7**, 790 (1939).
4. W. Mund, *L'Action Chimique des Rayons Alpha en phase Gazeuze*, Hermann et Cie, Paris, 1935; W. Mund, *Bull. Soc. Chim. Belg.*, **43**, 49 (1934).

5. H. Eyring, J. O. Hirschfelder, and H. S. Taylor, *J. Chem. Phys.*, **4**, 479, 570 (1936).
6. C. Smith and H. Essex, *J. Chem. Phys.*, **6**, 188 (1938); A. D. Kolumban and H. Essex, *ibid.*, **8**, 450 (1940); N. T. Williams and H. Essex, *ibid.*, **16**, 1153 (1948), **17**, 995 (1949); H. Essex, *J. Phys. Chem.*, **58**, 42 (1954).
7. O. Risse, *Ergeb. Physiol.*, **30**, 242 (1930).
8. H. Fricke, *Cold Spring Harbor Symp.*, **2**, 241 (1934).
9. H. Fricke and E. J. Hart, *J. Chem. Phys.*, **4**, 418 (1936).
10. A. Debierne, *Ann. Phys. (Paris)*, (9) **2**, 97 (1914).
11. I. G. Draganić and Z. D. Draganić, *The Radiation Chemistry of Water*, Academic Press, New York, 1971.
12. A. O. Allen, *Radiat. Res.*, **17**, 255 (1962).
13. *Actions Chimiques et Biologiques des Radiation* (Éd. M. Haissinsky), Masson et Cie, Paris, Vol. 5, P. 10 (1961).
14. Mechanisms in Radiobiology (Éd. M. Errera), Academic, New York, Vol. 1, pp. 181 and 93, 1961.

Additional references on the history of radiation chemistry:

15. M. Burton, *Chem. Eng. News*, **47**, 87 (February 10, 1969).
16. A. O. Allen, *J. Chem. Ed.*, **45** (5), 291 (1968).

CHAPTER 2

Sources of Radiation:
The Interaction of Radiation with Matter

Radioactive Isotope Sources • α-Ray • β-Ray • γ-Ray • Radioactive Cobalt • Nuclear Reactors • Spent Fuel Rods and Fission Products • Radioactive Decay • x-Ray Generators and High-Energy Particle Accelerators • x-Ray Machines • Resonant Transformer • Cockcroft Walton Accelerator • Van de Graaf Accelerator • Betatron • Cyclotron • Linear Electron Accelerator • Linear Ion Accelerator • Febetron • Radiation from Accelerators • Neutron Sources • Interaction of Radiation with Matter • Electrons • Energy Loss by Radiation • Energy Loss Through Inelastic Collisions • Elastic Scattering • Range of Electrons • Heavy Charged Particles • Range of Heavy Particles • Neutrons Elastic Scattering • Inelastic Scattering • Nuclear Reactions • Capture • Electromagnetic Radiation • The Photoelectric Effect • The Compton Effect • Pair-Production • Coherent Scattering • Photonuclear Reactions • Attenuation of Monoenergetic Radiations in Matter • Combination of Absorption Coefficients • Broad-Beam Attenuation • Build-Up of Secondary Radiation • Energy Absorption from Monoenergetic Radiations • Attenuation of x-Rays • Ionization and Excitation Produced by Radiation • Specific Ionization • Linear Energy Transfer • References •

The sources of radiation used in radiation-chemical studies can be divided into two groups, those employing natural or artificial radioactive isotopes, and those that employ some form of particle accelerator. The first group includes the classical radiation sources, radium and radon, and also isotopes such as cobalt-60, caesium-137, and strontium-90. The second group includes the familiar x-ray tube, the linear and Van de Graaff accelerators, the

9

betatron, the cyclotron, and other high-energy machines. Nuclear reactors may be considered as rather special members of the first group.

RADIOACTIVE ISOTOPE SOURCES

Table 2.1 lists the chief radioactive isotopes presently used as sources of radiation and the type and the energy of the radiation that they emit.

T A B L E 2.1   *Radioactive Isotopes Commonly Used as Sources of Radiation*

Isotope	Half-Life	Type and Energy (in MeV) of Principal Radiation Emitted	
Natural isotopes			
Polonium-210	138 days	α, 5.304 (100%)	
		γ, 0.8 (0.0012%)	
Radium-226	1620 years	α, 4.777 (94.3%)	
		α, 4.589 (5.7%)	
		γ, 0.188 ($\sim$4%)	
		(+ radiation from decay	
		products if present)	
Radon-222	3.83 days	α, 5.49	
		(+ radiation from decay	
		products if present)	
Artificial isotopes			
Caesium-137	30 years	β, 1.18 (max) (8%) $\Big\}$	0.24 (av)
		β, 0.52 (max) (92%)	
		γ, 0.6616 (82%)	
Cobalt-60	5.27 years	β, 0.314 (max)	0.093 (av)
		γ, 1.332	
		γ, 1.173	
Hydrogen-3 (tritium)	12.26 years	β, 0.018 (max)	0.0055 (av)
Phosphorus-32	14.22 days	β, 1.710 (max)	0.70 (av)
Strontium-90 + Yittrium-90	28.0 years (Y^{90}, 64 hr)	β, 0.544 (max) β, 2.25 (max)[a]	0.205 (av) 0.93 (av)[a]
Sulfur-35	87.2 days	β, 0.167 (max)	0.049 (av)

[a] From yttrium-90.

α-Ray

α-Particles are the nuclei of helium atoms, that is, helium atoms that have lost both electrons and hence have a double positive charge, $^4_2\text{He}^{2+}$. They are emitted by radioactive nuclei and have discrete energies that are characteristic of the radioisotope disintegrating. Polonium-210, for example, emits α-particles, each of which has an energy of 5.304 MeV.[1]

On passing through matter, α-particles lose energy principally by inelastic collisions with electrons lying in their path, leading to excitation and ionization of the atoms and molecules to which these electrons belong. The great difference in mass between the α-particle and the electron means that the α-particle loses only a small fraction of its energy and is virtually undeflected as a result of a collision. As a consequence α-particles are slowed down gradually as the result of a large number of small energy losses and travel in a very nearly straight path. Since each of the α-particles from a radioactive element has the same energy they will each have about the same range; the random nature of the collisions gives rise to small variations in the range of individual particles. This is shown by the solid curve in Fig. 2.1 where

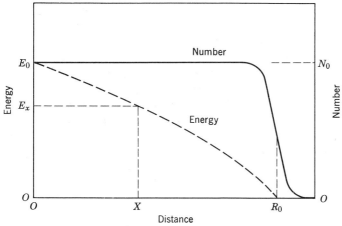

FIGURE 2.1 *Relationship between distance traveled and energy and number of α-particles.*

[1] One million electron-volts (1 MeV) is the kinetic energy an electron, or other singly charged particle, receives on being accelerated by a potential difference of one million volts. It should be noted that under these circumstances a helium nucleus, being doubly charged, would receive an energy increment of 2 MeV. The kilo electron-volt (keV) is equal to one thousandth of 1 MeV.

the number of α-particles is plotted against the distance traveled. If all particles had exactly the same range the curve would drop vertically (to $N = 0$) at the distance equal to the range. In practice some straggling occurs and the mean range (R_0) is found from the point of inflection of the curve.

The energy of the α-particles falls as the distance traveled increases. This is shown by the broken curve in Fig. 2.1 where the mean energy of all the α-particles is plotted against the distance they have traveled. If the α-particles are allowed to penetrate a thickness of material which is less than the mean range in that material they will be slowed down but not stopped. Advantage may be taken of this to obtain α-particles of lower energies than are normally available; for example, in Fig. 2.1 passage through an absorber of thickness X would lower the mean energy of the α-particles to E_x.

The energy lost by an α-particle in passing through matter produces considerable numbers of ions and excited molecules in the particle track. An α-particle from polonium-210, for example, gives a total of about 150,000 ion pairs and a rather greater number of excited molecules when slowed down in air. However, the extent of the chemical reaction that follows depends not only on the number of these active species formed but also upon their concentration in the particle track which, in turn, depends on the rate at which energy is lost as the particle is slowed down. The rate of energy loss is generally expressed in terms of the *linear energy transfer*, or LET, which is defined (1) as "the linear-rate of loss of energy (locally absorbed) by an ionizing particle traversing a material medium." The units in which LET values are expressed are usually kilo electronvolts per micron (keV μ^{-1}). Table 2.2 gives some values for the mean range in air and water and the average LET in water for some commonly used α-particles. The LET depends on the energy of the α-particle and increases as the particle is slowed down; however, a very rough approximation to the LET can be obtained by dividing the initial energy of the particle by its mean range. Values obtained in this way, and termed the "average LET," are included in Table 2.2. Such "average" values suffice to show the differences between the various types of radiation. A fuller discussion of LET is deferred until page 62.

T A B L E 2.2   *Range and LET Values for Alpha-Particles*

Isotope	Energy of α-Particle (MeV)	Mean Range in Air (15°C, 760 mm Hg) (cm)	Mean Range in Water (ref. 2) (μ)	Average LET in Water (keV μ^{-1})
Radium-226	4.795	3.3	33.0	145
Polonium-210	5.30	3.8	38.9	136
Radon-222	5.49	4.0	41.1	134

α-Particles are the least penetrating of the radiations from radioactive isotopes, but they have the highest LET values and are of particular value where a high LET is desired.

α-Particle sources may be constructed in various forms; one type, which employs polonium-210 as the active material, is shown in Fig. 2.2 (The Radiochemical Centre, Amersham, England). Polonium (1 mCi to 1 Ci)[2] is electrodeposited onto a gold foil disc (*A*) which is mounted in a stainless steel holder (*B*). A thin mica sheet (*C*) is cemented onto the front face of the holder to protect the polonium deposit, and a steel handle is provided that can be screwed into either the back or the side of the holder.

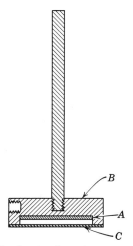

FIGURE 2.2   *Section through an α-particle source.*

Radon is also used as a source of α-radiation and may be used either mixed directly with the material to be irradiated or sealed in an extremely thin glass bulb which is surrounded by the material. The former has the advantage that it is possible to irradiate a relatively large bulk of material uniformly.

α-Particles can be produced in situ in the material to be irradiated by combining with it a compound of boron or lithium and irradiating the

---

[2] The *curie* (Ci) is the unit of radioactivity and is defined as the quantity of any radioactive nuclide in which the number of disintegrations per second is 3.700×10^{10}; this is very nearly equal to the rate of disintegration of 1 g of radium or of the radon in equilibrium with this weight of radium.

mixture with slow neutrons; the reactions $^{10}B(n, \alpha)^7Li$ and $^6Li(n, \alpha)^3H$ respectively occur with the liberation of α-particles and the formation of energetic nuclei.

Helium nuclei may also be accelerated to energies up to several hundred million electron volts by modern particle accelerators.

β-Ray

β-particles are fast electrons emitted by radioactive nuclei. In contrast to α-particles, the β-particles from a particular radioactive element are not all emitted with the same energy but with energies ranging from zero up to a maximum value (E_β) which is characteristic of the element.

The energy distributions, or energy spectra, of the β-particles emitted by phosphorus-32 and by an equilibrium mixture of strontium-90 and its daughter yttrium-90 are shown in Figs. 2.3 and 2.4, respectively. The maximum energy of the β-particles (E_β) determines the greatest range that the β-radiation will have in matter.

Monoenergetic electrons known as *conversion electrons* and *Auger electrons* are produced by radioactive decay processes distinct from that which gives the continuous β-energy spectrum. Conversion electrons may accompany γ-ray emission from the nucleus and may be treated as though they arise from interaction between the γ-ray and electrons in the atomic shells of the atom. Auger electrons arise from electron rearrangements after the production of an electron vacancy in an inner shell. The vacancy will immediately be filled by an electron from an outer shell and the excess energy lost either by x-ray emission or by the ejection of one of the other electrons

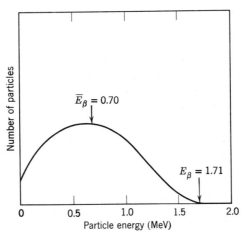

FIGURE 2.3   *Energy spectrum of phosphorus-32 β-particles (ref. 3).*

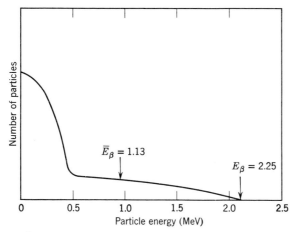

FIGURE 2.4 *β-Particle energy spectrum of strontium-90-yttrium-90 pair (ref. 3).*

from the atom (Auger electron). The process may be repeated if the electron which moved to fill the inner shell did not come from the outermost shell, and so a number of Auger electrons may arise from the original vacancy.

On passing through matter β-particles lose energy predominantly by inelastic collisions with electrons, in a similar manner to α-particles. However, because the β-particle and the electron with which it collides have the same mass, the β-particle can lose up to half of its energy in a single collision and may be deflected through a large angle. Deflection also occurs when the particle passes close to an atomic nucleus. As a result even β-particles that start with the same energy may come to rest at widely separated points. Thus β-particles have no fixed range in matter but show a maximum distance of penetration, or maximum range. The experimental number-distance curve for a beam of β-particles with energies ranging from zero to a maximum has the form shown in Fig. 2.5; the background count is due to bremsstrahlung, i.e., secondary x-rays (cf. p. 25). The maximum range, R_0, is taken as the point where the absorption curve merges with the background.

The range of β-rays and electrons is generally determined using aluminum absorbers; values for the maximum range of some β-rays in aluminum are given in Table 2.3 (3). Since the absorption and scattering of electrons is only slightly affected by changes in atomic number, ranges (in mg cm^{-2})[3] in

---

[3] The thickness of material penetrated by radiation is very often expressed as the *mass per unit area* (e.g., mg cm^{-2}) of the material. Expressed in this way the thickness is independent of the density of the material and is related more directly to the fundamental absorption process. There is also the practical advantage that the thickness of very thin absorbers is easier to determine by weighing than by direct measurement. Mass per unit area can be converted into units of length by dividing by the density of the material.

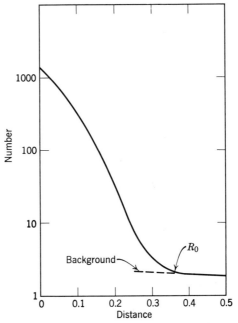

FIGURE 2.5   *Typical number-distance curve for β-rays. The number of β-particles transmitted through an absorber is plotted on a logarithmic scale against distance (the thickness of the absorber in g cm^{-2}) on a linear scale.*

other low-atomic number materials (e.g., water, tissue, organic compounds) are very nearly the same as the range in aluminum; the values for the maximum range in water in Table 2.3 were calculated by making the assumption that the ranges in water and aluminum were identical. Ranges in air are taken from ref. 4. It should be emphasized that these ranges are for β-particles with the maximum energy ($E_β$) alone and that only a very small fraction of the β-particles will actually penetrate to this distance; the average penetration for all the β-particles emitted is about one fifth of the maximum range. It should also be noted that the term range used in connection with β-particles and electrons is used here to denote the penetration of the radiation straight through the absorber. Since even the β-particles and electrons that penetrate furthest suffer some deflection from the straight path, the *path length*, measured along the curved path, will be rather greater than the distance to which the particle penetrates. Both path length and penetration are referred to as the range, but we reserve the term range for the penetration, and the path length is referred to as such. For heavy charged particles, such as α-particles, which travel along almost straight paths, the range (i.e., penetration) and path length are the same.

The average LET values given in Table 2.3 were calculated for β-particles with the maximum energy by dividing this energy by the path length (4) for the particles. Average LET values over the whole β-ray energy spectrum will be rather greater than those given because of the increased contribution from lower energy particles.

T A B L E 2.3   *Range and LET Values for β-Particles*

Isotope	Maximum Energy of β-Particles (MeV)	Path Length in Air (cm)	Maximum Range in Aluminum (cm)	Maximum Range in Water (cm)	Average LET in Water (keV μ^{-1})
Hydrogen-3	0.018	0.65	0.0002	0.00055	2.6
Sulfur-35	0.167	31.	0.012	0.032	0.52
Strontium-90	0.544	185.	0.066	0.18	0.27
Phosphorus-32	1.71	770.	0.29	0.79	0.21
Yttrium-90	2.25	1020.	0.40	1.1	0.20

β-Ray sources used at the present time employ the artificial radioisotopes shown in Table 2.1 which are used either as external sources of radiation or, mixed with the material to be irradiated, as internal radiation sources.

Strontium-90 β-ray sources similar in design to the α-ray source shown in Fig. 2.2 are available with activities ranging from 5 mCi to 1 Ci of strontium-90. In this case the active disc contains strontium-90 bonded in silver and protected by a surface layer of silver; the mica window is not needed. The strontium-90 is in equilibrium with its daughter, yttrium-90, and β-particles characteristic of each isotope are emitted. Though neither the strontium-90 nor the yttrium-90 emits any γ-radiation, secondary x-rays (bremsstrahlung) may be formed by interaction of the β-particles with high-atomic number materials in the source. The secondary radiation is unlikely to contribute appreciably to the energy absorbed by material being irradiated but, because of its greater penetrating power, may constitute a health hazard in the neighborhood of the source.

Phosphorus-32, sulfur-35, and hydrogen-3 (tritium) have been used as internal sources of β-radiation, Phosphorus-32 and sulfur-35 have the advantages of being readily available with high specific activities and of decaying sufficiently rapidly that there is no serious difficulty in disposing of the active residues.

γ-Ray

γ-Rays are electromagnetic radiation of nuclear origin with short wavelengths in the region of 3×10^{-9} cm to 3×10^{-11} cm. For the present purpose it is more convenient to describe the radiation in terms of energy than in terms of wavelength, since it is the energy absorbed from the radiation that is basically of interest. The relationship between wavelength and energy is

$$E = \frac{hc}{\lambda} \tag{2.1}$$

where h is Planck's constant, c is the velocity of light, and λ is the wavelength. Substituting for the constants

$$E \text{ (eV)} = \frac{12.4}{10^5 \, \lambda(\text{cm})} \tag{2.2}$$

In terms of energy the wavelength range 3×10^{-9} cm to 3×10^{-11} cm becomes approximately 40 keV to 4 MeV.

The γ-rays emitted by radioactive isotopes are either monoenergetic or have a small number of discrete energies as, for example, has cobalt-60 which gives equal numbers of γ-photons of energy 1.332 MeV and 1.173 MeV.

Unlike α- and β-particles, which lose their energy gradually through a number of small energy transfers, γ-rays tend to lose the greater part of their energy through a single interaction. The result is that whereas monoenergetic α-particles and electrons are slowed down by thin absorbers, rather than absorbed, in the same situation a part of the incident γ-rays are completely absorbed but the remainder are transmitted with their full initial energy. The number (N) of γ-photons transmitted through a sheet of absorbing material (under narrow-beam conditions; see p. 45) is given by

$$N = N_i e^{-\mu l} \tag{2.3}$$

where N_i is the number of incident photons, l is the thickness of the absorber, and μ is the total absorption coefficient of the material for γ-rays of the appropriate energy. This is shown graphically in Fig. 2.6.

γ-Rays do not have a definite range in matter; another term, the *half-thickness value* (or *half-value thickness*), is often used to relate the number of photons transmitted with the thickness of the absorber. The half-thickness value is the thickness of the absorber required to reduce the intensity of the γ-radiation (i.e., the number of photons transmitted) by one half. It can be calculated from Eq. 2.3 if the value of the absorption coefficient is known (the half-thickness value $= 0.693/\mu$), or read from a number distance curve (Fig. 2.6). Attenuation coefficients and half-thickness values are given in ref. 5.

Radioactive Cobalt

One of the most popular γ-radiation sources is cobalt-60. Cobalt-60 can be obtained with activities as high as 20 to 40 Ci g^{-1} though an activity of

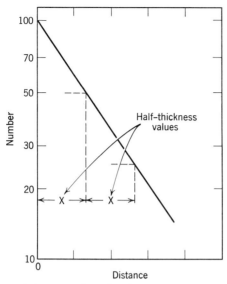

FIGURE 2.6   *Typical number-distance curve for γ-rays.*

1 to 5 Ci g^{-1} is more usual. The cobalt-59 from which it is prepared is generally irradiated in a reactor in the form of pellets, small slugs, or thin discs to give a uniformly active material, and these are assembled into radiation sources of the desired size. The containers in which the cobalt-60 is assembled also serve to filter out the β-rays emitted. Sources containing between 100 and 10,000 Ci are in common use in university laboratories and sources in the region of 1000 Ci for cancer therapy. Larger sources containing of the order of 100,000 to 1,000,000 Ci have been designed for irradiating materials on an industrial scale.

γ-Emitting isotopes must be surrounded by relatively thick shields of dense material to protect personnel working with them. Very many designs are available for cobalt-60 sources and the associated shielding, but, in general, they can be divided into two groups. In the first the cobalt is arranged to irradiate a cavity, in which the sample to be irradiated is placed inside a compact mass of shielding material, and some form of movable shielding is provided so that the sample can be introduced into the cavity without exposing the cobalt-60. The shielding is generally lead and such sources can be quite compact. An example is the Atomic Energy of Canada "Gammacell" which is shown in Fig. 2.7. In this the cobalt is in the form of a hollow cylinder into which the sample is introduced by means of the moving drawer; spiral tubes through the drawer allow wires and small tubes to be led into the

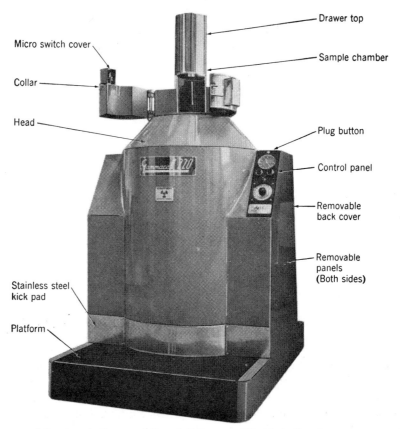

Drawer top

Micro switch cover

Sample chamber

Collar

Head

Plug button

Control panel

Removable back cover

Removable panels (Both sides)

Stainless steel kick pad

Platform

FIGURE 2.7   *Atomic Energy of Canada "Gammacell" cobalt-60 radiation source at the University of Saskatchewan. (Courtesy of Atomic Energy of Canada, Ltd., Commercial Prod. Div., Ottawa, Canada.)*

irradiation cavity without the escape of radiation. The "Gammacell" shown contains 1100 Ci of cobalt-60 and weighs about 7500 lb. A kilocurie source provides a total γ-ray power of 14.8 W.

The second group of cobalt-60 sources are the so-called cave-type in which the cobalt-60 is stored in a shielded container and, when required, moved out of the container and close to the sample in a small shielded room or cave. An example of this arrangement is shown in Fig. 2.8; in other facilities the cobalt-60 may be stored in the floor or walls of the cave itself. In this case it is necessary to provide adequate shielding for the room and also for the entrance, which is usually built in the form of a labyrinth. The common shielding material for cave-type sources is concrete, but the distance between

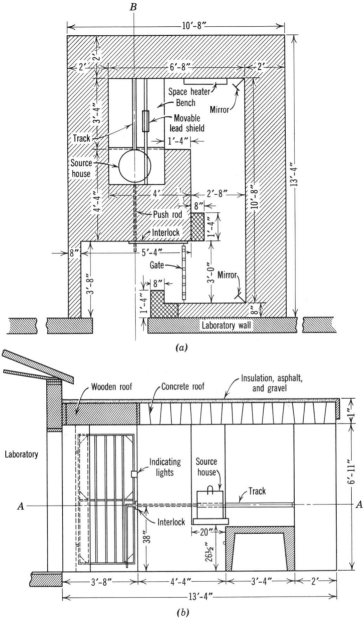

FIGURE 2.8 *Cave-type cobalt-60 irradiation facility at the University of Saskatchewan. A 90-Ci cobalt-60 slug is stored in the lead source house and in use is pushed out along the track by means of the push rod. (a) Horizontal cross-section; (b) Vertical cross-section. (Courtesy 1959 Nuclear Congress.)*

the active source and the outside of the shielding walls also contributes to the attenuation of the radiation because of the inverse-square relationship.

The cavity-type source has the advantage of compactness, so that it can be used in existing laboratories, but has the disadvantage that the size of samples and equipment that can be put into the cavity is limited. The intensity of the radiation inside the cavity is fixed. The cave-type of source offers a very much greater area for equipment and it is possible, by varying the distance between the cobalt source and the sample, to irradiate materials at different radiation intensities. A disadvantage of the cave-type source is the greater space required and the bulk and weight of the concrete shielding needed. All very large cobalt-60 sources (over about 10,000 curies) are of the cave type (6–9).

Nuclear Reactors

Fission of uranium in a nuclear reactor leads to the liberation of large amounts of energy of which about 80 to 85% appears as kinetic energy of the recoiling fission fragments, about 5 to 6% as the energy of neutrons and γ-rays produced during the fission reaction, and 5 to 6% as energy stored in radioactive fission products and liberated as β- or γ-radiation when these decay.

The range of the fission fragments is quite short, and the kinetic energy associated with them is normally dissipated as heat within the fuel elements of the reactor. Several proposals have been made for utilizing this energy to initiate chemical reaction, using intimate mixtures of chemicals and reactor fuel. Harteck and Dondes (10), for example, proposed incorporating uranium enriched with uranium-235 in glass-wool fibers a few microns in diameter and passing gaseous reactants through the fiber mat exposed to the neutron flux of a reactor. Whatever method is used the products will be contaminated with fission fragments and may also contain radioactive impurities formed by the absorption of neutrons; radioactive material from both sources must be removed before the product can be used.

Neutron and γ-radiation within, or near, the reactor core might also be used to initiate chemical reaction by pumping the material to be irradiated through these regions. Neutron absorption would again be a possible source of radioactive contamination. An alternative approach, in which the neutron flux in the core is used indirectly as a radiation source, is possible if the reactor is cooled with liquid sodium. The function of the sodium is to transfer heat from the reactor core to an external heat exchanger, after which it is recycled through the reactor. In the core the sodium will also absorb neutrons forming radioactive sodium-24 which emits γ-rays with energies of 1.37 and 2.75 MeV with a half-life of about 15 hr. The stream of radioactive sodium can be used to irradiate materials outside the reactor before being recycled through the core. Use of the sodium in this way will not affect the operation of the reactor and has the advantage that the material being irradiated is only exposed to γ-radiation and will not itself become radioactive. Other elements can, of course, be exposed to neutrons in the reactor to produce longer lived radioisotopes that can be used subsequently as radiation sources; the production of cobalt-60 from cobalt-59 is an example (11–15).

Spent Fuel Rods and Fission Products

Fuel rods that have been exposed in a nuclear reactor for long periods contain a large proportion of fission products and are extremely radioactive. When they are replaced by fresh fuel, the used fuel elements are allowed to stand for some time to let the short-lived activity decay; during this time they may be used directly as radiation sources, though the radioactivity decays rather rapidly, necessitating frequent replacement to maintain a reasonably constant radiation level (6, 16). After most of the short-lived radioactivity has decayed the fuel rods are dissolved and processed chemically to recover fissionable material, leaving a solution containing the fission fragments and various salts and impurities. The mixture, generally with water removed, is known as the gross fission product and may be processed further to separate groups of chemically similar radioisotopes, which are known as mixed fission products. The mixed fission products will probably have higher specific activities than the gross fission product, be more specific in the type and energy of the radiation emitted, and have a more stable half-life. A final step is to separate the mixed fission products into individual isotopes, such as caesium-137 and strontium-90; these are referred to as separated fission products. Both the mixed and separated fission products have found use as sources of β-and/or γ-radiation. One of the most attractive is caesium-137 which emits 0.66-MeV γ-rays and has a half-life of 33 years. The γ-radiation from 1 MCi of caesium-137 corresponds to 3.91 kW of radiation power (7). Strontium-90, in the form of $SrTiO_3$, has been advocated as a source of β-radiation.

Radioactive Decay

The activity of all radioactive radiation sources falls as the source grows older, and it is often necessary to make allowance for this. The activity (C_t) after a period of decay (t) is related to the original activity (C_0) by the expression

$$\frac{C_t}{C_0} = e^{-\lambda t} \tag{2.4}$$

where λ is the decay constant ($\lambda = 0.693$/half-life). If a table of exponential functions is not available C_t/C_0 can be calculated from

$$\log_{10}\left(\frac{C_t}{C_0}\right) = \bar{1}.(1.0 - 0.4343\lambda t). \tag{2.5}$$

For example, for cobalt-60, which has a half-life of 5.27 years or 63.24 months, λ is found to be 0.01096 month^{-1} and, by substitution in Eq. 2.5, the ratio C_t/C_0 to be 0.989 after 1 month, 0.947 after 5 months, 0.896 after 10 months, etc. Thus the activity of a cobalt-60 source falls by roughly 1% month^{-1}.

The expressions derived for the radioactive decay of a source apply equally to the fall in the intensity of the radiation from the source, so that Eqs. 2.4 and 2.5 can be rewritten

$$\frac{I_t}{I_0} = e^{-\lambda t} \tag{2.6}$$

and

$$\log_{10}\left(\frac{I_t}{I_0}\right) = \bar{1}.(1.0 - 0.4343\lambda t) \tag{2.7}$$

where I_0 and I_t are the initial intensity and the intensity after time t respectively.

X-RAY GENERATORS AND HIGH-ENERGY PARTICLE ACCELERATORS

Many of the high-energy particle accelerators developed for nuclear research have been used to study particular radiation-chemical problems. However, the very high energy radiation from these machines is not normally necessary, or even desirable, and the radiation chemist is more likely to be concerned with accelerators that are able to provide a very intense beam of radiation at moderate energies (the intensity may be several orders of magnitude greater than the intensity of the radiation available from a radioactive source). The main characteristics of x-ray machines and several particle accelerators are very briefly described below; fuller descriptions of the machines and their mode of operation are available in many physics textbooks and in almost all books on nuclear physics. The x-ray machine, cyclotron, Van de Graaff accelerator, and linear electron accelerator have been employed most frequently for radiation-chemical studies, but other accelerators are being developed for, and adapted to, this work as radiation becomes accepted as a useful industrial tool.

x-Ray Machines

x-Rays are electromagnetic radiation with wavelengths less than about 10^{-6} cm, i.e., with energies greater than about 0.1 keV. They are produced whenever high-speed electrons are rapidly decelerated, as when they pass through the electric field of an atomic nucleus. The radiation produced in this way is known as *bremsstrahlung* (i.e., braking radiation), and its production offers an alternative means of energy loss by electrons to loss by collision, mentioned earlier in connection with the stopping of β-rays.

The energy of the bremsttrahlung radiation ranges from near zero to the maximum energy of the incident electrons. The energy of an individual bremsstrahlung photon depends on the amount by which the electron producing it is slowed down; if the electron is brought to rest (i.e., to thermal energy), the photon will have the full energy of the electron. A typical energy spectrum for x-rays is shown in Fig. 2.9.

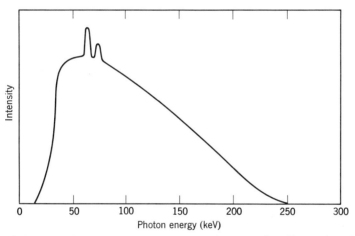

FIGURE 2.9   *Typical x-ray spectrum. (Energy distribution produced by 250-keV electrons striking a tungsten target; the characteristic radiation is not shown to scale.)*

Resonant Transformer

The resonant transformer is a form of x-ray generator in which the heavy iron core of the high-voltage transformer is absent and the secondary, high-voltage, circuit of the transformer, which is in series with the x-ray tube, is tuned to resonance. High voltages can be obtained with a great saving in weight and size in the transformer, and compact x-ray generators built on this principle are available with peak energies ranging from 100 to 3.5 MeV.

Cockcroft-Walton Accelerator

The Cockcroft-Walton voltage-multiplier circuit uses a series of rectifiers and condensers to rectify and multiply the output voltage of a transformer, each rectifier-condenser unit doubling the voltage. The potential difference generated in this way (up to about 1.5 MeV) may be used to accelerate positive ions or, though this is not common, electrons.

Van de Graaff Accelerator

In the Van de Graaff accelerator an electrostatic charge, which may be either negative or positive, is carried to a high-voltage electrode by means of a rapidly moving belt and by the potential difference between the electrode and the ground used to accelerate electrons or positive ions (protons, deuterons, helium nuclei, etc.) to high velocities.

The Van de Graaff accelerator is able to accelerate electrons or positive ions to any energy within the range of the machine (the maximum energy is generally between 1 and 5 MeV) and to maintain the energy constant to within about 0.1%, or better.

Betatron

The betatron is a circular accelerator into which electrons are injected as a pulse and then continuously accelerated and held in a circular path by a changing magnetic field. When the electrons have reached their full energy they are allowed to impinge on a target, producing x-rays, or are extracted from the machine as a beam. The process is repeated to give a pulsed beam of radiation with several hundred pulses a second, each lasting of the order of 1 μsec. Betatrons operating at maximum energies from 10 to 300 MeV have been constructed. Lower energy radiation than the maximum for the machine can be obtained by extracting the electrons before they have completed the full number of circles.

Cyclotron

In the cyclotron light ions are introduced into the center of a flat, evacuated, hollow disc which is divided into two halves along a diameter and situated between the poles of an electromagnet designed to produce a uniform field across it. By rapidly alternating the potential applied to the two halves, or dees, the ions can be accelerated along a spiral path which increases in diameter as the energy of the ions increases. The accelerated ions are extracted as they near the outer edge of the dees as an essentially continuous beam. Deuterium ions and protons have been accelerated in the cyclotron to about 20 MeV and helium nuclei to about 40 MeV; at energies much above these the increase in mass of the particles causes them to get out of step with the alternating potential, and modified forms of the cyclotron, the frequency modulated cyclotron, and the proton synchrotron are necessary if it is desired to accelerate the particles to higher energies.

Linear Electron Accelerator

The linear electron accelerator is a traveling wave accelerator; electrons are injected in pulses into a straight, segmented, waveguide and accelerated by the electric field of an electromagnetic wave that travels down the tube. Electron energies up to 630 MeV have been obtained in this way; lower energy linear electron accelerators are available commercially.

The accelerated electrons are delivered in pulses of a few nano- or microseconds duration, with a repetition rate of the order of 500 pulses $\sec^{-1}$.

Linear Ion Accelerator

In the linear ion accelerator positive ions are accelerated by passing through a series of tubes of increasing length to which an alternating potential is applied, the particles receiving an increment of energy each time they traverse a gap between two tubes.

Accelerators of this type have been used to accelerate the nuclei of light elements (helium to argon) to energies of about $10A$ MeV, where A is the atomic weight of the nucleus. These accelerated nuclei are of particular interest, since they give LET's an order of magnitude greater than those usually available with accelerated particles.

Febetron

See p. 192.

The Radiation from Accelerators

Table 2.4 gives the main characteristics of the radiation from the particle accelerators mentioned above. All the accelerators that give high-energy electrons can be used as x-ray sources by stopping the electrons with a heavy-metal target, e.g., tungsten. The x-rays produced will have a continuous distribution of energies from zero up to the energy of the incident electrons; the x-ray energies given in the table refer to this maximum, or peak, energy.

The terms pulsed and continuous beam refer to the fact that the radiation may be received either as a beam of constant intensity (i.e., continuous) or broken up into a number of pulses that are separated by periods in which no radiation is being received from the machine. Pulsed beams often give

TABLE 2.4   *Particle Accelerators*

Accelerator	Particles Accelerated or Radiation Produced	Energy (MeV)	Remarks
x-Ray machine	x-Rays	0.05–0.3	Pulsed beam unless constant potential power supply used, continuous energy spectrum.
Resonant transformer	x-Rays	0.1–3.5	Pulsed beam, continuous energy spectrum.
Cockcroft-Walton accelerator	Positive ions	0.1–1.5	Continuous beam, monoenergetic radiation.
Van de Graaff accelerator	x-Rays	1–5	Continuous beam, continuous energy spectrum.
	Electrons and positive ions	1–5	Continuous beam, monoenergetic radiation.
Betatron	x-Rays	10–300	Pulsed beam, continuous energy spectrum.
	Electrons	10–300	Pulsed beam, monoenergetic radiation.
Cyclotron	Positive ions	10–20	Essentially continuous beam, monoenergetic radiation.
Linear electron accelerator	x-Rays	3–630	Pulsed beam, continuous energy spectrum.
	Electrons	3–630	Pulsed beam, essentially monoenergetic radiation.
Linear ion accelerator	Positive ions	4–400	Pulsed beam, essentially monoenergetic radiation.
Radioactive sources			
	α-Ray	0–5.5	Monoenergetic.
	β-Ray	0.02–2.25	Continuous energy spectrum.
	γ-Ray	0.5–2	Monoenergetic or small number of discrete energies.

instantaneous radiation intensities much greater than are obtained with a continuous beam.

When positive ions are being accelerated the energies given in the table refer to singly charged ions; e.g., protons and deuterons. Ions with more than one positive charge will receive multiples of the energy shown, in proportion to their charge. A helium nucleus with two positive charges would, for example, be accelerated to twice the energy of a proton by the same accelerating potential. The properties of some accelerated particles are given in Table 2.5.

T A B L E 2.5   *Range and LET Values for Accelerated Particles*

Particle	Energy (MeV)	Range in Air (cm; 15°C, 760 mm Hg)	Range in Aluminum (mm)	Range in Water (mm)	Average LET in Water (keV μ^{-1})
Electron	1	405	1.5	4.1	0.24
	3	1400	5.5	15	0.20
	10	4200	19.5	52	0.19
Proton	1	2.3	0.013	0.023	43
	3	14	0.072	0.14	21
	10	115	0.64	1.2	8.3
	30	820	4.3	8.5	3.5
Deuteron	1	1.7	0.0096		
	3	8.8	0.049	0.088	34
	10	68	0.37	0.72	14
	30	480	2.65	5.0	6
Helium nucleus	1	0.57	0.0029	0.0053	190
	3	1.7	0.0077	0.017	180
	10	10.5	0.057	0.11	92
	30	71	0.375		

Neutron Sources

Neutrons are uncharged nuclear particles with a mass of one mass unit. They are not stable in the free state and, unless captured by a nucleus, decay spontaneously into a proton and electron with a half-life of about 13 min. Reaction with a nucleus is, however, their normal fate.

There are no long-lived[4] radioisotopes that emit neutrons directly with the exception of a few isotopes of heavy elements which undergo spontaneous

---

[4] Rather less than 1% of the neutrons released during a nuclear chain reaction are produced by the decay of radioactive fission products and appear up to about a minute after the fission event. These are known as *delayed neutrons*.

fission. Neutrons can, however, be produced by stopping accelerated positive particles or high-energy electromagnetic radiation with suitable target materials. A number of these nuclear reactions are listed in Table 2.6.

T A B L E 2.6   *Nuclear Reactions Used to Produce Neutrons*

Incident Radiation	Target	Threshold Energy (MeV)	Product
α-Particles	Beryllium-9		Carbon-12
	Boron-10 (18.8%)[a]		Nitrogen-13
	Boron-11 (81.2%)[a]		Nitrogen-14
	Lithium-7 (92.5%)[a]		Boron-10
Deuterons	Deuterium		Helium-3
	Tritium		Helium-4
	Beryllium-9		Boron-10
Protons	Tritium	1.19	Helium-3
	Beryllium-9		Boron-9
	Lithium-7 (92.5%)[a]	1.88	Beryllium-7
x- or γ-Rays	Deuterium	2.23	Hydrogen-1
	Beryllium-9	1.67	Beryllium-8
	Uranium-238	6.0	Uranium-237

[a] Natural abundance.

The threshold energy is the minimum energy that the incident radiation needs to initiate the nuclear reaction. Accelerated electrons produce similar reactions to x-rays via the bremsstrahlung radiation produced in the target.

The energy of the neutrons produced in nuclear reactions is governed solely by the energy balance of the reaction and depends on the type and energy of the incident radiation and on the target material; by selecting the appropriate reaction and conditions monoenergetic neutrons with any desired energy, within a broad range, can be obtained. Generally the efficiency of the nuclear reactions is not high, a large part of the incident radiation energy being lost by ionization and excitation in the target.

Small neutron sources may use radioisotopes as sources of α- or γ-radiation, or small ion sources producing accelerated deuterons in conjunction with a target containing tritium (17). The most frequently used radioisotope-target pairs are intimate mixtures of ^{226}Ra-Be, ^{210}Po-Be, and ^{124}Sb-Be, employing (α, n), (α, n), and (γ, n) reactions, respectively. Of these the polonium-beryllium source has the advantage that very little γ-radiation accompanies the neutrons. Larger neutron fluxes can be obtained by using particle accelerators to provide the incident radiation or from nuclear reactors.

Both radioisotopes and particle accelerators are widely used as radiation sources, and they are, to a certain extent, complementary. The radioisotope sources have the advantage of relatively low cost and of adaptability where small sources are required or large volumes are to be irradiated. The particle accelerators have a much greater range of energies and are able to produce much greater radiation intensities than are normally available from isotope sources. In addition the accelerators are able to accelerate particles that are not available naturally from radioactive isotopes.

INTERACTION OF RADIATION WITH MATTER

Some knowledge of the processes by which radiation interacts with matter is essential to an understanding of radiation-chemical phenomena, since the chemical effects are a direct consequence of the absorption of energy from the radiation. In this section the principal interactions of electrons, heavy charged particles, neutrons, and electromagnetic radiation with matter are outlined.

Electrons

Electrons interact with matter by a number of processes of which the most important are the emission of electromagnetic radiation and inelastic and elastic collisions. The relative importance of these processes varies strongly with the energy of the incident electrons and, to a smaller extent, with the nature of the absorbing material; at high energies energy is lost predominantly by radiation emission and at low energies through inelastic collisions. Elastic scattering (change in the direction of motion without conversion of kinetic energy to any other form of energy) is of greatest importance at low energies.

ENERGY LOSS BY RADIATION. High-speed charged particles passing close to the nucleus of an atom may be decelerated and, according to classical physics, will radiate electromagnetic energy (*bremsstrahlung*) with a rate, $-dE/dl$, proportional to $z^2 Z^2/m^2$, where z and Z are the charge on the particle and the nucleus respectively and m is the mass of the particle. Thus the energy loss by radiation will be greatest for light particles and for stopping materials of high atomic number.

For electrons, bremsstrahlung emission is negligible below 100 keV but increases rapidly with increasing energy, becoming the predominant mode of energy loss at an electron energy between 10 and 100 MeV (the exact energy depends on the stopping material). The bremsstrahlung energy spectrum extends from zero to the energy of the incident electrons and is, in fact, the continuous x-ray spectrum (Fig. 2.9), (see Bethe and Ashkin (18) and Heitler (19)).

ENERGY LOSS THROUGH INELASTIC COLLISIONS.   Charged particles can also lose energy in matter through Coulomb interactions with electrons of the stopping material. Interaction in this way, which produces excitation and ionization in the stopping material, is the dominant process whereby electrons are slowed down at electron energies below those at which bremsstrahlung emission occurs.

Bethe (18, 20) derived an expression for the energy loss from electrons by excitation and ionization

$$-\left(\frac{dE}{dl}\right)_{col} = \frac{2\pi N e^4 Z}{m_e v^2}\left[\ln\frac{m_e v^2 E}{2I^2(1-\beta^2)} - (2\sqrt{1-\beta^2} - 1 + \beta^2)\ln 2\right.$$

$$\left. + 1 - \beta^2 + \tfrac{1}{8}(1-\sqrt{1-\beta^2})^2\right]\text{ergs cm}^{-1} \qquad (2.8)$$

where v is the velocity of the electron in cm sec^{-1}, β is v/c, I is the mean excitation potential for the atoms of the stopping material in ergs, N is the number of atoms per cubic centimeter, e is the charge on the electron in electrostatic units, m_e is the rest mass of an electron in grams, c is the velocity of light in cm sec^{-1}, and Z is the atomic number of the stopping material.

The energy loss per unit path, $-(dE/dl)_{col}$, is known as the *specific energy loss* or *stopping power*. It should be noted that this is a function of the electron velocity, and changes as the electron is slowed down. A quantity which will be referred to later is the *mass stopping power*, S/ρ, given by

$$S/\rho = -\left(\frac{dE}{dl}\right)_{col} \times \frac{1}{\rho}\text{ ergs cm}^2\text{ g}^{-1} \qquad (2.9)$$

where ρ is the density of the material.

The mean excitation potential, I, takes into account the effect of the electron binding energies on the energy loss, and is given by $I = kZ$, where k is a "constant" that can be determined experimentally for different elements and which lies between 8 and 16. Experimental values of I are determined using heavy particles, since the strong scattering of electrons in matter make them unsuitable for this determination.

For an electron of energy E MeV the ratio of the energy loss by radiation to the loss by collision is given roughly (in the MeV range of energies) by

$$\frac{(dE/dl)_{rad}}{(dE/dl)_{col}} \approx \frac{EZ}{1600 m_e c^2} \qquad (2.10)$$

ELASTIC SCATTERING.   Charged particles may be deflected by the Coulomb (electrostatic) field of an atomic nucleus. This is particularly important in the case of electrons because of their small mass, and they frequently experience such elastic scattering. Scattering is greatest for low electron energies and for high atomic number materials.

RANGE OF ELECTRONS.   The range, or penetration, of β-particles (groups of electrons of mixed energy) has been discussed briefly on p. 15. Monoenergetic electrons are also characterized by an indefinite range due to frequent scattering and the possibility of large energy losses; a typical number-distance curve for monoenergetic electrons is shown in Fig. 2.10. The extrapolated or practical range, R_p, is found by extrapolating the more or less linear portion of the curve to intersect the background and the maximum range, R_0, from the point where the curve merges with the background. Both ranges are characteristic of the original electron energy.

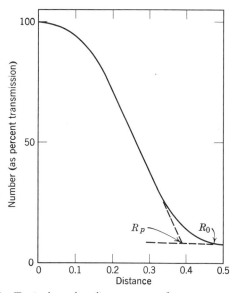

FIGURE 2.10   *Typical number-distance curve for monoenergetic electrons.*

Several empirical formulae have been used to relate the range and energy of electrons. Katz and Penfold (21) for example found that in aluminum both the maximum range of β-rays and the extrapolated range of monoenergetic electrons could be represented by

$$\text{range (mg cm}^{-2}) = 412E^n \tag{2.11}$$

where E (MeV) is the maximum β-ray energy or the energy of the mono-energetic electrons and $n = 1.265 - 0.0954 \ln E$, for energies from 0.01 to 2.5 MeV. For energies from 2.5 to about 20 MeV

$$\text{range (mg cm}^{-2}) = 530E - 106 \tag{2.12}$$

Though these formulae fit the data for aluminum best they also apply reasonably well to other materials, especially other light elements, since the range (in mg cm^{-2}) varies only slightly with atomic number.

The mean energy loss due to ionization and excitation and the range for electrons and positrons in a number of materials have been tabulated by Nelms (22), and others (22a, b, c).

Heavy Charged Particles

Heavy charged particles (protons, deuterons, α-particles, etc.) interact with matter in the same way as electrons, by bremsstrahlung emission, inelastic collisions, and elastic scattering. However, bremsstrahlung emission is only important at very high energies (of the order of 1000 MeV), and elastic scattering is relatively unimportant, so that energy loss is principally by inelastic collisions with the electrons of the stopping material. The rate of energy loss is inversely related to the velocity of the particle so that if two particles of equal energy but different mass are compared, the heavier will have the smaller velocity and hence a larger linear rate of energy loss. The ion density along the track of an α-particle, for example, is several hundred times greater than that along the track of an electron of the same energy.

RANGE OF HEAVY PARTICLES.   A typical number-distance curve for heavy particles is shown in Fig. 2.11. The broken line is obtained by differentiating the solid curve with respect to distance and represents the distribution of the range of individual particles about the mean range, R_0. The mean range can be found from the position of the maximum of the differential curve or the point of inflection of the solid, integral, curve. In practice the extrapolated range, R_e, is more readily determined (by drawing a tangent to the steepest part of the number-distance curve and extrapolating it to cut the distance axis) and the mean range calculated from this (18, 23). The difference between the mean and extrapolated range is a measure of the straggling of the particles and usually amounts to about 1 or 2% of the total range.

Data on the range of heavy particles are available from a number of sources, (18, 22a, 23–25). Frequently ranges are given for only one type of particle. However, the range of other heavy particles can be found quite readily, since for particles of the same velocity the linear rate of energy loss is proportional to the square of their charge (except at very low energies).

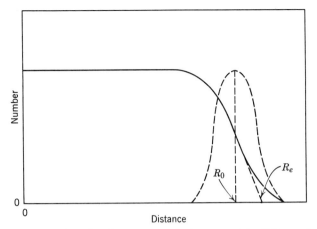

FIGURE 2.11   *Typical number-distance curve for heavy particles. (The number of particles left in the beam is plotted against the distance from the source.)*

The range is proportional to $E/(-dE/dl)$ and by considering particles with the same velocity the expressions

$$\text{range of particle } A = \frac{m_A Z_B^2}{m_B Z_A^2} \times \text{range of particle } B \qquad (2.13)$$

when

$$\text{energy of particle } A = \frac{m_A}{m_B} \times \text{energy of particle } B \qquad (2.14)$$

are obtained relating the range of particles A and B of mass and atomic number m_A and Z_A, and m_B and Z_B, respectively, in the same stopping material. For example, the range of an α-particle ($m = 4, Z = 2$) is related to the range of a proton ($m = 1, Z = 1$) of energy E_p by

$$\text{range of } \alpha\text{-particle} = \frac{4}{4} \times \text{range of proton} \qquad (2.15)$$

$$(\alpha\text{-particle energy} = 4E_p)$$

i.e., in a particular stopping material the range of an α-particle is equal to the range of a proton with one quarter the energy (both Eqs. 2.13 and 2.14, are required to obtain this result). Ranges for deuterons ($m = 2, Z = 1$) and α-particles in air calculated from the range of protons in air (6) using Eq. 2.13 and 2.14 are tabulated in Table 2.7. The calculated α-particle ranges are

TABLE 2.7   *Range (centimeters) of Protons,*
Deuterons, and α-Particles in Air (15°C, 760 mm Hg)

Energy (MeV)	Proton	Deuteron	α-Particle
0.5	0.86	—	—
1.0	2.30	1.72	—
2.0	7.20	4.61	0.86 (1.05)
4.0	23.1	14.4	2.30 (2.49)
8.0	77.3	46.2	7.20 (7.35)

smaller by a small constant value than the ranges found experimentally (shown in parentheses). This constant, found by experiment to be 0.20 cm, takes into account the capture and loss of electrons at low energies, which affects α-particles more than protons. Other multi-charged particles require a similar correction at low energies for exact results.

Neutrons

Neutrons, since they are uncharged, do not produce ionization directly in matter but interact almost exclusively with atomic nuclei. However, the products of neutron interactions often do produce ionization and thus give rise to typical radiation-chemical changes. The main ionizing species are protons, or heavier positive ions, and the chemical effects of neutron irradiation are similar to those produced by beams of these charged particles. Neutrons can, however, penetrate much greater thicknesses of material, and the consequences of neutron irradiation are not confined to the surface regions of the absorber.

The main processes by which neutrons and nuclei interact are elastic scattering, inelastic scattering, nuclear reaction, and capture. The type of interaction taking place depends to a large extent on the energy of the neutron, and for this reason it is convenient to divide neutrons into groups according to their energy. The groups are (26): thermal neutrons, in thermal equilibrium with their surroundings and having energies of the order of 0.025 eV at room temperature; intermediate neutrons, with energies from 0.5 to 10 keV; fast neutrons, with energies from 10 to 10 MeV; and relativistic neutrons with energies greater than 10 MeV.

ELASTIC SCATTERING.   Elastic scattering is the most probable interaction of fast neutrons with matter and is also important for neutrons in the intermediate range of energies. The energy of the incident neutron is shared between the recoiling neutron and nucleus according to the laws of conservation of energy and momentum, and, apart from gaining kinetic energy, the nucleus is unchanged. The fraction of the energy of the neutron that is transferred to the recoiling nucleus (atomic weight A) varies from zero up to a maximum which is given by

$$\left(\frac{\Delta E}{E_0}\right)_{max} = \frac{4A}{(A + 1)^2} \tag{2.16}$$

This fraction is greatest for hydrogen ($A = 1$) and decreases with increasing atomic weight. In biological tissues and other materials containing a large proportion of hydrogen, the most important interaction of fast neutrons is elastic scattering with hydrogen nuclei. The recoiling hydrogen nuclei (protons) subsequently produce excitation and ionization in the substrate.

The slowing down of neutrons in the moderator of a nuclear reactor is due mainly to elastic scattering and is accomplished most efficiently by low atomic weight materials. Thus the average number of collisions to reduce the energy of a neutron from 2 MeV, the average energy of neutrons produced by the fission process, to thermal energy (0.025 eV) is 18 in hydrogen, over 100 in graphite, and about 2000 in lead (27). The cross section for elastic scattering increases as the energy of the neutron decreases.

INELASTIC SCATTERING.   Inelastic scattering occurs if the neutron is absorbed by the nucleus and then a neutron with lower energy re-emitted, leaving the original nucleus in an excited state which eventually returns to the ground state by emitting one or more γ-rays. Inelastic scattering is not possible for neutron energies below that of the lowest excited state of the nucleus (generally a few hundred kiloelectron volts) but increases in importance as the neutron energy is increased; it may be as probable as elastic scattering at energies above about 10 MeV. Inelastic scattering can be considered as a (n, n) reaction.

NUCLEAR REACTIONS.   At high energies the colliding neutron may be incorporated into the nucleus and another particle such as a proton or α-particle emitted. Nuclear reactions (other than capture) only occur when the neutron has an energy in excess of the threshold energy for the reaction, and they are generally not significant at neutron energies below several megaelectron volts. The energy of the emitted particle depends not only on the neutron energy but also on the energy liberated or absorbed by the nuclear reaction taking place.

One nuclear reaction that can take place at thermal neutron energies is reaction with the mass-10 isotope of boron, $^{10}B(n, \alpha)^7Li$, and this is widely used in detectors for low-energy neutrons. The same reaction may also be used to produce α-particles and recoiling ^7Li nuclei in situ in a material containing boron, by irradiating it with low-energy neutrons. A second important thermal-neutron reaction is that with nitrogen-14 to give carbon-14, $^{14}N(n, p)^{14}C$.

CAPTURE.   At thermal energies the most probable interaction process is capture of the neutron by the nucleus to give an isotope of the target element.

Often the compound nucleus is formed in an excited state and returns to the ground state with the emission of one or more γ-rays; the reaction in this case can be classed as a (n, γ) reaction.

The cross section for capture is generally low at high neutron energies but frequently rises in a series of (resonance) peaks at energies within the intermediate range, increasing again in the thermal region in inverse proportion to the neutron velocity.

In biological tissue and similar systems (e.g., organic compounds containing only C, H, O, and N) the important reactions are $^1H(n, \gamma)^2H$ which yields a 2.2 MeV γ-ray, and $^{14}N(n, p)^{14}C$, which produces a 0.66 MeV proton; the capture cross sections of oxygen and carbon are small.

Electromagnetic Radiation

Unlike charged particles, which generally lose energy almost continuously through a large number of small energy transfers as they pass through matter, photons of electromagnetic radiation tend to lose a relatively large amount of energy whenever they interact with matter. However, not all incident photons will interact with any finite thickness of matter, and those that do not interact suffer no change and are transmitted with their original direction and energy. The effect of the absorbing matter (neglecting for the moment the photons that do interact) is therefore to reduce the number of photons transmitted, and thus to reduce the *intensity* of the radiation passing through it. By intensity is meant the radiation energy (the number of photons multiplied by their average energy) passing through a sphere of unit (maximum) cross-sectional area in unit time at the point of interest; the units of intensity are generally ergs per square centimeter per second.

The reduction in electromagnetic radiation intensity (dI) on passing through a small thickness (dl) of absorber is given by

$$dI = -I_i \mu \, dl \qquad (2.17)$$

where I_i is the intensity of the incident radiation and μ is the total *linear attenuation coefficient* (units cm^{-1}) of the material. The linear attenuation coefficient is thus the fraction of the incident photons diverted from the incident beam by unit thickness of absorber; it is a constant for a given material and for radiation of a given energy but varies from material to material and for different photon energies. Equation 2.17 only applies when dI and dl are very small but integration gives an expression which is not restricted in this manner

$$I = I_i e^{-\mu l} \qquad (2.18)$$

where I is the intensity of the radiation transmitted through a thickness l of absorber.

If the linear absorption coefficient is divided by density the *mass attenuation coefficient* (symbol μ/ρ, units $cm^2\ g^{-1}$) is obtained, which is independent of the density and physical state of the material. Also useful are the *atomic attenuation coefficient* (symbol $_a\mu$, units $cm^2\ atom^{-1}$) and the *electronic attenuation coefficient* (symbol $_e\mu$, units $cm^2\ electron^{-1}$) which are the absorption coefficients per atom and per electron respectively. They are related to the linear and mass attenuation coefficients in the following manner

$$_a\mu = \frac{\mu A}{\rho N_A}\ cm^2\ atom^{-1} \tag{2.19}$$

$$_e\mu = \frac{\mu A}{\rho N_A Z}\ cm^2\ electron^{-1} \tag{2.20}$$

where ρ is the density, A the atomic weight and Z the atomic number of the stopping material, and N_A is Avogadro's number. The atomic and electronic absorption coefficients have the dimensions of an area and are often referred to as *cross sections*. The term cross section is used interchangeably with attenuation coefficient to denote the probability of attenuation. Numerical values of the atomic and electronic attenuation coefficients or cross sections are of the order of $10^{-24}\ cm^2\ atom^{-1}$ or $electron^{-1}$ and are often given in terms of b $atom^{-1}$ or b $electron^{-1}$ where 1 b (barn) is $10^{-24}\ cm^2$.

The total attenuation coefficient is the sum of a number of partial coefficients representing various processes of attenuation. These processes are the photoelectric effect, the Compton effect, pair production, coherent scattering, and photonuclear reactions. The first three of these processes are the most important, though the relative importance of each process depends very much on the photon energy and the atomic number of the stopping material.

THE PHOTOELECTRIC EFFECT.   Low-energy photons interact mainly by the photoelectric effect. In this type of interaction the entire energy of the photon (E_0) is transferred to a single atomic electron, which is ejected from the atom with an energy equal to the difference between the photon energy and the binding energy (E_s) of the electron in the atom,

$$E_e = E_0 - E_s \tag{2.21}$$

At low photon energies the electrons are ejected predominantly at right angles to the direction of the incoming photon, but as the energy increases the distribution shifts increasingly toward the forward direction. Energy and momentum must be conserved, and this is made possible by the recoil of the remainder of the atom, so that photoelectric interaction is not possible with free electrons.

If the incident photon has sufficient energy it is generally the most tightly bound electron in the atom that is ejected, i.e., an electron from the K shell. Interactions with K-electrons account for about 80% of the photoelectric process for photons with energies greater than the K shell binding energy; most of the remainder of the interactions are with L-electrons. The vacancy created by loss of an electron from an inner atomic shell will be filled by an electron from an outer shell, with emission of characteristic x-radiation (fluorescent radiation) or of low-energy Auger electrons. For low atomic number (often referred to as "low-Z") materials the binding energy of the inner electron shells is relatively small (e.g., of the order of 500 eV for water), and the secondary x-rays and electrons will have low energies and will be absorbed in the immediate vicinity of the original interaction. On the other hand, materials of high atomic number may give moderately energetic secondary radiation (e.g., for tungsten, the secondary x-rays have energies up to 70 keV) which can travel some distance from the original interaction before being completely absorbed.

Photoelectric interactions are most probable for high atomic number materials and for low photon energies. The *electronic attenuation* coefficient varies from element to element approximately as Z^3 and for a given element decreases rapidly with increasing photon energy. Photoelectric absorption shows a marked increase as the photon energy increases from a value just below the K shell binding energy to a value just above this energy, a plot of the photoelectric attenuation coefficient against photon energy having a distinct edge at the binding energy (cf., Fig. 2.16).

THE COMPTON EFFECT.   In the Compton effect a photon interacts with an electron, which may be loosely bound or free, so that the electron is accelerated and the photon deflected with reduced energy. The energy and momentum of the incident photon are shared between the scattered photon and the recoil electron.

From the equations for the conservation of energy and momentum it is possible to calculate three of the four variables, θ, ϕ, E_γ, and E_e (Fig. 2.12) representing the Compton process if the energy of the incident photon is known and the remaining variable is assumed. For example, the energy of

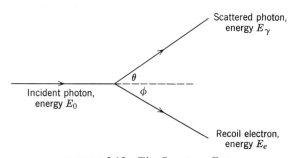

FIGURE 2.12   *The Compton effect.*

the scattered photon is related to the angle θ by

$$E_\gamma = \frac{E_0}{1 + (E_0/m_e c^2)(1 - \cos \theta)} \tag{2.22}$$

where E_0 and E_γ are the energies of the incident and scattered photon respectively and $m_e c^2$ is the rest energy of the electron. This equation shows that when the angle θ is small, the photon is scattered with little reduction in energy and that the greater the deflection θ the greater the energy loss from the photon.

The energy of the recoil electron is equal to the difference between the energy of the incident and scattered photon

$$E_e = E_0 - E_\gamma \tag{2.23}$$

and may have any value from zero up to a maximum which can be calculated from Eqs. 2.22 and 2.23 by putting $\theta = 180°$. The most probable electron energies are those near zero and near the maximum energy, energies near the maximum being particularly favored when the energy of the incident photon is high. The direction in which the electron recoils is predominantly that of the incident photon, and is more nearly so the higher the fraction of the photon energy it carries away and the higher the incident photon energy.

The probability of the photon being scattered with a definite energy or direction and the probability of Compton interaction as a whole were derived quantum mechanically by Klein and Nishina (28). The total electronic Compton absorption coefficient $_e\sigma$ gives the fraction of the photons of energy E_0 which interact by the Compton process, per electron cm^{-2}. An appreciable part of the energy of these photons is retained by the scattered photons, and the fraction of the incident photon energy, per electron cm^{-2}, that the scattered photons retain is given by the Compton *scattering coefficient*, $_e\sigma_s$. The fraction of the incident photon energy transferred to the recoil electrons, per electron cm^{-2}, is given by the *true* or *energy* Compton absorption coefficient, $_e\sigma_a$, where

$$_e\sigma_a = {_e\sigma} - {_e\sigma_s} \tag{2.24}$$

The ratio of the true to the scatter absorption coefficient varies with the energy of the incident photons (Fig. 2.13).

Compton electronic attenuation coefficients are independent of the atomic number of the stopping material and can be applied to any material, though separate coefficients must be determined for each different incident photon energy. Conversion of the electronic attenuation coefficients to the atomic, mass, or linear coefficients will involve one or more of the constants, atomic number, atomic weight, and density of the stopping material.

Compton interactions predominate for photon energies between 1 and 5 MeV in high atomic number (high-Z) materials and over a much wider

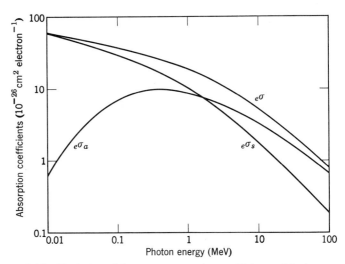

FIGURE 2.13   *Variation of Compton attenuation coefficients with photon energy.*

range of energies in low-Z materials. In water, for example, Compton interactions predominate from about 30 keV to 20 MeV.

Graphs of the angular and energy distributions for the Compton process have been published by Nelms (29).

PAIR-PRODUCTION.   Pair production involves the complete absorption of a photon in the vicinity of an atomic nucleus or, less frequently, an electron[5] with the production of two particles, an electron and a positron (Fig. 2.14). The energy of the photon less the rest energies of the two particles (each $m_e c^2$) is divided between the kinetic energy of the electron and positron (the small amount of energy transferred to the nucleus is nearly always neglected), i.e.,

$$E_0 = E_e + E_p + 2m_e c^2 \qquad (2.25)$$

Momentum is shared by the recoiling nucleus. The positron is slowed down in a similar manner to an electron and eventually combines with an electron, the two particles being replaced by two 0.51 MeV γ-rays (*annihilation radiation*) emitted in opposite directions.

Pair production cannot occur at photon energies less than 1.02 MeV (i.e., $2m_e c^2$). Above this energy the *electronic* attenuation coefficient or cross

[5] When the photon is absorbed in the field of an electron, this electron also is set in motion and hence the process is often referred to as *triplet production*. Triplet production is less common than pair production and will not be considered here, the discussion which follows applies particularly to pair-production in the field of a nucleus.

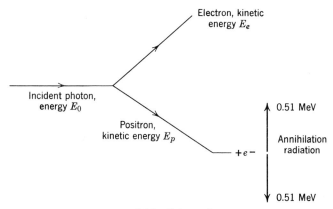

FIGURE 2.14 *Pair-production.*

section varies with the photon energy and is very nearly proportional to the atomic number (Z) of the stopping material. Since the atomic cross section is equal to the electronic coefficient multiplied by Z (Eqs. 2.19 and 2.20) the atomic cross section will vary as Z^2.

COHERENT SCATTERING. A photon may be scattered with little loss of energy by several processes, the chief process involving interaction with the atomic electrons (*Rayleigh scattering*). Rayleigh scattering is most probable at low photon energies (less than 0.1 MeV) and in high-Z materials; it is included in Fig. 2.15 for water and Fig. 2.16 for lead as *coherent scattering*. The term coherent refers to the fact that the effects combine coherently, i.e., by the addition of amplitudes, since there is a definite phase relationship between the incident and scattered radiation. Compton scattering effects combine incoherently by the addition of the intensities of the effects, since there is a random phase relationship between the incident and scattered radiation; hence the term *incoherent scattering* that is sometimes applied to Compton scattering.

Rayleigh scattering occurs in an energy range where the photoelectric cross section is large, and it can often be neglected without introducing a very large error. Furthermore, the angle of deflection is generally small and since it is not accompanied by any significant energy loss the scattered photon will only be distinguished from the primary beam under "narrow-beam" conditions (see p. 45).

In crystals, coherent scattering may be particularly intense in certain directions due to interference between waves scattered by different atoms; this is the basis for the determination of crystal structure by x-ray diffraction.

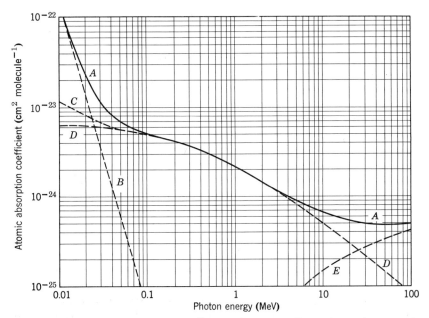

FIGURE 2.15 *Atomic attenuation coefficients for water. Curve A: total attenuation coefficient (with coherent scattering). B: photoelectric attenuation coefficient. C: total Compton coefficient (with coherent scattering). D: total Compton coefficient (without coherent scattering). E: pair-production coefficient.*

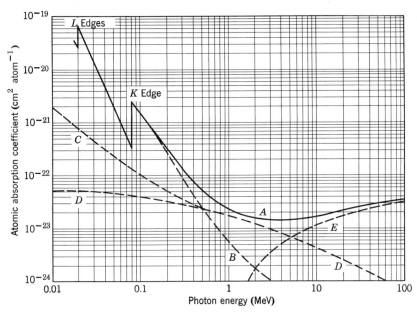

FIGURE 2.16 *Atomic attenuation coefficients for lead.*

PHOTONUCLEAR REACTIONS.   At energies above about 8 MeV for high-Z materials, and in the region of 10 to 20 MeV for low-Z materials, photons have sufficient energy to eject a proton or neutron from the nucleus of an atom. The cross section for these reactions is zero for photon energies below the binding energy of the particles and then generally rises to a maximum at energies two to four times the binding energy, falling again at higher energies. Photonuclear cross sections are generally small compared with the Compton and pair-production cross sections at the same energy, though the neutrons that are ejected may be significant in practice because of their considerable range.

Natural lead, for example, can undergo a (γ, n) reaction with a threshold energy of 7.9 MeV. The photonuclear cross section shows a maximum at 13.7 MeV and at this energy has a value about 4.3% of that for the combined Compton and pair-production processes (30).

Photonuclear reactions have been reviewed by Strauch (31).

ATTENUATION OF MONOENERGETIC RADIATION IN MATTER.   Attenuation coefficients for the processes outlined above may be added to give a total attenuation coefficient, i.e.,

$$\mu = \tau + \sigma + \kappa \tag{2.26}$$

where τ, σ, and κ are the linear attenuation coefficients for the photoelectric, Compton, and pair-production processes respectively, neglecting the small contributions from coherent scattering and photonuclear reaction. The total mass and the total atomic and total electronic attenuation coefficients can be obtained in the same way by adding the appropriate partial coefficients. It should be emphasized however that a value for a total attenuation co-efficient refers only to one material and one photon energy; separate values must be determined for every different material and for each photon energy. Figures 2.15 and 2.16 show the variation of the total and partial atomic attenuation coefficients for water and lead respectively with photon energy. The two curves are reasonably characteristic of those for other low-Z and high-Z materials respectively.

Figure 2.17 illustrates the experimental arrangement for measuring at-tenuation coefficients under *narrow-beam* conditions, i.e., where the radiation impinging on the absorber is restricted to a narrow beam by means of a collimator. Ideally, the radiation intensity measured by the detector under these conditions is due only to that part of the primary beam which has not interacted with the absorbing material in any way; scattered radiation due to both incoherent and coherent scattering being deflected away from the detector and not measured. Under narrow-beam conditions the intensity measured by the detector (I) is given by

$$I = I_i e^{-\mu l} \tag{2.18}$$

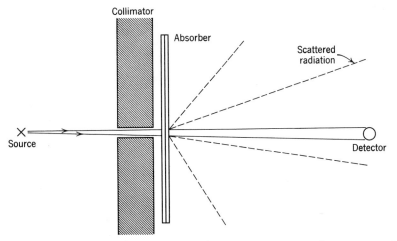

FIGURE 2.17   *Experimental arrangement for measuring narrow-beam attenuation coefficients.*

where I_i is the intensity measured in the absence of the absorber and l is the thickness of the absorber; μ is the total linear attenuation coefficient.

Tables of attenuation coefficients are given by Grodstein (32) and McGinnies (33), Hubbell (33a, b) and Evans (33c).

COMBINATION OF ATTENUATION COEFFICIENTS.   In general the probability of a photon interacting with an atom is independent of the environment of the atom. Attenuation coefficients for atoms of different elements present in an absorbing material can therefore be treated separately and, weighted in proportion to the abundance of the element (in appropriate units), added to give the attenuation coefficient for the material as a whole.

Electronic attenuation coefficients are weighted in proportion to the number of electrons each element contributes to the total number of electrons present in the compound. For example, a water molecule contains 10 electrons, one from each of the hydrogen atoms and 8 from the oxygen atom, and therefore

$$(_e\mu)_{H_2O} = \frac{2}{10}(_e\mu)_H + \frac{8}{10}(_e\mu)_O \tag{2.27}$$

Substituting numerical values from Table 2.8 for 1 MeV γ-rays

$$(_e\mu)_{H_2O} = 0.2 \times 0.211 + 0.8 \times 0.211 \tag{2.28}$$
$$= 0.211 \text{ b electron}^{-1}$$

TABLE 2.8   *Total Attenuation Coefficients for 1 MeV γ-Rays (ref. 32)*

Coefficient	Hydrogen	Oxygen	Water	Units
Electronic, $_e\mu$	0.211	0.211	0.211	b[a] electron^{-1}
Atomic, $_a\mu$	0.211	1.69	2.11	b[a] atom^{-1} or molecule^{-1}
Mass, μ/ρ	0.126	0.0636	0.0706	cm^2 g^{-1}
Linear, μ	—	—	0.0706	cm^{-1}

[a] 1 b $= 10^{-24}$ cm^2.

In this particular example attenuation is entirely by the Compton process and, since the Compton electronic attenuation coefficient is the same for all materials at any given photon energy, the electronic attenuation coefficient for water is the same as that for both hydrogen and oxygen. Electronic attenuation coefficients can, however, be combined in this way regardless of the process by which attenuation takes place.

Atomic attenuation coefficients are combined after weighting them in proportion to the number of atoms of each element present. For water, containing two atoms of hydrogen and one of oxygen, this is simply

$$(_a\mu)_{H_2O} = 2(_a\mu)_H + (_a\mu)_O \tag{2.29}$$

For 1 MeV γ-rays

$$(_a\mu)_{H_2O} = 2 \times 0.211 + 1.69 \tag{2.30}$$
$$= 2.11 \text{ b molecule}^{-1}$$

Mass attenuation coefficients are weighted in proportion to the weight of each element present. Water contains 2/18 part by weight of hydrogen and 16/18 part of oxygen and the mass attenuation coefficient of water is given by

$$\left(\frac{\mu}{\rho}\right)_{H_2O} = \frac{2}{18}\left(\frac{\mu}{\rho}\right)_H + \frac{16}{18}\left(\frac{\mu}{\rho}\right)_O \tag{2.31}$$

For 1 MeV γ-rays

$$\left(\frac{\mu}{\rho}\right)_{H_2O} = \frac{2}{18} \times 0.126 + \frac{16}{18} \times 0.0636 \tag{2.32}$$
$$= 0.0706 \text{ cm}^2 \text{ g}^{-1}$$

Linear attenuation coefficients cannot be combined quite as simply, since they depend also on the density of the elements and the compound. The linear attenuation coefficient (μ) of a compound (density ρ_c) containing two elements is given by

$$\mu = w_1\left(\frac{\rho_c}{\rho_1}\right)\mu_1 + w_2\left(\frac{\rho_c}{\rho_2}\right)\mu_2 \tag{2.33}$$

where w_1, ρ_1, and μ_1 are the weight fraction, density, and linear attenuation coefficient respectively for the first element. It is better to avoid using the densities of the elements altogether. This can be accomplished by adding their electronic, atomic, or mass attenuation coefficients to get the corresponding coefficient for the compound and converting this to the required linear attenuation coefficient. Only the density of the compound is needed.

The electronic, atomic, mass, and linear attenuation coefficients can be interconverted at any stage using the relationships

$$_a\mu = Z(_e\mu) \tag{2.34}$$

$$\frac{\mu}{\rho} = \frac{N_A Z}{A}(_e\mu) = \frac{N_A}{A}(_a\mu) \tag{2.35}$$

$$\mu = \left(\frac{\mu}{\rho}\right)\rho \tag{2.36}$$

where Z is the atomic number or, for a compound, the sum of the atomic numbers of the elements present (e.g., 10 for water), A is the atomic or molecular weight, N_A is Avogadro's number, and ρ is the density of the material.

Attenuation coefficients for mixtures of compounds are derived in the same way, conbining the attenuation coefficients for the elements present or, if known, for the compounds making up the mixture. Energy and scatter coefficients (see p. 53) can be combined in the same way as the total attenuation coefficients.

BROAD-BEAM ATTENUATION.   When a sheet of absorbing material is put in a beam of radiation, the intensity measured by a detector placed in the transmitted beam depends, in part, on the amount of scattered radiation that reaches the detector. Narrow-beam experiments (Fig. 2.17) are designed to keep to a minimum the amount of scattered radiation measured. If, however, the collimator is removed, as shown in Fig. 2.18, or the detector is moved close to the absorber, an appreciable amount of scattered radiation can enter the detector, which will then give a higher reading for the transmitted radiation intensity than under narrow-beam conditions; i.e., the measured attenuation coefficient under these broad-beam conditions is spuriously low.

Under broad-beam conditions Eq. 2.18 for the attenuation of the radiation is replaced by

$$I = I_i B e^{-\mu l} \tag{2.37}$$

where μ is the narrow-beam attenuation coefficient and B is the *build-up factor*. The build-up factor is the ratio of the measured (transmitted) intensity

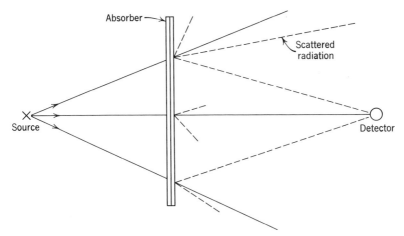

FIGURE 2.18 *Arrangement for measuring broad-beam attenuation coefficients.*

under broad-beam conditions (i.e., attenuated primary radiation plus scattered radiation entering the detector) to the measured intensity under narrow-beam conditions (attenuated primary radiation only). It is dependent on the kind and thickness of the absorber, the photon energy, and also on the type of detector used and its position relative to the absorber. Typical broad-beam and narrow-beam attenuation curves for monoenergetic radiation are shown in Fig. 2.19.

Data on build-up factors have been tabulated by Fano (34) and Goldstein and Wilkins (35), and are briefly discussed by Evans (33c) and Johns (35a).

BUILD-UP OF SECONDARY RADIATION.   When a beam of electromagnetic radiation penetrates an absorber, the secondary electron radiation builds up to a maximum at a distance below the surface that is roughly equal to the greatest range of the secondary electrons, and then falls as the primary beam is attenuated. The reason for the build-up is readily seen. Immediately below the surface a small volume of the absorber contains the electron tracks of secondary electrons generated in the volume and the tracks of electrons generated outside the volume but scattered into it (Fig. 2.20). The volume thus receives scattered electrons from all directions but the direction of the surface. As the distance of the volume element from the surface is increased, the volume of absorber from which it receives scattered electrons also increases, until the depth of the volume element below the surface corresponds to the greatest range of the secondary electrons. At this point it is receiving the maximum number of scattered electrons, since electrons produced at a greater distance from the volume element will not reach it. At greater depths

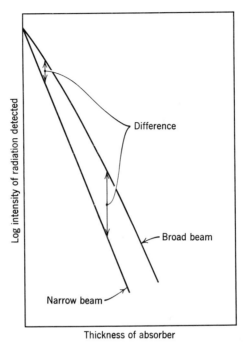

FIGURE 2.19 *Typical attenuation curves for monoenergetic γ-radiation determined under broad-beam and narrow-beam conditions.*

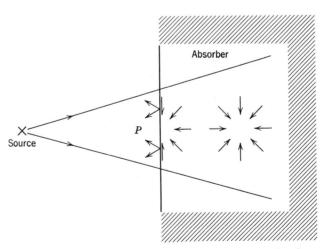

FIGURE 2.20 *Build-up of secondary electrons in an absorber.*

the number of secondary electrons falls as the primary beam of radiation is attenuated. Scattered electromagnetic radiation will build up in a similar manner but in general its effect will be small compared to that of the secondary electrons.

If the greatest range of the secondary electrons is small, the build-up effect can be neglected, through it may be quite large with high-energy radiation. Typical curves relating the energy absorbed per unit volume with the depth (*depth-dose curves*) are shown in Fig. 2.21 for radiation of different energies absorbed in water. The vertical axis gives the depth dose as a percentage of the maximum dose.

Part of the scattered electromagnetic radiation and secondary electrons will be directed back across the surface and will increase the measured intensity at a point *P* (Fig 2.20), in front of the surface, to a greater value than it would have had in the absence of the absorber. The amount of this *backscattered radiation* depends on the energy of the incident radiation, the nature of the absorbing material, and the area of the surface irradiated.

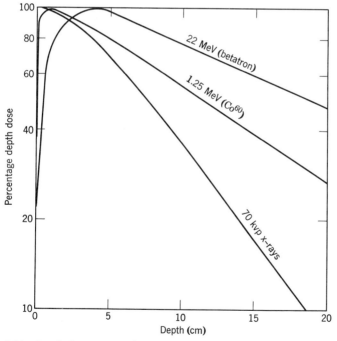

FIGURE 2.21 *Depth-dose curves for electromagnetic radiation in water.* (*Distance between radiation source and water surface, 80 cm; area of beam at surface of water, 100 cm².*)

Scattered electromagnetic radiation will generally be the most effective part of the backscattered radiation because of its greater range, and the term backscatter normally refers to this alone, the secondary electron contribution being neglected.

ENERGY ABSORPTION FROM MONOENERGETIC RADIATION.   The total attenuation coefficient (μ) relates to the attenuation of a narrow beam of monoenergetic radiation passing through an absorber. It is not directly a measure of the energy transferred to the absorbing material—the quantity that is primarily of interest in radiation chemistry—since part of the incident photon energy may be diverted from the primary beam as scattered or secondary radiation.

Processes which lead to photon scattering or to secondary electromagnetic radiation are:

(i)   Secondary x-ray (fluorescent radiation) emission from an atom that has lost an electron by the photoelectric process,
(ii) radiation scattering as a result of Compton interaction,
(iii) formation of annihilation radiation by the combination of a positron produced by pair-production and an electron,
(iv) coherent scattering,
(v)   bremsstrahlung emission by high-energy secondary electrons (cf., ref. 34, Table 5).

For moderate photon energies (0.1 to 5 MeV) and low-Z materials only Compton scattering (ii) is important. At lower energies and for high-Z materials processes (i) and (iv) may be significant, and at high energies processes (iii) and (v) become important. The energy actually absorbed is equal to the energy lost from the primary beam (found via the total attenuation coefficient) less the energy scattered by the five processes listed above; it is assumed that scattered and secondary electromagnetic radiation escape from the absorbing material without reacting further, which may often be true for small samples but is less likely to be true for large ones.

The energy absorbed may be equated with the energy transferred to the secondary electrons if these do not subsequently lose energy by bremsstrahlung emission (v) and if they remain within the absorbing material. However, if the range of the secondary electrons is greater than the dimensions of the sample, an appreciable number will escape and carry off part of the energy that would otherwise have been absorbed. This can be alleviated by surrounding the sample with similar material to a distance roughly equal to the greatest range of the secondary electrons. Under these conditions, as many electrons enter the sample from the surrounding material as escape from it, and the sample is said to be in a state of *electronic equilibrium*.

A series of coefficients are used to describe the scattered radiation and the energy transferred to secondary electrons. The first of these are the *scatter coefficients*, identified by the subscript *s* (e.g., μ_s), which relate to the fraction of the incident radiation energy lost from the beam in the form of scattered photons and secondary electromagnetic radiation. *Energy-transfer coefficients*, identified by the subscripts K or tr (e.g., μ_K or μ_{tr}), relate to the fraction of the energy transferred to secondary electrons (or other secondary particles if formed, e.g., if the primary radiation is not electromagnetic), while the *energy-absorption coefficients*, identified by the subscript en (e.g., μ_{en}), are concerned with the fraction of the energy which remains in the absorber. The latter coefficients are of most interest in radiation chemistry, since they give the fraction of the incident energy which becomes available to produce chemical and physical changes in the absorbing material. The energy absorption coefficients are smaller than the corresponding energy-transfer coefficients by the amount necessary to allow for the loss of bremsstrahlung by the secondary electrons. When such bremsstrahlung losses are small, e.g., when low to moderate energy photons are absorbed by low-Z materials, the energy-transfer and the energy-absorption coefficients are the same (e.g., Table 3.4, p. 78). This is often assumed to be true and the energy-transfer and energy absorption coefficients represented by a single *true*, or *energy*, *absorption coefficient* identified by a subscript *a* (e.g., μ_a). To avoid adding to a list of coefficients that is already long, this practice is followed in the remainder of this section and it is assumed that the energy lost from the primary beam can be described by a (total) attenuation coefficient (e.g., μ) which is the sum of a scatter coefficient (μ_s) and an (energy) absorption coefficient (μ_a). The symbols used to represent the various attenuation, scatter, and absorption coefficients are listed in Table 2.9. Note particularly that the subscripts *s* and *a* denoting scatter and energy absorption respectively are placed after the symbol (μ, σ, etc.) identifying the process involved; subscript *a* and *e before* the symbol indicate atomic and electronic coefficients and hence the units to be used.

The linear scatter (μ_s) and linear energy absorption (μ_a) coefficients are defined in a similar manner to the total linear attenuation coefficient, that is, by

$$dI_s = -I'\mu_s \, dl \tag{2.38}$$

and

$$dI_a = -I'\mu_a \, dl \tag{2.39}$$

where dI_s is the reduction in intensity due to scattering and dI_a the reduction due to energy absorption when radiation with intensity I' passes through a very small thickness dl of absorber. Since these two processes account for all the energy lost from the incident radiation, it is clear that the total reduction

TABLE 2.9   *Attenuation, Scatter, and Absorption Coefficients*

Coefficient	Linear	Mass	Atomic	Electronic
Units	cm^{-1}	$cm^2\,g^{-1}$	$cm^2\,atom^{-1}$	$cm^2\,electron^{-1}$
Units in which absorber thickness is measured	cm	$g\,cm^{-2}$	$atoms\,cm^{-2}$	$electrons\,cm^{-2}$
Total attenuation coefficient	μ	μ/ρ	$_a\mu$	$_e\mu$
scatter component	μ_s	μ_s/ρ	$_a\mu_s$	$_e\mu_s$
energy absorption component	μ_a	μ_a/ρ	$_a\mu_a$	$_e\mu_a$
Compton attenuation coefficient	σ	σ/ρ	$_a\sigma$	$_e\sigma$
scatter component	σ_s	σ_s/ρ	$_a\sigma_s$	$_e\sigma_s$
energy absorption component	σ_a	σ_a/ρ	$_a\sigma_a$	$_e\sigma_a$
Photoelectric attenuation coefficient[a]	τ	τ/ρ	$_a\tau$	$_e\tau$
Pair-production coefficient[a]	κ	κ/ρ	$_a\kappa$	$_e\kappa$

N.B. considering only the three main absorption processes

$$\mu = \tau + \sigma + \kappa$$
$$= \mu_s + \mu_a$$
$$\mu_s = \tau_s + \sigma_s + \kappa_s$$
$$\mu_a = \tau_a + \sigma_a + \kappa_a$$

[a] Scatter and energy absorption components are represented as shown for the Compton absorption coefficient.

in intensity (dI), which is given by

$$dI = -I'\mu dl \tag{2.17}$$

is also given by

$$dI = dI_s + dI_a \tag{2.40}$$

and that

$$\mu = \mu_s + \mu_a \tag{2.41}$$

The quantity with which we are mainly concerned is the energy absorbed by a sample placed in a beam of radiation of known intensity. This can be derived as follows: Consider an absorber made up of a large number of very thin layers, perpendicular to the direction of the incident radiation and each dl thick. Then the energy absorbed by any one layer is a function of the reduction in intensity as the radiation passes through it as given by Eq. 2.39; I' is now the radiation intensity at the position occupied by the layer. However the intensity at any depth in the absorber is given by Eq. 2.18, i.e.,

$$I' = I_i e^{-\mu l} \tag{2.18}$$

where I_i is the radiation intensity at the surface. Thus at any depth l, dI_a is given by

$$dI_a = -I_i\mu_a e^{-\mu l}\, dl \tag{2.42}$$

The total reduction in intensity due to energy absorption as the radiation passes through the entire thickness of absorber, l cm, is found by integrating Eq. 2.42,

$$I_a = \int_0^l -I_i\mu_a e^{-\mu l}\, dl \tag{2.43}$$

$$= \frac{I_i(1 - e^{-\mu l})\mu_a}{\mu} \text{ ergs cm}^{-2} \text{ sec}^{-1} \tag{2.44}$$

The reduction in intensity due to energy scattered is found in a similar manner to be

$$I_s = \frac{I_i(1 - e^{-\mu l})\mu_s}{\mu} \text{ ergs cm}^{-2} \text{ sec}^{-1} \tag{2.45}$$

The total reduction in intensity of the radiation beam as it passes through the absorber, due to both scattering and energy absorption, can be found by adding Eqs. 2.44 and 2.45

$$I_a + I_s = I_i(1 - e^{-\mu l}) \text{ ergs cm}^{-2} \text{ sec}^{-1} \tag{2.45a}$$

(since $\mu = \mu_a + \mu_s$) or by subtracting the intensity of the transmitted radiation (from Eq. 2.18) from the incident radiation intensity

$$I_i - I = I_i(1 - e^{-\mu l}) \text{ ergs cm}^{-2} \text{ sec}^{-1} \tag{2.45b}$$

The energy lost from the beam as it passes through the absorber is found by integrating Eq. 2.45a with respect to time, i.e.,

$$E_{\text{lost}} = E_i(1 - e^{-\mu l}) \text{ ergs cm}^{-2} \tag{2.46}$$

where E_i is the energy incident upon the absorber in units of ergs cm^{-2}. In the same way the energy absorbed (E_a) can be derived from Eq. 2.44

$$E_a = \frac{E_i(1 - e^{-\mu l})\mu_a}{\mu} \text{ ergs cm}^{-2} \tag{2.47}$$

A simplified form of Eq. 2.47 can be used if μl is small, i.e., if the radiation is not appreciably attenuated by passage through a thickness l of the stopping material. Under these conditions

$$1 - e^{-\mu l} \sim \mu l$$

and Eq. 2.47 becomes

$$E_a = E_i\mu_a l \text{ ergs cm}^{-2} \tag{2.48}$$

For some purposes it is convenient to express this in terms of the mass energy absorption coefficient, μ_a/ρ, when

$$E_a(\text{ergs g}^{-1}) = \frac{E_i(\text{ergs cm}^{-2})\mu_a}{\rho} \tag{2.49}$$

Values for the energy absorbed from a beam of radiation calculated using both Eq. 2.47 and Eq. 2.48 are compared in Table 2.10. As the thickness of absorber is increased the approximate value given by Eq. 2.48 deviates from the true value to an increasing extent. In using these equations it is assumed that none of the energy scattered is subsequently absorbed by the sample.

TABLE 2.10 *Absorbed Energy Calculations*[a]

Thickness of Absorber (Water) (cm)	Total Energy Lost from Beam (Eq. 2.46) (ergs cm^{-2})	Energy Absorbed by Water (ergs cm^{-2})	
		(Eq. 2.47)	(Eq. 2.48)
1	6.2	2.9	3
2	12	5.6	6
3	17.5	8.2	9
5	27	13	15
10	47	22	30
20	72	34	60

[a] Calculations of the energy absorbed from 1.25 MeV (Co^{60}) γ-rays when the energy incident upon the absorber (water) is 100 ergs cm^{-2}. For 1.25 MeV γ-rays in water $\mu = 0.064$ cm^{-1} and $\mu_a = 0.03$ cm^{-1}.

Equations 2.48 and 2.49 are only true for thin absorbers which do not attenuate the radiation appreciably. However they may be used to find the ratio of the energies absorbed by two different stopping materials, exposed to the same beam of radiation, over a much wider range of thicknesses, since the errors in the numerator and denominator largely nullify each other. Thus

$$\frac{(E_a)_A}{(E_a)_B} = \frac{E_i(\mu_a/\rho)_A}{E_i(\mu_a/\rho)_B} = \frac{(\mu_a/\rho)_A}{(\mu_a/\rho)_B} \tag{2.50}$$

The total energy absorption coefficient μ_a, like the total attenuation coefficient μ, is the sum of three partial coefficients relating to the three major absorption processes:

$$\mu_a = \tau_a + \sigma_a + \kappa_a \tag{2.51}$$

For the photoelectric process

$$\tau_a = \tau \frac{E_0 - fE_s}{E_0} \tag{2.52}$$

where f is the probability that the binding energy E_s of the electron ejected from the atom will be re-emitted as a secondary x-ray photon. If the energy of the incident photon, E_0, is much greater than the binding energy, then for all practical purposes $\tau_a = \tau$. The Compton electronic energy absorption coefficient $_e\sigma_a$ is related to the linear coefficient by

$$\sigma_a = {_e\sigma_a} \frac{\rho N_A Z}{A} \tag{2.53}$$

The pair-production linear energy absorption coefficient is

$$\kappa_a = \kappa \frac{E_0 - 2m_e c^2}{E_0} \tag{2.54}$$

where $2m_e c^2$ ($= 1.02$ MeV) is that part of the absorbed energy that reappears as annihilation radiation.

Values of the total energy absorption coefficient for a number of materials and photon energies are given by Hubbell (33a, b) and Evans (33c), and in the 1964 report of the ICRU (36). The values are corrected for energy loss through bremsstrahlung production by secondary electrons.

For low-Z materials there is a range of energies for which the only signifi-cant absorption process is the Compton effect; for water this range is about 0.1 to 4 MeV (Fig. 2.15). In this region

$$\mu = \sigma \tag{2.55}$$

and

$$\mu_a = \sigma_a \tag{2.56}$$

The absorption of cobalt-60 and caesium-137 γ-rays by aqueous, biological, and most organic systems comes in this region.

ATTENUATION OF X-RAYS.   In the preceding sections, factors governing the absorption of monoenergetic electromagnetic radiation have been described. In practice, radiation made up of photons with a wide range of energies may be encountered, as in the case of an x-ray beam. The absorption of such polyenergetic radiation can be treated in a manner similar to the absorption of monoenergetic radiation by considering it as a number of superimposed monoenergetic beams, whose radiation penetrates independently. Calcula-tion of the absorption characteristics of a polyenergetic beam involves integration over the range of energies represented in the beam and requires a knowledge of the number of photons of each energy present, i.e., of the energy spectrum of the beam.

Low-energy photons tend to have higher attenuation coefficients than those of higher energies and will be removed from the primary beam faster than the high-energy components. Thus as the beam penetrates to greater depths, the energy spectrum of the attenuated primary beam changes, the proportion of low-energy photons falling (neglecting any build-up of low-energy scattered radiation). At the same time the average attenuation co-efficient for the beam decreases. Consequently it is not possible to calculate a single average attenuation coefficient for an x-ray beam at all depths in an absorber, but only an average coefficient applicable to the energy spectrum as it exists at a particular point, and which will only be effective over a limited range of thicknesses. This is shown by an increase in the half-value layer for an x-ray beam as the beam penetrates further through the absorber.

The energy absorbed from an x-ray beam can be calculated in a similar manner to the energy absorption from a beam of monoenergetic γ-rays by using, e.g.,

$$E_a = \int_{E_m}^{E_0} EN_E(\mu_a)_E l \, \mathrm{d}l \qquad (2.51)$$

(based on Eq. 2.48) where N_E is the number of photons of energy E incident upon the absorber, $(\mu_a)_E$ is the energy absorption coefficient for photons of energy E, l is the thickness of the absorber, and E_0 and E_m are the highest and lowest photon energies present in the x-ray beam respectively. The integration can be solved graphically.

Ionization and Excitation Produced by Radiation

The energy lost when a moving charged particle is slowed down in matter gives rise to a trail of excited and ionized atoms and molecules in the path of the particle. Excited states are produced when bound electrons in atoms and molecules of the stopping material gain energy and are raised to higher energy levels, and ions when the energy gained is sufficient, and the transient excited states produced such that electrons are expelled. Electromagnetic radiation produces a similar result, since the energy absorbed is transferred to electrons and positrons and then dissipated along the paths of these particles. The overall result of the absorption of any type of ionizing radiation by matter is thus the formation of tracks of excited and ionized species. These species will, in general, be the same in a particular material regardless of the type or energy of the radiation responsible. All ionizing radiation will, therefore, give rise to qualitatively similar chemical effects. However, radiation of different types and energy will lose energy in matter at different rates, and consequently will form tracks that may be densely or sparsely populated with the active species. The differences observed in the chemical effect of different radiations—differences in the quantities and proportions of the chemical products—stem from the different density of active species in the particle tracks. Expressions that reflect this changing density, such as specific ionization and linear energy transfer, are therefore useful in evaluating the overall chemical effect. Track effects of this sort are more important in the case of liquids, where the active species are hindered from moving apart by the proximity of other molecules, than in gases, where they can move apart with relative ease. In gases the different types of radiation do not give the markedly different yields of products that they may in liquid systems.

Electrons ejected as a consequence of the ionization produced by radiation may themselves be sufficiently energetic to produce further ionization and excitation. If the energy of these secondary electrons is relatively small, less than about 100 eV, their range in liquid or solid materials will be short and any secondary ionizations that they produce will be situated close to the original ionization, giving a small cluster or spur[6] of excited and ionized species. Some of the secondary electrons will have enough energy to travel further from the site of the original ionization and will form tracks of their own, branching off from the primary track. Such electrons are known as δ-rays (Fig. 2.22); their tracks will be similar to those of other electrons with

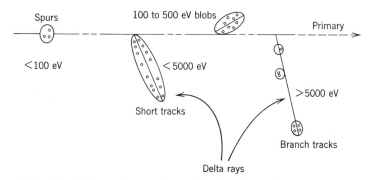

FIGURE 2.22 *Distribution of ions and excited molecules in the track of a fast electron (not to scale). Several types of energy deposition are associated with the passage of a primary energetic electron through a condensed medium. Only positive ions are indicated by the dots; neither the associated (germinate) electrons nor possible (associated) excited species present in spurs, blobs, and tracks are shown (38f).*

the same energy. Lea (37) has calculated that about one half of the total number of ionizations produced by a primary particle, whether electron, proton, or α-particle, are to be found in the tracks of δ-rays with energy exceeding 100 eV. The remaining ionization events are distributed along the primary track as isolated single ionizations or in small spurs. Wilson (38), using cloud-chamber photographs, counted the ion pairs formed along the track of fast electrons in water vapor and found the following distribution of

[6] We will use the term spur rather than the alternative, cluster, when referring to these small groups of excited and ionized species.

ion pairs in the spurs:

number of ion pairs in the spur:	1	2	3	4	over 4	Total
frequency of spurs of each size:	0.44	0.22	0.12	0.10	0.12	1.00

The average spur contains 2 to 3 ion pairs and corresponds to an energy loss of about 100 eV. The spurs will also contain excited species (Fig. 2.22). Beekman (38a) obtained a value of 0.60 rather than 0.44. The discrepancy has been discussed by Ore and Larsen (38b). Mozumder and Magee (38c, d, e) have suggested that for δ electrons in the energy range 100 to 500 eV the generated second generation electrons are not energetic enough to leave the site of birth—a large spur is produced called a "blob." δ-Electrons with larger energies may form short tracks (500 eV–5 keV) or branch tracks (> 5 keV) (Fig. 2.22, ref. 38f).

With densely ionizing radiation (e.g., α-particles) the spurs overlap and form a column of ions and excited species about the track (so-called *columnar ionization*); less densely ionizing radiation gives spurs at intervals along the track (Fig. 2.23). For fast particles, such as the secondary electrons formed when energetic γ-rays are absorbed in water and organic liquids, Samuel and Magee (39) have calculated that the spurs occur at intervals of about 10^4 Å and that they have an initial diameter of about 20 Å.

Once the electrons have slowed down to thermal energies they will probably either neutralize a positive ion directly or add to a neutral molecule to form a negative ion, which subsequently neutralizes a positive ion.

SPECIFIC IONIZATION. An indication of the linear rate of energy loss by a charged particle and the density of ionization in the particle track is given by the *specific ionization*, the total number of ion pairs produced in a gas per unit length of track. The specific ionization includes both ions produced in the primary particle track and those produced by δ-rays. In Fig. 2.24 the specific ionization for protons and α-particles are plotted as a function of the distance from the end of the particle track (*Bragg curve*). The most intense ionization is produced near the end of the track where the particle velocity is low, though the ion density falls when the velocity becomes low enough for the particle to capture electrons.

The energy loss is related to the specific ionization by the important quantity W, the mean energy loss per ion pair formed

$$\text{specific ionization} = \frac{\mathrm{d}E/\mathrm{d}l}{W} \qquad (2.58)$$

where $\mathrm{d}E/\mathrm{d}l$ is the specific energy loss or stopping power. The quantity W is the total energy consumed per ion pair formed; i.e., the energy to eject an electron plus the energy, per ion pair, used in producing excited molecules. Some experimentally determined values of W for gases are given in Table 2.11.

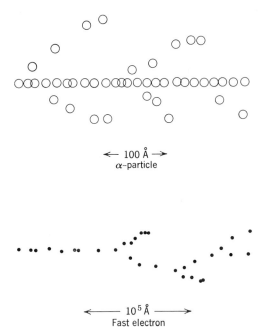

← 100 Å →
α-particle

←——— 10^5 Å ———→
Fast electron

FIGURE 2.23 *Distribution of spurs in particle tracks. The spurs will have a diameter of the order of 20 Å. (Not to scale.)*

TABLE 2.11   *Experimental Values of W for Gases (40)*

Gas	W (eV/ion pair)	
	Electrons	α-Particles
Air	33.73	34.98
Helium	41.5	46.0
Argon	26.2	26.3
Hydrogen	36.6	36.2
Nitrogen	34.6	36.39
Oxygen	31.8	32.3
Carbon dioxide	32.9	34.1
Methane	27.3	29.1
Ethane	24.6	26.6
Ethylene	26.3	28.03
Acetylene	25.7	27.3

Values of W are very nearly independent of the energy of the radiation and are practically the same for all types of radiation; electromagnetic radiation and electrons give the same values.

LINEAR ENERGY TRANSFER.   Linear energy transfer was defined earlier in this chapter as the linear-rate of loss of energy (locally absorbed) by an ionizing particle traversing a material medium (38 g), and a rough average value calculated by dividing the total energy of a particle by its path length. Several factors contribute to the roughness of this average value. In the first place, the rate of energy loss by a particle changes as it is slowed down, as is shown by the Bragg curves (Fig. 2.24), and the LET will vary at different positions along the track. Second, energy lost by the primary particle in a particular section of the track is not necessarily absorbed locally but may be transferred in part to δ-rays or to secondary electromagnetic radiation. γ- and x-rays transfer all their energy to the medium through secondary electrons having a wide range of energies and LET.

In practice some sort of mean value may be derived for the LET (41–44) or the complete distribution of LET values calculated for particles of all energies present (43, 44). In either case the nature of the calculations and

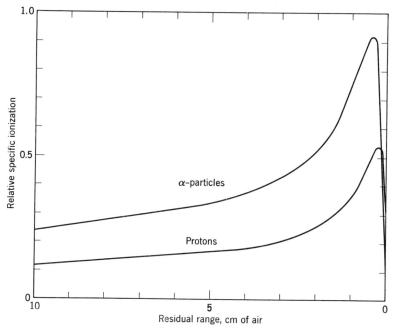

FIGURE 2.24   *Bragg curves for protons and α-particles.*

the assumptions made depend on the author, so that it is essential to use care when comparing LET values from different sources. Some track-average LET values in liquid water for various radiations with $\Delta = 0.1$ keV are given in Table 2.12 (38g). For ^{60}Co γ-rays, additional values are given for $\Delta = 1$ and 10 keV. $\bar{L}_{\Delta, T}$ is the track-average LET with cut-off Δ.

TABLE 2.12 *Track-Average Values of LET in Water Irradiated with Various Radiations*

Radiation	Cut-off Energy (Δ/eV)	$\bar{L}_{\Delta, T}$ (keV/μm)
^{60}Co γ-rays	Unrestricted	0.23_9
	10,000	0.23_2
	1,000	0.23_0
	100	0.22_9
22-MV x-rays	100	0.19
2-MeV electrons (whole track)	100	0.20
200-kV x-rays	100	1.7
^{3}H β-rays	100	4.7
50 kV x-rays	100	6.3
5.3 MeV α-rays (whole track)	100	43

For charged particle radiation, the LET values increase in the order shown below, assuming the protons and heavier particles to have about the same energy,

LET increases
- γ-rays, high-energy electrons
- low-energy x-rays, β-particles
- protons
- deuterons
- α-particles
- heavy ions (ionized N, O, etc.)
- fission fragments from nuclear reactions

The increase in LET for electrons in water as the electron energy decreases is shown in Fig. 2.25 (43, 46). The LET plotted here is the average energy dissipated per unit track length by electrons having the energy shown, and not the energy loss averaged over the entire path length; energy transfers giving δ-rays with more than 100 eV were not included as part of the energy locally dissipated. Above 1 MeV the LET is practically constant, and little difference in biological or chemical effect would be expected of radiations giving secondary electrons with energies in this range.

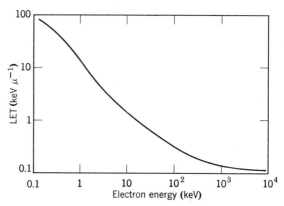

FIGURE 2.25   *Average linear energy transfer (LET) for electrons in water.*

REFERENCES

1. Report of the International Commission on Radiological Units and Measurements (ICRU), (1959), *National Bureau of Standards (U.S.) Handbook 78*, 1961.
2. D. E. Lea, *Actions of Radiations on Living Cells*, Cambridge University Press, Cambridge, p. 25, 1946.
3. L. Slack and K. Way, *Radiations from Radioactive Atoms in Frequent Use*, U.S. Atomic Energy Commission, 1959.
4. A. T. Nelms, *Energy Loss and Range of Electrons and Positrons*, National Bureau of Standards, Circular 577, 1956.
5. G. W. Grodstein, *x-Ray Attenuation Coefficients from 10 keV to 100 MeV*, National Bureau of Standards (U.S.) Circular 583, 1957, and R. T. McGinnies, Supplement to National Bureau of Standards (U.S.), Circular 583, 1959.
6. *Large Radiation Sources in Industry*, International Atomic Energy Agency, Vienna, Vol. 1, 1960, p. 180.
7. *CAPRI Documentation*, **1**, 1 (January 1974). C.E.N. Saclay, France.
8. *Gamma Irradiation in Canada*, Atomic Energy of Canada Limited, Report PP-19-60, p. 44.
9. *Seventh Hot Laboratories and Equipment Conference*, Cleveland, Ohio, 1959, p. 350.
10. P. Harteck and S. Dondes, *Nucleonics*, **14**, 22 (July 1956); P. Harteck, S. Dondes, and J. W. Michener, *Proc. 2nd Int. Conf. Peaceful Uses of Atomic Energy*, United Nations, Geneva, **7**, 544 (1958).
11. L. E. Brownell, *Radiation Uses in Industry and Science*, U.S. Atomic Energy Commission, 1961.
12. A. Charlesby, *Atomic Radiation and Polymers*, Pergamon Press, New York, 1960, p. 61.
13. B. Manowitz, Nuclear Engineering, Part II, *Chem. Eng. Progr., Symp. Ser.*, **50** (12) (1954).
14. R. C. Loftness, *Proc. 1955 Conf. Nuclear Eng.*, University of California, Los Angeles (1955).
15. W. Wild and J. Wright, *Symp. Utilization of Radiation from Fission Products, A.E.R.E.* (Gt. Britain), Report C/R 1231, 3 (1953).
16. Industrial Uses of Large Radiation Sources, *IAEA*, **2**, 175 Vienna (1963).
17. *Nucleonics*, **18**, 69, 72, 75 (December 1960).

The topics included in the second half of the chapter are discussed more fully in ref. 18 and by G. N. Whyte, *Principles of Radiation Dosimetry*, Wiley, New York, 1959.

18. H. A. Bethe and J. Ashkin in *Experimental Nuclear Physics* (ed. E. Segrè), Vol. 1, Wiley, New York, pp. 166, et seq., 1953.
19. W. Heitler, *Quantum Theory of Radiation*, 3rd Ed., Oxford University Press, 1954.
20. H. A. Bethe, *Handbuch der Physik*, Julius Springer, Berlin, Vol. 24, 1933, p. 273; *Ann. Phys.*, **5**, 325 (1930); *Z. Phys.*, **76**, 293 (1932).
21. L. Katz and A. S. Penfold, *Rev. Mod. Phys.*, **24**, 28 (1952).
22. A. T. Nelms, *Energy Loss and Range of Electrons and Positrons*, National Bureau of Standards (U.S.), Circular 577, 1956.
22a. H. Bichsel in *Radiation Dosimetry* (eds. F. H. Attix and W. C. Roesch), Vol. I, Academic Press, New York, Chap. 4, 1968, p. 157.
22b. M. J. Berger and S. M. Seltzer, *Tables of Energy Losses and Ranges of Positrons and Electrons*, NASA Tech. Report SP 3012 (1964).
22c. M. J. Berger and S. M. Seltzer, *Additional Stopping Power and Range Tables for Protons, Mesons and Electrons*, NASA Tech. Report SP 3036 (1966).
23. M. S. Livingston and H. A. Bethe, *Rev. Mod. Phys.*, **9**, 245 (1937).
24. W. A. Aron, B. G. Hoffman, and F. C. Williams, *Range-Energy Curves*, AECU 663 (1951).
25. *American Institute of Physics Handbook* (ed. D. E. Gray), McGraw-Hill, New York, Section 8c, 1957.
26. Measurement of Absorbed Dose of Neutrons, and of Mixtures of Neutrons and Gamma Rays, *National Bureau of Standards (U.S.) Handbook 75*, 1961.
27. G. N. Whyte, *Principles of Radiation Dosimetry*, John Wiley and Sons, New York, 1959.
28. O. Klein and Y. Nishina, *Z. Phys.*, **52**, 853 (1929).
29. A. T. Nelms, *Graphs of the Compton Energy-Angle Relationship and the Klein-Nishina Formula from 10 keV to 500 MeV*, National Bureau of Standards (U.S.), Circular 542, 1953.
30. R. Montalbetti, L. Katz, and J. Goldemberg, *Phys. Rev.*, **91**, 659 (1953).
31. K. Strauch, *Ann. Rev. Nucl. Sci.*, **2**, 105 (1953).
32. G. W. Grodstein, *x-Ray Attenuation, Coefficients from 10 keV to 100 MeV*, National Bureau of Standards (U.S.), Circular 583, 1957.
33. R. T. McGinnies, *x-Ray Attenuation Coefficients from 10 keV to 100 MeV*, Supplement to National Bureau of Standards (U.S.), Circular 583, 1959.
33a. J. H. Hubbell and M. J. Berger, *Engineering Compendium Radiation Shielding* (ed. R. G. Jaeger), Springer–Verlag, New York, 1968, pp. 167–184 and 185–202.
33b. J. H. Hubbell, *Photo Cross Sections, Attenuation Coefficients, and Energy Absorption Coefficients from 10 keV to 100 GeV*, NSRDS–29 (1969).
33c. R. D. Evans in *Radiation Dosimetry* (eds. F. H. Attix and W. C. Roesch), Vol. I, Academic, New York, 1968, Chap. 3, p. 94.
34. U. Fano, *Nucleonics*, **11** (August), 8 (September), 55 (1953).
35. H. Goldstein and J. E. Wilkins, USAEC Report NYO–3075, 1954.
35a. H. E. Johns in *Radiation Dosimetry* (eds. F. H. Attix and E. Tochilin), Vol. 3, Academic, New York, 1969, Chap. 17, p. 1.
36. ICRU 64; Physical Aspects of Radiation, *National Bureau Standards Handbook 85*, 1964.
37. D. E. Lea, *Actions of Radiations on Living Cells*, Cambridge University Press, Cambridge, 1946, p. 27.
38. C. T. R. Wilson, *Proc. Roy. Soc. (London)*, Ser. A, **104**, 192 (1923).

38a. W. J. Beekman, *Physica*, **15**, 327 (1949).
38b. A. Ore and A. Larsen, *Radiat. Res.*, **21**, 331 (1964).
38c. A. Mozumder and J. L. Magee, *Radiat. Res.*, **28**, 203, 215 (1966); *J. Chem. Phys.*, **45**, 3332 (1966).
38d. A. Mozumder, A. Chatterjee, and J. L. Magee, *Advan. Chem. Ser.*, **81**, 27 (1968).
38e. A. Mozumder in *Advances in Radiation Chemistry* (eds. M. Burton and J. L. Magee), Vol. I, Wiley–Interscience, New York, 1969, pp. 1–102.
38f. M. Burton, *Chem. Eng. News*, **46**, 86 (February 10, 1969).
38g. *Linear Energy Transfer*, ICRU Report 16, ICRU, Washington, U.S.A., June 1970.
39. A. H. Samuel and J. L. Magee, *J. Chem. Phys.*, **21**, 1080 (1953).
40. I. T. Myers in *Radiation Dosimetry* (eds. Attix, Roesch, Tochelin), 2nd ed., Academic, New York Vol. I, 1968, Ch. 7, p. 317.
41. L. H. Gray, *Brit. J. Radiol. Suppl.*, **1**, 7 (1947).
42. D. V. Cormack and H. E. Johns, *Brit. J. Radiol.*, **25**, 369 (1952).
43. P. R. J. Burch, *Brit. J. Radiol.*, **30**, 524 (1957).
44. H. E. Johns in *Radiation Dosimetry* (eds. G. J. Hine and G. L. Brownell), Academic, New York, 1956, Chap. 12.
45. D. V. Cormack, personal communication.
46. P. R. J. Burch, *Radiat. Res.*, **6**, 289 (1957).

CHAPTER 3

Radiation Dosimetry

Dosimetric Terms and Units • Calorimetry • Ionization Measurements (Electromagnetic Radiation) • Standard Free-Air Chamber • Thimble Ionization Chambers • Calculation of Absorbed Dose from Ionization Measurements • Dosimetry of Charged Particles • Charge-Collection Measurements • Ionization Measurements • Other Measurements • Neutron Dosimetry • Ionization Measurements • Dosimetry of Internal Radiation Sources • Chemical Dosimetry • Fricke (Ferrous Sulfate) Dosimeter • Ceric Sulfate Dosimeter • Other Chemical Dosimeters • Dosimetry in Pulse Radiolysis • Solid-State Dosimetry • Calculation of Absorbed Dose in Specimen • Electromagnetic Radiation • Charged Particle Radiation • Energy Partition in Mixtures • Personnel Dosimetry • References •

Quantitative studies in radiation chemistry require a knowledge of the amount of energy transferred from the radiation field to the absorbing material and, if possible, the distribution of the absorbed energy within the material. The determination of these quantities, in more or less detail, constitutes radiation dosimetry.

DOSIMETRIC TERMS AND UNITS

The terms and units described below are discussed in more detail by Roesch and Attix (3) and in the publications of the International Commission on Radiation Units and Measurements (e.g., ref. 4).

The *absorbed dose* is the quantity generally sought and is the amount of energy absorbed per unit mass of irradiated material. The official unit of absorbed dose is the *rad*, which is defined as an energy absorption of 100 erg g^{-1} or 10^{-2} J kg^{-1}, although units of *electron volts per gram* or *electron volts per cubic centimeter* are also used (1 rad = 6.241×10^{13} eV g^{-1} or $6.241\rho \times 10^{13}$ eV cm^{-3}, where ρ is the density of the material in g cm^{-3}). The absorbed dose is a direct measure of the energy transferred to the irradiated material and capable of producing chemical or physical change in it; it is determined both by the composition of the material and characteristics of the radiation field.

The *absorbed dose rate* is the absorbed dose per unit time and has the units rads, eV g^{-1}, or eV cm^{-3}, per unit time, e.g., rads min^{-1} or eV g^{-1} sec^{-1}.

The *exposure of x- or γ-radiation* (formerly exposure dose) at a certain place is a measure of the radiation based on its ability to produce ionization in air. The unit of exposure is the *roentgen* (symbol R). By definition (4), exposure of air to 1 R produces ions of one sign (i.e., either the $-$ve or the $+$ve ions are considered, not both together) carrying a total electrical charge of exactly 2.58×10^{-4} C kg^{-1} of air when all the secondary electrons (both $-$ve and $+$ve) are stopped in air (1 R = 2.58×10^{-4} C kg^{-1}). This is identical with the previous definition in which the charge collected was defined as 1 esu of charge per 0.001293 g air, i.e., per 1 cm^3 dry air at 0°C and 760 mm Hg pressure. By definition, the roentgen can only be used for photon irradiation.

Neither the rad nor the roentgen are SI units although they have been accepted by the International Committee on Weights and Measures as units to be used with the SI for a limited time (5, 6) and are used in the subsequent discussion. However, SI units for quantities commonly met in radiation chemistry are listed in Table 3.1 with conversion factors from current to SI units. The SI units for absorbed dose rate and exposure rate may be expressed in two forms, since 1 J sec^{-1} = 1 W and 1 C sec^{-1} = 1 A, but confusion is likely to be minimized, at least during the transition to SI units, if the units J kg^{-1} sec^{-1} and C kg^{-1} sec^{-1} are used for absorbed dose rate and exposure rate, respectively.

Other terms met in radiation dosimetry are fluence, flux density, and kerma. *Fluence* (symbol Φ, units m^{-2}) is a measure of the number of particles entering a sphere of unit cross sectional area at the point of interest while *energy fluence* (Ψ, J m^{-2}) gives the sum of the energies of the particles entering the sphere, exclusive of their rest energies. The *fluence rate* or *flux density* ($\varphi = d\Phi/dt$, m^{-2} sec^{-1}) gives the number of particles entering the sphere per second and the *energy fluence rate* or *energy flux density* ($\psi = d\Psi/dt$, J m^{-2} sec^{-1} or erg cm^{-2} sec^{-1}) the sum of the energies of the particles entering the sphere per second; the latter quantity is also described as the *intensity* of the radiation. *Kerma* (K, rad or J kg^{-1}; from *kinetic energy*

T A B L E 3.1 *Current and SI Units Used in Radiation Chemistry (refs. 4–6)[a]*

Quantity (Symbol)	Commonly Used Unit (1975)	Equivalent in SI Units
Activity (A)	1 curie (Ci)[b]	3.7×10^{10} sec^{-1} (exactly)
Absorbed dose (D)	1 rad[b]	0.01 J kg^{-1} (exactly)
	1 electron volt per gram (eV g^{-1})	1.6022×10^{-16} J kg^{-1}
	1 electron volt per cubic centimeter (eV cm^{-3})	$1.6022 \times 10^{-16}/\rho$ J kg^{-1}
Absorbed dose rate ($\dot{D}$)	1 rad per second (rad sec^{-1})	0.01 J kg^{-1} sec^{-1} (exactly)
	1 electron volt per gram per second (eV g^{-1} sec^{-1})	1.6022×10^{-16} J kg^{-1} sec^{-1}
	1 electron volt per cubic centimeter per second (eV cm^{-3} sec^{-1})	$1.6022 \times 10^{-16}/\rho$ J kg^{-1} sec^{-1}
Exposure (X)	1 roentgen (R)[b]	2.58×10^{-4} C kg^{-1} (exactly)
Exposure rate ($\dot{X}$)	1 roentgen per second (R sec^{-1})	2.58×10^{-4} C kg^{-1} sec^{-1} (exactly) 2.58×10^{-4} A kg^{-1} (exactly)
Mass energy absorption coefficient (μ_{en}/ρ)	1 cm^2 g^{-1}	0.1 m^2 kg^{-1} (exactly)
Mass stopping power (S/ρ)	1 erg cm^2 g^{-1}	10^{-8} J m^2 kg^{-1} (exactly)
	1 MeV cm^2 g^{-1}	1.6022×10^{-14} J m^2 kg^{-1}
Mean energy to form an ion-pair (W)	1 electron volt (eV)	1.6022×10^{-19} J
Radiation chemical yield ("G" value)	1 G unit (equivalent to one molecule of the specified material formed or destroyed per 100 eV energy absorbed)	1.0364×10^{-7} mol J^{-1} (the equivalent of 1 G unit is 1.0364×10^{-7} mole material changed per joule energy absorbed = 0.10364 μmol J^{-1})
	$M/N = 1$ (an ion-pair, or ionic, yield of one represents the formation or destruction of one molecule of the specified material for each ion-pair formed by irradiation)	$1.6606 \times 10^{-24}/W$ (joules) mol J^{-1} $1.0364 \times 10^{-5}/W$ (eV) mol J^{-1} (W is the mean energy to form an ion-pair; units joules or eV)

[a] The symbol ρ used in the table is the density of the material in g cm^{-3}.
[b] Unit recognized by the ICRU (4).

*r*eleased in *ma*terial) is the sum of the initial kinetic energies of all the charged particles (electrons and ions) produced per unit mass of material by uncharged particles (e.g., photons, neutrons). The term was introduced to allow a more exact description of the absorption of photons and neutrons, which deposit energy in the absorber in two steps: (*a*) transfer of energy to charged particles, and (*b*) ionization and excitation of the substrate by these secondary particles. Kerma describes the results of process *a* at a point, and absorbed dose the results of process *b*. The kerma and absorbed dose are the same if bremsstrahlung losses are negligible, and an equilibrium is established in which energy lost to the sample by escaping charged particles is balanced by energy gained from charged particles entering the system though produced outside it.

The *mass energy-transfer coefficient* (μ_K/ρ, cm^2 g^{-1} or m^2 kg^{-1}) is the ratio of the kerma to the energy fluence of photons or neutrons at any point, i.e., $\mu_K/\rho = K/\Psi$, The *mass energy-absorption coefficient* (μ_{en}/ρ, cm^2 g^{-1} or m^2 kg^{-1}), is smaller than μ_K/ρ by the amount necessary to allow for the loss of bremsstrahlung by the secondary charged particles produced by the photons or neutrons. The *mass attenuation coefficient* (μ/ρ, cm^2 g^{-1} or m^2 kg^{-1}), or mass absorption coefficient, described in Chapter 2 relates to the total loss of energy from the primary beam by all absorption and scattering processes. These three coefficients are used when the primary radiation is uncharged (i.e., photon or neutron radiation) and interaction between the radiation and molecules of the absorbing material is relatively infrequent.

An alternative approach based on the average rate of energy loss or stopping power is used when the primary radiation consists of charged particles, which interact much more strongly with the absorber. The *total mass stopping power* (S/ρ, MeV cm^2 g^{-1} or J m^2 kg^{-1}) of a material for charged particles is defined by $S/\rho = (dE/dl)/\rho$, where dE is the energy lost by a charged particle of specified energy in traversing a distance dl, and ρ is the density of the medium. For energies at which nuclear interactions can be neglected, the total mass stopping power is the sum of two partial stopping powers that relate to energy lost by collision (i.e., transferred to molecules of the absorber in the form of ionization or excitation) and to energy lost by radiative processes (i.e., bremsstrahlung formation). It is the first of these partial stopping powers, the *collisional mass stopping power*, $(S/\rho)_{col}$, that is of primary interest in radiation chemistry.

Techniques for measuring ionizing radiation can be divided into absolute and secondary methods. Absolute methods involve direct determination of exposure or absorbed dose from physical measurements of, for example, the energy absorbed (by calorimetry), the ionization produced in a gas, or the charge carried by a beam of charged particles of known energy. The absolute methods are often not suited to routine use and, in practice, secondary dosimeters (e.g., thimble ionization chambers and chemical dosimeters), whose response to radiation is known from comparison with an absolute dosimeter, are generally used.

CALORIMETRY

The most direct way of determining the amount of energy carried by a beam of radiation is to measure the increase in temperature of a block of material placed in the beam, the method originally used by Curie and Laborde to measure the rate of energy release by the radioactive decay of radium (7). The material must be such that all the absorbed energy is converted to heat, none, for example, being used to initiate chemical reaction, Good thermal conductivity is also necessary and in practice graphite or metals are generally used for this purpose. If the block is of sufficient size to completely absorb the radiation (Fig. 3.1a) the rate of temperature increase is related directly to the *energy flux density* or *intensity* (erg cm^{-2} sec^{-1}) of the beam (8). With low intensity radiation, such as that normally available from x- and γ-ray sources, the temperature rise is very small and calorimetry is not a convenient method of dosimetry for routine use, although it is important as a check on other, less direct, methods, since the results are obtained directly in energy units.

The *absorbed dose* at any depth in the material can be calculated from the measured intensity or may be measured directly by calorimetry (8, 9). The direct calorimetric measurement of absorbed dose is rather more difficult than the measurement of radiation intensity and a smaller sample of material is used, so that the radiation is only partially absorbed, and the sample is surrounded by the same, or similar, material so that it is in electronic equilibrium with its surroundings (Fig. 3.1b). Radak and Marković (9) give the range of absorbed dose rate that can be measured in this way as 10^{-7} W g^{-1} (36 rads hr^{-1}) to 10 W g^{-1} (10^6 rads sec^{-1}). Absorbed dose measurements with calorimeters in which water is the absorbing material have been used to calibrate the Fricke and other aqueous chemical dosimeters described later in this chapter.

A novel application of calorimetry to determine the three dimensional absorbed dose distribution in transparent liquids has been described by Hussmann and McLaughlin (10, 11) who used holographic interferometry to observe changes in refractive index caused by the local heating produced by a short radiation pulse.

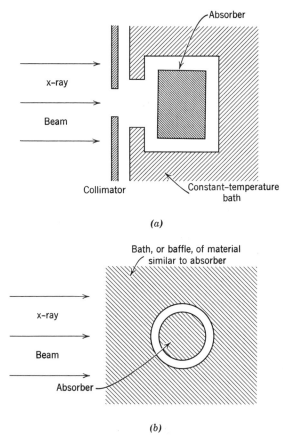

FIGURE 3.1   *Principle of calorimeters to measure (a) radiation intensity and (b) absorbed dose.*

IONIZATION MEASUREMENTS (ELECTROMAGNETIC RADIATION)

Measurements of the ionization produced in a gas by radiation have been used for the purposes of dosimetry since the discovery of x-rays and radio-activity and are still favored means of dosimetry when x- and γ-rays are used for medical purposes.

Ionization is measured by means of an ionization chamber (which may be either an absolute or a secondary instrument) consisting of two electrodes separated by a gas-filled space in which the incident radiation produces ionization. A potential is applied to the electrodes to attract the ions to them, and the resulting current, or the discharge of the electrodes, is measured by some appropriate means. The quantity obtained in this way is the exposure, from which the absorbed dose can be calculated knowing the composition of the medium and the energy, or energy spectrum, of the radiation. Factors affecting the design of ionization chambers are discussed by Boag (12) and by Burlin (13).

Standard Free-Air Chamber

For absolute ionization measurements a free-air chamber, represented by Fig. 3.2, is used. The photon beam (generally an x-ray beam) enters the air-filled chamber through a diaphram, which defines the cross section of the beam inside the chamber, and then passes between the parallel-plate electrodes. The lower plate consists of a collecting section, in the center, and guard plates on either side to define the (shaded) *sensitive volume*—the volume of air irradiated by the beam for which the ionization will be measured. All ions of the appropriate sign (negative in Fig. 3.2) produced in the *collecting volume*, i.e., between the verticals through A and B, will be attracted to the collecting electrode. Some of the ions produced in this volume may arise from secondary electrons produced outside of the sensitive volume. However, these will be compensated for by electrons which are scattered from the sensitive volume outside the collecting volume. To ensure that this electronic equilibrium is set up the distance between the walls of the chamber and the

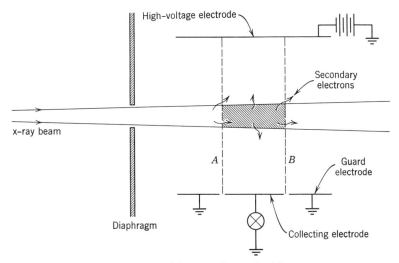

FIGURE 3.2   *Principal features of a standard free-air chamber.*

nearest edge of the sensitive volume must be more than the greatest range of the secondary electrons. The distance between the x-ray beam and the electrodes must also be greater than this range, so that the secondary electrons are not collected before they have completed their tracks and produced their full complement of ions. If these conditions are satisfied, the charge collected (in esu), divided by the sensitive volume (cm^3), is numerically equal to the exposure dose in roentgens. After applying a number of corrections, including corrections to bring the temperature and pressure to NTP, the exposure dose should be accurate to better than 1%.

Free-air chambers are used as standards of exposure dose for x-rays in the range 10 to 300 kV. At higher energies the long range of the secondary electrons entails the use of very large chambers, or of chambers operating under pressures of several atmospheres; both of which present practical difficulties.

Thimble Ionization Chambers

An alternative to the use of air itself to produce electronic equilibrium in the air in the sensitive volume is the use of a much thinner layer of solid material with the same chemical composition as air. Such a material, whose physical properties (except density) are the same as those of air for the purposes of radiation absorption, is referred to as an *air-equivalent* material. Materials having mean atomic numbers close to that of air approximate an air-equivalent material, and include such organic polymers as Bakelite (a phenolformaldehyde resin), Lucite (polymethylmethacrylate), and nylon (adipic acid-hexamethylenediamine polyamide). Thimble ionization chambers in which a small volume of air (the sensitive volume) is surrounded by a thin shell of polymer are available in a wide variety of forms, are both rugged and portable, and offer a convenient and sensitive means of measuring exposure with x- and γ-radiation.

An example of a thimble ionization chamber is the Victoreen condenser r-meter, shown in Fig. 3.3. The small air volume is enclosed by a Bakelite cap, which is made conducting on the inside by a coating of colloidal carbon, and in the center is an aluminum electrode, insulated from the cap. The instrument acts as a small condenser and before use is charged to an appropriate voltage with the device shown. When the chamber is irradiated, the charge is partly neutralized and the drop in voltage, multiplied by the capacitance of the instrument, gives a measure of the ionization produced by the radiation. The charging device is also used to measure the voltage drop, and is calibrated in roentgens; corrections are applied to the observed exposure for temperature and barometric pressure, and also an instrument correction found by comparing the thimble chamber with a standard air chamber. The calibration is generally independent of energy over a considerable range of x-ray energies.

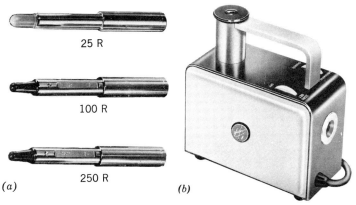

FIGURE 3.3   *Victoreen condenser r-meter. (a) Ionization chambers for measuring doses up to 25R, 100R, and 250R respectively (the dose which can be measured is increased as the ionization chamber, the plastic cap on the left of the instrument, is made smaller). (b) The charging and reading device.*

At high photon energies the thickness of the thimble chamber wall is not sufficient to ensure electronic equilibrium, and the chamber must be surrounded by sufficient air-equivalent material to fulfill this condition (Table 3.2, column 2). Large thicknesses of the air-wall material will attenuate the radiation (Table 3.2, column 3), and a correction for this must be applied to obtain the exposure in free air at the point of interest. Up to energies of a few megaelectron volts the correction may be made by plotting the observed

TABLE 3.2   *Typical Values of Equilibrium Thickness and Wall Attenuation for X-Rays (ref. 14)*

Tube Potential (MV)	Equilibrium Thickness (g cm^{-2})	Approximate Wall Attenuation (%)
0.2	<0.05	<0.2
1.0	0.2	0.6
2.0	0.4	1
5.0	1.0	3
10	2.0	7
20	4	10
50	7	20
100	11	30
Co^{60} γ-rays (1.25 MeV)	0.4	1

Courtesy Miss E. M. Whyte, Ottawa, Ont.

ionization against the thickness of the absorber surrounding the thimble chamber, using various thicknesses, and extrapolating the curve back to zero thickness. The necessity for electronic equilibrium makes it increasingly difficult to determine the exposure in roentgens as the energy of the x- or γ-radiation approaches very high values. For this reason 3 MeV is sometimes arbitrarily taken as the upper limit of the energy range over which the roentgen should be used.

Ritz and Attix (15) have described a graphite cavity ionization chamber suitable for use with kilocurie cobalt-60 sources which allows the absolute determination of exposure dose rates to within 2% at dose rates up to 10^7 R/hr.

Calculation of Absorbed Dose from Ionization Measurements

The absorbed dose in air, or air-equivalent material, exposed to x- or γ-radiation and in electronic equilibrium can be derived using the definition of the roentgen. By definition, 1 R of x- or γ-radiation produces in 1-kg air electrons, or positive ions, carrying 2.58×10^{-4} C of charge, and therefore,

$$1 \text{ R} \equiv 2.58 \times 10^{-4} \frac{(C)}{(kg)} \times 6.241 \times 10^{18} \frac{\text{(electron charges)}}{(C)}$$

$$\times 33.73 \frac{(eV)}{\text{(electron, or ion-pair)}} \times 1.602 \times 10^{-19} \frac{(J)}{(eV)} \times 100 \frac{\text{(rad kg)}}{(J)}$$

$$= 0.870 \text{ rad} \quad or \quad 8.70 \times 10^{-3} \text{ J kg}^{-1} \text{ (in air)} \tag{3.1}$$

The main uncertainty in this calculation is the value of W_{air}, for which we have used 33.73 ± 0.15 eV per ion-pair as recommended by Myers (16) for high-energy electrons, x- and γ-rays; the uncertainty quoted leads to a value of 0.870 ± 0.004 for the conversion factor rad per roentgen. Other values of W_{air}, notably 32.5 and 34.0 eV per ion-pair, have been widely used in the past and give slightly different values for the conversion factor. However, results obtained using these values of W_{air} can be brought into line with current calculations by multiplying by $33.73/W$, where W is the value originally used.

The relationship expressed in Eq. 3.1 is true for x- and γ-radiation of all energies greater than 20 keV (i.e., photon energies for which W is assumed to be constant) when absorbed in air. For other materials the value of the constant varies with the atomic composition of the material and the energy of the radiation.

Ionization measurements can be used to determine the absorbed dose in media other than air in either of two ways: first, by measuring the exposure dose in air at a point, using an ionization chamber, and then substituting the material to be irradiated at the same position, so that it is exposed to the same beam of radiation; second, by taking measurements of the ionization with a thimble chamber actually inside the material. The two problems differ somewhat and will be treated separately.

Case I. EXPOSURE DOSE MEASURED IN AIR. It will be assumed that:

(i) both the ionization chamber and the sample are small enough not to attenuate the radiation appreciably,
(ii) both the ionization chamber and the sample are in electronic equilibrium,
(iii) the scattered electromagnetic radiation is not absorbed by either the chamber or the sample,
(iv) the thimble chamber has been calibrated by comparison with a standard chamber for radiation of the appropriate energy, and the exposure correctly measured at the point to be occupied by the sample.

Given these conditions the energy absorbed by the sample and by air are proportional to their mass energy absorption coefficients (Eq. 2.50) and

$$D_M = D_A \times \frac{(\mu_{en}/\rho)_M}{(\mu_{en}/\rho)_A} \text{ rads} \tag{3.2}$$

where D_M and D_A are the absorbed dose in the medium and in air respectively and $(\mu_{en}/\rho)_M$ and $(\mu_{en}/\rho)_A$ are the corresponding mass energy absorption coefficients. From Eq. 3.1

$$D_A = 0.870 X_A \text{ rads} \tag{3.3}$$

where X_A is the exposure in roentgens, and therefore

$$D_M = 0.870 X_A \times \frac{(\mu_{en}/\rho)_M}{(\mu_{en}/\rho)_A} = f X_A \text{ rads} \tag{3.4}$$

Table 3.3 lists representative values of mass energy absorption coefficients for monoenergetic photons and elements of increasing atomic number. In Table 3.4 these values are compared with the corresponding mass attenuation and mass energy-transfer coefficients for several of the elements. Values of the conversion factor, f rad/R, calculated using Eq. 3.4 are given in Table 3.5 for elements with a range of atomic numbers and for water. Mass energy-absorption coefficients for the latter can be obtained by adding the appropriate coefficients for the elements present, hydrogen and oxygen, weighted in proportion to their mass in the molecule, i.e., $(\mu_{en}/\rho)_{H_2O} = 2.02/18.02(\mu_{en}/\rho)_H + 16.00/18.02(\mu_{en}/\rho)_O$; alternatively the f factors for hydrogen and oxygen may be combined in the same manner. Other compounds and mixtures may be treated in the same way. For a given material, the value of f is constant in the range of photon energies where only Compton interactions occur, but varies with the energy of the incident radiation in the regions where photoelectric

TABLE 3.3 *Mass Energy-Absorption Coefficients for Monoenergetic Photons* $(\mu_{en}/\rho; cm^2 g^{-1})^a$

Photon Energy (MeV)	$_1$H	$_6$C	$_7$N	$_8$O	$_{14}$Si	$_{16}$S	$_{26}$Fe	$_{82}$Pb	air[b]
0.01	0.00986	1.97	3.38	5.39	33.3	49.7	142	130.7	4.61
0.05	0.0271	0.0233	0.0319	0.0437	0.241	0.372	1.64	6.54	0.0406
0.10	0.0406	0.0215	0.0224	0.0237	0.0459	0.0609	0.219	2.28	0.0234
0.5	0.0593	0.0297	0.0296	0.0297	0.0298	0.0300	0.0295	0.0951	0.0296
1.0	0.0555	0.0279	0.0279	0.0278	0.0277	0.0278	0.0262	0.0377	0.0278
1.5	0.0507	0.0255	0.0255	0.0254	0.0253	0.0253	0.0237	0.0271	0.0254
5	0.0317	0.0171	0.0173	0.0175	0.0187	0.0192	0.0198	0.0259	0.0174
10	0.0225	0.0138	0.0143	0.0148	0.0175	0.0184	0.0209	0.0310	0.0145

[a] Values are selected from the comprehensive tables prepared by J. H. Hubbell and quoted by Evans (17; see also Refs. 18, 19)

[b] 75.5% N_2, 23.2% O_2, 1.3% Ar by weight.

TABLE 3.4 *Values of Mass Attenuation* (μ/ρ), *Mass Energy-Transfer* (μ_K/ρ), *and Mass Energy-Absorption* (μ_{en}/ρ) *Coefficients* $(cm^2 g^{-1})^a$.

Material	Photon Energy (MeV)	μ/ρ With Coherent Scattering	μ/ρ Without Coherent Scattering	μ_K/ρ	μ_{en}/ρ
Water	0.01	5.21	4.99	4.79	4.79
	0.1	0.171	0.168	0.0256	0.0256
	1	0.0706	—	0.0311	0.0309
	10	0.0222	—	0.0162	0.0157
$_6$C	0.01	2.32	2.16	1.97	1.97
	0.1	0.152	0.149	0.0215	0.0215
	1	0.0635	—	0.0279	0.0279
	10	0.0196	—	0.0143	0.0138
$_{26}$Fe	0.01	172.6	171.6	142	142
	0.1	0.370	0.342	0.219	0.219
	1	0.0600	0.0596	0.0264	0.0262
	10	0.0299	—	0.0238	0.0209
$_{82}$Pb	0.01	136.6	132.1	131.0	130.7
	0.1	5.78	5.62	2.28	2.28
	1	0.0708	0.0689	0.0396	0.0377
	10	0.0496	—	0.0419	0.0310

[a] Values are selected from the tables prepared by J. H. Hubbell and quoted by Evans (17; see also Refs. 18, 19).

TABLE 3.5   *Absorbed Dose per Unit Exposure for Monoenergetic Photons*

Photon Energy (MeV)	$f = 0.870 \times \dfrac{(\mu_{en}/\rho)_M}{(\mu_{en}/\rho)_A}$ rad/R[a]					
	$_1$H	$_6$C	$_8$O	$_{26}$Fe	$_{82}$Pb	Water
0.01	0.00186	0.372	1.017	26.8	24.7	0.904
0.05	0.581	0.499	0.937	35.2	140	0.896
0.10	1.51	0.800	0.881	8.14	84.8	0.952
0.5	1.74	0.873	0.873	0.867	2.80	0.970
1.0	1.74	0.873	0.870	0.820	1.18	0.967
1.25 (^{60}Co γ-rays)	1.74	0.873	0.870	0.816	1.06	0.967
5.0	1.59	0.855	0.875	0.990	1.30	0.955
10.0	1.35	0.828	0.888	1.25	1.86	0.942

[a] Values of μ_{en}/ρ were taken from Evans (17). $W_{air} = 33.73 \pm 0.15$ eV per ion-pair (16).

absorption and pair-production predominate. If the radiation is not mono-energetic, values of f must be calculated for all energies present and combined in proportion to the energy fluence for photons of each energy to give a mean value, $\bar{f}$.

If photon absorption is entirely by the Compton process, Eq. 3.4 can be simplified to

$$D_M = 0.870 X_A \times \frac{(Z/A)_M}{(Z/A)_A} = f X_A \text{ rads} \qquad (3.5)$$

which does not require knowledge of the mass energy-absorption coefficients, since Z/A is simply the ratio of atomic number to atomic weight for the material. This follows because Compton absorption is proportional to the number of electrons in the medium and not to the way in which they are bound in atoms, and the number of electrons per unit mass of material is proportional to Z/A (this is discussed more fully in Chapter 2). For materials other than elements a mean value of Z/A is used, calculated from

$$\overline{Z/A} = \sum w_i \left(\frac{Z}{A}\right)_i \qquad (3.6)$$

w_i being the fraction by weight of the ith element in the medium. For a compound this is simply the sum of the atomic numbers of the atoms present divided by the molecular weight of the compound, e.g., for water $\overline{Z/A} = (2 \times 1 + 8)/18.02 = 0.555$. Table 3.6 lists values of f calculated using both Eqs. 3.5 and 3.4 (assuming a photon energy of 1 MeV); the agreement ranges from excellent for the low-Z materials to poor for such high-Z materials as lead. Comparison of Tables 3.5 and 3.6 shows that Eq. 3.5 (which gives a single value of f irrespective of photon energy) can be used over a fairly

T A B L E 3.6   *Absorbed Dose per Unit Exposure for Photon Absorption in the Compton Absorption Region*

Material	Z/A	f, rad/R^a Calc. Using Eq. 3.5	Calc. Using Eq. 3.4[b]
Air[c]	0.4992	0.870	0.870
$_6C$	0.4995	0.871	0.873
$_8O(=O_2)$	0.5000	0.871	0.870
$_{26}Fe$	0.4656	0.812	0.820
$_{82}Pb$	0.3958	0.690	1.18
Water	0.5551	0.968	0.967
0.4 M H_2SO_4[d]	0.5533	0.964	0.964
CH_4	0.6233	1.09	1.09
CH_3COOH	0.5329	0.929	0.930
H_2NCONH_2	0.5328	0.929	0.930
$CaCO_3$	0.4996	0.871	0.870

[a] $W_{air} = 33.73 \pm 0.15$ eV per ion-pair (16).
[b] Calculated for 1 MeV photons using μ_{en}/ρ values from Evans (17).
[c] 75.5% N_2, 23.2% O_2, 1.3% Ar by weight.
[d] 10.84% H_2, 87.91% O_2, 1.25% S by weight.

wide range of photon energies if the material is composed predominantly of low-Z elements, but that it will introduce significant errors if medium and high-Z elements are major components of the system. The latter interact mainly by photoelectric absorption at low photon energies and by pair-production at high energies. The same arguments are employed when comparing absorbed dose in liquid and solid systems exposed to the same x- or γ-ray field (cf., p. 111 and Fig. 3.5).

In large samples of material all the scattered and secondary electromagnetic radiation may not escape, and allowance must be made for the additional energy absorbed. The nature of the corrections and the use of backscatter tables are discussed by Johns (20). With large samples the exposure is best measured at the position to be occupied by the surface of the sample, and the exposure at different depths in the material calculated from attenuation data or depth-dose curves.

Case II. ABSORBED DOSE BY CAVITY IONIZATION. Ionization measurements made inside the irradiated medium with a gas-filled thimble chamber, or other cavity ionization chamber, can be used to calculate the absorbed dose by means of the *Bragg-Gray cavity principle* (21, 22). This is expressed by the equation

$$E_M = J_G W_G (s_m)_{gas}^{medium} \tag{3.7}$$

where E_M is the energy absorbed by the medium in erg g^{-1}, J_G is the ionization produced in the gas-filled cavity in esu g^{-1}, W_G is the mean energy expended by the secondary electrons crossing the cavity per ion pair formed in the gas (ergs ion-pair^{-1}), and $(s_m)_{\text{gas}}^{\text{medium}}$ is the ratio of the mass stopping power of the medium to that of the gas for these secondary electrons.[1] The principle follows from the assumption that if the cavity is sufficiently small it will not alter the electron fluence in the medium when it is introduced. For the principle to be valid, (1) the cavity must be small compared to the range of the ionizing particles (i.e., electrons for electromagnetic radiation) in the gas, (2) direct interaction between the incident radiation and the gas should contribute little to the ionization in the cavity, and (3) the incident radiation must not be appreciably attenuated in the medium over a distance equal to the dimensions of the cavity. The conditions for validity have been discussed by a number of authors (12, 13, 25, 26). From the definition of the rad it follows that the absorbed dose in the medium is

$$D_M = 0.01 E_M \text{ rads} \tag{3.8}$$

If the cavity gas is air ($W = 33.73$ eV ion-pair^{-1}) and the ionization produced is expressed in terms of Q, the charge in esu carried by ions of either sign per 0.001293 g of air (i.e., per cubic centimeter air at $0°$C and 760 mm Hg)

$$D_M = 0.01 \frac{(\text{rad})}{(\text{erg g}^{-1})} \times \frac{Q}{0.001293} \frac{(\text{esu})}{(\text{g air})} \times 2.082 \times 10^9 \frac{(\text{electrons})}{(\text{esu})}$$

$$\times 33.73 \frac{(\text{eV})}{(\text{electron, or ion-pair})} \times 1.602 \times 10^{-12} \frac{(\text{erg})}{(\text{eV})} \times (s_m)_{\text{air}}^{\text{medium}}$$

$$= 0.870 Q \times (s_m)_{\text{air}}^{\text{medium}} \text{ rads} \tag{3.9}$$

Electronic equilibrium, which is so important when measuring the exposure dose, is no longer essential when the absorbed dose is determined by cavity ionization. The cavity ionization measurements are determined by the secondary electron fluence at the point of interest, and this also determines the energy absorbed at that point. Near the surface of an absorber, where electronic equilibrium has not yet been reached, the electron fluence will be smaller than at a greater depth in the medium, but both the observed ionization[2] and the energy absorbed will also be smaller. For the same reason,

---

[1] The mass stopping power ratio can be calculated approximately from the stopping power equation derived by Bethe (Eq. 2.8 for electrons):

$$(s_m)_{\text{gas}}^{\text{medium}} \sim \frac{(S/\rho)_{\text{medium}}}{(S/\rho)_{\text{gas}}}$$

The quantity is more complex than this equation would suggest, however, and it is influenced by other factors such as the size and shape of the cavity (13, 23, 24).

[2] An ionization chamber with a very thin wall, or a wall of material similar in composition to the medium, would be used.

scattered and secondary electromagnetic radiation absorbed in the medium need not be corrected for separately, since electrons produced by such radiation, and passing through the point of interest, will contribute to the measured ionization. It might be pertinent to point out also that cavity ionization measurements can be used to measure other ionizing radiation provided that the requirements for the Bragg-Gray principle to be valid can be met and that the principle is not limited to gas-filled cavities, the same theoretical approach being valid for liquid and solid "cavities" (13, 26).

Equation 3.9 is derived making the assumption that the secondary electrons crossing the cavity originate in the medium. That is to say, either the walls of the ionization chamber are of the same composition as the medium or else are very thin, compared to the range of the secondary electrons crossing them. If this assumption is not true, and the secondary electrons crossing the cavity are produced entirely within the walls of the ionization chamber itself, then

$$D_M = 0.870Q \times (s_m)_{\text{air}}^{\text{wall}} \times \frac{(\mu_{\text{en}}/\rho)_M}{(\mu_{\text{en}}/\rho)_{\text{wall}}} \text{ rads} \qquad (3.10)$$

When, under these circumstances, the walls of the ionization chamber are made of an air-equivalent material, the stopping power ratio $(s_m)_{\text{air}}^{\text{wall}}$ has the value 1, and Eq. 3.10 becomes

$$D_M = 0.870Q \times \frac{(\mu_{\text{en}}/\rho)_M}{(\mu_{\text{en}}/\rho)_A} \text{ rads} \qquad (3.11)$$

In this equation Q is equal to the exposure in roentgens measured at the point of interest in the medium[3], apart from this the equation is similar to Eq. 3.4. It should be noted, however, that because of the attenuating and scattering effect of the medium the exposure Q will not be the same as the exposure in air (X_A) at the same point if the medium were removed. The thickness of absorber in front of the ionization chamber may attenuate the radiation reaching the chamber to a very much greater extent than an equal thickness of air, in which case Q would be very much smaller than X_A, though both could be termed an exposure dose at that particular point. A distinction is sometimes made between the two types of exposure by referring to the exposure measured in air as an *air-dose*.

Between the limits where the secondary electrons crossing the cavity are produced entirely within the medium or entirely within the walls of the

[3] Q is also equal to the exposure dose in roentgens, even if the wall material is not air-equivalent, provided the ionization chamber has been calibrated against a standard chamber. In this case $(s_m)_{\text{air}}^{\text{wall}}$ is included in the calibration correction. The secondary electrons crossing the cavity must, however, still be produced entirely within the walls of the ionization chamber.

ionization chamber, the observed ionization will lie between that found for the two limiting cases, and the absorbed dose will lie between that given by Eq. 3.9 and that given by Eqs. 3.10 or 3.11. Whether or not a given ionization chamber can be classed as thin or thick walled depends, of course, on the energy of the incident radiation. A chamber with walls of air-equivalent material of thickness about 0.5 g cm^{-2} would be thick walled with respect to ^{60}Co γ-rays but thin walled with respect to 100 MVp x-rays (Table 3.2).

Tables of mass stopping-power ratios relative to air, and other data useful in determining the absorbed dose from cavity-ionization measurements with air-filled chambers have been published (13, 25–28). The range of values to be expected is illustrated in Table 3.7; more precise values, applicable to specific ranges of electron and photon energies, are given by Burlin (13, 26).

TABLE 3.7 *Mean Mass Stopping-Power Ratios Relative to Air,*
$(s_m)_{air}^M$ or $(s_m)_{air}^{wall}$ *(Ref. 14).*

Photon Energy (MeV)	Graphite	Polystyrene	Lucite	Bakelite	Aluminum	Water
0.1	1.01	1.14	1.12	1.09	0.85	1.17
0.5	1.01	1.12	1.11	1.08	0.88	1.16
1.0	1.00	1.11	1.11	1.08	0.88	1.15
3.0	0.99	1.09	1.09	1.05	0.89	1.13
Co^{60} γ-rays (1.25 MeV)	1.00	1.11	1.11	1.07	0.89	1.15

DOSIMETRY OF CHARGED PARTICLES

Two features distinguish charged-particle dosimetry from that of x- and γ-radiation. First, the fact that the particles are charged means that information about the beam can be gained by collecting their charge and measuring it. The second feature arises from the greater rate of energy loss by charged particles which leads to large variations in the absorbed dose over comparatively small distances, and to the complete stopping of a charged-particle beam by a thickness of absorber that would barely attenuate an x- or γ-ray beam. Regardless of other considerations, the latter must obviously influence the design of instruments for use with these less-penetrating radiations.

Charge-Collection Measurements

The particle flux density (φ) of a beam of charged particles from an accelerator can be determined by measuring the total charge carried by the beam by means of a Faraday cup (or Faraday cage) (29–31) consisting of a metal block thick enough to stop the beam completely, supported on insulators inside an evacuated chamber (Fig. 3.4). The block acquires the charge of every particle absorbed, and the current flowing from it is a direct measure of the number of particles entering the chamber; if necessary, a correction is applied for loss of charge due to particles scattered out of the block.

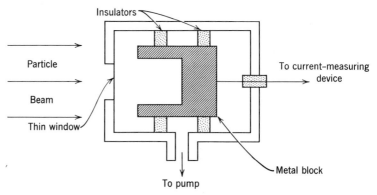

FIGURE 3.4   *Faraday cage.*

In order to obtain the energy flux density (ψ) or intensity of the radiation, the energy (E) of the particles is also required. This will often be known from the characteristics of the accelerator, or it can be determined from the range of the particles or from magnetic-deflection measurements (29, 31). Then

$$\psi(\text{J m}^{-2} \text{ sec}^{-1}) = \varphi(\text{m}^{-2} \text{ sec}^{-1}) \times E(\text{J}) \qquad (3.12)$$

The absorbed dose in a thin target through which the beam passes is given by

$$D_M = 100 \times \Phi(\text{particles m}^{-2}) \times \frac{S}{\rho}(\text{J m}^2 \text{ kg}^{-1}) \text{ rads} \qquad (3.13)$$

$$= 1.602 \times 10^{-8} \times \Phi(\text{particles cm}^{-2}) \times \frac{S}{\rho}(\text{MeV cm}^2 \text{ g}^{-1}) \text{ rads} \quad (3.14)$$

where Φ is the particle fluence (the time integral of the particle flux density) and S/ρ the mass stopping power of the absorbing material (the value of the collision mass stopping power depends on the nature of the particle and absorber and the particle energy; data for electrons are given in refs. 31 to 33). For other than very thin samples the particle fluence and mass stopping power should be integrated over the range of particle energies found in the absorber. If the target is thick enough to absorb the beam completely, the absorbed dose in the target is the total energy of the beam less energy scattered by bremsstrahlung emission; ignoring bremsstrahlung losses, the average absorbed dose in an absorber of thickness d (cm) and density ρ (g cm^{-3}) will be

$$D_M = 1.602 \times 10^{-8} \times \Phi(\text{particles cm}^{-2}) \times \frac{E}{\rho d} \text{ rads} \qquad (3.15)$$

where E (MeV) is the particle energy.

Ionization Measurements

Ionization chambers are generally used as secondary instruments with charged-particle radiation and calibrated against a Faraday cup. They may be either parallel-plate chambers, similar to the standard free-air chamber used with x-rays, or cavity chambers; in either case the design is influenced by the need to keep absorbing material in the path of the beam to a minimum. Chambers for use with electron beams have been described by Boag (12) and Laughlin (29), and for heavy charged-particle beams by Raju et al. (30). When the sample is sufficiently thick and the range of the particles great enough, the absorbed dose is generally measured by cavity ionization using a very thin-walled ionization chamber. The Bragg-Gray principle is valid provided the conditions concerning the size of the cavity are met (12, 25, 26), and the absorbed dose is given by

$$D_M = 6.241 \times 10^{20} \times Q(\text{C kg}^{-1}) \times W_G(\text{J}) \times (s_m)_{\text{gas}}^{\text{medium}} \text{ rads} \qquad (3.16)$$

$$= \frac{10^5 \times Q'(\text{C}) \times W_G(\text{eV}) \times (s_m)_{\text{gas}}^{\text{medium}}}{\rho_G V_G} \text{ rads} \qquad (3.17)$$

where Q and Q' are the charge collected in coulomb per kilogram cavity gas and coulomb respectively, W_G (J or eV) is the energy necessary to form an ion-pair in the cavity gas (density ρ_G g cm^{-3}), V_G (cm^3) the collecting volume, and $(s_m)_{\text{gas}}^{\text{medium}}$ the ratio of the mass stopping powers of the medium and the cavity gas for the ionizing particles crossing the cavity. An additional factor may be included in the expression to correct for perturbations in the electron fluence in the medium caused by introducing the ionization chamber (31). The mass stopping-power ratio will vary according to the type and mean energy of the particles producing ionization in the cavity (26, 30). W_G is also dependent on the nature of the ionizing particles but is the same for electron, hard x-ray, and γ-radiation, thus $W_{\text{air}} = 33.73 \pm 0.15$ eV per ion-pair for electrons having energies up to 50 MeV (16, 31).

Other Measurements

Calorimetric determination of the energy absorbed from beams of charged particles is simpler than for x- or γ-radiation because of the greater radiation intensities involved, and a number of authors have described calorimeters for this purpose (31, 34–37).

At energies above about 20 MeV, flux densities of heavy particles can be estimated by means of nuclear reactions giving radioactive products that the particles initiate in suitable stopping materials. The number of radioactive nuclei formed can be measured by standard counting techniques and then used to calculate the particle flux density (30).

Solid state and chemical dosimeters can also be used with charged-particle radiation and are described later in this chapter.

Electron-beam monitors, which give a continuous indication of electron fluence rate or other characteristic of the beam, are often used with electron accelerators in conjunction with dosimetric measurements and are described in ref. 31.

It should be noted in connection with the dosimetry of heavy charged particles that the rate of energy deposition, or LET, is not constant along the particle track but increases sharply toward the end of the track (i.e., the Bragg curves in Fig. 2.24, p. 62). Most dosimeters give the average dose received by the active volume of the dosimeter, masking the fact that the dose may not be absorbed uniformly throughout the sample.

NEUTRON DOSIMETRY

Neutrons do not, themselves, produce ionization in matter, and they are detected by means of their interactions with nuclei. The kerma from neutron irradiation can be calculated if the number and energy distribution of the incident neutrons are known; the kerma is equal to the absorbed dose if secondary particle equilibrium exists. A discussion of the methods by which these quantities can be determined is beyond the scope of this book but is included in many that deal specifically with dosimetry (e.g., refs. 38 to 42) and in others concerned with nuclear physics and reactor technology. Ionization measurements can also be used to determine the absorbed dose, and these are briefly described below.

Ionization Measurements

The Bragg-Gray principle (Eq. 3.7) is applicable to cavity ionization measurements made in neutron fields, and it enables the absorbed dose to be calculated from such measurements.

In practice, cavity ionization measurements have been used most frequently to determine the absorbed dose for fast neutrons in biological tissue, where most of the neutron energy is transferred to hydrogen nuclei (protons). Ionization chambers for this purpose are made with walls of hydrogen-rich material so that ionization in the gas-filled cavity, like the energy deposition in the tissue, is brought about by recoil protons. In these circumstances W_G, the mean energy expended in producing an ion-pair in the cavity gas, will not be the same for neutron irradiation as for x- and γ-ray irradiation, since the particles causing ionization, protons, and electrons, respectively, are not the same. When the cavity gas is air, W_{air} (protons) $= 36.0 \pm 0.4$ eV per ion-pair (16). The stopping power ratio, $(s_m)_{gas}^{medium}$, will also be different when ionization in the cavity gas is brought about by protons rather than electrons. However, uncertainty regarding the stopping power ratio may be avoided by choosing gas and wall materials with the same atomic composition, e.g., ethylene and polyethylene, when $(s_m)_{gas}^{medium} = 1$. This has the further advantage that restrictions on the cavity size, which are particularly stringent for neutrons because of the short range of the recoil protons, are removed.

If the gas filling the cavity has the same composition as the wall material, and the ions, of either sign, produced in the cavity carry a charge of Q esu cm^{-3} at NTP, the absorbed dose in the wall material is given by

$$D_{wall} = 0.01 \frac{(\text{rads})}{(\text{erg g}^{-1})} \times Q \frac{(\text{esu})}{(\text{cm}^3)} \times \frac{1}{\rho} \frac{(\text{cm}^3)}{(\text{g})}$$

$$\times 2.082 \times 10^9 \frac{(\text{electrons})}{(\text{esu})} \times W_G \frac{(\text{eV})}{(\text{electron, or ion-pair})}$$

$$\times 1.602 \times 10^{-12} \frac{(\text{erg})}{(\text{eV})}$$

$$= 3.34 \times 10^{-5} Q W_G / \rho \text{ rads} \tag{3.18}$$

where ρ is the density of the gas (g cm^{-3}).

The absorbed dose in the medium surrounding the ionization chamber is equal to D_{wall} if both the medium and the wall material have the same atomic composition. However, if their composition is different the absorbed dose in the medium is

$$D_{medium} = D_{wall} \times \frac{(\mu_K/\rho)_{medium}}{(\mu_K/\rho)_{wall}} \tag{3.19}$$

where μ_K/ρ is a mass energy transfer coefficient averaged, if necessary, over the types of atoms and neutron energies present (42).

Neutron dosimetry is complicated by the fact that neutrons are very often accompanied by electromagnetic radiation which, for many purposes, must be estimated separately. When the incident radiation is a mixture of neutrons and γ-rays, cavity ionization chambers of the type considered above (i.e., with gas and walls of similar hydrogenous material) will respond to both types of radiation. The measured ionization, Q, will be related to the total absorbed dose by Eq. 3.18, though W_G should now be a mean value, since both protons and electrons will be ionizing the cavity gas. The appropriate mean value for W_G can be estimated if the relative contribution to the total absorbed dose from neutrons and γ-rays is known approximately. If the wall material and the surrounding medium have the same composition $D_{\text{medium}} = D_{\text{wall}}$; otherwise the absorbed dose in the wall due to neutron interactions can be converted to the absorbed dose in the medium using Eq. 3.19, and the absorbed dose due to the γ-rays similarly converted using the ratio of the mass energy absorption coefficients for the medium and wall material (Eq. 3.2). Some knowledge of the relative neutron and γ-ray contributions to the absorbed dose is needed to make this conversion from D_{wall} to D_{medium}.

The γ-ray contribution to the absorbed dose can be estimated by using a second cavity ionization chamber, similar in size and shape to the first, but with thick walls of a material that contains no hydrogen (e.g., aluminum, graphite, or teflon). The second chamber will have a lower response to neutrons than the first but a roughly similar response to γ-rays, so that it is possible to discriminate between the effects due to these two types of radiation (41).

DOSIMETRY OF INTERNAL RADIATION SOURCES

The short range of α- and β-particles limits the quantity of material that can be irradiated by an external α- or β-source, and it is sometimes convenient to mix the radioactive material with the material to be irradiated, and so achieve uniform irradiation of a large sample. The most frequently used isotopes for this purpose are the β-emitting phosphorus-32 and sulfur-35 and, rather less often, α-emitting radon and β-emitting tritium; it is usually a prerequisite that the isotope decay quickly, and that the decay products be stable, so that radioactive residues from the experiment do not present a disposal problem.

Generally the sample irradiated will be large, compared to the range of the ionizing particles in the medium, and the absorbed dose in the sample can be assumed to be the total energy liberated by the disintegrating nuclei.[4] Loss of energy by the escape of secondary electromagnetic radiation from

---

[4] Actually that part of the total energy liberated that is transferred to ionizing particles; energy liberated in the form of energetic γ-rays or antineutrinos does not contribute to the absorbed energy being considered in this section.

the sample can usually be disregarded. If the concentration of the radioactive isotope present in the sample is C mCi g^{-1} (equivalent to $3.7 \times 10^7 C$ disintegrations g^{-1} sec^{-1}) and the mean energy released by each disintegration and transferred to the ionizing particles is $\bar{E}$ MeV, then the absorbed dose rate, $\dot{D}_M$, is given by

$$\dot{D}_M = 3.7 \times 10^7 C \frac{(\text{disintegrations})}{(\text{g sec})} \times \bar{E} \frac{(\text{MeV})}{(\text{disintegration})}$$

$$\times 1.602 \times 10^{-8} \frac{(\text{g rad})}{(\text{MeV})}$$

$$= 0.5927 \times C \times \bar{E} \text{ rads sec}^{-1} \tag{3.20}$$

The concentration of the radioisotope (C) can be found by counting the number of particles emitted per unit time by a small aliquot of the radioactive material. Any counting technique can be used, but the result must be corrected to give an absolute value of the activity present; i.e., the total number of nuclei disintegrating per gram-second rather than the number of disintegrations per gram-second that happen to be observed. Allowance must also be made for radioactive decay in the period between counting the activity and mixing the active material with the sample.

Radioactive decay will also occur during the irradiation itself. For irradiations that last a much shorter time than the half-life of the isotope being used, it is sufficient to calculate the dose rate ($\dot{D}_M$) from the activity present at the midpoint of the irradiation and to use this value to calculate the total absorbed dose. For longer irradiations, for example longer than the half-life of the isotope, the absorbed dose rate should be integrated over the period of the irradiation (t) to give the true absorbed dose (D_M), i.e.,

$$D_M = \int_0^t \dot{D}_M \, dt \tag{3.21}$$

Substituting for $\dot{D}_M$ from Eq. 3.20,

$$D_M = 0.5927\bar{E} \int_0^t C \, dt \text{ rads} \tag{3.22}$$

For radioactive decay,

$$C_t = C_0 e^{-\lambda t} \tag{3.23}$$

where C_0 and C_t (millicuries per gram) are the concentration of radioactive material present in the sample at the start of the irradiation and after time t (sec) respectively, and λ (sec^{-1}) is the decay constant for the isotope being used. Substituting for C the integration (Eq. 3.22) becomes

$$D_M = 0.5927 \times \bar{E} \times C_0 \int_0^t e^{-\lambda t} \, dt \text{ rads} \tag{3.24}$$

and, integrating,

$$D_M = \frac{0.5927 \times \bar{E} \times C_0}{\lambda} (1 - e^{-\lambda t}) \text{ rads} \qquad (3.25)$$

for an irradiation lasting t seconds. If λ and t are expressed in min^{-1} and minutes respectively, the numerical constant in Eq. 3.25 becomes 35.56, and if the units are days^{-1} and days, the constant has the value 5.121×10^4. The greatest absorbed dose is obtained if the mixture is left until the radioactivity has decayed completely. In this case $t \to \infty$ and Eq. 3.25 becomes

$$D_M = \frac{0.5927 \times \bar{E} \times C_0}{\lambda} \text{ rads} \qquad (3.26)$$

Values for the energy of the particles emitted by radioactive nuclei can be found in most collections of nuclear data; a few values are given in Table 2.1. For β-particles the average β-particle energy is used in Eqs. 3.20 to 3.26. When more than one particle is expelled for each disintegration of the parent isotope (i.e., when the daughter elements are radioactive and have much shorter half-lives than the parent element), $\bar{E}$ is the mean energy carried by all these particles, per disintegration of the parent. An example of this is the energy released by radon when it is in equilibrium, in the sample, with its daughter elements (Table 3.8). The energy released was calculated using Eq. 3.20, taking $\bar{E}$ as the sum of the energies of the individual particles.

T A B L E 3.8 *Energy Release by Radon in Equilibrium with Its Daughter Elements*

Element	Half-Life	Particle Emitted	Particle Energy (MeV)	Rate of Energy Release Per Millicurie of Radon per Gram (rads sec^{-1})
Radon	3.83 days	α	5.49 ⎫	
RaA (^{218}Po)	3.05 min	α	6.00 ⎬ Total Energy	11.4
RaC' (^{214}Po)	10^{-4} sec	α	7.68 ⎭ $\bar{E} = 19.17$	
RaB (^{214}Pb)	26.8 min	β	0.23 (av) ⎫ Total Energy	
RaC (^{214}Bi)	19.7 min	β	0.76 (av) ⎭ $\bar{E} = 0.99$	0.59

Energy transferred to the recoiling nucleus by the expulsion of an α- or β-particle ought also to be included in the mean energy of the disintegration ($\bar{E}$), though in practice this may not be very significant. The energy transferred to the nucleus by loss of an α-particle, calculated from the equation for the conservation of momentum, is

$$E_{\text{nucleus}} = E_\alpha \times \left(\frac{\text{mass } \alpha\text{-particle}}{\text{mass recoiling nucleus}} \right) \qquad (3.27)$$

In the example given in Table 3.8 the energy transferred to the nuclei by the loss of the three α-particles totals 0.36 MeV; inclusion of this energy raises the absorbed dose rate for α-particle absorption from 11.4 to 11.6 rads sec^{-1}. The fraction of the energy transferred to the nucleus by the loss of a β-particle will be less than that resulting from the loss of an α-particle, and can be disregarded.

CHEMICAL DOSIMETRY

In chemical dosimetry the radiation dose is determined from the chemical change produced in a suitable substrate. Calculation of the dose requires a knowledge of the G value for the reaction or product estimated, which is found by comparing the chemical system with some form of absolute dosimeter. Chemical dosimeters are therefore secondary dosimeters and are used because of their greater convenience. The quantity that they measure is the average absorbed dose in the material composing the dosimeter, which can be converted to the absorbed dose in other materials as described later in this chapter (p. 110). In order to facilitate this conversion and to reduce errors, the dosimeter system is usually chosen so as to have the same atomic composition and density as the sample to be irradiated, as far as this is possible. Aqueous dosimeters, for example, are used if the sample is an aqueous solution, biological material, or organic substance.

In view of the large number of chemical systems that might conceivably be used for dosimetry, it is useful to list some of the features desirable in a dosimeter. The response of the dosimeter should be:

(i) proportional to the radiation dose over a wide range of dose (the range of interest in radiation chemistry is about 10 to 10^8 rads, though no one dosimeter is likely to cover more than a part of this range),

(ii) independent of dose rate (this may range from a few rads per minute to as high as 10^{13} rads sec^{-1} during the individual radiation pulses from an accelerator),

(iii) independent of the energy and LET of the radiation,

(iv) independent of temperature,

(v) reproducible; for many purposes a precision of between $\pm 1\%$ and $\pm 5\%$ is acceptable.

In addition the dosimeter should be:

(vi) stable under normal conditions, such as exposure to light and air, both before and after irradiation,

(vii) simple to use.

Two other desirable characteristics are applicable to chemical dosimeters:

(viii) the dosimeter should be easy to prepare from standard shelf reagents and solvents (i.e., the system should be insensitive to small amounts of impurities and not require elaborate purification of the reagents used) and, for convenience, should not require degassing,

(ix) the response should not be dependent on minor changes in the composition of the dosimeter, for example, upon small changes in the concentration of reagents or pH of the solution.

In practice, no one dosimeter meets all these demands. Of the chemical systems the Fricke, or ferrous sulfate, dosimeter probably comes closest and at the present time is the best understood and most widely used chemical dosimeter. Other systems may be superior for some purposes and a number of these, together with the Fricke dosimeter, are described below after outlining the method of calculating absorbed dose from measured chemical change.

For a dosimeter in which radiation induces a chemical change, the mean absorbed dose (D_D) over the volume occupied by the dosimeter is derived as follows. For any system, by definition, $G(\text{product})$ is the number of molecules of product formed per 100 eV energy absorbed and 1 rad corresponds to an energy absorption of 0.01 J kg^{-1}. Then

$$D_D = \text{moles product formed per kg} \frac{(\text{mol})}{(\text{kg})} \times 6.022 \times 10^{23} \frac{(\text{molecules})}{(\text{mol})}$$

$$\times \frac{100}{G(\text{product})} \frac{(\text{eV})}{(\text{molecule})} \times 1.602 \times 10^{-19} \frac{(\text{J})}{(\text{eV})} \times 100 \frac{(\text{kg rad})}{(\text{J})}$$

$$= 9.647 \times 10^8 \times \frac{\text{moles product formed per kg}}{G(\text{product})} \text{ rads} \qquad (3.28)$$

or

$$D_D = 9.647 \times 10^8 \times \frac{\text{moles product formed per liter}}{\rho G(\text{product})} \text{ rads} \qquad (3.29)$$

where ρ is the density of the system (g cm^{-3}). Very often the yield of product will be determined spectrophotometrically when, assuming Beer's Law to be obeyed,

$$\text{moles product formed per liter} = \frac{\Delta A}{\Delta \varepsilon l} \qquad (3.30)$$

and

$$D_D = 9.647 \times 10^8 \times \frac{\Delta A}{\Delta \varepsilon l \rho G(\text{product})} \text{ rads} \qquad (3.31)$$

where ΔA is the difference in absorbance (or optical density) between the irradiated and nonirradiated solution, $\Delta \varepsilon$ is the difference in molar extinction coefficient (liter mol^{-1} cm^{-1}) of reactant and product at the wavelength being used, and l (cm) is the optical path length (i.e., sample thickness) used when determining the absorbances.

Fricke (Ferrous Sulfate) Dosimeter

The reaction involved in the Fricke dosimeter (28, 31, 43, 44) is the oxidation of an acid solution of ferrous sulfate to the ferric salt, in the presence of oxygen and under the influence of radiation.

Fricke (45) proposed this system as an x-ray dosimeter as early as 1929, and chose 0.4 M sulfuric acid as the solvent so that the response to x-rays should be the same as that of standard air ionization chambers over a range of x-ray energies. A more dilute solution of acid gives a solution closer in composition to biological materials and 0.05 M H_2SO_4 is sometimes used as solvent. However, the standard dosimeter solution is one containing about 10^{-3} M ferrous sulfate or ferrous ammonium sulfate and 10^{-3} M sodium chloride in air-saturated ($\sim 2.5 \times 10^{-4}$ M O_2) 0.4 M sulfuric acid (pH 0.46). The quantities required to prepare such a solution are 0.28 g $FeSO_4 \cdot 7H_2O$ [or 0.39 g $Fe(NH_4)_2(SO_4)_3 \cdot 6H_2O$], 0.06 g NaCl and 22 ml concentrated (95 to 98%) H_2SO_4 per liter of solution; the solution slowly oxidizes and should not be stored longer then a few days. Law and his colleagues (46, 47) have found that traces of reducing impurities present in sulfuric acid affect the yield of the dosimeter and recommend preirradiating the acid or treating it with hydrogen peroxide before making up the dosimeter solution. Chloride is added to the solution to inhibit the oxidation of ferrous ions by organic impurities (48) and is unnecessary if both the water and the reagents used are purified exhaustively. At very high dose rates, such as those used in pulse radiolysis studies, the response of the dosimeter is improved if chloride is *not* present. The purity of the water used as a solvent is a major consideration whenever aqueous solutions are irradiated. Ordinary distilled water is commonly purified by redistillation from an alkaline permanganate solution, and then from an acid dichromate solution, to reduce the amount of organic impurities present; the product is "triply distilled" water. It is even better to reflux the alkaline and acid solutions for extended periods and to finally pass the vapor through a red-hot silica tube, though even this does not suffice to remove all organic material, and final traces are best removed by preirradiating the triply distilled water and then removing traces of peroxide by exposure to ultraviolet light (49). The residual hydrogen peroxide may also be removed by purging the solution with hydrogen (to remove O_2 and CO_2) and then irradiating for a further period in the presence of hydrogen, when the hydrogen peroxide is converted to water by a radiation-induced chain reaction (50). Water deionized by passage through ion-exchange resins generally contains traces of dissolved resin and is not an acceptable substitute for carefully distilled water for radiation-chemical purposes.

The care taken in purifying solvents and reagents should be equalled by the attention given to cleaning glassware which will be in contact with aqueous solutions before, and during, their irradiation. The glassware is often freed of organic matter by treatment with a strong acid-oxidizing agent mixture (e.g., conc. $H_2SO_4 + CrO_3$, $Na_2Cr_2O_7 \cdot 2H_2O$, etc., or HNO_3 + oxidizing agent), and then thoroughly washed with tap, distilled, and triply distilled water, though it may be better to avoid chemical cleaning altogether and rely on baking for an hour at about 550°C, or preirradiation, to remove organic films (43). Contact between the solution to be irradiated and any organic material (e.g., rubber stoppers or tubing, or plastic tubing) is best avoided. Nevertheless, carefully washed polyethylene (Polythene) and poly-methylmethacrylate (Lucite, Perspex) containers have been used successfully in ferrous sulfate dosimetry, though polystyrene appears more likely to give spuriously high results (e.g., 51–54).

To determine the absorbed dose (in 0.4 M sulfuric acid) using the Fricke dosimeter, a sample of the dosimeter solution in a container thick enough to ensure electronic equilibrium[5] is placed in the radiation field for a measured length of time, and then the yield of ferric ions measured. To avoid undue wall effects (i.e., so that practically all the secondary electrons contributing to the energy absorption originate in the solution) glass containers for the solution should have an inner diameter of at least 8 mm when γ-radiation is being determined (55, 56); Burlin (13, 57), using a modified cavity theory, has calculated that with a silica cell and ^{60}Co γ-rays a diameter of 6 cm is needed to reduce the wall effect to below 0.1%. The most common

---

[5] This applies particularly to high energy radiation, since at low and moderate energies (e.g., for ^{60}Co γ-rays) the walls of most glass vessels represent an adequate thickness; for true electronic equilibrium the walls should have the same atomic composition as the dosimeter solution.

Electronic equilibrium and wall effects can be ignored if a liquid dosimeter is chosen which has essentially the same atomic composition as the material to be irradiated and both dosimeter and sample are irradiated in the same vessel.

method of measuring the ferric ion formed is by spectrophotometric analysis, comparing the absorbance of the irradiated and nonirradiated dosimeter solutions at the wavelength at which ferric ions show maximum absorption (about 304 nm). The optical readings should be taken soon after the irradiation, so that adventitious oxidation of the solutions is minimized. The mean absorbed dose (D_D) for the volume occupied by the dosimeter solution is given by Eq. 3.31.

The value of ε should be determined (46, 58, 59) with the spectrophotometer used for the optical measurements, making the determination at the wavelength at which the absorption by ferric ions is greatest (302 to 305 nm, depending on the instrument calibration), and, at the same time, the fact that the measured absorbance is proportional to the amount of ferric ion present confirmed for the instrument being used [Scharf (60) has described procedures that may be followed if the response is not linear, i.e., if Beer's law is not obeyed exactly]. The extinction coefficient has a rather large temperature coefficient of $+0.69\%$ per °C (51, 58, 61) so that the extinction coefficient and absorbances should be measured at the same temperature or, if this is not possible, the absorbed dose corrected using the expression

$$\text{absorbed dose (corr.)} = \frac{\text{absorbed dose (from absorbance at } t_2 \text{°C)}}{1 + 0.007(t_2 - t_1)} \quad (3.32)$$

where t_1°C is the temperature at which the extinction coefficient was determined and t_2 the temperature of the absorbance measurements. Published values for the ferric and ferrous ion molar extinction coefficients at the ferric ion absorption maximum (304 nm) and 25°C are (31, 59): $\varepsilon(Fe^{3+}) = 2205 \pm 3 \ M^{-1} \ cm^{-1}$ and $\varepsilon(Fe^{2+}) = 1 \ M^{-1} \ cm^{-1}$.

The density of the dosimeter solution is essentially that of 0.4 M sulfuric acid, i.e., $1.024_5 \pm 0.001_5$ between 15°C and 25°C(62); the density used is that at the temperature at which the optical measurements are made, rather than the temperature of the solution during irradiation.

Substituting $\Delta\varepsilon_{304} = 2204$ (25°C), $\rho = 1.024$, and $G(Fe^{3+}) = 15.5$ (for ^{60}Co γ-rays) in Eq. 3.31,

$$D_D = 2.76 \times 10^4 \times \frac{\Delta A}{l} \text{ rads} \quad (3.33)$$

$$= 1.72 \times 10^{18} \times \frac{\Delta A}{l} \text{ eV g}^{-1} \quad (3.34)$$

and, assuming the density of the dosimeter solution during irradiation to be 1.024,

$$D_D = 1.76 \times 10^{18} \times \frac{\Delta A}{l} \text{ eV cm}^{-3} \quad (3.35)$$

These equations should be accurate to within a few percentage for radiation for which $G(Fe^{3+})$ is 15.5; more accurate values are obtained by substituting the appropriate values of $\Delta\varepsilon$, ρ, $G(Fe^{3+})$ in Eq. 3.31 and, if necessary, correcting the resulting absorbed dose for temperature using Eq. 3.32.

The $G(Fe^{3+})$ values depend on the LET of the radiation and are greatest when the LET is low; a rather arbitrary selection of values from the current literature is given in Table 3.9; more complete lists are given by Fricke and Hart (43) and in the ICRU publications (28, 44). The values shown in the

TABLE 3.9   *G Values for the Ferrous Sulfate Dosimeter*[a]

Radiation	$G(Fe^{3+})$	Ref.
160-MeV protons	16.5 ± 1	63
1 to 30 MeV electrons	15.7 ± 0.6[b]	31
11 to 30 MV x-rays	15.7 ± 0.6[b]	28
5 to 10 MV x-rays	15.6 ± 0.4[b]	28
4-MV x-rays	15.5 ± 0.3[b]	28
^{60}Co γ-rays (1.25 MeV)	15.5 ± 0.2[b,c]	28
2-MV x-rays ($\bar{E} = 0.44$ MeV)	15.4 ± 0.3[b]	28
^{137}Cs γ-rays (0.66 MeV)	15.3 ± 0.3[b]	28
250-kV x-rays ($\bar{E} = 48$ keV)	14.3 ± 0.3	64
50-kV x-rays ($\bar{E} = 25$ keV)	13.7 ± 0.3	64
^{3}H β-rays (E_{max} 18 keV, E_{av} = 5.7 keV)	12.9 ± 0.3	65
12-MeV deuterons	9.81	66
14.3-MeV neutrons	9.6 ± 0.6	67
1.99-MeV protons	8.00	66
3.47-MeV deuterons	6.90	66
^{6}Li $(n, \alpha)^{3}$H recoil nuclei	5.69 ± 0.12	68
^{210}Po α-rays (5.3 MeV; internal source)	5.10 ± 0.10	69
^{10}B $(n, \alpha)^{7}$Li recoil nuclei	4.22 ± 0.08	66
^{235}U fission fragments	3.0 ± 0.9	63
Limiting yield at an infinite LET (extrapolated from results obtained with accelerated ^{12}C, ^{16}O and ^{14}N ions)	2.9	70

[a] The values given are for the standard dosimeter solution made up in 0.4 M H_2SO_4 (pH 0.46) and for a temperature of 25°C; G values for solutions in 0.05 M H_2SO_4 are about 2% lower. x-Ray energies are maximum (peak) values; $\bar{E}$ values are effective mean energies as given by the authors.

[b] Mean value recommended by the International Commission on Radiation Units and Measurements.

[c] A value of 15.6 is also widely used for ^{60}Co γ-radiation (Eg. 43).

table were obtained by comparing the Fricke dosimeter with some form of absolute dosimeter utilizing calorimetry, ionization measurement, or charge counting. The G values quoted for the heavy charged particles normally assume that the particles are completely stopped in the dosimeter solution; thus they are, in effect, averaged along the particle track. This is significant for the heavy particle radiations because the rate of energy loss by the particle (i.e., the LET) increases very considerably toward the end of the particle track (cf., the Bragg curves in Fig. 2.24, p. 62) and $G(Fe^{3+})$ will not be the same for the plateau region of the Bragg curve and for the peak region (71).

The processes by which ferrous ions are oxidized to ferric consume oxygen and exhuastion of the oxygen present in solution mark an upper limit beyond which the yield of the dosimeter falls (cf., Fig. 7.5). The highest dose that an air-saturated solution can register accurately is about 50,000 rads, and in practice the absorbed dose is usually kept between 4000 and 40,000 rads with low-LET (e.g., x, γ, or electron) radiation; the maximum dose is about four times as great if dissolved air is displaced by pure oxygen before the solution is irradiated, although the ferrous ion concentration should also be increased to between 0.02 and 0.05 M under these conditions (72). The lower limit is the dose that produces sufficient oxidation to be accurately measured. Using a 1-cm absorption cell ($l = 1$ in Eq. 3.31), this is about 4000 rads, but rather lower doses can be measured with the same accuracy by using longer cells (54, 73), a more sensitive measuring technique (74), or by measuring the ferric absorbance at 224 nm where Fe^{3+} absorbs more strongly than at 304 nm; at 224 nm, $\varepsilon(Fe^{3+}) = 4565$ and $\varepsilon(Fe^{2+}) = 20\ M^{-1}\ cm^{-1}$ [these values can be used over the temperature range 20 to 30°C, since the molar extinction coefficient varies only 0.13% per °C (54, 61, 73)]. Lower doses can also be measured by increasing the sensitively of the analytical method. In the spectrophotometric determination this can be achieved, at the expense of convenience, by making a derivative of ferric iron that has a higher extinction coefficient than the ferric ion alone. Ehrenberg and Sæland, for example, used the thiocyanate complex ($\varepsilon = 8500$ at 465 nm) for this purpose (63). By adding ^{59}Fe to the dosimeter solution as a radioactive tracer, and subsequently isolating and counting the ferric ions formed, Rudstam and Svedberg (75) and Gal (76) were able to measure absorbed dose in the region 20 to 1000 rads.

The response of the standard dosimeter solution (containing NaCl) is independent of dose rate up to 2×10^8 rad sec^{-1} (35, 77, 78), but falls at higher dose rates. Thus Rotblat and Sutton (77) found $G(Fe^{3+}) = 13.1$ for electron irradiation at a dose rate of 8×10^9 rad sec^{-1}, compared with a value of 15.45 at lower dose rates, and Anderson (35) a G value of 11.4 for the range from 2 to 3×10^9 rad sec^{-1}. However, a modified dosimeter saturated with oxygen and containing 0.01 M ferrous sulfate, but containing *no* sodium chloride, is dose rate independent to dose rates of the order of 10^{10} rad sec^{-1} (43, 79).

For low-LET radiation, the response is essentially independent of ferrous ion concentration between 5×10^{-2} and 10^{-4} M (56, 80), and of sulfuric acid concentration between 0.75 and 0.05 M (81, 82), although Day and Law (83) have suggested reducing $G(Fe^{3+})$ by 2% when 0.05 M H_2SO_4 is used in place of the more usual 0.4 M acid. A slight dependence of $G(Fe^{3+})$ on ferrous ion concentration has been reported for high-LET radiation (68, 71, 84). It should also be noted that altering the acid concentration alters the density of the solution (e.g., in Eq. 3.31) and, by changing the absorption coefficient or mass stopping power for the solution, the amount of energy absorbed.

The yield of the dosimeter is reduced slightly (about 0.1% per °C) by increasing the temperature of the solution during irradiation (85, 86). Increasing the temperature of the solution will also lower its density, resulting in a lower energy absorption per unit volume, but it will not directly affect the energy absorption per unit mass, upon which the absorbed dose is based.

Barr, Geisselsoder, and Laughlin (87) have suggested adding high-Z solutes such as zinc or cadmium sulfate to the dosimeter solution when low-energy radiation (e.g., low energy x-rays), which interacts mainly by the photoelectric process, is to be measured; more recently, Frankenberg (88) has suggested cesium sulfate as a suitable solute. The solutes cause no significant change in $G(Fe^{3+})$ but increase the mean atomic number of the solution, which can be matched to the mean atomic number of the system being studied.

In common with other liquid dosimeters, the ferrous sulfate dosimeter has the advantage that it can be irradiated in a container of any size or shape, duplicating in these respects the sample to be irradiated.

Ceric Sulfate Dosimeter

The use of ceric sulfate in dosimetry is based on the reduction of ceric ions to cerous under the influence of radiation (43, 89–92).

The response of this dosimeter depends only slightly on the presence of oxygen and consequently exhaustion of the dissolved oxygen does not impose an upper limit on the dose that can be measured. Acid ceric sulfate solutions can, in fact, be used to measure absorbed doses as large as 10^8 rads (93), the upper limit being set by the solubility of ceric salts and the complete reduction of the ceric ions present.

The yield of cerous ions is measured by the difference in ceric ion concentration before and after irradiation. At low ceric ion concentrations ($<2 \times 10^{-4}$ M) this is determined from the change in absorbance at the ceric absorption maximum at 320 nm (the exact wavelength should be determined with the instrument being used for dosimetry); the molar extinction coefficient for ceric ions at this wavelength, 5610 M^{-1} cm^{-1} at 25°C (43), is not temperature dependent, $\varepsilon(Ce^{3+})_{320} = 2.7$ M^{-1} cm^{-1} (58). At higher concentrations of ceric ion the change can be measured using standard analytical techniques, for example by titration with ferrous sulfate (93) or, more rapidly, potentiometrically (91, 92). The absorbed dose is calculated using Eqs. 3.29 or 3.31.

Since the cerous ion concentration is determined by difference, it is advantageous to start with a ceric ion concentration that is of the same order as, but slightly larger than, the expected change in concentration. Reduction of between 20 and 80% of the ceric ions originally present is generally desirable, and concentrations of ceric sulfate between 10^{-5} and 0.4 M are used. Solutions are made up in 0.4 M sulfuric acid using triply distilled water and high-purity reagents, since the system is sensitive to traces of impurities [e.g., 20 ppm of formic acid is reported to increase $G(Ce^{3+})$ from 2.50 to 3.55 (94)]. Alternatively, the effect of organic contaminants can be reduced by adding at least 3×10^{-3} M cerous salt to the dosimeter solution (90).

Representative $G(Ce^{3+})$ values for 10^{-5} to 0.03 M ceric sulfate solutions are given in Table 3.10; the dosimeter is generally calibrated against the

TABLE 3.10 *G Values for the Ceric Sulfate Dosimeter*

Radiation	$G(Ce^{3+})$	Assumed $G(Fe^{3+})$, or Method of Energy Measurement	Ref.
8–14-MeV electrons	2.5 ± 0.18	Calorimetry	94
^{60}Co γ-rays (1.25 MeV)	2.5 ± 0.026	15.6	94
	2.33 ± 0.03	15.45	95
	2.45 ± 0.08	15.5	96
	2.44 ± 0.03^a	15.6	97
	2.32 ± 0.02	15.6	92
200-kvp x-rays	3.15 ± 0.10	15.5	96
10-MeV deuterons	2.80 ± 0.04	Charge collection	95
11-MeV helium ions	2.90 ± 0.06	Charge collection	95
^{210}Po α-particles (5.3 MeV; internal source)	3.20 ± 0.06	Absolute counting	98
^{210}Po α-particles (3.4 MeV; filtered through mica)	2.88 ± 0.02	4.7	99
^{10}B $(n, \alpha)^7$Li recoil nuclei	2.94 ± 0.12	4.22	95

^{a} This is the average of six independent literature values normalized to a standard set of conditions, taking 5609 M^{-1} cm^{-1} as the molar extinction coefficient for ceric ion at 320 nm.

Fricke dosimeter. In contrast to the ferrous sulfate dosimeter, $G(Ce^{3+})$ values increase slightly with increasing LET.

The $G(Ce^{3+})$ values are independent of the initial ceric concentration from 10^{-5} to 0.05 M (56, 94), but are concentration dependent above about 0.05 M. Cerous ions may also affect the yield as they build up in the irradiated solution, necessitating the use of a calibration curve when concentrated ceric solutions are given large doses of radiation; the response is essentially linear with absorbed dose in the region of 10^5 to 10^7 rads.

Taimuty, Towle, and Peterson (94) found the yield to be independent of dose rate up to 2×10^8 rad sec^{-1} and of temperature from 7°C to 35°C, though Rotblat and Sutton (77) and Pikaev and Glazunov (100) give a rather lower dose-rate independent limit of about 10^7 rad sec^{-1}. Matthews (92) has given an empirical equation that corrects for the effect of temperature and also the effect of initial cerous ion and oxygen concentration on $G(Ce^{3+})$ for ^{60}Co γ-radiation.

Nicksic and Wright (101) have drawn attention to the need to avoid unnecessary exposure of ceric solutions to light in order to avoid photo-reduction.

Other Chemical Dosimeters

Many other systems have been proposed and, to a limited extent, used as chemical dosimeters (2, 43, 102, 103), but at the present time none has been adopted as widely as the Fricke dosimeter and, to a lesser degree, the ceric sulfate dosimeter. A few of these systems, primarily chosen to illustrate the range available, are listed in Table 3.11.

When the ferrous sulfate dosimeter is modified by the inclusion of cupric sulfate, the ferrous ions are oxidized without the consumption of oxygen, and the range of the dosimeter is extended up to 10^7 rads. The water, oxalic acid, and glucose dosimeters are also intended for use in the high-dose range encountered in reactor technology and commercial applications employing radiation. They have the advantage over systems containing iron or cerium that they do not become radioactive when exposed to neutrons and also that they are less corrosive than the dosimeters containing sulfuric acid. Solid oxalic acid (106, 124) and solid succinic acid (125) have also been proposed for use in reactor dosimetry where their thermal stability is an added advantage; the absorbed dose is calculated from the loss in weight caused by the formation of gaseous products.

Color changes or bleaching very often occur when solutions of organic dyes are exposed to ionizing radiation, and the bleaching of methylene blue solutions has been applied in dosimetry for many years. More recently the development of color in colourless solutions of triphenylmethane dye cyanides has been investigated and it has been shown (111, 126, 127) that by choosing suitable solvent-dye combinations, absorbed doses from 10 to 10^8 rads can be measured; solid solutions in gelatin are used for the higher dose ranges. The development of color in the visible region means that the change can be measured using relatively unsophisticated equipment or even estimated visually.

Low absorbed doses can be estimated using the polyacrylamide system which is also largely unaffected by fast and thermal neutrons. This dosimeter is particularly simple and inexpensive and has been suggested for use in personnel monitoring and for Civil Defence purposes. The calcium benzoate dosimeter employs extremely sensitive fluorescence measurements in order to determine the small amounts of chemical change produced by low radiation doses.

Organic halogen compounds liberate acid products on irradiation and have formed the basis for several chemical dosimeters. The simplest of these is merely water saturated with pure chloroform. Hydrochloric acid is formed on irradiation and can be determined by titration with dilute alkali, from the conductivity of the solution, or by any other of the many standard analytical methods for strong acids and chloride ions. $G(HCl)$ is relatively independent of the dose rate and energy of the radiation, but since the radiolysis reaction consumes oxygen, exhaustion of the dissolved oxygen imposes an upper limit to the dose that can be measured. A similar restriction applies to the two-phase chloroform-water system, which consists of a layer of pure chloroform covered with a layer of water. Most of the radiation-induced reaction here takes place in the organic layer, and the acid products are subsequently extracted into the water layer and estimated. Radiation initiates a chain reaction in pure chloroform and consequently very high yields of acid can be obtained and rather small doses of radiation detected. However, chain reactions have the disadvantage of being generally very dependent on the radiation dose rate, and on such factors as the temperature and the presence of impurities. These undesirable features can be overcome in part by adding inhibitors, generally organic alcohols or phenols, which limit the chain length of the reaction. By adding carefully controlled amounts of inhibitors the characteristics of the two-phase chloroform-water dosimeter can be improved, and it may be used to measure doses in the range 50 to 1000 rads. Two-phase systems of tetrachloroethylene and water, containing varying amounts of inhibitor, can be prepared so as to have a useful range over some part of the region from 1 to 10^6 rads. The response of the two-phase dosimeters is energy dependent at low x- and γ-ray energies, since they contain a high proportion of the relatively high-Z element chlorine, and

TABLE 3.11 *Systems Used in Chemical Dosimetry*

Dosimeter	Chemical Change Measured and G Value	Method of Measurement	Dose Range (rads)	Ref.
Aqueous ferrous sulfate (Fricke) (aerated; 10^{-3} M $FeSO_4$, 10^{-3} M NaCl, 0.4 M H_2SO_4)	$Fe^{2+} \rightarrow Fe^{3+}$ $G(Fe^{3+}) = 15.5$ (^{60}Co γ-rays) (see also Table 3.9)	Spectrophotometry (at 304 nm)	4×10^3 to 4×10^4 ($\pm 1\%$)	43 p. 93
Aqueous ceric sulfate [10^{-5} M to 0.4 M $Ce(SO_4)_2$, 0.4 M H_2SO_4]	$Ce^{4+} \rightarrow Ce^{3+}$ $G(Ce^{3+}) = 2.32$ (^{60}Co γ-rays) (see also Table 3.10)	Spectrophotometry (at 320 nm), titration, or potentiometry	10^4 to 10^8 ($\pm 2\%$)	43 p. 98
Aqueous ferrous + cupric sulfates (aerated; 10^{-3} M $FeSO_4$, 10^{-2} M $CuSO_4$, 5×10^{-3} M H_2SO_4)	$Fe^{2+} \rightarrow Fe^{3+}$ $G(Fe^{3+}) = 0.66$ (^{60}Co γ-rays) $= 2.0$ [^{10}B $(n, \alpha)^7$Li recoil nuclei]	Spectrophotometry (at 304 nm)	10^5 to 10^7 or 10^8 ($\pm 2\%$)	43 104
Water + 10^{-4} M I$^-$ (aerated; pH 7)	$H_2O \rightarrow H_2 + 1/2\ O_2$ $G(H_2 + O_2) = 0.575$ (^{60}Co γ-rays)	Volume gas produced	5×10^6 to 10^{10} (± 2 to $\pm 5\%$)	43 105
Aqueous oxalic acid (aerated; 0.025 − 0.6 M)	Loss of oxalic acid $G(-$oxalic acid$) = 4.9 \pm 0.4$ (^{60}Co γ-rays)	Spectrophotometry or titration with base or permanganate	10^6 to 2×10^8 ($\pm 3\%$)	106 107 108
Aqueous D(+) glucose (aerated; 10 or 20% solution)	Degradation of the sugar $G(-$glucose$) \cong 2.5$ (^{60}Co γ-rays)	Polarimetry	10^6 to 4×10^8 ($\pm 2\%$)	109 110
Pararosaniline cyanide (aerated; 10^{-3} to 5×10^{-2} M in acidified, polar, solvent)	Formation colored dye. $G(dye) = 0.038$ to 0.27 depending on solvent.	Spectrophotometry	10^3 to 10^6	111

102

System	Observation	Method of measurement	Range	Ref.
Aqueous polyacrylamide (aerated; 0.078% material with mol. wt. 5 to 6 × 10^6)	Degradation of the polymer to smaller molecules	Viscosity	50 to 7500	112
Aqueous calcium benzoate (aerated; 6 × 10^{-4} M)	COO^- → COO^- OH (structure) G(salicylic acid) = 0.6 (^{60}Co γ-rays)	Fluorescence	5 to 5000 ($\pm 5\%$)	113
Aqueous chloroform (single phase) (aerated; saturated solution, i.e., about 0.07 M)	Formation HCl G(HCl) = 28.4 ± 1.0 (^{60}Co γ-rays) G(total acid) = 30.1 ± 1.1	Titration with base, conductivity or pH measurements	10^3 to 4 × 10^4 ($\pm 5\%$)	114 115 116
Chloroform + water (two phase; air-saturated)	Formation HCl G(HCl) approx. 6000 or, if inhibitor (alcohol or phenol) present, approx. 35–90	Titration with base or (+ indicator) spectrophotometry	(+ inhibitor) 75–1000	115
Cyclohexane	Formation hydrogen $G(H_2)$ = 5.25	Volume gas produced	10^4 to 10^8 (± 3 to 6%)	117 118 119
Nitrous oxide (100–1000 torr, 290–300 K)	$N_2O \rightarrow N_2, O_2, NO, NO_2$ $G(N_2)$ = 10.0 ± 0.2 increasing to 12.3 ± 0.3 at extremely high dose rates	Pressure measurement after trapping condensable products at 77K; gas chromatography or mass spectrometry	5 × 10^4 to 2 × 10^6 ($\pm 5\%$)	37 120 121
Ethylene	Formation of hydrogen $G(H_2)$ = 1.35 ± 0.05	Pressure measurement after trapping condensable products	5 × 10^5 to 2 × 10^7 ($\pm 5\%$)	121 122 123

they are of limited use in the laboratory. Taplin (115) and Fricke and Hart (43) have published comprehensive reviews of chlorinated-hydrocarbon dosimeters and their applications.

Gels containing water-soluble chlorine compounds and a pH-sensitive dye give a three-dimensional picture of the dose distribution when they are irradiated (128, 129). The gel can be cut into sections after irradiation and the local doses estimated from the color change, or from conductivity measurements, at the point of interest.

Cyclohexane is not widely used in dosimetry but is of interest as a non-aqueous liquid dosimeter containing only low-Z elements. Nitrous oxide and ethylene are useful dosimeters for gaseous systems, although ionization measurements are probably more reliable in this instance (121, 123). Nitrous oxide is stable at temperatures up to about 200°C and may prove a useful general dosimeter at temperatures in the range -80 to 200°C at which other chemical dosimeters cannot be used.

Dosimetry in Pulse Radiolysis

The dosimetric procedures used in pulse radiolysis (36) generally differ from those used in steady state radiolysis because of the very much higher instantaneous dose rates involved, typically 10^8 to 10^{12} rad sec^{-1} during a pulse lasting 5×10^{-9} to 10^{-6} sec compared with continuous dose rates of the order of 10 to 10^3 rad sec^{-1} from ^{60}Co γ-ray sources, and because of the possibility of observing and measuring the yields of species having lifetimes of the order of 10^{-6} to 10^{-3} sec. Furthermore, the very high concentrations of radicals produced by pulse irradiation leads to increased radical-radical reaction at the expense of radical-solute reactions so that G values determined at lower dose rates are often not applicable in pulse radiolysis experiments.

The Fricke dosimeter may be used with pulse irradiation if sodium chloride is *not* added to the dosimeter solution; $G(Fe^{3+}) = 15.6$ with instantaneous dose rates up to 5×10^8 rad sec^{-1} (36). A modified dosimeter containing no sodium chloride, 10^{-2} M ferrous sulfate, and saturated with oxygen can be used up to 7×10^9 rad sec^{-1} with $G(Fe^{3+}) = 16.1$ (reported in ref. 36). It is very often convenient to measure the absorbance of the dosimeter in situ immediately following the radiation pulse using the analyzing light and optical system that will subsequently be used with the sample solutions. In this event, the absorbance may be measured either after about 30 msec, when reaction of HO_2 and OH with Fe^{3+} is complete and $G(Fe^{3+}) = 6.62$ $(= G_H + G_{e_{aq}^-} + G_{OH})$, or after about 30 sec when H_2O_2 will also have reacted and $G(Fe^{3+}) = 15.6$; $\varepsilon(Fe^{3+}) = 2205$ M^{-1} cm^{-1} at 304 nm, and the absorbed dose is given by Eq. 3.31. The two measurements may differ, since the first gives the absorbed dose along the light path in the irradiation cell immediately following the pulse while the second gives the absorbed dose after the contents of the cell have undergone some mixing by diffusion.

A complex ion containing ferrous iron is oxidized in the ferrocyanide dosimeter where the chemical change is the oxidation of ferrocyanide, $Fe(CN)_6^{4-}$, to ferricyanide, $Fe(CN)_6^{3-}$. The dosimeter solution consists of a neutral $5 \times 10^{-3} M$ potassium ferrocyanide solution saturated with either oxygen or nitrous oxide, and the ferricyanide absorbance is measured either at 420 nm, $\varepsilon(\text{ferricyanide}) = 1000 \ M^{-1} \ cm^{-1}$, or at 440 nm where $\varepsilon(\text{ferricyanide}) = 620 \ M^{-1} \ cm^{-1}$. Oxidation is brought about by a rapid reaction with OH followed by a very much slower reaction with HO_2 (or O_2^-) and H_2O_2. If the absorbance is measured 2 to 100 μsec after the pulse, only the former reaction need be considered and $G(\text{ferricyanide}) = 3.2 (= G_{OH})$ (130) in the oxygen-saturated solution or 5.5 ($= G_{e_{aq}^-} + G_{OH}$) with the nitrous oxide saturated solution; a filter to remove wavelengths shorter than 300 nm should be used between the analyzing light source and the irradiation cell to minimize photochemical oxidation. There is some uncertainty in the values of $G_{e_{aq}^-}$ and G_{OH} to be used in neutral solution (131) and the G values reported here for the ferricyanide and thiocyanate dosimeters should be checked when this point is resolved; radical yields in acid solutions are well established and only small changes might be expected in the G values given for the Fricke dosimeter.

The thiocyanate dosimeter consists of a neutral 2 to $5 \times 10^{-3} M$ potassium thiocyanate solution saturated with either oxygen or nitrous oxide. Upon irradiation, hydroxyl radicals react with CNS^- to form a species $(CNS)_2^-$ which absorbs at 475 nm with $\varepsilon = 7600 \ M^{-1} \ cm^{-1}$ (132) and is formed with $G[(CNS)_2^-] = 2.9 (= G_{OH})$ in oxygen-saturated solution and with $G = 5.8 (= G_{e_{aq}^-} + G_{OH})$ in nitrous oxide-saturated solution. With this dosimeter the transient decays relatively rapidly by a second order process with a rate constant of $3 \times 10^9 \ M^{-1} \ sec^{-1}$, and the absorption should be extrapolated back to midpulse or corrected mathematically (36) to allow for the decay and obtain the total absorbed dose in the solution. In order to do this, the absorption is monitored as soon as possible after the pulse, ignoring any rapid decay due to e_{aq}^- during the first few hundred nanoseconds.

The hydrated electron dosimeter utilizes the intense absorption of e_{aq}^- in the region from 600 to 800 nm (36, 43, 133), and consists of either $10^{-2} M$ sodium hydroxide solution saturated with O_2-free hydrogen or an oxygen-free $10^{-4} M$ sodium hydroxide solution containing $10^{-2} M$ ethanol. The role of the hydrogen and ethanol is to convert hydroxyl radicals to hydrogen atoms by reaction with hydrogen, or to remove them by reaction with the alcohol. For the hydrogen purged solution, $G(e_{aq}^-) = G_{e_{aq}^-} + G_H + G_{OH}$ and $G(e_{aq}^-)\varepsilon = 7.12 \times 10^4$ at 578 nm and 12.1×10^4 at 700 nm (133). The dosimeter containing ethanol is less sensitive and $G(e_{aq}^-)\varepsilon = 2.94 \times 10^4$ at 578 nm and 5.0×10^4 at 700 nm (133). By choosing the appropriate conditions, doses from 10 to 10^4 rad pulse^{-1} can be measured. As with the thiocyanate dosimeter, the absorption is measured immediately following the radiation pulse and corrected for radical decay by extrapolation back to midpulse, or mathematically (36). The hydrogen-saturated dosimeter is interesting in that no permanent chemical change is produced so that the solution can be sealed in a quartz cell and used repeatedly.

SOLID-STATE DOSIMETRY

Under the general term "solid state dosimeter" are gathered a diverse group of dosimeters that often have little more in common than that they are solids, as the rather arbitrary selection described in Table 3.12 shows. Generally the radiation-induced changes are cumulative and the dosimeter indicates absorbed dose, though sometimes the changes are transitory and the system measures absorbed dose rate as does, for example, the *n* on *p* solar cell. All the solid state devices are secondary dosimeters and must be calibrated against an absolute dosimeter or well-characterized secondary dosimeter such as the Fricke dosimeter. In laboratory work in radiation chemistry the liquid and gas phase dosimeters described in the previous section are generally preferred. However, by sectioning a solid dosimeter the dose distribution through the sample can be determined. Solid systems are often preferred in monitoring industrial irradiations, which involve high dose rates and the need for dosimeters that can be easily read and that will withstand rough treatment.

The anthracene dosimeter described in Table 3.12 is representative of several systems in which absorbed dose is determined by the degree to which a luminescent organic compound (e.g., anthracene or *p*-quaterphenyl) is degraded to nonluminescent products. The dosimeter is made up in the form of a thin wafer or film and is read using a fluorimeter; it is primarily intended for use in industrial irradiations. Solids frequently become discolored on exposure to ionizing radiation and the development of color centers is used to estimate absorbed dose in the polymethylmethacrylate and polystyrene dosimeters. The organic polymers are relatively soft and can be cut into sections to map the dose distribution within the sample. Polystyrene is resistant to radiation damage and can be used as a solvent or support for more radiation-sensitive materials as in the *trans*-stilbene-polystyrene dosimeter; the change in this instance is conversion of *trans*-stilbene to the *cis*-isomer and is followed by the decrease in the *trans*-absorption peak at 324 nm. In relating absorbed dose in the dosimeter to that in the material being irradiated, the organic dosimeters are advantageous if the material is biological or organic, since they have a similar atomic composition (i.e., effective atomic number) and density. The same advantage may be claimed for photographic films, which are also organic based, but these dosimeters are more complex and generally require careful processing before the absorbed dose can be determined. Nevertheless, by choosing appropriate film and processing conditions, absorbed doses within a very wide range can be measured by photographic methods.

Coloration induced in inorganic materials by exposure to radiation has been used to estimate absorbed dose for many years, one of the earliest dosimeters to achieve widespread use being based on color changes of barium platinocyanide tablets, giving rise to the so-called pastille dose applied, particularly, in the dermatological use of x-rays. Subsequently attention turned to other materials and the cobalt glass dosimeter is an example of a system in which rather large absorbed doses are estimated from the color change in an inorganic glass. These have some advantages over organic polymers, since they are more rigid, so that smaller dosimeters can be fabricated, chemically more inert, harder, and stronger so that they are less likely to be damaged by careless handling. Low absorbed doses can be measured using silver-doped phosphate glasses in which irradiation produces fluorescent centers. Originally these contained silver phosphate in a mixture of aluminum, barium, and potassium phosphates, but more recently a mixture of lithium and aluminum phosphates has been used as the base, since this gives an effective atomic number closer to that of organic and biological materials and helps avoid large correction factors at low x-ray energies.

The silver-doped phosphate glasses are examples of radiophotoluminescent (RPL) dosimeters in which radiation produces defects that cause fluorescence when the irradiated glass is exposed to ultraviolet light. Related to these are the thermoluminescent (TL) dosimeters in which the radiation-induced defects (trapped electrons and positive holes) recombine with the production of fluorescence when the irradiated solid is heated, the absorbed dose being related to the number of fluorescence photons emitted. The major thermoluminescent dosimeters at the present time are lithium fluoride, calcium fluoride both in its natural state (CaF_2: natural) and doped with manganese (CaF_2:Mn), and lithium borate doped with manganese ($Li_2B_4O_7$:Mn). The lithium-based dosimeters have an energy response (i.e., effective atomic number) close to that of organic and biological materials. Lithium fluoride is available containing the natural isotopic distribution of lithium isotopes (7.5% ^6Li and 92.5% ^7Li) and as ^6LiF and ^7LiF; the natural and ^6LiF detect thermal neutrons but the ^7LiF is insensitive to them.

The p on n silicon solar cell is analogous to the anthracene dosimeter in the sense that the absorbed dose is estimated from the extent to which a component or characteristic of the system is degraded on irradiation. In this

TABLE 3.12 *Solid State Dosimeters*

Material	Effect of Irradiation	Method of Measurement	Absorbed Dose Range (rad)	Maximum Dose Rate Reported (rad sec⁻¹)	Ref.
Anthracene; pressed powder or in gelatin	Degradation	Loss of anthracene fluorescence at 440 nm	5×10^5 to 5×10^7	4×10^3	134 135
Polymethylmethacrylate (PMMA)	Degradation and formation color centers	Spectrophotometry (260 to 345 nm)	10^5 to 10^7 ($\pm 2\%$)	3×10^8	134 136
Polystyrene	Formation color centers	Spectrophotometry (~ 420 nm)	5×10^5 to 2×10^8 ($\pm 2\%$)		137
Polystyrene doped with trans-stilbene	Isomerization trans- to cis-stilbene	Spectrophometric detmn. trans-isomer at 324 nm	2×10^5 to 10^8	10^{14}	138
Photographic films	Formation latent image or, at high doses, darkening	Densitometry, generally after development	10^{-4} to 10^8 Using various films and methods of processing		139 140 141
Cobalt glass	Formation color centers	Spectrophotometry	10^4 to 2×10^6 ($\pm 2\%$)	10^6	134
Silver doped Li-Al phosphate glasses (RPL, radiophoto-luminescent dosimeter)	Creation fluorescent centers	Measurement fluorescence at ~ 650 nm excited by ultraviolet (~ 320 nm) light	10^{-2} to 10^4 ($\pm 2\%$)	10^5	134 141 142
	Formation color centers	Spectrophotometry or densitometry	10^3 to 10^6		

Lithium fluoride (TL, thermoluminescent dosimeter)	Formation trapped electrons and positive holes	Measurement fluorescence (~400 nm) emitted on heating to 195°C	10^{-3} to 10^5 (± 1 to 3%)	10^{11}	134 141 142 143
Manganese doped calcium fluoride (TL dosimeter)	Formation trapped electrons and positive holes	Measurement fluorescence (~500 nm) emitted on heating to 260°C	10^{-3} to 3×10^5 (± 2 to 3%)		134 141 142 144
p on n solar cell (integrating dosimeter)	Degradation cell response	Photocurrent	10^4 to 10^8 (± 5%)	3×10^3	145 146
n on p solar cell (dose-rate meter)	Generation ionization current	Current measurement		0.1 to 3×10^3 (± 5%)	145 147
Stimulated exoelectron emission from insulator, e.g., LiF, BeO, CaSO$_4$	Changes in surface structure such that electrons are released upon stimulation	Electrons released on thermal or photo stimulation counted			141

instance the property followed is the ability of the cell to produce a photo-current on exposure to monochromatic light (1200 nm). The operation of the *n* on *p* silicon solar cell is different and this functions as a solid state ionization chamber to give a continuous current upon irradiation which is related to the absorbed dose rate; the current produced is measured using a suitable microammeter or galvanometer. Stimulated exoelectron emission is the release of electrons (which are counted) from the surface of an irradiated insulator such as natural or doped LiF, BeO, or $CaSO_4$, upon exposure to light or upon heating. Dosimeters based on this phenomena are under study.

CALCULATION OF DOSE ABSORBED IN SPECIMEN

A problem that frequently arises is the determination of the absorbed dose in a specimen from the dose absorbed by a dosimeter irradiated under the same conditions. If the dosimeter and specimen are homogeneous, have the same size, density, and atomic composition (i.e., the same proportions of elements by weight), the absorbed dose will be the same. This condition is met, for example, if both dosimeter and sample are dilute aqueous solutions, and they are irradiated in turn in the same container in the same position in the radiation field, or if the Fricke dosimeter and a solution in 0.4 *M* sulfuric acid are compared. Often, however, dosimeter and sample differ and a calculation must be performed to obtain the absorbed dose in the specimen. The necessary corrections are described below for differences in composition and density for electromagnetic and charged particle irradiation, covering the situations met most often in radiation chemistry. The equations given do not allow for differences in size between dosimeter and sample and apply particularly to the irradiation of equal volumes of liquid dosimeter and sample under the same conditions. Dosimeters give the average absorbed dose over the volume of the dosimeter, and with large samples, or radiation with a short range in the sample, this may mask significant differences in absorbed dose between the front and back of the specimen (e.g., the depth dose curves in Fig. 2.21). The distribution of dose in large specimens and other problems related to sample size and inhomogeneity are of particular concern in radiobiology and radiation therapy and are described in publications dealing with these subjects.

The radiochemical yield of a product formed by irradiation, $G(P)_M$, is related to the absorbed dose in the sample, D_M, by Eq. 3.29 and by

$$G(P)_M = G(P')_D \times \frac{[P]_M}{[P']_D} \times \frac{\rho_D}{\rho_M} \times \frac{D_D}{D_M} \quad \text{molecules product per} \atop \text{100 eV energy absorbed} \qquad (3.36)$$

where $G(P')_D$ is the radiochemical yield of product P' formed in the dosimeter, $[P]_M$ and $[P']_D$ are the measured yields of P and P' (moles per unit volume, e.g., mole liter^{-1}) when sample and dosimeter are exposed in the same radiation field, ρ_M and ρ_D are the densities of sample and dosimeter, and D_D/D_M is the ratio of absorbed dose per unit mass in specimen and dosimeter calculated as described below.

Electromagnetic Radiation

When both dosimeter and specimen are exposed to the same radiation field and are in electronic equilibrium, the absorbed dose in the dosimeter (D_D) and specimen (D_M) are related by

$$D_M = D_D \times \frac{(\mu_{en}/\rho)_M}{(\mu_{en}/\rho)_D} \qquad (D \text{ in rads or eV g}^{-1}) \qquad (3.37)$$

where $(\mu_{en}/\rho)_D$ and $(\mu_{en}/\rho)_M$ are the mass energy absorption coefficients for dosimeter and specimen respectively. Mass energy absorption coefficients for some common elements and several photon energies are listed in Table 3.3; more extensive tabulations are given by Hubbell and Berger (18, 19) and by Evans (17). For compounds and mixtures the mass absorption coefficients are combined in proportion to their weight in the sample (Eq. 2.31, cf., Eq. 3.6) to obtain an average value which can be substituted in Eq. 3.37. The values shown for water in Table 3.4 were calculated in this way by summing $2.02/18.02 \ (\mu_{en}/\rho)_H + 16.00/18.02(\mu_{en}/\rho)_O$. When photons of more than one energy are present in the incident radiation, mean absorption coefficients must be calculated for dosimeter and sample based on energy absorption coefficients for each range of energies present and the energy fluence due to photons in each energy range.

In the range of photon energies in which absorption is predominantly by the Compton process equation 3.37 can be simplified to

$$D_M = D_D \times \frac{(Z/A)_M}{(Z/A)_D} \qquad (D \text{ in rads or eV g}^{-1}) \qquad (3.38)$$

where Z/A is the ratio of atomic number (Z) to atomic weight (A) for an element and the ratio of the sum of the atomic numbers of the elements present to molecular weight for a compound (cf. Table 3.6); mean values are calculated for mixtures by adding Z/A for each component multiplied by its fraction by weight in the mixture (Eq. 3.6).

For Eq. 3.38 to be valid it is necessary to assume that the incident radiation does not interact appreciably by photoelectric absorption or pair-production in either material. This assumption can be verified by examination of graphs (e.g., Figs. 2.15 and 2.16) or tables (19, 148) which give the attenuation coefficients for these three processes (Compton, photoelectric, and pair-production) over the energy range concerned. If the material being irradiated contains a number of elements, energy absorption must be by the Compton process for those elements which are major constituents. Fortunately the absorption of cobalt-60 and caesium-137 γ-radiation by aqueous, biological, and most organic systems is well within the Compton region for these materials. Values of the ratios $(\mu_{en}/\rho)_M(\mu_{en}/\rho)_{Fricke}$ and $(Z/A)_M(Z/A)_{Fricke}$ for various materials M and the Fricke dosimeter solution (0.4 M H_2SO_4) are compared in Fig. 3.5, which shows that the "Compton region" for the low-Z materials water and polyethylene extends over the energy ranges 0.15 to 10 MeV and 0.25 to 2 MeV, respectively. The Compton region extends over a more limited range of energies for aluminum which has $Z = 13$, and is nonexistent for lead ($Z = 82$). Equation 3.38 can be used with polyenergetic x-rays if essentially all the photon energies present in the beam are within the Compton region for the elements present in dosimeter and specimen.

Equations 3.37 and 3.38 must be modified to include the densities of the media if the energy absorbed per unit volume of absorber, D_v (in, for example, eV cm^{-3}), is required. The modified equations are

$$(D_v)_M = (D_v)_D \times \frac{(\mu_{en}/\rho)_M}{(\mu_{en}/\rho)_D} \times \frac{\rho_M}{\rho_D} \qquad (3.39)$$

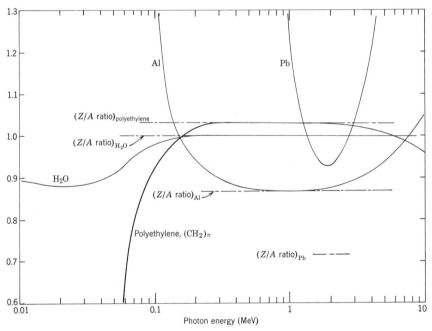

FIGURE 3.5   *Comparison of* $(\mu_{en}/\rho)_M/\mu_{en}/\rho)_{Fricke}$ (————) *and* $(Z/A)_M/(Z/A)_{Fricke}$ (— — — —) *for various materials, M, and the Fricke dosimeter solution* (0.4 M H_2SO_4).

and

$$(D_v)_M = (D_v)_D \times \frac{(Z/A)_M}{(Z/A)_D} \times \frac{\rho_M}{\rho_D} \tag{3.40}$$

The equations given in this section (Eqs. 3.37 to 3.40) are strictly true only when the γ- or x-rays are not appreciably attenuated by passing through the absorbing media (this is the condition for Eq. 2.50 and hence Eq. 3.2 to be true), although in practice they are used even when an appreciable fraction of the incident radiation is absorbed. It should be born in mind, however, that significant errors may result due to reabsorption of the secondary (scattered) photon radiation when samples which are more than several centimeters thick are irradiated (149, 150). This effect will be most important when dosimeter and specimen differ in atomic composition.

Charged Particle Radiation

If the incident radiation is a beam of charged particles (electrons, protons, etc.) and the samples are relatively thick, the incident radiation may be completely absorbed by both dosimeter and specimen so that the total energy absorbed will be essentially the same in each system. Small differences in absorbed energy will arise if the energy lost due to back-scattering of the charged particles and bremsstrahlung are not the same for each system, though these differences can be minimized by using a dosimeter that has an atomic composition close to that of the specimen.

When the samples are thin compared with the range of the incident particles, the absorbed dose in dosimeter and specimen is proportional to their collisional mass stopping powers, $(S/\rho)_{col}$, and

$$D_M = D_D \times \frac{(S/\rho)_{col, M}}{(S/\rho)_{col, D}} \quad (D \text{ in rads or eV g}^{-1}) \tag{3.41}$$

Collisional mass stopping powers for electrons have been calculated by Berger and Seltzer (31, 32) for a number of elements and electron energies using a modification of the Bethe equation (Eq. 2.8) incorporating a density correction; the values given in Table 3.13 are from this source. Total mass stopping powers for electrons, which include energy losses by collisional and radiative processes, are given by Berger and Seltzer and, for several additional elements, in earlier publications by Nelms (151). Mass stopping powers for compounds can be estimated by combining values for the elements present weighted in proportion to the mass of the element in the compound (Bragg additivity law) (cf., Eqs. 2.31 and 3.6), although at high electron energies the values obtained deviate from those calculated directly using the Bethe equation and data for the compound (Fig. 3.6), since there is an appreciable "density effect" due to polarization in condensed media.

TABLE 3.13   *Collisional Mass Stopping Powers for Electrons*
$[(S/\rho)_{col}; MeV\ cm^2\ g^{-1}]$ (32)

Electron Energy (MeV)	$_1H$	$_6C$	$_7N$	$_8O$	$_{13}Al$	$_{26}Fe$	$_{82}Pb$	Water	Poly-ethylene
0.01	51.47	20.15	19.81	19.64	16.57	14.07	8.419	23.20	24.65
0.05	14.29	5.909	5.834	5.795	5.059	4.455	2.997	6.747	7.113
0.10	8.766	3.685	3.642	3.621	3.191	2.838	1.964	4.197	4.415
0.5	4.205	1.801	1.806	1.798	1.603	1.449	1.059	2.061	2.148
1.0	3.826	1.634	1.665	1.658	1.473	1.337	1.002	1.876	1.937
5	4.112	1.692	1.837	1.831	1.561	1.430	1.135	1.931	1.982
10	4.400	1.761	1.983	1.977	1.637	1.505	1.217	2.000	2.056
50	5.089	1.899	2.330	2.324	1.792	1.656	1.380	2.144	2.209
100	5.258	1.953	2.440	2.425	1.849	1.711	1.436	2.204	2.270

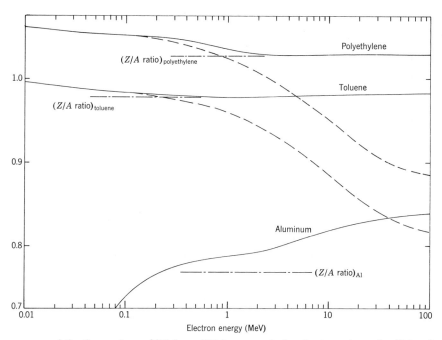

FIGURE 3.6   *Comparison of $(S/\rho)_{col,\ M}/(S/\rho)_{col,\ H_2O}$ calculated using values of collisional mass stopping powers for the compounds (——————) and for the elements present (– – –). Values of the ratio $(Z/A)_M/(Z/A)_{H_2O}$ are also shown. Collisional mass stopping powers for elements and compounds were taken from Berger and Seltzer (32).*

For low-Z materials such as the organic compounds polyethylene and toluene and electron energies above about 0.1 MeV, an empirical approach using Eq. 3.38 rather than Eq. 3.41 may be more successful than attempting to combine the mass stopping powers of the elements present, since Z/A ratios approximate more closely to the true mass stopping power ratios (Fig. 3.6). In any event, errors will be reduced if the dosimeter duplicates as far as possible the size, density, and atomic composition of the specimen in which the absorbed dose is required. If electron mass stopping powers are used, they should be those for the mean energy of the primary electrons ($\bar{E}$) at the centre of the sample; this is given by (31)

$$\bar{E} = E_0 \left(1 - \frac{d}{R_p} \right) \tag{3.42}$$

where E_0 is the energy of the electrons incident on the sample, d(cm) the distance to the centre of the sample, and R_p(cm) the practical or extrapolated range of the incident electrons (cf., Chapter 2).

The calculation of mass stopping powers for heavy charged particles is discussed by Bichsel (33).

Equation 3.41 applies if the absorbed dose is measured in units of energy absorption per unit mass (e.g., rads or eV g^{-1}), if the units used are energy absorption per unit volume (e.g., eV cm^{-3}) a density correction must also be applied as in Eqs. 3.39 and 3.40.

Energy Partition in Mixtures

If a mixture of elements or compounds is irradiated with electromagnetic or charged particle radiation, the fraction of the total absorbed dose absorbed by each component of the mixture is proportional to the fraction of the component present by weight and to the mean collisional mass stopping power of the component for the primary and secondary ionizing particles of various energies present in the medium. The latter is not readily determined and in practice various approximations have been used, the simplest and most widely adopted being to assume that the mass stopping powers are proportional to the Z/A ratio for the component, when

$$\frac{D_1}{D_{mix}} = w_1 \times \frac{(Z/A)_1}{(Z/A)_{mix}} \qquad (D \text{ in rads or eV g}^{-1}) \tag{3.43}$$

$$= \varepsilon_1 \text{ (the electron fraction of 1 in the mixture)}$$

where

$$\left(\frac{Z}{A} \right)_{mix} = w_1 \left(\frac{Z}{A} \right)_1 + w_2 \left(\frac{Z}{A} \right)_2 + \cdots w_n \left(\frac{Z}{A} \right)_n \tag{3.44}$$

for a mixture with n components; w_1 is the weight of the first component divided by the weight of the mixture, and D_1/D_{mix} the fraction of the total absorbed dose absorbed by the first component. Several authors have pointed out that a large proportion of the energy transferred to the medium will be by relatively low-energy secondary particles which will not have sufficient energy to excite inner-shell electrons and have therefore proposed that Z in the above expression should be replaced by the number of outer-shell or valence electrons in the materials present (e.g., 6 for oxygen and sulfur, 5 for nitrogen and phosphorus) (e.g., 152, 153). Klots (154) has derived a table of "effective atomic numbers" that can replace Z in Eq. 3.43 by examining ion yields in binary gas mixtures irradiated with α-particles. The results support an earlier conclusion (155, 156) that unsaturated hydrocarbons, containing π electrons, absorb rather more energy in an irradiated mixture than saturated compounds containing only σ electrons.

PERSONNEL DOSIMETRY

Doses below 1 rad, down to the dose absorbed due to natural background radiation, are of concern to all radiation workers, since they represent the most common hazard in this field. Consequently, some knowledge of the dose absorbed by the worker, over fairly long periods of time, and of the dose-rate in working areas, are essential to health and safety.

Low exposure rates, of the order of milliroentgens per hour, that may be found in working areas, are controlled by means of survey instruments. These include portable instruments based on ionization measurements, and capable of indicating dose rates in ranges varying from 0 to 25 mr/hr to (for Civil Defense purposes) 0 to 500 r/hr, portable Geiger counters, which are rather more sensitive and have operating ranges varying from 0 to 0.2 mr/hr to 0 to 10 r/hr, and scintillation counters, which can be made even more sensitive than the Geiger counters.

Low exposures, accumulated over a period of time, can be measured with sensitive ionization chambers reading up to a few hundred milliroentgens. The sensitive volume of these chambers is generally only one or two cubic centimeters and the whole instrument no larger than a fountain pen, so that they are aptly named *pocket dosimeters*. Some form of separate charging device is necessary, but the dosimeters are often designed to include a calibrated scale that can be read without other instrumentation. The fogging produced in photographic film by exposure to ionizing radiation is also used to determine the accumulated dose to persons working with radiation sources (157, cf., 158). The films are subsequently developed under carefully controlled conditions and the degree of darkening provides a measure of the exposure. Careful control of the developing procedure is essential, and this is usually carried out at a central laboratory devoted to this purpose. Both the survey instruments and the pocket dosimeters will be energy dependent, and their response may vary with such conditions as the temperature and barometric pressure, or the direction in which they are exposed to the radiation. Great accuracy cannot be expected from them unless they are calibrated and used under controlled conditions. Generally, for the purposes for which they are intended, an accuracy of 25% is adequate. Survey instruments and pocket dosimeters are reviewed by Bemis (157), Kiefer, Maushart and Mejdahl (159), and Oliver (160).

REFERENCES

Many of the topics included in this section are discussed more fully by G. N. Whyte, *Principles of Radiation Dosimetry*, Wiley, New York, 1959, and in ref. 1 and 2 below. ICRU reports are available from the International Commission on Radiation Units and Measurements, 7910 Woodmont Avenue, Washington, D.C., 20014.

1. *Radiation Dosimetry* (eds. F. H. Attix, W. C. Roesch, and E. Tochilin), Vol. 1, Academic, New York, 1968; Vol. 2, 1966; Vol. 3, 1969.
2. *Manual on Radiation Dosimetry* (eds. N. W. Holm and R. J. Berry), Dekker, New York, 1970.
3. W. C. Roesch and F. H. Attix in ref. 1, Vol. 1, p. 1.
4. *Radiation Quantities and Units*, ICRU Report 19 (1971).
5. K. Liden, *Phys. Med. Biol.*, **18**, 462 (1973).
6. M. L. McGlashan, *Ann. Rev. Phys. Chem.*, **24**, 51 (1973).
7. P. Curie and A. Laborde, *C. R.*, **136**, 673 (1903).
8. J. S. Laughlin and S. Genna in ref. 1, Vol. 2, p. 389.
9. B. Radak and V. Marković in ref. 2, p. 45.
10. E. K. Hussmann, *Appl. Optics*, **10**, 182 (1971).
11. E. K. Hussmann and W. L. McLaughlin, *Radiat. Res.*, **47**, 1 (1971).
12. J. W. Boag in ref. 1, Vol. 2, p. 1.
13. T. E. Burlin in ref. 2, p. 13.
14. G. N. Whyte, *Principles of Radiation Dosimetry*, Wiley, New York, 1959, p. 64.
15. V. H. Ritz and F. H. Attix, *Radiat. Res.*, **16**, 401 (1962).
16. I. T. Myers in ref. 1, Vol. 1, p. 317.
17. R. D. Evans in ref. 1, Vol. 1, p. 93.
18. J. H. Hubbell and M. J. Berger, *Engineering Compendium Radiation Shielding* (ed. R. G. Jaeger), Springer-Verlag, New York, 1968, pp. 167 and 185.
19. J. H. Hubbell, *Photo Cross Sections, Attenuation Coefficients, and Energy Absorption Coefficients from 10 keV to 100 GeV*, NSRDS-29 (1969), U.S. Dept Commerce-National Bureau of Standards, Washington, D.C.
20. H. E. Johns, *The Physics of Radiology*, 2nd ed., Charles C. Thomas, Springfield, Ill., 1961; H. E. Johns in ref. 1, Vol. 3, pp. 1 and 677.
21. W. H. Bragg, *Studies in Radioactivity*, Macmillan, London, 1912, p. 94.

22. L. H. Gray, *Proc. Roy. Soc. (London), Ser. A*, **156**, 578 (1936); *Brit. J. Radiol.*, **10**, 600, 721 (1937).
23. P. R. J. Burch, *Radiat. Res.*, **3**, 361 (1955).
24. L. V. Spencer and F. H. Attix, *Radiat. Res.*, **3**, 239 (1955).
25. F. W. Spiers in *Radiation Dosimetry* (eds. G. J. Hine and G. L. Brownell), Academic, New York, 1956, Chap. 1.
26. T. E. Burlin in ref. 1, Vol. 1, p. 331.
27. G. N. Whyte, *Principles of Radiation Dosimetry*, Wiley, New York, 1959, p. 64, p. 68 (Table 5.2).
28. *Radiation Dosimetry: X-rays and Gamma Rays with Maximum Photon Energies between 0.6 and 50 MeV*, ICRU Report 14 (1969).
29. J. S. Laughlin in ref. 1, Vol. 2, p. 91.
30. M. R. Raju, J. T. Lyman, T. Brustad, and C. A. Tobias in ref. 1, Vol. 3, p. 151.
31. *Radiation Dosimetry: Electrons with Initial Energies between 1 and 50 MeV*, ICRU Report 21 (1972).
32. M. J. Berger and S. M. Seltzer, *Tables of Energy Losses and Ranges of Positrons and Electrons*, NASA Tech. Report SP 3012 (1964), and *Additional Stopping Power and Range Tables for Protons, Mesons and Electrons*, NASA Tech. Report SP 3036 (1966).
33. H. Bichsel in ref. 1, Vol. 1, p. 157.
34. S. I. Taimuty, *Nucleonics*, **15**, 182 (November 1957).
35. A. R. Anderson, *J. Phys. Chem.*, **66**, 180 (1962).
36. E. M. Fielden and N. W. Holm in ref. 2, p. 261.
37. C. Willis, A. W. Boyd, and A. O. Miller, *Radiat. Res.*, **46**, 428 (1971).
38. J. A. Auxier, W. S. Snyder, and T. D. Jones, in ref. 1, Vol. 1, p. 275.
39. J. De Pangher and E. Tochilin in ref. 1, Vol. 3, p. 309.
40. J. Moteff in ref. 1, Vol. 3, p. 201.
41. E. Tochilin and B. W. Shumway in ref. 1, Vol. 3, p. 247.
42. *Neutron Fluence, Neutron Spectra and Kerma*, ICRU Report 13 (1969).
43. H. Fricke and E. J. Hart in ref. 1, Vol. 2, p. 167.
44. *Radiation Dosimetry: X-rays Generated at Potentials of 5 to 150 kV*, ICRU Report 17 (1970).
45. H. Fricke and S. Morse, *Am. J. Roentgenol*, **18**, 430 (1927); *Phil. Mag.*, **7**, 129 (1929); H. Fricke and E. J. Hart, *J. Chem. Phys.*, **3**, 60 (1935).
46. J. V. Davies and J. Law, *Phys. Med. Biol.*, **8**, 91 (1963).
47. J. Law and A. T. Redpath, *Phys. Med. Biol.*, **13**, 371 (1968).
48. H. A. Dewhurst, *J. Chem. Phys.*, **19**, 1329 (1951); *Trans. Faraday Soc.*, **48**, 905 (1952).
49. H. Fricke, E. J. Hart, and H. P. Smith, *J. Chem. Phys.*, **6**, 229 (1938).
50. R. A. Sassé, *Health Phys.*, **19**, 675 (1970).
51. J. L. Haybittle, R. D. Saunders, and A. J. Swallow, *J. Chem. Phys.*, **25**, 1213 (1956).
52. S. A. Ahmed, J. Law, and P. W. Walton, *Phys. Med. Biol.*, **15**, 311 (1970).
53. J. Law, *Phys. Med. Biol.*, **15**, 117 (1970).
54. M. Kartha, *Radiat. Res.*, **42**, 220 (1970).
55. J. Weiss, A. O. Allen and H. A. Schwarz, *Proc. Int. Conf. Peaceful Uses Atomic Energy*, United Nations, New York, **14**, 179 (1956).
56. Jerome Weiss, *Nucleonics*, **10**, 28 (July 1952).
57. T. E. Burlin and F. K. Chan, *J. Appl. Radiat. Isotop.*, **20** 767 (1969).
58. C. M. Henderson and N. Miller, *Radiat. Res.*, **13**, 641 (1960).
59. R. K. Broskiewicz and Z. Bulhak, *Phys. Med. Biol.*, **15**, 549 (1970).
60. K. Scharf, *Phys. Med. Biol.*, **16**, 77 (1971).
61. K. Scharf and R. M. Lee, *Radiat. Res.*, **16**, 115 (1962).

62. *International Critical Tables*, (United States) National Research Council, 1928, Vol. 3, p. 56.
63. L. Ehrenberg and E. Sæland, JENER *Publications*, No. 8 (1954).
64. J. Law, *Phys. Med. Biol.*, **14**, 607 (1969).
65. A. O. Fregene, *Radiat. Res.*, **31**, 256 (1967).
66. E. J. Hart, W. J. Ramler, and S. R. Rocklin, *Radiat. Res.*, **4**, 378 (1956).
67. J. Law, R. C. Lawson, and D. Porter, *Phys. Med. Biol.*, **19**, 643 (1974).
68. R. H. Schuler and N. F. Barr, *J. Am. Chem. Soc.*, **78**, 5756 (1956).
69. C. N. Trumbore and E. J. Hart, *J. Phys. Chem.*, **63**, 867 (1959).
70. M. Imamura, M. Matsui, and T. Karasawa, *Bull. Chem. Soc. Jap.*, **43**, 2745 (1970).
71. A. Appleby and E. A. Christman, *Radiat. Res.*, **60**, 34 (1974).
72. A. O. Allen and W. G. Rothschild, *Radiat. Res.*, **7**, 591 (1957).
73. J. Law, *Int. J. Appl. Radiat. Isotop.*, **22**, 701 (1971).
74. R. G. Waggener, C. W. Hiatt, L. F. Rogers, and P. Zanca, *Radiat. Res.*, **45**, 244 (1971).
75. G. Rudstam and T. Svedberg, *Nature*, **171**, 648 (1953).
76. O. S. Gal, *Int. J. Appl. Radiat. Isotop.*, **13**, 304 (1962).
77. J. Rotblat and H. C. Sutton, *Proc. Roy. Soc. (London)*, *Ser. A*, **255**, 490 (1960).
78. P. Y. Glazunov and A. K. Pikaev, *Dokl. Akad. Nauk SSSR*, **130**, 1051 (1960).
79. J. K. Thomas and E. J. Hart, *Radiat. Res.*, **17**, 408 (1962).
80. C. J. Hochanadel and J. A. Ghormley, *J. Chem. Phys.*, **21**, 880 (1953).
81. H. A. Dewhurst, *Trans. Faraday Soc.*, **49**, 1174 (1953).
82. A. O. Allen, V. D. Hogan, and W. G. Rothschild, *Radiat. Res.*, **7**, 603 (1957).
83. M. J. Day and J. Law, *Phys. Med. Biol.*, **14**, 665 (1969).
84. R. H. Schuler and A. O. Allen, *J. Am. Chem. Soc.*, **77**, 507 (1955); **79**, 1565 (1957).
85. C. J. Hochanadel and J. A. Ghormley, *Radiat. Res.*, **16**, 653 (1962).
86. J. Law, *Phys. Med. Biol.*, **15**, 741 (1970).
87. N. F. Barr, J. Geisselsoder, and J. S. Laughlin, *Radiat. Res.*, **14**, 291 (1961).
88. D. Frankenberg, *Phys. Med. Biol.*, **14**, 597 (1969).
89. E. Bjergbakke in ref. 2, p. 323.
90. R. W. Matthews, *Int. J. Appl. Radiat. Isotop.*, **22**, 199 (1971).
91. R. W. Matthews, *Ibid.*, **23**, 179 (1972).
92. R. W. Matthews, *Radiat. Res.*, **55**, 242 (1973).
93. J. T. Harlan and E. J. Hart, *Nucleonics*, **17**, 102 (August 1959).
94. S. I. Taimuty, L. H. Towle, and D. L. Peterson, *Nucleonics*, **17**, 103 (August 1959).
95. N. F. Barr and R. H. Schuler, *J. Phys. Chem.*, **63**, 808 (1959).
96. G. R. A. Johnson and J. Weiss, *Proc. Roy. Soc. (London)*, *Ser. A*, **240**, 189 (1957).
97. J. W. Boyle, *Radiat. Res.*, **17**, 427 (1962).
98. M. Lefort and X. Tarrago, *J. Phys. Chem.*, **63**, 833 (1959).
99. J. Weiss and N. Miller, *J. Phys. Chem.*, **63**, 888 (1959).
100. A. K. Pikaev and P. Ya. Glazunov, *Izv. Akad. Nauk SSSR, Otd. Khim. Nauk*, **940** (1960).
101. S. W. Nicksic and J. R. Wright, *Nucleonics*, **13**, 104 (November 1955).
102. I. G. Draganić and Z. D. Draganić, *The Radiation Chemistry of Water*, Academic, New York, 1971, Chap. 8.
103. I. G. Draganić and B. L. Gupta in *Dosimetry in Agriculture, Industry, Biology and Medicine*, International Atomic Energy Agency, Vienna, 1973, p. 351.
104. E. Bjergbakke in ref. 2, p. 319.
105. E. J. Hart in ref. 2, p. 327.
106. I. G. Draganić and O. Gal, *Radiat. Res. Rev.*, **3**, 167 (1971).
107. E. Hara, *Int. J. Appl. Radiat. Isotop.*, **20**, 395 (1969).
108. N. W. Holm in ref. 2, p. 337.

120   An Introduction to Radiation Chemistry

109. V. V. Generalova, M. B. Kishinevskaya, and B. A. Vainshtok, *Atomnaya Energiya*, **26**, 31 (1969).
110. R. D. Russell, *Int. J. Appl. Radiat. Isotop.*, **21**, 143 (1970).
111. W. L. McLaughton and M. Kosanić, *Ibid.*, **25**, 249 (1974).
112. A. L. Boni, *Radiat. Res.*, **14**, 374 (1961).
113. W. A. Armstrong and D. W. Grant, *Nature*, **182**, 747 (1958); *Can. J. Chem.*, **38**, 845 (1960); W. A. Armstrong, R. A. Facey, D. W. Grant, and W. G. Humphreys, *Radiat. Res.*, **19**, 120 (1963).
114. J. Teply and J. Bednár, *Proc. 2nd Int. Conf. Peaceful Uses Atomic Energy*, United Nations, Geneva, **29**, 71 (1958).
115. G. V. Taplin in *Radiation Dosimetry* (eds. G. H. Hine and G. L. Brownell), Academic, New York, 1956, Chap. 8.
116. B. J. Rezansoff, K. J. McCallum, and R. J. Woods, *Can. J. Chem.*, **48**, 271 (1970).
117. R. H. Schuler and A. O. Allen, *J. Am. Chem. Soc.*, **77**, 507 (1955).
118. H. A. Dewhurst, *J. Phys. Chem.* **61**, 1466 (1957).
119. J. Prévé and G. Gaudemaris, *C. R.*, **250**, 3470 (1960).
120. G. R. A. Johnson, *Radiation Chemistry of Nitrous Oxide Gas. Primary Processes, Elementary Reactions, and Yields*, NSRDS-NBS 45 (1973), U.S. Department Commerce-National Bureau of Standards, Washington, D.C.
121. K. G. McLaren, *Int. J. Appl. Radiat. Isotop.*, **25**, 87 (1974).
122. G. R. Freeman, *Radiat. Res. Rev.*, **1**, 1 (1968).
123. D. W. Huyton and T. W. Woodward, *Radiat. Res. Rev.*, **2**, 205 (1970).
124. O. S. Gal and I. G. Draganić, *Int. J. Appl. Radiat. Isotopes*, **22**, 753 (1971).
125. O. S. Gal and P. Premović, *Ibid*, **23**, 541 (1972).
126. W. L. McLaughlin in ref. 2, p. 377.
127. W. L. McLaughlin, E. K. Hussmann, H. H. Eisenlohr and L. Chalkley, *Int. J. Appl. Radiat. Isotop.*, **22**, 135 (1971).
128. H. L. Andrews, R. E. Murphy and E. J. LeBrun, *Rev. Sci. Instrum.*, **28**, 329 (1957).
129. L. H. Gevantman, *Radiat. Res. Suppl.*, **2**, 608 (1960).
130. G. E. Adams, J. W. Boag and B. D. Michael, *Trans. Faraday Soc.*, **61**, 492 (1965).
131. G. V. Buxton, *Radiat. Res. Rev.*, **1**, 209 (1968).
132. K. N. Jha, G. L. Bolton and G. R. Freeman, *J. Phys. Chem.*, **76**, 3876 (1972).
133. E. J. Hart and E. M. Fielden in ref. 2, p. 331.
134. J. F. Fowler and F. H. Attix in ref. 1, Vol. 2, p. 241.
135. J. Ilic-Popovic and P. E. Hjortenberg, *Int. J. Appl. Radiat. Isotop.*, **20**, 541 (1969).
136. C. G. Orton in ref. 2, p. 357.
137. W. W. Parkinson, C. D. Bopp, D. Binder, and J. E. White, *J. Phys. Chem.*, **69**, 828 (1965).
138. L. A. Harrah, *Radiat. Res.*, **39**, 223 (1969).
139. R. A. Dudley in ref. 1, Vol. 2, p. 325.
140. W. L. McLaughlin in ref. 2, pp. 129, 387.
141. K. Becker, *Solid State Dosimetry*, CRC Press, Cleveland, Ohio, 1973.
142. J. R. Cameron in ref. 2, p. 105.
143. J. R. Cameron in ref. 2, p. 405.
144. F. H. Attix in ref. 2, p. 397.
145. J. F. Fowler in ref. 1, Vol. 2, p. 291.
146. A. C. Muller in ref. 2, p. 429.
147. A. C. Muller in ref. 2, p. 423.
148. G. W. Grodstein, *X-ray Attenuation Coefficients from 10 keV to 100 MeV*, National Bureau of Standards Circular 583 (1957); R. T. McGinnies, *National Bureau of Standards Supplement to Circular 583* (1959).

149. A. Brnjolfsson, *Adv. Chem. Series*, **81**, 550 (1968).
150. J. Weiss and F. X. Rizzo in ref. 2, p. 231.
151. A. T. Nelms, *Energy Loss and Range of Electrons and Positrons*, National Bureau of Standards, Circular 577 (1956), Supplement to National Bureau of Standards Circular 577 (1958), National Bureau of Standards, Washington, D.C.
152. A. MacLachlan, *J. Am. Chem. Soc.*, **82**, 3309 (1960).
153. A. J. Swallow, *Discuss. Faraday Soc.*, **36**, 273 (1963).
154. C. E. Klots, in *Fundamental Processes in Radiation Chemistry* (ed., P. Ausloos), Interscience, New York, 1968, pp. 1–57; *J. Chem. Phys.*, **39**, 1571 (1963); **44**, 2715 (1966); **46**, 3468 (1967).
155. M. Inokuti, *J. Phys. Soc. Jap.*, **13**, 537 (1958); *Isotop. Radiat.* (*Tokyo*), **1**, 82 (1958).
156. J. Lamborn and A. J. Swallow, *J. Phys. Chem.*, **65**, 920 (1961).
157. E. A. Bemis in *Radiation Dosimetry* (eds., G. H. Hine and G. L. Brownell), Academic Press, New York, 1956, Chap. 10.
158. R. A. Dudley in *Radiation Dosimetry, ibid.*, Chap. 7.
159. H. Kiefer, R. Maushart, and V. Mejdahl in ref. 1, Vol. 3, p. 557.
160. R. Oliver in ref. 2, p. 181.

CHAPTER 4

Ions and Excited Molecules:
Free Radicals

Excited Molecules • Internal Conversion • Intersystem Crossing • Non-radiative Energy Transfer • Unimolecular Reaction • Bimolecular Reaction • Electron Transfer • Ions • The Mass Spectrometer • Ion Recombination • Dissociation • Ionization Potentials • Charge Transfer • Ion-Molecule Reactions • Electron Addition • Subexcitation Electrons • Summary of the Reactions of Ions and Excited Molecules Excited Molecules • Ions • Free Radicals • Formation • Thermal Dissociation • Photodissociation • Oxidation-Reduction Processes • Properties and Reactions • Radical Rearrangement • Radical Dissociation • Radical Attack on Substrates • Radical-Destroying Processes • Disproportionation • Electron Transfer • "Hot" Radicals • Reaction with Oxygen • Detection • Magnetic Susceptibility • Electron Paramagnetic Resonance • Flash Photolysis • Summary of the Reactions of Free Radicals • References •

In the earlier chapters ions and excited molecules were introduced as the first species formed when ionizing radiation is absorbed by matter. Their formation is related to the absorption of energy by the material and their numbers are directly proportional to the absorbed dose; for most molecules in the gas phase, approximately equal numbers of ionized and excited molecules are produced (1). At a later period these primary species may break down and (or) react with the substrate to bring about chemical change. Free radicals formed from the excited and ionized molecules are largely responsible for these changes and generally dominate mechanisms postulated for radiation-induced reactions. In subsequent chapters

122

we of necessity concentrate more on the free radicals than on their precursors but, to emphasize that these precursors are not unimportant, brief descriptions of the properties and reactions of excited and ionized molecules are given here.

EXCITED MOLECULES

The properties of excited atoms and molecules are known from spectroscopic and photochemical studies in which the excited species are produced by the absorption of visible or ultraviolet light. The same excited species are formed by the interaction of ionizing radiation with matter, though ionizing radiation may also produce more highly excited states, with more intrinsic energy, and states (e.g., triplet) which are not formed directly by the absorption of light. Reactions observed in photochemistry, therefore, are not necessarily the only ones stemming from excited molecules which occur upon radiolysis. The distribution of excited species in media exposed to light and to ionizing radiation is also different (Fig. 1.1); photochemically produced species are distributed randomly in any plane perpendicular to the direction of the incident light but those produced by ionizing radiation are concentrated along the tracks of charged particles.

Ionizing radiation can produce excited molecules in matter directly (Eq. 4.1) and also indirectly by neutralization of the ions formed (Eq. 4.2).

$$A \rightsquigarrow A^* \tag{4.1}$$

$$A \rightsquigarrow A^+ + e^- \ (\text{or} \ (A^+)^* + e^-) \tag{4.2}$$

$$A^+ + e^- \rightarrow A^{**} \rightarrow A^*$$

(An asterisk will be used to designate a molecule or ion in an excited state.) The highly excited molecules formed on neutralization (A^{**}) often lose rapidly part of their energy through collisions with other molecules and drop to lower excited states (A^*) similar to those formed photochemically.

The absorption of ultraviolet and visible light by matter can be represented by

$$A + hv \rightarrow A^* \tag{4.3}$$

where hv (Planck's constant multiplied by the frequency of the radiation) is the energy of a light quantum (photon). This is similar to photoelectric absorption of x- and γ-radiation in the sense that the photon disappears completely and its energy is transferred to the molecule. However, it differs from photoelectric absorption in two important respects. First, the incident photon does not have sufficient energy to eject an electron from the molecule and an electron is merely moved to an orbital, farther from the nucleus, corresponding to a higher energy than the original orbital. (The electron excited will be a loosely bound electron in an outer orbital rather than one from an inner shell as is generally the case with photoelectric absorption.) Second, in photoelectric absorption an electron is liberated which can carry off energy of the photon in excess of the binding energy of the electron. This is not so in the process represented by Eq. 4.3, and all the photon energy must be transferred to the excited molecule. Since the possible energy states of the molecule are strictly regulated by quantum-mechanical considerations, it also follows that only light quanta can be absorbed whose energy is equal to the difference between the energies of two of these states, and then only if the transition between these two states is allowed. Light absorption is therefore selective, and the energy of the light absorbed depends on the molecular structure of the absorbing matter. A corollary to this is that absorption of light of a given wavelength will raise the absorbing molecule to a particular, well-defined, energy state. Furthermore, in complex molecules light absorption is often limited to one of the functional groups present. Ionizing radiation, on the other hand, is not selective and can excite any part of the molecule.

The formation and properties of excited molecules can be illustrated by considering a simple diatomic molecule. Figure 4.1 represents the potential

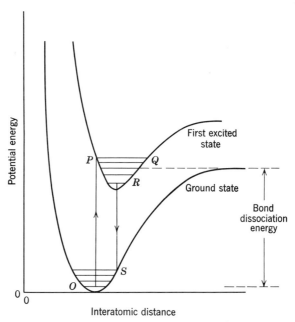

FIGURE 4.1   *Potential-energy curves for a diatomic molecule in its ground state and in an electronically excited state.*

energy of such a molecule plotted against the distance between the two nuclei. Potential energy in this context comprises the electronic energy associated with electron orbitals other than the ground orbitals, vibrational energy of the two nuclei as they move together and apart along the line joining their centers, and rotational energy of the molecule as a whole; translational (thermal) energy associated with the movement of the molecule from place to place is not included. Quanta of rotational energy are small and will not concern us further, since they do not contribute materially to the energy available for bringing about chemical change. The figure shows the stable ground state of the molecule in which the electrons are in their lowest orbitals and the first (singlet) excited state in which one of the electrons is raised to a higher orbital. More energetic electronic states also exist and would be represented in the figure by other curves lying above the first excited state. The minima in the curves in Fig. 4.1 show that in this instance a stable (covalent) bond is formed between the two nuclei in both ground and excited states. The vertical distance between the horizontal portion of the ground state curve and the x axis (or, more accurately, the horizontal through O) is a measure of the bond dissociation energy in the ground state, while the interatomic distance at the minimum of the curve is the equilibrium bond length. Excited states that have potential-energy curves similar to the ground state curve, as does the excited state shown in Fig. 4.1, generally have a smaller dissociation energy than the ground state, corresponding with a weaker bond.

The normal state of the molecule is represented by the lowest vibrational level of the ground state, that is, by the lowest horizontal line within the ground state curve, where the vibrational quantum number is zero and the molecule has only zero-point vibrational energy. For the sake of brevity, states in which the molecule has only zero-point vibrational energy will be called vibrationless states. Absorption of light with an energy corresponding to the vertical distance between O and P can raise the molecule to the upper curve at P. The most probable transitions between different electronic energy states are those that can be represented as a vertical line on a potential-energy diagram. This restriction, which is known as the Franck-Condon principle, can also be expressed by saying that the two nuclei are unlikely to move, relative to one another, during the transition; the transition from the ground to the excited state takes place in about 10^{-15} sec whereas a complete vibration of the molecule takes about 10^{-13} sec. Other near-vertical transitions are not completely excluded, but they occur with a smaller probability. Futhermore, transitions are possible with a lower probability from positions other than the center of the lowest vibrational level so that a series of absorption bands are obtained rather than a single absorption line.

At P the excited molecule is compressed, since P lies to the left of the minimum in the excited curve, which represents the stable internuclear distance for the excited state, and the molecule subsequently vibrates with an amplitude equal to the distance PQ. The vibrational energy that the molecule gains in this way is also quantized; that is, it can only have a certain number of definite values, and these are represented in Fig. 4.1 by the horizontal lines (e.g., PQ) contained in the two curves. Vibrational energy can be lost quite readily by transfer to other molecules as thermal (translational) energy when these collide with the vibrationally excited molecule. On the potential energy diagram this is shown by the molecule dropping from the vibrationally excited state PQ to lower vibrational levels and eventually to the vibrationless state R.

Electronic excitation energy is not lost as readily as vibrational energy by collisional processes and the excited molecule may remain at R for a relatively long time (e.g., with a half-life of about 10^{-8} sec) before emitting a photon and returning to the ground state at S. A vertical transition is again most probable, and the energy of the emitted radiation (*fluorescence*) is given by the distance RS. Since RS is shorter than OP the fluorescent radiation has a longer wavelength than the radiation originally absorbed, or, in other words, the emission (fluorescent) radiation bands occur at longer wavelengths than the absorption bands. At S the molecule has excess vibrational energy which will rapidly be lost by collisions with other molecules, returning the molecule to its original state, O.

Fluorescence is only one of several ways in which an electronically excited molecule can return to the ground state. Including fluorescence these are:

(i) fluorescence (radiative conversion to the ground state),
(ii) internal conversion (nonradiative conversion to a lower state of the same multiplicity),[1]
(iii) intersystem crossing (nonradiative conversion to a state of different multiplicity),
(iv) nonradiative energy transfer to a neighboring molecule,
(v) unimolecular reaction (dissociation or isomerization of the excited molecule),
(vi) bimolecular reaction (chemical reaction involving a second molecule).

Influences external to the excited molecule may favor one or another of the alternatives. If they restrict or prevent fluorescence, the fluorescence is is said to be quenched.

---

[1] The multiplicity of a state is a type of degeneracy that is determined by the total electron spin. Most stable molecules have singlet ground states and singlet and triplet excited states.

Internal Conversion

If the potential-energy curve of the excited state crosses or comes close to that of the ground state at a vibrational level lower than that to which the molecule is originally excited (Fig. 4.2), there is a finite possibility that the molecule will cross from the excited to the ground state at this point. In the ground state the molecule will be highly vibrationally excited but will rapidly lose this energy by collision with other molecules; the degraded energy will appear as thermal energy. Internal conversion to the ground state may not be complete, but any such conversion which does take place will reduce the intensity of the fluorescence observed.

More common than conversion from an excited state to the ground state is internal conversion from higher electronically excited states to the lowest excited state of the same multiplicity (e.g., singlet or triplet). This is particularly true for polyatomic molecules because the potential-energy surfaces corresponding with the many excited states will have numerous intersections. Conversion to the lowest excited state takes place rapidly within 10^{-13} sec (2). Fluorescence is therefore characteristic of the lowest electronically excited state of a given multiplicity even though higher excited states may be formed initially. Chemical reactions of electronically excited molecules are also those of the lowest excited state unless the first step is rapid unimolecular dissociation of the excited molecule. Higher excited states may dissociate before they have lost energy through internal conversion and in this case the extra energy is carried away by the dissociation products.

Intersystem Crossing

Figure 4.3 shows the potential-energy curve of a second electronically excited state of different multiplicity crossing the original excited state curve. The second curve corresponds to a lower energy than the first, and since the curves cross at a point (T) whose vibrational level is below that to which the molecule is originally excited, the excited molecule is in a position to cross to the second curve at T. Some of the excited molecules may do this, thereby decreasing the intensity of the fluorescence, but allowing the possibility of radiative transfer from the second curve to the ground state. This phenomenon has been found in molecular spectroscopy, and the original and second (lower energy) excited curves have been identified as singlet and triplet excited states, respectively.

Electronically excited states formed directly by the absorption of light are normally singlet states while triplet excited states are formed from these by intersystem crossing, the process shown in Fig. 4.3. In general, triplet excited

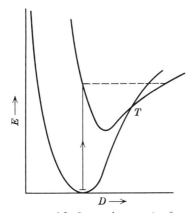

FIGURE 4.2 *Internal conversion from an electronically excited state to the ground state.*

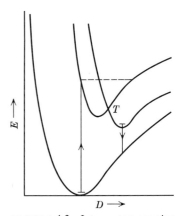

FIGURE 4.3 *Intersystem crossing from singlet to triplet excited state.*

states are not formed directly from the ground state by the absorption of light, and the transition from singlet ground state to triplet excited state is said to be forbidden.

Triplet excited states have longer lifetimes than singlet excited states because the triplet → singlet emission is highly forbidden. Lifetimes from 10^{-4} to 10^{-3} sec are usual, but the lifetime may be as long as several seconds. The radiation emitted when an excited triplet state drops to the ground state has a longer wavelength than the corresponding fluorescence and is known as *phosphorescence*. Some of the triplet excited molecules may gain enough energy through collisions with other molecules to raise them to the vibrational level at T (Fig. 4.3), so that they can return to the singlet excited state and fluoresce in the normal way. Fluorescence due to molecules which have spent some time in the triplet state will be delayed beyond the normal half-life of the singlet excited molecules (about 10^{-8} sec) and is known as *delayed* or *slow fluorescence*; the effect is enhanced by heating the triplet excited medium.

The lifetime of triplet excited states is influenced by the physical state of the medium and is longest in very viscous or solid systems; singlet excited states differ in this respect and their lifetimes are not affected by the nature of the surrounding medium. Phosphorescence is not observed at all in non-viscous solutions where, in the absence of other nonradiative processes such as bimolecular reaction, the triplet state is deactivated in a radiationless process involving both the triplet excited molecule and the solvent.

Structurally the triplet state is distinguished by inversion of the spin of the excited electron (Fig. 4.4). In stable molecules the electrons are normally

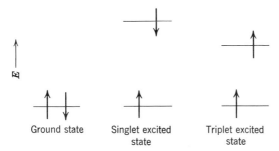

FIGURE 4.4   *Electron distribution and spin in molecules in the ground (singlet) state, excited singlet, and excited triplet states.*

grouped in pairs in each of which the two electrons have opposite spins. Excitation to an excited singlet state results in the translation of one of the electrons to a higher-energy orbital, without change in the direction of spin. Excitation to an excited triplet state involves both transfer of an electron to a higher orbital and inversion of the spin of the excited electron. The triplet state will therefore contain two electrons whose spins are parallel, and for many purposes it can be treated as a diradical. The two unpaired electrons associated with the triplet state have been detected by magnetic susceptibility and electron spin resonance measurements similar to those used to detect the single unpaired electron present in free radicals. In Fig. 4.4 the energy of the molecule in the triplet state is shown as rather less than the energy of the excited molecule in the singlet state, since a system in which the electron spins are parallel has a lower energy than one in which they are all paired (Hund's rule); a triplet level is normally the lowest electronically excited state of the molecule.

Interconversion of singlet and triplet states is facilitated by the presence of paramagnetic atoms or heavy nuclei. The lifetime of triplet excited states for example is shorter when the excited compound contains heavy elements which facilitate conversion to the singlet ground state.

Triplet excited molecules are important in radiation chemistry both because of their relatively long lifetimes and diradical character, which favor chemical reaction with the substrate rather than energy loss by phosphorescence and internal conversion, and because they are formed in the tracks of charged particles in rather greater numbers than might be anticipated by analogy with photochemistry. Swiftly moving charged particles can produce only optically allowed transitions—from singlet ground states only singlet excited states—but slowly moving charged particles, particularly electrons, can induce optically forbidden transitions, e.g., give triplet excited states from singlet ground states. Furthermore, neutralization of positive ions with electrons (Eq. 4.2) in the presence of other ions, i.e., under the conditions present in the clusters or spurs found along the track of a charged particle, can give both singlet and triplet excited molecules (3). Finally, slow secondary electrons that do not have quite enough energy to excite a molecule to its lowest singlet excited state may have sufficient energy to raise it to the lowest triplet state.

Representative lifetimes for the processes described above are given in the following diagram. The lifetimes are in seconds and refer to condensed (liquid) systems excited photochemically.

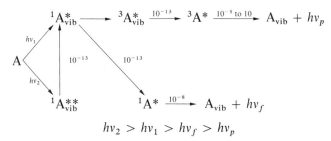

$$hv_2 > hv_1 > hv_f > hv_p$$

A is the molecule in the ground state; the superscripts 1 and 3 denote the singlet and triplet states respectively, the subscript "vib" a molecule with excess vibrational energy, the asterisk (*) indicates the lowest electronic excited states (either singlet or triplet), and the double asterisk (**) any higher electronic excited state.

Nonradiative Energy Transfer

The transfer of electronic excitation energy from one molecule to another can take place in several ways, but the overall process is represented simply by,

$$A* + B \rightarrow A + B* \qquad (4.4)$$

The electronically excited molecule produced, B*, may revert to the ground state by any of the processes listed earlier in this chapter. Transfer of excitation energy is only possible if the energy necessary to excite B is equal to or less than the excitation energy of A*, i.e., if the excitation potential of B is equal to or less than the excitation potential of A. Energy of A* in excess of that necessary to excite B appears as vibrational or translational energy of the two molecules.

In its simplest form reaction 4.4 occurs when an excited molecule collides directly with a molecule able to accept the excitation energy, so-called collisions of the second kind. Photochemistry offers many examples of this process including a large number of reactions, mainly in the gas phase, in which the first step is collision of an excited mercury atom with another molecule. When this second molecule is hydrogen, for example, the energy transferred is sufficient to cause it to dissociate into two hydrogen atoms; light of wavelength 2537 Å raises mercury to a triplet excited state with an energy of 133 kcal mole^{-1} while the bond dissociation energy of molecular hydrogen is 103 kcal mole^{-1}. Similar reactions can take place in liquid media, but in condensed systems the two processes described below may play a significant part in the energy transfer. A well-known example of energy transfer in liquid media occurs with liquid scintillators in which energy is absorbed by a solvent and transferred to a dissolved organic compound which subsequently fluoresces.

Transfer of excitation energy also occurs if the excited molecule A* fluoresces and the fluorescent radiation is absorbed by the molecule B, raising it to the excited state B*. Qualitatively similar to this picture but more efficient is nonradiative resonance, or dipole-dipole, transfer of excitation energy between the two molecules (4). The requirements are similar to those for the emission and reabsorption of fluorescent radiation, i.e., that there be maximum overlap of the emission spectrum of the sensitizing molecule (A) and the absorption spectrum of the accepting molecule (B), but the process is pictured as involving both molecules simultaneously. The particular interest of the resonance-transfer process is that it is not restricted to collisions of the two molecules concerned but can take place when these are separated by distances large by comparison with molecular dimensions, i.e., by distances of the order of 50 to 100 Å. It is applicable to gas, liquid, and solid systems.

A further energy transfer process is possible if the excited molecule is one of an orderly array of similar molecules, e.g., one molecule in a molecular crystal. In this case it is possible to treat the system as though the electronic excitation energy (or exciton) moved rapidly from one molecule to another, spending a very short time (less than the period of one vibration) on each. The energy of the exciton is that of the vibrationless lowest excited state, since extra electronic or vibrational energy will be lost quite rapidly, and its lifetime will be about the same as if it were located on a single molecule, i.e., about 10^{-8} sec. If the crystal contains an impurity molecule which has an excited state with a lower energy than the host molecules, the exciton, if it becomes located on the impurity, will raise it to an excited state with both electronic and vibrational energy. If now the vibrational energy is lost the exciton will no longer have sufficient energy to return to the host molecules and will be trapped on the impurity, and, if the excited impurity fluoresces, the fluorescence will be characteristic of the impurity and not of the host molecules which make up the bulk of the crystal. Trapping of the absorbed radiation energy in this way and the subsequent emission of fluorescence from a minor component in the solid is frequently observed with the solids used as organic scintillators.

Energy transfer by exciton migration is most likely in crystals and in large molecules such as polymers, but it appears that it may also occur in liquids. Burton and his colleagues (5) have proposed, in order to explain high rates of excitation transfer in certain scintillator solutions, that in liquid hydrocarbons small groups of ordered molecules (domains) may exist. Exciton transfer can occur within these groups of perhaps 10 or 15 molecules, so that collision with any molecule of a domain containing an excited molecule is equivalent to collision with the initially excited molecule itself.

A review of energy transfer in radiation chemistry has been given by Matheson (6).

Unimolecular Reaction

Polyatomic excited molecules may reach a stable state through some form of molecular rearrangement (7). A simple example is the conversion of a *trans* compound into an equilibrium mixture of the *cis* and *trans* forms on irradiation with ultraviolet light, i.e.,

$$\underset{trans}{\overset{\displaystyle \text{H}}{\underset{\text{H}}{\diagdown}\text{C}=\text{C}\diagup}} \quad \xrightarrow{h\nu} \quad \underset{cis}{\overset{\displaystyle}{\underset{\text{H}}{\diagdown}\text{C}=\text{C}\underset{\text{H}}{\diagup}}} \tag{4.5}$$

If the excited state has sufficient energy the molecule may break at a covalent bond to give two free radical fragments,

$$(R{:}S)^* \rightarrow R{\cdot} + {\cdot}S \tag{4.6}$$

The two electrons comprising the bond are divided one to each radical, and the least excitation energy that will produce dissociation is equal to the dissociation energy of the bond. Frequently rather more energy than this is required and the excess appears as kinetic energy of the fragments. Two situations that lead to dissociation are illustrated by the potential-energy curves in Fig. 4.5. In Fig. 4.5a the molecule is excited directly into a repulsive state and dissociation occurs in less time than the period of one vibration (less than about 10^{-13} sec); the excited state involved may be the lowest or a higher electronic excited state. In *b* the molecule is raised to a stable excited state, but it may cross to a repulsive state at *T*. The excited molecules will not necessarily all pass into the repulsive state and dissociate; some may drop past *T* and fluoresce in the normal manner. Dissociation of the excited molecule after crossing into a second excited state, as in *b*, is called *predissociation*.

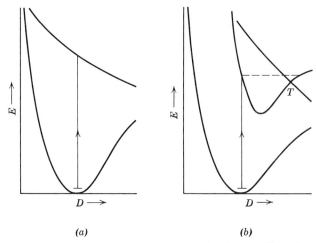

(a)   *(b)*

FIGURE 4.5   (*a*) *Transition to an unstable* (*repulsive*) *state*; (*b*) *predissociation.*

Dissociation is very often the first step in photochemical reaction, the chemical changes being brought about subsequently by the free radicals produced. With light of short wavelength, i.e., of relatively high energy, the radicals themselves may be excited or may have an appreciable amount of kinetic energy. Radicals formed with very little extra energy may not, in liquid media, be able to escape through the close-packed molecules that surround them. In this case they recombine, dissipating the energy of re-combination as heat and producing no overall chemical change. The caging and recombination of free radicals from an excited molecule before they have become separated is often referred to as the *Franck-Rabinowitch* or *cage* effect (8). A good example of the cage effect has been provided by Lampe and Noyes (9) who determined the quantum yields for the photodissociation of iodine in various solvents. They found that the quantum yield decreased in the series hexane, carbon tetrachloride, hexachlorobutadiene as the mass of the solvent molecules increased. In all solvents the quantum yields were below that found in the gas phase where the dissociation products can move apart without hindrance. More recently Meadows and Noyes (10) have drawn attention to the fact that the cage effect will operate in the opposite sense once a dissociation product is past the first layer of solvent molecules; thereafter the cage will act to keep the fragments apart.

Another feature of photochemical dissociation is the possibility of intra-molecular energy transfer prior to dissociation. The bond breaking is not necessarily the site at which light was absorbed; often it is the weakest bond in the molecule which breaks. Aldehydes and ketones, for example, absorb light at a wavelength characteristic for the carbonyl group, but subsequently dissociate in such a manner that a carbon–carbon bond breaks and not the carbon–oxygen bond responsible for the light absorption. Acetone dissociates to give an acetyl and a methyl radical,

$$CH_3COCH_3 \xrightarrow{h\nu} CH_3CO\cdot + \cdot CH_3 \tag{4.7}$$

Aldehydes and ketones with longer hydrocarbon chains break between the α and β carbon atoms; thus methyl butyl ketone decomposes almost entirely into acetone and propylene,

$$CH_3(CH_2)_3COCH_3 \xrightarrow{h\nu} CH_3CH{=}CH_2 + CH_3COCH_3 \tag{4.8}$$

The redistribution of electronic energy within a molecule is a rapid process and often it is necessary to consider the molecule as a whole to be excited rather than associate excitation with a particular location in the molecule.

Dissociation of excited molecules into molecular products is possible (reaction 4.8 is an example) though less common than dissociation into free radicals. Another example is the dehydrogenation of excited ethane to give ethylene,

$$C_2H_6{}^* \rightarrow C_2H_4 + H_2 \tag{4.9}$$

Rearrangement and dissociation may involve either singlet or triplet excited states.

Bimolecular Reaction

Excited molecules may react directly with a second molecule to form new chemical products. Reactions of this type can be classified under the headings, electron transfer, abstraction, addition, and Stern-Volmer reactions.

ELECTRON TRANSFER.   An electron may be transferred between the excited molecule and a second molecule or ion;

$$A^* + B \rightarrow A^+ + B^- \text{ (or } A^- + B^+) \tag{4.10}$$

Ferrous ions quench the fluorescence of methylene blue in aqueous solution by a mechanism that is believed to include transfer of an electron from a ferrous ion to an excited dye molecule (D*) on collision,

$$D^* + Fe^{2+} \rightarrow D\cdot^- + Fe^{3+} \tag{4.11}$$

$$D\cdot^- + H^+ \rightarrow DH\cdot \tag{4.12}$$

$$2DH\cdot \rightarrow D + DH_2 \tag{4.13}$$

The excited dye is converted by electron transfer to the radical-ion $D\cdot^-$, and then, by combination with a hydrogen ion, to the radical $DH\cdot$. Disproportionation between two of these radicals regenerates a molecule of the dye and gives a reduced form of the dye (the leuco dye), DH_2 (11). Oxidizing agents (e.g., Fe^{3+}, Ce^{4+}, O_2) may also quench the fluorescence, accepting an electron in the electron transfer step and being reduced.

Reaction frequently takes place between two molecules or ions that form a complex and are already loosely bound together when one of the species is excited. Photochemical electron transfer reactions between transition metal ions and anions or water molecules are well established, e.g.,

$$Fe^{3+}X^- \xrightarrow{h\nu} Fe^{2+} + X\cdot \qquad (4.14)$$

and

$$M^{z+}H_2O \xrightarrow{h\nu} M^{(z+1)+}OH^- + H\cdot \qquad (4.15)$$

where $X = F$, Cl, Br, OH, N_3, C_2O_4 and $M = Fe$, Cr, V, or Ce, etc. In both reactions one of the products is a free radical.

ABSTRACTION. Abstraction reactions generally involve reduction of the excited molecule through hydrogen abstraction from the substrate,

$$A^* + RH \rightarrow AH\cdot + R\cdot \qquad (4.16)$$

Both products are free radicals. Bolland and Cooper (12) have investigated the oxidation of ethanol in neutral or acid aqueous solution photosensitized by the anthraquinone-2,6-disulfonate ion. The singlet excited quinone abstracts hydrogen from the ethanol and in the absence of oxygen the process is represented as

$$Q^* + CH_3CH_2OH \rightarrow QH\cdot + CH_3\dot{C}HOH \qquad (4.17)$$

$$CH_3\dot{C}HOH + Q \rightarrow CH_3CHO + QH\cdot \qquad (4.18)$$

$$2QH\cdot \rightarrow Q + QH_2 \qquad (4.19)$$

The quinone is reduced to the hydroquinone and the ethanol oxidized to acetaldehyde. Oxygen, when present, adds to the free radical intermediates to give peroxy-radicals which interact to give hydrogen peroxide, acetaldehyde, and acetic acid as the final products; the initial step, hydrogen abstraction by the singlet excited quinone, remains the same.

ADDITION. Addition reactions appear to be characteristic of the triplet rather than the singlet excited state. Typical of these reactions is the addition of oxygen, which itself has a triplet ground state, to triplet excited linear polynuclear aromatic hydrocarbons (except naphthalene) to form transannular peroxides

$$\text{anthracene} + O_2 \xrightarrow{h\nu} \qquad (4.20)$$

Schenck and his collaborators (13) have shown that photochemical oxidations sensitized by dyes (e.g., eosin and methylene blue) can produce transannular peroxides from cyclic dienes and hydroperoxides from mono-olefins. The hydroperoxides formed are very often isomers of those formed by autoxidation, which is known to proceed through a free radical mechanism, and a different process must be involved in the sensitized oxidations. Schenck has suggested that the triplet excited sensitizer (D*) adds oxygen to form an intermediate diradical peroxide, DO_2, which subsequently reacts with the substrate (P)

$$D^* + O_2 \rightarrow DO_2 \tag{4.21}$$

$$\dot{D}O_2 + P \rightarrow D + PO_2 \tag{4.22}$$

Addition reactions are not limited to photo- or photosensitized oxidations. Dainton and Ivin (14) have shown that excited sulfur dioxide will add to both saturated and unsaturated aliphatic hydrocarbons. Excited sulfur dioxide apparently adds directly to the saturated compounds to give sulfinic acids

$$SO_2^* + RH \rightarrow R-SO_2H \tag{4.23}$$

Addition to the unsaturated hydrocarbons is believed to give an intermediate diradical which in the gas phase rearranges to give a sulfinic acid,

$$SO_2^* + R-CH{=}CH_2 \rightarrow R-\dot{C}H-CH_2-SO_2\cdot \tag{4.24}$$

$$R-\dot{C}H-CH_2-SO_2\cdot \rightarrow R-CH{=}CH-SO_2H \tag{4.25}$$

In the liquid phase the diradical initiates polymerization to give a polysulfone with the repeating unit—$(RCH-CH_2-SO_2)$—.

STERN-VOLMER REACTIONS. These reactions involve an interchange of atoms between two molecules, one or both of which are excited,

$$2A^* \rightarrow P + Q \tag{4.26}$$

It has been suggested (15) that a reaction of this type may account in part for the production of hydrogen and hydrogen peroxide in water irradiated with densely ionizing radiation:

$$2H_2O^* \rightarrow H_2 + H_2O_2 \tag{4.27}$$

IONS

Ionization is the distinctive consequence of the absorption of ionizing radiation by matter. The ions are produced by the process,

$$A \rightsquigarrow A^+ + e^- \ [\text{or } (A^+)^* + e^-] \tag{4.28}$$

In this section we consider the fate of both the positive ion and the electron and the results that follow recombination of a positive ion with either an electron or a negative ion. Much of the available information about ions is the result of gas phase studies with the mass spectrometer and the principle of this instrument is first described briefly.

The Mass Spectrometer

The mass spectrometer enables ions of different mass and different charge to be separated and collected. In the instrument (Fig. 4.6) the gas or vapor to be examined is ionized and the ions accelerated by a potential difference

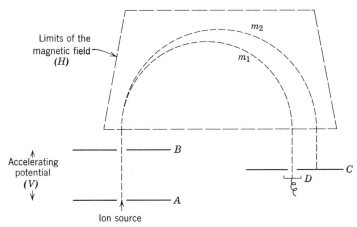

FIGURE 4.6 *Principle of the mass spectrometer.*

V and subjected to a magnetic field (strength H) perpendicular to the direction of motion of the ions. In the magnetic field the ions follow a circular path of radius r given by

$$r^2 = \frac{2Vm}{H^2 e} \tag{4.29}$$

where m and e are the mass and charge of the ion respectively. When V and H are constant, r depends on the ratio m/e or, for singly charged ions (which predominate), upon m only.

Mass spectrometry is used extensively to identify organic compounds using for this purpose ionizing potentials in the region of 50 to 75 V. Potentials in this range are great enough to ionize any organic material and to bring about dissociation reactions so that the mass spectra of complex organic compounds are made up of a large number of peaks (each due to an ion of different mass/charge ratio) of varying intensity.

Other developments in mass spectrometry of particular interest to radiation chemists include the construction of instruments capable of operating at ionization chamber pressures of up to 1 mm mercury (16), and the use of coupled mass spectrometers (17). In the latter case, ions produced by one mass spectrometer are allowed to interact with neutral molecules, and the ionic products are analyzed by a second mass spectrometer. Both these developments allow a more detailed study of the reactions between ions and neutral molecules (ion-molecule reactions) than is possible with conventional instruments.

The mass spectra of organic compounds have also been correlated with the products obtained from them by irradiation in the gas phase (18–20). The results confirm that ions are important intermediates in gas-phase radiolysis, and the correlation is surprisingly good considering that the mass spectra tell nothing about the contribution of excited molecules to the radiolysis reactions. It is more debatable whether the mass spectrum of an organic vapor is closely related to the products and yields obtained by the irradiation of the same compound in the liquid phase. In the mass spectrometer the ions formed are essentially isolated from one another and from the parent molecules. This condition is approached, though not closely, in gas phase radiolyses at low to moderate pressures but is quite different from the situation in liquid systems. In the condensed phase bimolecular collisions (e.g., ion-molecule) will be many times more probable than in the gas and may lead to far more bimolecular reaction and to the stabilization by energy loss of ions that do not survive in the mass spectrometer. Unimolecular reactions (e.g., fragmentation) will be favored in the mass spectrometer by retention by the ion of its original energy and by the longer lifetime of the ions (about 10^{-5} sec compared to a lifetime of about 10^{-13} sec for ions formed by irradiation in the liquid phase (21)).

Ion Recombination

The normal fate of ions is to react with an ion of opposite sign to reform a neutral entity:

$$A^+ + e^- \rightarrow A^{**} \tag{4.30}$$

or

$$A^+ + A^- \rightarrow A^* + A \tag{4.31}$$

The electron ejected during ionization may have considerable energy, and this will be lost through interactions with other molecules before the electron neutralizes a positive ion; in other words the election is reduced to thermal or near thermal energy before recombination occurs. The immediate product of the recombination is a neutral but highly excited molecule which may be in either the singlet or the triplet state and which contains sufficient energy for spontaneous re-ionization. However, re-ionization is unlikely except perhaps in the case of a gas at very low pressures where the excited molecule is effectively isolated. In other cases internal conversion will rapidly convert the molecule to the lowest (singlet or triplet) excited state. Dissociation of the excited molecule often occurs to give either molecular products (Eq. 4.32) or, more frequently, free radicals (Eq. 4.33).

$$A^* \ (\text{or} \ A^{**}) \rightarrow M^* + N \tag{4.32}$$

$$A^* \ (\text{or} \ A^{**}) \rightarrow R\cdot^* + S\cdot \tag{4.33}$$

One or other of the products will most probably be excited (22).

Highly excited radicals can be formed by the neutralization of an ion which is also a free radical,

$$R\cdot^+ + e^- \rightarrow R\cdot^{**} \tag{4.34}$$

Excited radicals are often more reactive than those with only thermal energy and they are distinguished from the latter by the name *hot radicals*. The term also applies to radicals that are formed with a large amount of kinetic energy.

If the ionized molecule is closely associated with another molecule, as in a molecular complex or a transient ion-molecule collision complex, and neutralization occurs, chemical reaction between the two molecules may result:

$$A\cdot B^+ + e^- \rightarrow C + D \tag{4.35}$$

The initial ion-molecule combination is similar to the ion-clusters postulated by Lind and Mund as intermediates in gas-phase radiolysis reactions, though smaller and possibly transient. However reaction 4.35 is likely to be only one of several simultaneous reactions that contribute to the overall chemical change. The reaction is similar to the Stern-Volmer reaction (Eq. 4.26) mentioned in the section on excited molecules.

Other processes, dissociation, charge transfer, ion-molecule reaction, or electron addition may intervene between the formation of an ion pair and ultimate recombination of two oppositely charged ions but, in all cases, neutralization with an ion of opposite sign also occurs and may lead to further chemical reaction.

Dissociation

Positive ions formed in the mass spectrometer and by ionizing radiation are nearly always vibrationally excited, since they are formed by vertical Franck-Condon transitions from the ground state of the neutral molecule, and internuclear distances in the ion will in general be different to those in the neutral molecule. This is similar to the formation of vibrationally excited states of electronically excited molecules (Fig. 4.1). The positive ions may also be electronically excited if the radiation initiating ionization has an energy greater than the lowest ionization potential of the molecule; a condition that is true in the mass spectrometer when ionizing potentials in the usual range of 50 to 75 V are used and is also true for a large proportion of the ionizing events produced by ionizing radiation. It must be emphasized, however, that the mass spectrometer does not distinguish between excited and nonexcited ions of the same mass.

In common with electronically excited molecules, excited polyatomic ions may dissociate and (or) rearrange. Dissociation can be represented by

$$(A^+)^* \rightarrow M^+ + N \tag{4.36}$$

or

$$(A^+)^* \rightarrow R^{\cdot +} + S^{\cdot} \tag{4.37}$$

The fragments may be either both molecules or both free radicals but in each case one has a positive charge. The charge normally remains with the fragment having the lowest ionization potential, which is generally the larger of the two fragments. Excited ionized ethane dissociates into the molecular products ethylene and hydrogen,

$$(C_2H_6^+)^* \rightarrow C_2H_4^+ + H_2 \tag{4.38}$$

In this instance the charge remains with the ethylene which has an ionization potential of 10.5 V rather than with the hydrogen for which the ionization potential is 15.4 V. (Ionization potentials are listed in Table 4.1.) Large ions may break down in stages to give progressively smaller fragment ions, though this is a feature of mass spectrometry rather than radiation chemistry, since the ion and its breakdown products will need to escape neutralization for a relatively long time.

As with excited molecules, the bond breaking in the dissociation of an ion is not necessarily the site of the original interaction. The electrons remaining in the ion may readily change their position, resulting in a migration of charge within the molecule. Furthermore, the excitation energy is also redistributed within the molecule among the various vibrational and electronic excited states in an essentially random fashion. If at any instant sufficient energy is concentrated in a single bond, the molecule can dissociate at that bond. Dissociation is therefore not immediate and not specific (a group of similarly excited molecules will not all break at the same bond) though it will be favored with respect to weak bonds that require less than the average amount of energy to break them. Aliphatic halogen compounds for example frequently break preferentially at the relatively weak carbon–halogen bond

T A B L E 4.1 *Ionization Potentials*[a]

Atoms (ref. 23)

H·	13.6	He	24.6
F·	17.4	Ne	21.6
Cl·	13.0	Ar	15.8
Br·	11.8	Kr	14.0
I·	10.4	Xe	12.1

Molecules (ref. 24)

H_2	15.4	Cl_2	11.5
N_2	15.6	Br_2	10.6
O_2	12.1	I_2	9.3
CO	14.0	HCl	12.7
CO_2	13.8	HBr	11.6
SO_2	12.3	HI	10.4
H_2O	12.6	NO	9.2
NH_3	10.2	N_2O	12.9
CH_4	13.0	$CH_2{=}CH{-}CH_2{-}CH_3$	9.6
C_2H_6	11.7	C_2H_2	11.4
C_3H_8	11.1	C_6H_6	9.2
$n\text{-}C_4H_{10}$	10.6	$C_6H_5{-}CH_3$	8.8
$iso\text{-}C_4H_{10}$	10.6	$C_6H_5{-}C_2H_5$	8.8
C_2H_4	10.5	Naphthalene	8.1
C_3H_6	9.7	Cyclohexane	9.9
$CHCl_3$	11.4	CH_3OH	10.9
CH_3Cl	11.3	HCHO	10.9
CH_3Br	10.5	HCOOH	11.1
CH_3I	9.5	CH_3COCH_3	9.7

Radicals (ref. 25)

·OH	13.7	$\cdot C_2H_5$	8.7
$\cdot O_2H$	11.5	$\cdot C_4H_9(t)$	6.9
$\cdot CH_3$	10.0	$\cdot C_6H_5$	9.9

[a] In volts.

both in the mass spectrometer and upon irradiation. Studies with the mass spectrometer have shown that fragmentation of complex ions may be delayed as long as 10^{-5} sec which, particularly in condensed systems, would allow ample time for other processes to compete with the dissociation.

Ions may rearrange before or during the fragmentation process and give products that cannot be obtained from the parent ion by fission of a single bond; the formation of ethylene from the ion $C_2H_6^+$ for example involves the migration of a hydrogen atom. Hydrogen is particularly liable to migrate and the mass spectra of many organic compounds show evidence of this. Isobutane for example gives a peak at mass 29 due to the ion $C_2H_5^+$, which is accounted for by migration of hydrogen and the subsequent splitting of two carbon–carbon bonds. Alcohols (R—CH_2OH) frequently give a peak at mass 19 due to the hydronium ion, H_3O^+, which is formed in a rearrangement involving two hydrogen atoms. Other groups may also migrate. Rearrangement is very common in the mass spectra of unsaturated hydrocarbons; for example, the four compounds shown below have almost indistinguishable mass spectra, as though the molecular structure were lost on ionization and the bonds redistributed in a random manner before fragmentation occurs (26).

$$CH_3—CH\!=\!CH—CH_2—CH_3 \qquad CH_3—\overset{\overset{\displaystyle CH_3}{|}}{C}\!=\!CH—CH_3$$

2-pentene   2-methyl-2-butene

$$CH_2\!=\!\overset{\overset{\displaystyle CH_3}{|}}{C}—CH_2—CH_3 \qquad CH_2\!=\!CH—\overset{\overset{\displaystyle CH_3}{|}}{CH}—CH_3$$

2-methyl-1-butene   3-methyl-1-butene

Charge Transfer[2]

In charge-transfer reactions (27, 28) a positive ion removes an electron from an adjacent neutral molecule

$$A^+ + B \rightarrow A + B^+ \tag{4.39}$$

For the reaction to take place the ionization potential of the neutral molecule must be equal to or less than the ionization potential of the neutral counterpart of the ion, i.e., $I_B \leqslant I_A$. The reaction

$$He^+ + Ne \rightarrow He + Ne^+ \tag{4.40}$$

for example, is possible because the ionization potential of helium (24.6V) is greater than that of neon (21.6 V). When the ionization potentials of the two neutral molecules differ, energy will be liberated in the process and may be retained by the products as vibrational or electronic energy, though the process is most favored when the energy difference is small. In some cases

---

[2] Charge transfer is frequently treated as a special case of energy transfer, which in this case embraces both the transfer of excitation energy and of an electron. In this book, however, we have preferred to make a clear distinction between the two processes, and therefore limit the term energy transfer to the transfer of excitation energy, using "charge transfer" where an electron moves from one molecule to another.

the energy liberated may be sufficient to bring about dissociation of the newly formed ion, e.g.,

$$Ar^+ + CH_4 \rightarrow Ar + \cdot CH_3^+ + H\cdot \tag{4.41}$$

Charge transfer between like molecules will not produce any observable change in liquid systems where the molecules are free to diffuse, but it is a form of energy migration that may be significant in the solid state.

It may be pertinent to point out in connection with the table of ionization potentials (Table 4.1) that the values quoted are for materials in the gas phase and may not be true in liquid media. However, the relative values are believed to be the same in both phases and so the individual values can be used to decide the likelihood of charge transfer in liquid systems.

Ion-Molecule Reactions

Ion-molecule reactions are those between an ion and a neutral molecule,

$$A^+ + B \rightarrow C^+ + D \tag{4.42}$$

Reactions of this type are not normally detected by the mass spectrometer, since these operate at low pressures at which ion-molecule collisions will be scarce. However, by increasing the gas pressure in the instrument this sort of collision can be rendered more probable and the products of ion-molecule reactions detected. As the pressure is increased the ion-current due to the ions formed directly will increase in proportion to the pressure while the ion-current due to ions formed by bimolecular ion-molecule reactions will increase in proportion to the square of the pressure and can thereby be identified.

Ion-molecule reactions are believed to be important steps in the radiolysis of many compounds, and they can take a number of different courses. Hydrogen abstraction by an ion is common and leads to ions one mass unit heavier than the parent compound, e.g.,

$$HBr^+ + HBr \rightarrow H_2Br^+ + Br\cdot \tag{4.43}$$

$$H_2O^+ + H_2O \rightarrow H_3O^+ + \cdot OH \tag{4.44}$$

$$CH_4^+ + CH_4 \rightarrow CH_5^+ + \cdot CH_3 \tag{4.45}$$

Other ion-molecule reactions involve transfer of a proton (H^+) or (29) a hydride ion (H^-),

$$H_2^+ + O_2 \rightarrow HO_2^+\cdot + H\cdot \tag{4.46}$$

$$C_3H_5^+ + \text{neo-}C_5H_{12} \rightarrow C_3H_6 + C_5H_{11}^+ \tag{4.47}$$

or the formation and rupture of carbon—carbon bonds,

$$CH_3^+ + CH_4 \rightarrow C_2H_5^+ + H_2 \qquad (4.48)$$

$$C_2H_4^+ + C_2H_4 \rightarrow C_3H_5^+ + \cdot CH_3 \qquad (4.49)$$

It should perhaps be pointed out that ions are not bound by the laws of valency in quite the same way as neutral molecules and ions such as CH_5^+ are stable.

Exothermic ion-molecule reactions with very few exceptions require no activation energy, reaction taking place at the first collision, and consequently the reactions are very fast even at low temperatures. Collisions between polyatomic ions and molecules are facilitated by the attraction that the ion exerts on the molecule by virtue of its charge. The collision complex formed when the molecular ion and molecule collide is also likely to be longer lived than that between two neutral molecules (e.g., 10^{-5} sec compared to 10^{-13} sec); the ion-molecule complex is an example of so-called *sticky collisions*.

Electron Addition

In most systems the only negatively charged entities with which we are concerned are the electrons produced by ionization. However, a minority of chemical compounds are able to form negative ions by capture of a slow electron (30), accompanied in some cases by dissociation,

$$A + e^- \rightarrow A^- \qquad (4.50)$$

or

$$A + e^- \rightarrow B^- + C \qquad (4.51)$$

Compounds showing such electron affinity include the halogens and organic halogen compounds, oxygen, liquid water, and alcohols, e.g.,

$$I_2 + e^- \rightarrow I^- + I\cdot \qquad (4.52)$$

$$C_2H_5I + e^- \rightarrow \cdot C_2H_5 + I^- \qquad (4.53)$$

$$O_2 + e^- \rightarrow O_2^- \qquad (4.54)$$

In liquid water

$$H_2O_{aq} + e^- \rightarrow e_{aq}^- \rightarrow OH_{aq}^- + H\cdot \qquad (4.55)$$

The dissociation is endothermic and possible only in liquid water where solvation of the ions provides the necessary energy.

Reaction 4.50 differs from simple ionization in having no means of dissipating any excess energy that the electron might have. Since there is no process of energy removal in reaction 4.50 the captured electron can only have energies that can accommodate to the various energy levels in the negative ion formed (i.e., electron capture is a resonance process). Reaction (4.51) occurs if attachment of the electron to A, or to a fragment formed by the dissociation of A, liberates enough energy to break one of the bonds in the molecule. Capture of electrons with kinetic energy much above thermal energy always leads to dissociation.

A third process, detected in the mass spectrometer, results in the formation of an ion pair and is analogous to simple ionization,

$$A + e^- \rightarrow B^+ + C^- + e^- \tag{4.56}$$

This is not a resonance process and is possible over a much wider range of electron energies than electron capture. Reactions of this type have been detected in the mass spectra of organic halogen compounds and are attributed to the large electron affinities of the halogens. Ethyl chloride for example undergoes the following reaction at electron energies near 9 eV,

$$C_2H_5Cl + e^- \rightarrow C_2H_5^+ + Cl^- + e^- \tag{4.57}$$

though the ionization potential of ethyl chloride, the lowest electron energy for reaction 4.58 to take place, is 11.2 eV.

$$C_2H_5Cl + e^- \rightarrow C_2H_5Cl^+ + 2e^- \tag{4.58}$$

Magee and Burton (30) have pointed out that electron capture by a neutral molecule competes with capture by a positive ion, and the relative probability of these two processes in the track of densely ionizing particles in condensed systems will change as the track gets older and the ions diffuse further apart. Soon after the track is formed, the positive ions and electrons will be concentrated near the track and electrons will be captured predominantly by positive ions.[3] Later, as the ions diffuse outwards and get further apart, capture by neutral molecules (assuming that this is possible) will become increasingly important.

However it is formed the negative ion will eventually neutralize a positive ion (31)

$$A^+ + A^- \rightarrow A^* + A \tag{4.31}$$

One or both of the neutral molecules formed will be excited, though the excitation energy will be less than that gained by a positive ion neutralized by an electron and will decrease as the electron affinity of A increases.

Electron capture has been demonstrated experimentally by Hamill and his colleagues (32) by irradiating frozen solutions of biphenyl or naphthalene in organic solvents (e.g., hydrocarbons and ethers). Ultraviolet absorption

---

[3] When an electron recombines with its parent ion this is referred to as geminate recombination (see also p. 181).

spectra characteristic of the products of electron attachment, the ions $C_{12}H_{10}^-$ and $C_{10}H_8^-$ respectively from biphenyl and naphthalene, are observed. Furthermore, when a second solute is present which competes with the biphenyl or naphthalene for electrons, the decrease in $G(C_{12}H_{10}^-)$ or $G(C_{10}H_8^-)$ gives a measure of the electron-attaching efficiency of the second solute. Of a number of solutes tested by Hamill, those showing the greatest relative efficiency were iodine (the most efficient), carbon tetrabromide, sulfur dioxide, benzaldehyde, bromoform, carbon tetrachloride, and benzyl acetate.

Electron attachment by a minor component in an irradiated mixture may lead to a disproportionately high rate of decomposition. When dilute solutions of carbon tetrachloride in benzene are irradiated, for example, the carbon tetrachloride decomposes to a greater extent than would be anticipated from the radiolysis of the two components separately. The effect can be accounted for by assuming that the carbon tetrachloride captures electrons produced by ionization of both components (33).

Subexcitation Electrons

Platzman (34) has shown that electrons lose energy in a medium much more slowly once they have fallen to an energy (about 0.5 to 4 eV) below that of the lowest excitation energy of the medium. The rates of energy loss $(-dE/dt)$ for an electron with energy 10 or 20 eV and one with an energy slightly less than E_p, the minimum excitation energy of the molecules of the medium, are approximately 10^{16} and 10^{13} eV sec^{-1} respectively in liquid water. On the basis of their longer lifetimes Platzman suggested treating these *subexcitation electrons* separately from the more energetic electrons.

Subexcitation electrons are likely to be most important in systems containing one or more minor components which have a lower excitation energy (E_m) than that of the principal component (E_p). In this case practically all the subexcitation electrons with energies between E_m and E_p may cause excitation of the minor component, which might therefore play a larger part in the radiolysis reactions than would be anticipated from its concentration. In these circumstances (and in the event of electron capture by a minor component) the equations given earlier for calculating the fraction of the total absorbed energy absorbed by a single component in a mixture (Eqs. 3.43 and 3.44) will not be true because part of the energy will be channeled selectively to the minor components, and the energy actually absorbed by these will be greater than that calculated from the equations. Platzman estimated that for high-energy radiations about 15 to 20% of the absorbed energy might be dissipated by subexcitation electrons.

SUMMARY OF THE REACTIONS OF IONS AND EXCITED MOLECULES

Excited Molecules

$$A \rightsquigarrow A* \qquad (4.1)$$
(excitation to singlet and triplet excited states)

$$A* \rightarrow A$$
(radiative or nonradiative conversion to the ground state; no chemical reaction)

$$A* + B \rightarrow A + B* \qquad (4.4)$$
(nonradiative energy transfer)

$$A* \rightarrow R\cdot + S\cdot \qquad (4.6)$$
(dissociation into free radicals)

$$A* \rightarrow M + N \qquad (4.9)$$
(dissociation into molecular products)

$$A* + B \rightarrow A^+ + B^- \; [\text{or } A^- + B^+] \qquad (4.10)$$
(electron transfer)

$$A* + RH \rightarrow AH\cdot + R\cdot \qquad (4.16)$$
(hydrogen abstraction)

$$A* + B \rightarrow AB \qquad (4.20)$$
(addition)

$$A* + B \rightarrow P + Q \qquad (4.26)$$
(Stern-Volmer reaction)

Ions

$$A \rightarrow A^+ + e^- \; [\text{or } (A^+)* + e^-] \qquad (4.28)$$
(ionization)

$$A^+ + e^- \rightarrow A** \rightarrow A* \qquad (4.30)$$
(neutralization producing both singlet and triplet excited states)

$$A^+ + A^- \rightarrow A* + A \qquad (4.31)$$
(neutralization by a negative ion)

$$A* \;(\text{or } A**) \rightarrow M* + N \qquad (4.32)$$

(dissociation into molecular products following neutralization)

$$A^* \text{ (or } A^{**}) \rightarrow R\cdot^* + S\cdot \qquad (4.33)$$

(dissociation into free radicals following neutralization)

$$A\cdot B^+ + e^- \rightarrow C + D \qquad (4.35)$$

(neutralization of complex with reaction)

$$(A^+)^* \rightarrow M^+ + N \qquad (4.36)$$

(dissociation of an excited ion into an ion and a molecule)

$$(A^+)^* \rightarrow R\cdot^+ + S\cdot \qquad (4.37)$$

(dissociation of an excited ion into a radical-ion and a radical)

$$A^+ + B \rightarrow A + B^+ \qquad (4.39)$$

(charge transfer)

$$A^+ + B \rightarrow C^+ + D \qquad (4.42)$$

(ion-molecule reaction)

$$A + e^- \rightarrow A^- \qquad (4.50)$$

(electron capture)

$$A + e^- \rightarrow B^- + C \qquad (4.51)$$

(electron capture with dissociation)

A more extensive discussion of excited states is given by C. Reid, *Excited States in Chemistry and Biology*, Butterworths Scientific Publications, London, 1957, and of mass spectrometry by J. H. Beynon, *Mass Spectrometry and Its Applications to Organic Chemistry*, Elsevier, Amsterdam, 1960. D. P. Stevenson and D. O. Schissler have reviewed "Mass Spectrometry and Radiation Chemistry" in *Actions Chimiques et Biologiques des Radiations* (ed. M. Haissinsky), Masson et Cie, Paris, Vol. 5, 1961. Also pertinent to this chapter is *Electron Impact Phenomena* by F. H. Field and J. L. Franklin, Academic, New York, 1957. See also P. F. Knewstubb, *Mass Spectrometry and Ion-Molecule Reactions*, Cambridge University Press, 1969; and *Ion-Molecule Reactions in the Gas Phase*, American Chemical Society Advances in Chem. Series 58, 1966.

FREE RADICALS

Free radicals are atoms or molecules that have one or more unpaired electrons available to form chemical bonds.[4] They are formed for instance

---

[4] The presence of an unpaired electron is sometimes indicated by adding a dot to the chemical formula, e.g., R·. The dot is often omitted for simple radicals such as H, Cl, and OH.

when a molecule is divided at a covalent bond in such a manner that one of the bonding electrons remains with each fragment,

$$R:S \rightleftharpoons R\cdot + \cdot S \qquad (4.59)$$

The process is reversible, the back reaction representing combination of two radicals to give a stable molecule. Free radicals are generally electrically neutral, although this is not necessarily so. Examples of charged radicals are the negative radical-ion O_2^- and the positive radical-ion $\cdot CH_3^+$. Triplet excited states, which were discussed in the last chapter, often react as though they are biradicals, with two separate unpaired electrons in the molecule.

Relatively stable free radicals have been known since the beginning of the century, the earliest being triphenylmethyl, $Ph_3C\cdot$,[5] discovered by Gomberg (35). More recently, reactive radicals have been shown to be transient intermediates in many reactions, including such diverse processes as autoxidation, electrolysis, polymerization, and pyrolysis. Consequently the properties and reactions of free radicals are well established, having been investigated by a variety of techniques and under many different conditions.

In radiation chemistry, free radicals are produced by the dissociation of excited molecules and by ion reactions (e.g., dissociation, ion-molecule reaction, neutralization) in or near the tracks of ionizing particles. Radicals that do not undergo radical-radical reactions in this region of high radical concentration diffuse into the bulk of the medium and generally react with the substrate. However, the initial high concentration of radicals close to the particle tracks can lead to radical phenomena in radiation chemistry different from those where the radicals are formed by other means (e.g., chemically or photochemically) and are more randomly distributed.

In this section the formation, properties, and reactions, and, with particular reference to radiation chemistry, detection of radicals are discussed.

Formation

Methods used to produce free radicals are of more than theoretical interest in radiation chemistry, since, judiciously applied, they allow the effect on a system of a single radical species (normally one of the several formed upon irradiation) to be studied. They may also be applied to convert relatively large amounts of material to products similar to those produced by irradiation.

Preparative methods, other than the use of ionizing radiation, involve either the transfer of sufficient energy to a molecule to cause unimolecular dissociation (e.g., thermal or photodissociation) or an oxidation-reduction

---

[5] Ph is used to represent the phenyl group, C_6H_5.

process. The latter entails transfer of a single electron, or an atom or group with an unpaired electron, from one atom or molecule to another. Mechanical stress (e.g., ultrasonic vibration) can sometimes disrupt molecules and produce radicals.

THERMAL DISSOCIATION.   Covalent bonds can be broken if a molecule is heated to a sufficiently high temperature. A well-known instance is the thermal dissociation of molecular iodine:

$$I_2 \rightleftharpoons 2I \cdot \qquad (4.60)$$

The reaction is reversible and the equilibrium lies completely to the left at low temperatures (below about 700°C at normal pressures) and to the right at high temperatures (above about 1700°C). At intermediate temperatures both molecular and atomic iodine are present in the equilibrium mixture. Organic molecules dissociate in a similar manner, and free radicals are important intermediates in the high-temperature pyrolysis of organic compounds. With organic compounds, however, the reverse, recombination reaction is only one of a number of possible radical-radical and radical-molecule reactions, and pyrolysis will give a variety of products.

Compounds with relatively weak covalent bonds may break down at moderate temperatures and provide a useful source of radicals. Organic peroxides, for example, contain weak O—O bonds and are convenient radical sources both in the gaseous state and in solution. In the gas phase di-tert-butyl peroxide is used as a source of methyl radicals:

$$(CH_3)_3CO—OC(CH_3)_3 \xrightarrow{120-200°C} 2(CH_3)_3CO \cdot \rightarrow$$
$$2 \cdot CH_3 + 2CH_3COCH_3 \quad (4.61)$$

Azo compounds (R—N=N—R) and certain organometallic compounds (e.g., dimethyl mercury and tetraethyl lead) are also useful thermal sources of radicals.

Though a variety of different radicals are available from thermal decompositions, these suffer as radical sources from the disadvantage that experiments must be conducted at temperatures high enough to bring about fission of the radical precursor, restricting the range of possible reaction conditions.

PHOTODISSOCIATION.   Several examples were given in the previous section of molecules which are excited by the absorption of light and then dissociate into free radical fragments. In some cases these provide a convenient source of radicals. Acetone, for example, is frequently used as a photochemical source of methyl radicals, both in the vapor state and in solution:

$$CH_3COCH_3 \xrightarrow{h\nu} CH_3CO \cdot + \cdot CH_3 \qquad (4.62)$$
$$CH_3CO \cdot \rightarrow \cdot CH_3 + CO \qquad (4.63)$$

Molecules that do not absorb light in a convenient range of wavelength can often be induced to dissociate by energy transfer from an atom that can be excited with the available light. Mercury is a common sensitizing atom and has been employed to induce dissociation in a variety of systems, mainly gaseous. An example is the photosensitized dissociation of hydrogen, where

$$Hg + h\nu \rightarrow Hg^* \tag{4.64}$$

$$Hg^* + H_2 \rightarrow Hg + 2H\cdot \tag{4.65}$$

With more complicated molecules more than one type of radical may be produced, limiting the usefulness of photosensitization in these cases.

OXIDATION-REDUCTION PROCESSES. Oxidation-reduction reactions that involve the transfer of a single electron between two reactants may produce free radicals. Probably the best known radical source of this nature is the Fenton reagent (36), a mixture of hydrogen peroxide and ferrous ion which produces hydroxyl radicals,

$$H_2O_2 + Fe^{2+} \rightarrow Fe^{3+} + \cdot OH + OH^- \tag{4.66}$$

Similar reactions occur with persulfate ($S_2O_8^{2-}$) and with the organic analogues of hydrogen peroxide, hydroperoxides (ROOH), and peroxides (ROOR). Organic hydroperoxides for instance give alkoxy radicals with ferrous ion

$$RO_2H + Fe^{2+} \rightarrow Fe^{3+} + RO\cdot + OH^- \tag{4.67}$$

Ions of other metals of variable valency, in their lower valence states, can often be substituted for ferrous ion in these reactions. Hydroxyl radicals generated by Fenton's reagent can produce similar effects to irradiation in aqueous solution (37), where hydroxyl radicals are among the active intermediates formed from the water.

Photo-initiated electron transfer has been mentioned earlier in this chapter and occurs between a water molecule and an anion or cation, or between loosely bound anions and cations, on irradiation with ultraviolet light; for example,

$$M^{z+}H_2O \xrightarrow{h\nu} M^{(z+1)+}OH^- + H\cdot \tag{4.68}$$

$$Fe^{3+}X^- \xrightarrow{h\nu} Fe^{2+} + X\cdot \tag{4.69}$$

where M = Fe, Cr, V, Ce, etc., and X = F, Cl, Br, OH, N_3, C_2O_4, etc.

Electrolysis can also produce free radicals. Solutions containing salts of carboxylic acids for example form radicals at the anode upon electrolysis

$$RCOO^- \xrightarrow{-e^-} RCOO\cdot \tag{4.70}$$

The radicals formed normally lose carbon dioxide and dimerize to give a hydrocarbon (R—R) (Kolbe reaction). Radical-radical reactions are favored by the high radical concentration at the electrode and predominate over other radical reactions. Nevertheless, products obtained from a compound by electrolysis often resemble those obtained upon irradiation.

The preceding processes involve one-electron transfer steps, but reactions in which atoms are transferred may also lead to radicals. The sodium-flame technique originated by Polanyi (cf., 38) is an interesting example and occurs in the vapor phase between sodium and organic halides, e.g.,

$$Na + RCl \rightarrow NaCl + R\cdot \tag{4.71}$$

The same reaction has been adapted to the preparation of specific hydrocarbon radicals for electron paramagnetic resonance study (39) by condensing alternately very thin layers of sodium and an organic halide (preferably an iodide) on a rotating surface at liquid nitrogen temperature. The radicals formed are trapped as the frozen matrix is built up and thereby preserved for subsequent examination. A somewhat similar atom-transfer reaction takes place in solution between cobaltous chloride and organic halides:

$$CoCl + RCl \rightarrow CoCl_2 + R\cdot \tag{4.72}$$

Properties and Reactions

The most characteristic property of free radicals is the instability associated with the presence of an unpaired electron. Radicals are often extremely reactive, reacting in such a manner that the odd electron is paired with a similar electron in another radical or eliminated by an electron-transfer reaction. Alternatively, the radical may react so as to produce a second, more stable, free radical. The reactivity and, in the opposite sense, the stability of radicals depends to a great extent upon their structure. Taking the two extremes, radicals may be either very reactive under normal conditions or else unreactive and relatively stable. Some examples of these two types are listed below:

Reactive radicals: $H\cdot$, $\cdot OH$, $Cl\cdot$, $\cdot CH_3$, $Ph\cdot$
Stable radicals: $Ph_3C\cdot$ and substituted triphenylmethyl radicals,

$$^-O \langle \bigcirc \rangle O\cdot \text{ (semiquinone),}$$

$$O_2N \langle \bigcirc \rangle \overset{NO_2}{\underset{NO_2}{}} N{-}NPh_2 \text{ (diphenylpicrylhydrazyl, DPPH)}$$

NO, NO_2, O_2

The radicals in the first group have only a transient existence under normal reaction conditions and are generally atoms or are formed from relatively small molecules. However, they can be trapped and preserved in solids, usually at low temperatures, where they are unable to diffuse and react together (40). Organic radicals in the second, stable, group are generally formed from large molecules where the odd electron is distributed over a much greater volume; in some cases steric factors may enhance the radical stability by hindering dimerization. Nitric oxide and nitrogen dioxide are included, since each has a single unpaired electron and behaves as a rather unreactive free radical. Oxygen has a triplet ground state and behaves in radical reactions as a diradical. Needless to say the two groups contain radicals of varying stability and many other radicals are known which fall between the two extremes.

In Table 4.2 a number of reactive radicals are arranged in order of decreasing reactivity (increasing stability) as the table is descended. The relative positions are only approximate[6], and no clear distinction can be made between radicals in adjacent levels. Nevertheless the divisions, rough as they are, should help to clarify radical reactions.

T A B L E 4.2 *Relative Stability of Reactive Free Radicals[a]*

R = Alkyl Radical	Atoms and Inorganic Radicals	Alkyl Radicals	Halogen Derivatives	Unsaturated and Oxygenated Radicals	Aromatic Compounds
—	H·, ·OH	—	—	$CH_2{=}CH·$	Ph·
					$o, m, p,$ $CH_3C_6H_4·$
R·, $RCO_2·$	F·	$·CH_3, ·C_2H_5$	—	$(CH_3CO_2·)$	—
RCO·	—	$n\text{-}C_3H_7·$	$·CF_3$	$CH_3CO·$	—
		$n\text{-}C_4H_9·$	$ClCH_2CH_2·$		
RO·	Cl·	$i\text{-}C_3H_7·$	$ClCH_2·$	$(CH_3O·)$	—
—	—	$t\text{-}C_4H_9·$	$Cl_2CH·$	—	—
—	—	—	$·CCl_3, Br_2CH·$	—	PhCO·
—	Br·	—	$·CBr_3$	$CH_2{=}CH{-}CH_2·$	$PhCH_2·$
RS·, $RO_2·$	I·, $(HO_2·)$	—	—	$(CH_3S·)(CH_3O_2·)$	—

[a] The radicals are arranged from top to bottom in order of increasing stability and decreasing reactivity. The relative positions are approximate and, for the radicals in parentheses, questionable.

[6] The radicals (R) were grouped by comparing the bond dissociation energies in the compounds R—H and R-halogen and assuming that the largest dissociation energies signified the most reactive radicals.

The stability of organic radicals is increased when hydrogen attached to the carbon atom carrying the free electron is replaced by any other atom or group. Radical stability increases, for example, in the series primary ($-CH_2\cdot$) < secondary ($=CH\cdot$) < tertiary ($\equiv C\cdot$). Stability will also be affected by the nature of the substituents. Fluorine has only a small effect when it replaces hydrogen, but the effectiveness of the halogens increases in the series F < Cl < Br < I; stability increases also as the number of halogen substituents is increased. Unsaturated substituents increase stability when they involve carbon atoms adjacent to that carrying the odd electron, but not when this carbon atom itself is part of the unsaturated grouping. For instance $CH_2{=}CH{-}CH_2\cdot$ is very much more stable than $CH_2{=}CH\cdot$ or Ph$\cdot$. In all cases the effectiveness of the stabilizing substituent falls as it is moved farther from the carbon with the odd electron.

Radicals that are high in Table 4.2 tend to be less selective in their reaction with stable molecules than those which are lower in the table. Hydrogen atoms, for example, tend to react with the first molecule with which they collide. Chlorine atoms are more selective in their attack on the substrate than hydrogen atoms, but are less selective than bromine atoms. In general radicals are less selective at high temperatures but, conversely, even very reactive radicals may be quite selective at very low temperatures. Other factors affecting radical reactions with a substrate besides temperature include the reactivities of the attacking and displaced radicals, the dissociation energies of the bonds broken and formed, and polar effects.

Representing radical attack on a substrate by the reaction

$$A\cdot + B{-}C \rightarrow A{-}B + C\cdot \tag{4.73}$$

is is generally true that the newly formed radical is less reactive than the attacking radical. To take an extreme example, hydrogen atoms will react with molecular iodine to give hydrogen iodide and an iodine atom,

$$H\cdot + I_2 \rightarrow HI + I\cdot \tag{4.74}$$

but the back reaction,

$$I\cdot + HI \rightarrow H\cdot + I_2 \tag{4.75}$$

which would replace a relatively stable iodine atom by a very reactive hydrogen atom, is most unlikely. The relative radical reactivities (Table 4.2) were estimated by comparing bond dissociation energies in representative compounds; the same conclusions can be reached by considering these energies directly. The overall energy change in reaction 4.74 is the difference between the energy released in forming a hydrogen-iodine bond and that absorbed in breaking an iodine-iodine bond, i.e., the difference between the bond dissociation energies for these two bonds. Thus, from Table 4.3, the

TABLE 4.3 *Bond Dissociation Energies*[a]

Bond	Bond Dissociation Energy kcal mole^{-1}	eV molecule^{-1}	Bond	Bond Dissociation Energy kcal mole^{-1}	eV molecule^{-1}
H—H	103.2	4.5	F—F	36	1.6
(H—H)$^+$	61	2.6	Cl—Cl	57.1	2.5
HO—H	117.5	5.1	Br—Br	45.5	2.0
(HO—H)$^+$	142	6.2	I—I	35.6	1.5
·O—H	101.5	4.4	H—F	134	5.8
HO—OH	51	2.2	H—Cl	102.2	4.4
HO$_2$—H	90	3.9	(H—Cl)$^+$	108	4.7
·O$_2$—H	47	2.0	H—Br	86.5	3.8
O=O	118	5.1	(H—Br)$^+$	93	4.0
HS—H	~90	~3.9	H—I	70.5	3.1
H$_2$N—H	102	4.4	(H—I)$^+$	74	3.2
CH$_3$—H	101	4.4	CH≡CH	230	10.0
(CH$_3$—H)$^+$	31	1.3	CH≡C—H	<121	<5.2
·CH$_2$—H	88	3.8	CH$_2$=CH$_2$	140	6.1
CH$_3$—CH$_3$	83	3.6	CH$_2$=CH—H	104–122	4.5–5.3
C$_2$H$_5$—H	96	4.2	Benzene C⋯C	193	8.4
(C$_2$H$_5$—H)$^+$	25	1.1	C$_6$H$_5$—H	102	4.4
n-C$_4$H$_9$—H	101	4.4	(C$_6$H$_5$—H)$^+$	110	4.8
t-C$_4$H$_9$—H	89	3.9	C$_6$H$_5$CH$_2$—H	83	3.6
CH$_3$—F	107	4.6	Cl$_3$C—H	90	3.9
CH$_3$—Cl	81	3.5	Cl$_3$C—Cl	68	2.9
CH$_3$—Br	67	2.9	Cl$_2$CH—Cl	72	3.1
CH$_3$—I	53	2.3	Br$_3$C—Br	49	2.1
n-C$_4$H$_9$—I	49	2.1	CH$_3$—CN	103	4.5
t-C$_4$H$_9$—I	45	2.0	CH$_3$—NH$_2$	80	3.5
C$_6$H$_5$—I	57	2.5	CH$_3$—SH	74	3.2
C$_6$H$_5$CH$_2$—I	39	1.7	CH$_3$S—H	89	3.9
			CH$_3$S—SCH$_3$	73	3.2
OHC—H	76	3.3	CH$_3$CO$_2$—H	112	4.9
HCO—OH	~90	~3.9	CH$_3$CO—OH	90	3.9
CH$_3$—OH	91	3.9	CH$_3$—CO$_2$·	−17	−0.7
CH$_3$O—H	100	4.3	CH$_3$O—OCH$_3$	37	1.6
CH$_3$CO—H	~85	~3.7	CH$_3$CO—OCCH$_3$	57	2.5
CH$_3$—CO·	~17	~0.7	CH$_3$CO$_2$—O$_2$CCH$_3$	30	1.3
CH$_3$CO—CH$_3$	72	3.1			

[a] The values are largely taken from *The Strengths of Chemical Bonds* by T. L. Cottrell, 2nd ed., Butterworths Scientific Publications, London, 1958. Values for ions are given assuming that the charge remains with the fragment having the lower ionization potential; they are taken from *Electron Impact Phenomena and the Properties of Gaseous Ions* by F. H. Field and J. L. Franklin, Academic, New York, 1957. (1 eV molecule^{-1} = 23.06 kcal mole^{-1})

energy change is

$$\text{D(H—I)} - \text{D(I—I)} = 70.5 - 35.6 = +34.9 \text{ kcal mole}^{-1}$$

Energy is in fact released and the reaction is perfectly feasible in the direction shown. The reverse reaction between iodine atoms and hydrogen iodide (Eq. 4.75) is very much less likely because 35 kcal of energy per mole must be supplied from the thermal energy of the system, or from an external source, for the reaction to take place.

Though the difference between the bond dissociation energies gives the overall energy change for a radical-molecule reaction such as 4.74, a certain amount of activation energy may be required to initiate the reaction (Fig. 4.7).

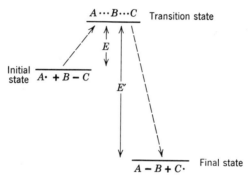

FIGURE 4.7 *Energy diagram for the reaction* $A \cdot + B\text{–}C \rightarrow A\text{–}B + C \cdot$ (*Potential energy increases with increasing distance from the bottom of the figure*). $E =$ *activation energy for forward reaction* $E' =$ *activation energy for back reaction*

$$E' - E = D(A - B) - D(B - C)$$

Activation energies may be an important factor if the attacking radical can react with the substrate in more than one way, when the reaction with the lowest activation energy will be favored. Normally the activation energy is provided from the thermal energy of the system, but in the case of radicals with extra energy (i.e., hot radicals), it can be provided by the radical itself and this will be more reactive than similar radicals in thermal equilibrium with the system. For most radical reactions the activation energies are appreciably lower than for nonradical reactions.

Polar factors, that is, effects due to differences in electron density in the reactants and media, such as the type of solvent, pH, inductive, and hyper-conjugation effects, may influence radical reactions, though to a smaller extent than heterolytic reactions, which involve transfer of electron pairs in bond breaking and bond formation. Radicals can be classed as electron-acceptors or electron-donors according to their tendency to gain or lose electrons respectively. Halogen atoms tend to gain electrons and attack preferentially, though generally not exclusively, points of high electron density in the substrate. Methyl radicals on the other hand are relatively nucleophilic and tend rather to donate electrons and seek electron deficient centers in the substrate. This is illustrated by the attack of chlorine atoms and methyl radicals on compounds $\cdots CH_2 \cdots CH_2$—R where R is an electron attracting group (either Cl or COOH) (e.g., 41); the reactions are

$$Cl\cdot + \cdots CH_2 \cdots CH_2-R \rightarrow HCl + \cdots \dot{C}H \cdots CH_2-R \quad (4.76)$$

and

$$\cdot CH_3 + \cdots CH_2 \cdots CH_2-R \rightarrow CH_4 + \cdots CH_2 \cdots \dot{C}H-R \quad (4.77)$$

The reactions of radicals can be divided into unimolecular (rearrangement and dissociation) and bimolecular, and the latter subdivided into attack on the substrate and radical-destroying processes according as the products do, or do not, include a free radical.

RADICAL REARRANGEMENT. Reactive radicals can sometimes attain more stable structures by rearrangement. Aromatic nuclei and halogen atoms appear to migrate most frequently, but other atoms (e.g. hydrogen) and groups (e.g., methyl) may also shift. Migration of phenyl is illustrated by the radical β,β,β,-triphenylethyl (I) (formed by decarbonation of β,β,β-triphenyl-propionaldehyde) which rearranges completely to the radical (II) (42),

$$Ph_3C-CH_2\cdot \rightarrow Ph_2\dot{C}-CH_2Ph \quad (4.78)$$
$$\text{(I)} \qquad\qquad \text{(II)}$$

A number of reactions have been reported by Nesmeyanov and Freidlina (43) in which a chlorine atom migrates in a radical of general structure (III) to give the more stable radical (IV),

$$Cl_3C-\dot{C}= \rightarrow Cl_2\dot{C}-CCl= \quad (4.79)$$
$$\text{(III)} \qquad\quad \text{(IV)}$$

RADICAL DISSOCIATION. Radicals may dissociate, characteristically breaking down into a smaller radical and an unsaturated compound, e.g.,

$$Br_2CH-\dot{C}HBr \rightarrow Br\cdot + BrCH{=}CHBr \quad (4.80)$$

and

$$(CH_3)_3CO\cdot \rightarrow \cdot CH_3 + CH_3COCH_3 \quad (4.81)$$

Acetoxy and acetyl radicals lose carbon dioxide and carbon monoxide respectively,

$$CH_3CO_2\cdot \rightarrow \cdot CH_3 + CO_2 \quad (4.82)$$

$$CH_3CO\cdot \rightarrow \cdot CH_3 + CO \quad (4.63)$$

The decarboxylation of acetoxy radicals (Eq. 4.82) occurs very readily, aided by the high heat of formation of carbon dioxide. In every case dissociation increases in importance as the temperature is raised.

In radiation chemistry, excited radicals formed by the neutralization of a radical-ion may dissociate, in the same way that excited molecules and ions can.

RADICAL ATTACK ON SUBSTRATES. *Addition.* Addition to unsaturated compounds is a characteristic reaction of free radicals,

$$A\cdot + \quad \overset{\diagdown}{\underset{\diagup}{C}}=\overset{\diagup}{\underset{\diagdown}{C}} \quad \longrightarrow \quad \overset{\diagdown}{\underset{\diagup}{C}}-\overset{\overset{\displaystyle A}{\displaystyle |}}{\underset{\diagdown}{C}}\diagup \qquad (4.83)$$

This is the reverse of radical dissociation (e.g., Eq. 4.80), and addition may in fact be a reversible reaction. Many examples of addition reactions are known and include radical-induced halogenation, with steps such as

$$Br\cdot + R-CH=CH_2 \rightarrow R-\dot{C}H-CH_2Br \qquad (4.84)$$

and the radical-initiated polymerization of unsaturated compounds, where the original radical grows by the stepwise addition of unsaturated molecules, e.g.,

$$-CH_2-\dot{C}HR + CH_2=CHR \rightarrow$$
$$-CH_2-CHR-CH_2-\dot{C}HR \text{ etc.} \qquad (4.85)$$

Initiation of polymerization in an unsaturated compound is sometimes used as a diagnostic test for free radicals.

Abstraction. Abstraction reactions are represented by the process,

$$A\cdot + BC \rightarrow AB + C\cdot \qquad (4.86)$$

and are the normal reaction of radicals with saturated organic compounds. Abstraction reactions may also compete with radical-addition in the case of unsaturated compounds. The species abstracted (B in the example above) is generally a univalent atom, either hydrogen or halogen, e.g.,

$$\cdot OH + CH_3OH \rightarrow H_2O + \cdot CH_2OH \qquad (4.87)$$
and
$$H\cdot + CH_3I \rightarrow HI + \cdot CH_3 \qquad (4.88)$$

Generally the radical formed is more stable than the attacking radical. Thus most radicals attack tertiary in preference to secondary or primary hydrogen atoms, though other factors (e.g., polar and energy considerations) will also influence the reaction.

RADICAL-DESTROYING PROCESSES. *Combination.* Combination is the reverse of molecular dissociation to give radicals,

$$R\cdot + S\cdot \rightarrow RS \qquad (4.89)$$

Reactions of this type are favored energetically, since they have little or no activation energy and energy is liberated equal to the dissociation energy of the bond formed. Combination of methyl radicals to form ethane for example takes place in the gas phase at almost every radical-radical collision (44),

$$2\cdot CH_3 \rightarrow C_2H_6 \qquad (4.90)$$

The readiness with which radicals combine accounts for their very short lifetimes, even under conditions of low radical concentration.

The energy liberated as a consequence of combination is sufficient, if it is not delocalized or lost, to allow immediate redissociation into the original radicals. In complex molecules it is distributed in a number of internal degrees of freedom and is unlikely to be concentrated in any one bond (so as to cause rupture) before it is lost through collisions with other molecules. Smaller molecules have fewer possibilities for distributing the energy, and diatomic molecules formed by the combination of two free atoms will normally dissociate again unless they can lose energy quickly. In condensed phases the newly formed diatomic molecule can lose energy rapidly to adjacent molecules, but in a gaseous system combination will only give a stable molecule if it takes place on a surface (e.g., a wall of the reaction vessel) or in the presence of a third body which can share the energy liberated,[7] e.g.,

$$O\cdot + O\cdot + M \rightarrow O_2 + M \qquad (4.91)$$

Surfaces may therefore affect gas-phase radiolyses in which free atoms are intermediates but will have little effect (in this sense) upon the radiolysis of liquids, where there are ample opportunities for rapid energy loss by collisional processes. It might be pointed out that the energy released by radical combination is less than that liberated by recombination of an ion pair, as the following example shows:

$$H\cdot + \cdot OH \rightarrow H_2O + 116 \text{ kcal mole}^{-1} \text{ (5.0 eV molecule}^{-1}) \qquad (4.92)$$

$$H_2O^+ + e^- \rightarrow H_2O + 291 \text{ kcal mole}^{-1} \text{ (12.6 eV molecule}^{-1}) \qquad (4.93)$$

The energy released by radical combination may be retained by the newly formed molecule long enough to influence its reactions. Alexander and Rosen

[7] The possibility that the newly formed molecule will lose energy by fluorescence is ignored here, since it is not normally significant.

(45) have suggested that hydrogen peroxide formed by the combination of two hydroxyl radicals in α-particle tracks in water is more reactive toward amino acids than ordinary hydrogen peroxide. They suggest that the enhanced reactivity, which has a relatively long lifetime, is due to the formation of the peroxide in a triplet excited state:

$$2 \cdot OH \rightarrow H_2O_2^* \qquad (4.94)$$

In the gas phase, combination of hydrogen atoms with ethyl radicals gives an excited form of ethane which, at low pressures, may dissociate into two methyl radicals before deactivation can occur:

$$H \cdot + \cdot C_2H_5 \rightarrow C_2H_6^* \rightarrow 2 \cdot CH_3 \qquad (4.95)$$

Dissociation of the excited molecule to methyl radicals rather than the original fragments is favored by the lower bond dissociation energy of the carbon-carbon bond (Table 4.3). In other instances molecular dissociation products may result, e.g. (46),

$$\cdot C_2H_5 + I \cdot \rightarrow C_2H_4 + HI \qquad (4.96)$$

in a process resembling radical disproportionation (see below).

When two radicals are formed in a liquid by dissociation of a single molecule, they may sometimes be prevented from moving apart by the surrounding molecules. In this event they will eventually recombine with one another and, if the original molecule is reformed,[8] there will be no chemical change produced in the system. Recombination of radicals in this way, before they have diffused apart, is known as the Franck-Rabinowitch or cage effect.

Disproportionation.   An alternative to the combination of two radicals to give a single stable molecule is transfer of an atom (generally hydrogen) from one radical to the other, giving two stable molecules, one of which is unsaturated:

$$2RH \cdot \rightarrow RH_2 + R \qquad (4.97)$$

Ethyl radicals can react together in this manner to give a molecule each of ethane and ethylene,

$$2 \cdot C_2H_5 \rightarrow C_2H_6 + C_2H_4 \qquad (4.98)$$

and the radicals $CH_3\dot{C}HOH$, derived from ethanol by loss of an α-hydrogen atom, to give ethanol and acetaldehyde,

$$2CH_3\dot{C}HOH \rightarrow CH_3CH_2OH + CH_3CHO \qquad (4.99)$$

---

[8] It need not be since disproportionation can also occur.

Unlike radicals may also disproportionate, e.g.,

$$\cdot CH_3 + \cdot C_2H_5 \rightarrow CH_4 + C_2H_4 \qquad (4.100)$$

Disproportionation is less common than simple radical combination, and it is rare with aromatic radicals, which generally dimerize. Often both disproportionation and combination will be found in the same system.

Electron Transfer. Just as radicals may be formed by one-electron transfer processes so they may be destroyed in the same way, e.g.,

$$\cdot OH + Fe^{2+} \rightarrow Fe^{3+} + OH^- \qquad (4.101)$$

However, electron transfer need not always result in radical destruction, for instance in aqueous solutions electron transfer between halide ions and hydroxyl radicals gives halogen atoms, e.g.,

$$\cdot OH + Br^- \rightarrow Br\cdot + OH^- \qquad (4.102)$$

"HOT" RADICALS. In some situations atoms and radicals may be formed with more kinetic or excitational[9] energy than the average thermal energy of the surrounding molecules; in other words they are "hot" relative to their surroundings.[10]

Energetic radicals (distinguished here by italic type) may be formed by the neutralization of molecular or radical ions, i.e., by

$$A^+ + e^- \rightarrow A^{**} \rightarrow R\cdot + S\cdot \qquad (4.103)$$

or

$$R\cdot^+ + e^- \rightarrow R\cdot \qquad (4.104)$$

respectively. They may also be formed by the dissociation of excited molecules if the excitation energy is greater than the dissociation energy of the bond

---

[9] Electronically excited radicals are "hot" in the sense used here.

[10] A distinction ought, perhaps, to be made here. Hot radicals, in the sense that the term is used in this book, will certainly not exceed thermal energy by more than a few electron volts. Their reactions are essentially those of similar radicals with thermal energies, though the rates of their reactions may be enhanced by the extra energy that they possess. Hot atoms produced by nuclear reactions, for instance the hot tritium atoms produced by the reactions ^{3}He$(n, p)^3$H and ^{6}Li$(n, \alpha)^3$H, can have very much higher energies in the kiloelectron volt range and can bring about quite different types of reaction. The hot tritium atoms formed in these reactions, for example, can displace other atoms or groups from molecules with which they collide (47), whereas hot hydrogen atoms in our sense, and thermal hydrogen atoms, will only undergo abstraction and addition reactions. The chief difference between the "low-energy" hot hydrogen atoms and thermal hydrogen atoms is that the former will often react with the first molecule they collide with and are therefore immune to materials present in the medium in low concentration. Thermal hydrogen atoms, on the other hand, may collide a number of times before they react and will often react preferentially with a reactive solute, even though the concentration of the solute is quite low.

broken,

$$A^* \rightarrow R\cdot + S\cdot \qquad (4.105)$$

All three processes can occur in the tracks of ionizing particles.

Exothermic chemical reactions can also produce energetic radicals, an example being the addition of hydrogen atoms to olefins, e.g.,

$$H\cdot + CH_3-CH{=}CH_2 \rightarrow CH_3-\dot{C}H-CH_3 + energy \qquad (4.106)$$

Propyl radicals formed by this reaction dissociate to an appreciable extent at room temperature,

$$\cdot C_3H_7 \rightarrow C_2H_4 + \cdot CH_3 \qquad (4.107)$$

whereas thermally equilibrated propyl radicals are stable (48).

Energetic radicals formed in a liquid by reactions 4.103 and 4.105 are less likely to be trapped in a solvent cage than similar radicals formed with only thermal energy, and the possibility of reaction between the radicals and the substrate is therefore enhanced. Furthermore the extra energy that the hot radical possesses also favors rapid reaction with the substrate, so that hot radicals generally react in this way rather than with other radicals. However, if the hot radical should escape almost immediate reaction, it will lose energy by collisions with other molecules and will eventually be reduced to thermal energy.

The reactions of hot radicals are likely to be less temperature dependent than those of thermal radicals. For example, hot radicals formed when a frozen material is irradiated at liquid nitrogen temperatures may react with the substrate, though the same radicals formed under these conditions with only thermal energy fail to react because the necessary activation energy is not available.

REACTION WITH OXYGEN. Oxygen readily adds to free radicals and, if present, will almost invariably play a part in radiation-induced reactions. The combination of oxygen with radicals is easily understood if molecular oxygen is written as a diradical, $\cdot O{-}O\cdot$, in recognition of its triplet ground state; thus

$$R\cdot + \cdot O{-}O\cdot \rightarrow R{-}O{-}O\cdot \qquad (4.108)$$

The product is a relatively stable peroxy radical.

In some cases peroxy radicals are able to abstract hydrogen from the substrate:

$$RO_2\cdot + RH \rightarrow RO_2H + R\cdot \qquad (4.109)$$

Reactions 4.108 and 4.109, taken together, constitute a chain reaction of the type responsible for the *autoxidation* (self-oxidation under mild conditions) of a wide variety of organic compounds; everyday examples include the drying of linseed oil based paints, the production of rancidity in fats, and the deterioration of rubber and plastic materials exposed to air and light. One of the earliest examples of autoxidation to be recognized was the slow, light accelerated, oxidation of benzaldehyde to benzoic acid on exposure to air (49). Later work has shown that the oxidation is a chain reaction that is described by equations 4.108 and 4.109 when R is the group PhCO. The overall chemical change produced by the chain reaction is the conversion of benzaldehyde to perbenzoic acid,

$$PhCHO + O_2 \rightarrow Ph—C\overset{\displaystyle O}{\underset{\displaystyle O—OH}{\diagdown}} \qquad (4.110)$$

The latter subsequently reacts with another molecule of benzaldehyde giving benzoic acid,

$$PhCO_3H + PhCHO \rightarrow 2PhCOOH \qquad (4.111)$$

It might be noted in connection with this oxidation that hydrogen is generally more weakly bound in an aldehyde group than in an unsubstituted saturated hydrocarbon, and his facilitates the hydrogen-abstraction step (Eq. 4.109); $D(PhCO—H)$ is 79 kcal mole^{-1}.

Since oxygenated radicals are of very great importance in radiation chemistry some of the reactions attributed to them are given below, though these are not as well established as the reactions of the parent alkyl radicals; the list is by no means complete.

Peroxy radicals $(RO_2 \cdot)$ are the initial products formed by the addition of oxygen to an organic radical, and they can form hydroperoxides (RO_2H) by hydrogen abstraction (Eq. 4.109) or by reaction with a perhydroxyl radical $(HO_2 \cdot)$:

$$RO_2 \cdot + HO_2 \cdot \rightarrow RO_2H + O_2 \qquad (4.112)$$

In some cases hydroperoxides can be formed directly by addition of oxygen to an excited molecule. Combination of two peroxy radicals may give an organic peroxide,

$$2RO_2 \cdot \rightarrow R—O—O—R + O_2 \qquad (4.113)$$

but the products of the reaction are more often two alkoxy radicals $(RO \cdot)$,

$$2RO_2 \cdot \rightarrow 2RO \cdot + O_2 \qquad (4.114)$$

Other reactions which have been suggested whereby a peroxy radical is degraded to an alkoxy radical include

$$RO_2 \cdot + R \cdot \rightarrow 2RO \cdot \qquad (4.115)$$

$$RO_2 \cdot + O_2 \rightarrow RO \cdot + O_3 \qquad (4.116)$$

and, at higher temperatures,

$$RO_2 \cdot \overset{RH}{\to} RO_2H \to RO \cdot + \cdot OH \qquad (4.117)$$

Alkoxy radicals are more reactive than peroxy radicals and can abstract hydrogen from most organic compounds, resembling hydroxyl radicals in this respect,

$$RO \cdot + RH \to ROH + R \cdot \qquad (4.118)$$

Metal ions able to undergo one-electron oxidation-reduction reactions can react with alkoxy and peroxy radicals and with hydroperoxides and peroxides (cf., Eq. 4.67), and may play an important part if they are present during radical reactions carried out in the presence of oxygen.

Finally it might be remarked that the molecular products from these reactions with oxygen may include chemically unstable hydroperoxides and peroxides which can continue to react slowly long after they have been formed. Radical reactions carried out in the presence of oxygen may therefore show delayed effects.

Detection

Several methods based on physical measurements are available for the detection and identification of free radicals. The most important from the point of view of radiation chemistry is electron paramagnetic resonance, but others, such as magnetic susceptibility measurements and mass spectroscopy (50), are also useful on occasion. Flash photolysis enables the absorption spectra of momentarily high concentrations of radicals and other transient intermediates to be obtained.

MAGNETIC SUSCEPTIBILITY. All molecules containing unpaired electrons are paramagnetic, that is, they are drawn from a region of low magnetic field strength into a region of high field strength. With suitable equipment the very small forces exerted on paramagnetic materials by a magnetic field can be measured, though application of this to the detection of free radicals is limited because of the relatively high radical concentration needed to make the measurement possible. As a consequence magnetic susceptibility measurements are restricted to relatively stable radicals; they have been used for example to study the dissociation of substituted hexaphenyl-ethanes into triphenylmethyl-type radicals in solution. The technique has also been used to demonstrate the presence of unpaired electrons in the phosphorescent (triplet) state of fluorescein (51).

ELECTRON PARAMAGNETIC RESONANCE. Paramagnetic resonance measurements provide the most sensitive method for detecting unpaired electrons and may be used both for stable free radicals and, at low temperatures, for reactive radicals that have been stabilized in frozen media. Further advantages are the possibility of identifying the radicals from their spectra and of estimating their numbers by comparison with a stable paramagnetic substance (e.g., DPPH or a paramagnetic salt). Radicals formed by the irradiation of solids or frozen materials can readily be studied by this technique.

Electron paramagnetic resonance (epr) or electron spin resonance (esr) as it is variously called is, like magnetic susceptibility, based on the magnetic properties associated with an unpaired electron. Measurements are made by placing the paramagnetic material in a uniform magnetic field, which causes the unpaired electron to orientate itself with respect to the field. The orientation will be either parallel or antiparallel to the field, and all unpaired electrons will take up one or other of these two possible states. The two states have slightly different energies, and at normal temperatures there will be rather more electrons in the lower-energy state, the number in the lower state increasing as the temperature is lowered. The energy difference between the two states is given by

$$\Delta E = g\beta H \qquad (4.119)$$

where H is the strength of the magnetic field (gauss, symbol G), β is a constant (the Bohr magneton) equal to 0.927×10^{-20} erg G^{-1}, and g is the spectroscopic splitting factor, a dimensionless number whose value depends on the environment of the unpaired electron (for a free electron g has the value 2.0023). Simultaneously the sample is subjected to electromagnetic radiation with a frequency such that

$$h\nu = g\beta H \qquad (4.120)$$

causing the unpaired electrons to reverse their orientation with respect to the magnetic field. Since there are more electrons in the lower-energy state this results in a net absorption of radiation energy, which with appropriate equipment can be observed as a spectroscopic absorption line.

Substituting numerical values into Eq. 4.120, taking $g = 2$ and setting the magnetic field strength at a reasonable level (about 3500 G), the frequency required is found to be about 10,000 Mc, i.e., is in the microwave region of the electromagnetic spectrum. In practice a wavelength of 3.2 cm (about 9400 Mc) is frequently used as microwave components designed for this wavelength are readily available. The experimental results are obtained in the form of an absorption curve by maintaining the frequency constant and varying the magnetic field by means of small secondary coils, so that the magnetic field varies about the value that corresponds to maximum absorption. A curve of the type obtained by plotting energy absorption against magnetic field strength for a system containing isolated free electrons is shown in Fig. 4.8.

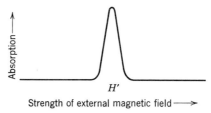

Strength of external magnetic field ⟶
FIGURE 4.8   *Epr absorption spectrum of an isolated electron.*

In addition to the external magnetic field, the unpaired electron may experience magnetic fields due to neighboring nuclei which possess nuclear magnetic moments. Nuclei having this property include hydrogen, deuterium, nitrogen-14, and chlorine-35, while carbon-12, oxygen-16, and sulfur-32 are examples of nuclei with zero magnetic moment. In the external magnetic field the nuclear magnetic fields become orientated, the number of different orientations possible for a particular nucleus being given by the expression $(2I + 1)$, where I is the spin of the nucleus (Table 4.4). Nuclear magnetic fields for nuclei close enough to an unpaired electron to affect it can either reinforce the external field or detract from it, and it is the resultant field from these two sources (external and nuclear) that governs the energy absorption. Energy absorption is dependent therefore not only on the external conditions but also on the molecular environment of the unpaired electron.

TABLE 4.4   *Nuclear Spins (I)*

0	$\frac{1}{2}$	1	$\frac{3}{2}$	$\frac{5}{2}$
Helium-4	Hydrogen-1	Deuterium	Chlorine-35	Oxygen-17
Carbon-12	Carbon-13	Nitrogen-14	Chlorine-37	
Oxygen-16	Fluorine-19			
Sulfur-32				

Consider the simplest example of the effect of a nuclear magnetic field, that of an unpaired electron adjacent to a nucleus of spin $\frac{1}{2}$ (e.g., hydrogen) and subjected to an external magnetic field H. Then the nuclear magnetic field (a) can, from the expression given above, assume either of two orientations, which can be represented by $+a$ and $-a$. The magnetic field experienced by the electron is therefore $H + a$ or $H - a$ (in the system as a whole, half the electrons present experience the higher field and half the lower field). Thus there will be two values of the external field (H) which satisfy Eq. 4.120, and each will give maximum energy absorption. The absorption spectrum (Fig. 4.9) does in fact show two peaks arranged sym-

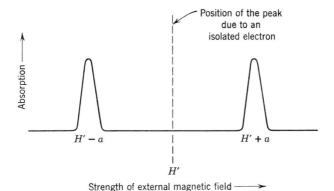

FIGURE 4.9   *Epr absorption spectrum of a hydrogen atom.*

metrically about the position occupied by the single (isolated free electron) peak in Fig. 4.8. If H' is the value of the external magnetic field for maximum absorption by the isolated electron, then $H' - a$ and $H' + a$ are the values corresponding to the two peaks in Fig. 4.9.

When two nuclei of spin $\frac{1}{2}$ are equidistant from the unpaired electron the two nuclear magnetic fields may either reinforce or cancel each other. Four combinations of the external and nuclear fields are possible, $H + a + a$, $H + a - a$, $H - a + a$, and $H - a - a$, but two coincide and the absorption curve (Fig. 4.10) will show three peaks, the center peak having twice the area of the outer peaks. When more than one nuclei with spin $\frac{1}{2}$ are equidistant from the odd electron, the number of peaks in the spectrum is given by the number of terms in the expansion of $(1 + x)^n$, n being the number of equivalent nuclei. The relative areas of the peaks are given by the numerical

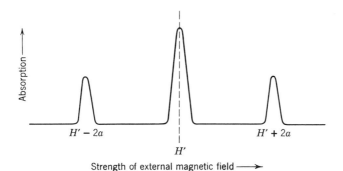

FIGURE 4.10   *Epr absorption spectrum due to two equidistant protons.*

coefficients of each term, i.e., $1:2:1$ when $n = 2$ (Fig. 4.10). It must be emphasized, however, that this only applies when the spins of the nuclei are $\frac{1}{2}$ and when they are equidistant from the unpaired electron. To take a practical example, the radical V would be expected to give an epr spectrum with seven lines with relative intensities $1:6:15:20:15:6:1$ [from the expansion of $(1 + x)^6$], since there are six hydrogen atoms equidistant from the carbon carrying the odd electron. Carbon and oxygen have zero spin and therefore do not contribute to the nuclear field while the hydrogen of the hydroxyl group is probably too far away to exert much effect. An epr spectrum with seven peaks in roughly the right ratio has in fact been obtained with irradiated crystalline α-amino isobutyric acid(VI), suggesting that irradiation results in loss of the amino group (52).

$$
\begin{array}{cc}
\mathrm{H_3C} \diagdown \mathrm{O} & \mathrm{NH_2} \\
\overset{\cdot}{\mathrm{C}}\!-\!\mathrm{C} & \mathrm{H_3C}\diagdown |\diagup\mathrm{O} \\
\mathrm{H_3C} \diagup \diagdown \mathrm{O\!-\!H} & \mathrm{C}\!-\!\mathrm{C} \\
& \mathrm{H_3C}\diagup\diagdown\mathrm{O\!-\!H}
\end{array}
$$

(V) (VI)

More complex spectra are obtained when the nuclei concerned are not equidistant from the unpaired electron. For instance, two hydrogen nuclei at different distances from the electron produce four peaks (Fig. 4.11); the nuclear fields at the electron due to the two protons are not now the same and are represented in Fig. 4.11 as a_1 and a_2. The second proton has thus split into two each of the lines due to the first.

Nuclei with spins greater than $\frac{1}{2}$ can give rather more absorption lines. Deuterium has spin 1 and so its nuclear magnetic field can assume three different orientations with respect to the external magnetic field; using the

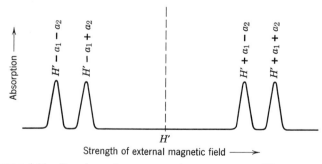

FIGURE 4.11   *Epr absorption spectrum due to two nonequidistant protons.*

notation above these can be represented as $+a$, 0, and $-a$. The spectrum of the deuterium atom therefore consists of three equally intense lines (Fig. 4.12). By listing the various combinations for the magnetic field at the unpaired electron it can be seen that two equidistant deuterium nuclei give a five-line spectrum, the intensities of the lines being in the ratio $1:2:3:2:1$. Two deuterium nuclei at different distances give nine equally intense lines since none of the lines coincide. The different nuclear spins of deuterium and hydrogen can aid in the interpretation of epr spectra if deuterium is substituted for selected hydrogen atoms in the compound being studied and the resulting changes in the spectrum are analyzed (53).

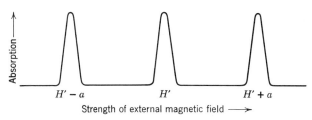

FIGURE 4.12 *Epr absorption spectrum of a deuterium atom.*

Diphenylpicrylhydrazyl (DPPH) (VII) is a stable free radical that is of interest in radiation chemistry as a radical scavenger. The structure of DPPH suggests that it would give a nine line epr spectrum, since the nearest nuclei to the odd electron are two nitrogen nuclei

$$O_2N \overbrace{\hspace{1em}}^{NO_2} —\dot{N}—NPh_2$$
$$NO_2$$
(VII)

(spin $= 1$). In fact the major features of the DPPH spectrum are five lines with relative intensities of $1:2:3:2:1$, suggesting that the electron is not localized on one of the nitrogen atoms but is spread over them both, so that both nitrogen nuclei are equivalent as far as the epr spectrum is concerned.

In radiation chemistry, electron paramagnetic resonance is used to identify the radicals formed on irradiation, by analysis of the fine structure of their spectra. This may be accomplished more readily if the absorption is plotted as a derivative curve, plotting the slope of the normal absorption curve against the magnetic field. An absorption peak is shown by the derivative curve crossing the zero axis; e.g., the derivative curve for an isolated free electron would appear as in Fig. 4.13, the normal absorption curve is shown in Fig. 4.8. Actual derivative curves found for the radicals H·, D·, ·OH, and ·OD in irradiated ice are shown in Fig. 7.2, p. 270.

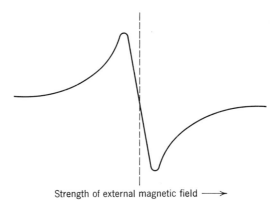

Strength of external magnetic field ⟶

FIGURE 4.13   *Derivative curve for an isolated free electron.*

FLASH PHOTOLYSIS.   Flash photolysis is a technique whereby a very high radical concentration is produced momentarily by photolysis, and the absorption spectrum of the reaction mixture determined immediately afterwards or after very short intervals (10^{-5} to 1 sec) (54, 55). The apparatus consists of a reaction vessel (A in Fig. 4.14) in the form of a quartz tube with

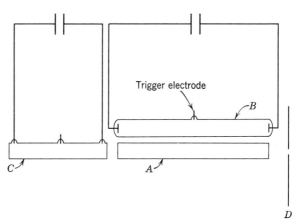

FIGURE 4.14   *Apparatus for flash photolysis.*

plane ends and, lying next to it, a quartz discharge lamp (*B*), also in the form of a tube but with tungsten electrodes and filled with about 10-cm pressure of inert gas. The reaction vessel and lamp are surrounded by a reflector to concentrate the light. A bank of condensers discharge through the lamp and with suitably high potential and capacity give a very intense flash lasting about 100 μsec (the flash may be sufficiently intense to cause complete dissociation of materials in the reaction tube). A second discharge lamp (*C*) is arranged end-on to the reaction tube and discharged a predetermined interval after the photoflash. Light from this second lamp travels along the reaction tube and into the spectrograph (*D*) which records the absorption spectrum of the reaction mixture. In practice the experiment is generally repeated a number of times, each time increasing the interval between the photo and spectroscopic flashes. In this way the decay of transient intermediates can be followed and their lifetimes estimated.

The material investigated may be either gaseous or in solution. However, the absorbed light is rapidly degraded to thermal energy and in gases at low pressures temperatures of several thousand degrees are generated, so that under these conditions the reaction mixture may contain the products of thermal as well as photolytic dissociation. In solution or in the presence of a relatively large excess of inert gas the temperature rise can be kept to about 10°C.

Flash photolysis has been used to identify free radical intermediates, to study the spectra (56) and chemical reactions of radicals, and, particularly, to study the kinetics of radical reactions. The technique has also been used to investigate nonradical transient intermediates (e.g., triplet excited states). Matheson and Dorfman (57) and McCarthy and MacLachlan (58) have detected radicals produced by ionizing radiation using an essentially similar technique, but replacing the photoflash by an intense pulse of electrons from an accelerator. This technique of pulse radiolysis is being increasingly exploited at the present time (59–61) both to identify transient radiolysis intermediates and, more particularly, to determine absolute rate constants for their reactions. It is discussed in detail on p. 191.

Summary of the Reactions of Free Radicals

FORMATION

$$RS \rightarrow R\cdot + S\cdot \qquad (4.59)$$
(thermal or photo dissociation)

$$M^{z+} + RO_2H \rightarrow M^{(z+1)+} + RO\cdot + OH^- \quad \text{etc.} \qquad (4.67)$$
(oxidation-reduction processes)

REACTIONS

$$AB\cdot \rightarrow BA\cdot$$
(rearrangement)

$$AB\cdot \rightarrow A\cdot + B$$
(dissociation)

$$R\cdot + \underset{/}{\overset{\backslash}{C}}=\underset{\backslash}{\overset{/}{C}} \longrightarrow \overset{R}{\underset{/}{\overset{|}{C}}}-\underset{\backslash}{\overset{/}{\dot{C}}} \qquad (4.83)$$
(addition)

$$A\cdot + BC \rightarrow AB + C\cdot \qquad (4.86)$$
(abstraction)

$$R\cdot + O_2 \rightarrow R{-}O{-}O\cdot \qquad (4.108)$$
(oxygen addition)

DESTRUCTION

$$R\cdot + S\cdot \rightarrow RS \qquad (4.89)$$
(combination)

$$2RH\cdot \rightarrow RH_2 + R \qquad (4.97)$$
(disproportionation)

$$M^{z+} + R\cdot \rightarrow M^{(z+1)+} + R^- \qquad (4.101)$$
(electron transfer)

For further reading on free radical chemistry *Free Radicals in Solution* by C. Walling, Wiley, New York, 1957, is recommended, and on electron paramagnetic resonance *Free Radicals as Studied by Electron Spin Resonance* by D. J. E. Ingram, Butterworths Scientific Publications, London, 1958, and *Electron Spin Resonance Spectra of Radiation-Produced Radicals* by R. W. Fessenden and R. H. Schuler in *Advances in Radiation Chemistry* (ed. M. Burton and J. L. Magee), Vol. 2, Wiley-Interscience, New York, 1970. Other valuable sources of information about free radicals are *Atomic and Free Radical Reactions* by E. W. R. Steacie, 2nd ed., Reinhold Publishing, New York, 1954; *Some Problems in Chemical Kinetics and Reactivity* by N. N. Semenov (translated by M. Boudart), Princeton University Press, Princeton N.J., 1958; *The Spectra and Structures of Simple Free Radicals* by G. Herzberg, Cornell University Press, Ithaca, 1971; *Wolne Rodniki w Chemii Radiacyjnej* (*Free Radicals in Radiation Chemistry*) by Jerzy Kroh, Warsaw, 1967, Panstwowe Wydawnictwo Naukowe, and *Electron Para-*

magnetic Resonance in Chemistry by L. A. Blumenfeld, V. V. Voevodskii, and A. G. Semenov, Academy of Science, USSR, 1962.

REFERENCES

1. R. L. Platzman, *Int. J. Appl. Radiat. Isotop.*, **10**, 116 (1961).
2. M. Kasha, *Discuss. Faraday Soc.*, **9**, 14 (1950).
3. J. L. Magee in *Comparative Effects of Radiation* (ed. M. Burton, J. S. Kirby-Smith, and J. L. Magee), Wiley, New York, 1960, p. 130.
4. Th. Förster, *Discuss. Faraday Soc.*, **27**, 7 (1959).
5. S. Lipsky and M. Burton, *J. Chem. Phys.*, **31**, 1221 (1959); J. M. Nosworthy, J. L. Magee, and M. Burton, *J. Chem. Phys.*, **34**, 83 (1961).
6. M. S. Matheson, *Ann. Rev. Phys. Chem.*, **13**, 77 (1962).
7. P. De Mayo and S. T. Reid, *Quart. Rev.*, **15**, 393 (1961).
8. J. Franck and E. Rabinowitch, *Trans. Faraday Soc.*, **30**, 120 (1934).
9. F. W. Lampe and R. M. Noyes, *J. Am. Chem. Soc.*, **76**, 2140 (1954).
10. L. F. Meadows and R. M. Noyes, *J. Am. Chem. Soc.*, **82**, 1872 (1960).
11. J. Weiss, *Trans. Faraday Soc.*, **35**, 48 (1939).
12. J. L. Bolland and H. R. Cooper, *Proc. Roy. Soc. (London)*, *Ser. A*, **225**, 405 (1954).
13. G. O. Schenck, *Angew. Chem.*, **64**, 12 (1952); G. O. Schenck, K. G. Kinkel, and H-J. Mertens, *Annalen*, **584**, 125 (1953); G. O. Schenck, *ibid.*, p. 156; G. O. Schenck, H. Eggert, and W. Denk, *ibid.*, p. 177.
14. F. S. Dainton and K. J. Ivin, *Trans. Faraday Soc.*, **46**, 374, 382 (1950); *Proc. Roy. Soc. (London)*, *Ser. A*, **212**, 96 (1952).
15. H. A. Dewhurst, A. H. Samuel, and J. L. Magee, *Radiat. Res.*, **1**, 62 (1954); P. Kelly, T. Rigg, and J. Weiss, *Nature*, **173**, 1130 (1954).
16. T. W. Martin, R. E. Rummel, and C. E. Melton, *Science*, **138**, 77 (1962).
17. H. Von Koch and E. Lindholm, *Arkiv Fys.*, **19**, 123 (1961); cf. refs. cited by P. Wilmenius and E. Lindholm, *Arkiv Fys.*, **21**, 97 (1962).
18. G. G. Meisels, W. H. Hamill, and R. R. Williams, *J. Phys. Chem.*, **61**, 1456 (1957).
19. L. M. Dorfman and M. C. Sauer, *J. Chem. Phys.*, **32**, 1886 (1960).
20. L. Reinisch, *J. Chim. Phys.*, **57**, 1064 (1960).
21. A. H. Samuel and J. L. Magee, *J. Chem. Phys.*, **21**, 1080 (1953).
22. J. L. Magee and M. Burton, *J. Am. Chem. Soc.*, **72**, 1965 (1950).
23. *American Institute of Physics Handbook* (ed. D. E. Gray), McGraw-Hill, New York, 1957, Section 7c.
24. K. Watanabe, *J. Chem. Phys.*, **26**, 542 (1957).
25. F. H. Field and J. L. Franklin, *Electron Impact Phenomena*, Academic Inc., New York, 1957, Table 10a, p. 106.
26. J. H. Beynon, *Mass Spectrometry and Its Applications to Organic Chemistry*, Elsevier, Amsterdam, 1960, p. 263.
27. J. L. Magee, *J. Phys. Chem.*, **56**, 555 (1952).
28. M. Burton and J. L. Magee, *J. Phys. Chem.*, **56**, 842 (1952).
29. F. H. Field and F. W. Lampe, *J. Am. Chem. Soc.*, **80**, 5587 (1958).
30. J. L. Magee and M. Burton, *J. Am. Chem. Soc.*, **73**, 523 (1951).
31. J. L. Magee, *Discuss. Faraday Soc.*, **12**, 33 (1952).
32. P. S. Rao, J. R. Nash, J. P. Guarino, M. R. Ronayne, and W. H. Hamill, *J. Am. Chem. Soc.*, **84**, 500 (1962); J. P. Guarino, M. R. Ronayne, and W. H. Hamill, *Radiat. Res.*, **17**, 379 (1962).

33. W. Van Dusen and W. H. Hamill, *J. Am. Chem. Soc.*, **84**, 3648 (1962).
34. R. L. Platzman, *Radiat. Res.*, **2**, 1 (1955).
35. M. Gomberg, *Ber.*, **33**, 3150 (1900); *J. Am. Chem. Soc.*, **22**, 757 (1900).
36. H. J. H. Fenton, *J. Chem. Soc.*, **65**, 899 (1894); H. J. H. Fenton and H. Jackson, *ibid.*, **75**, 1 (1899); H. J. H. Fenton and H. O. Jones, *ibid.*, **77**, 69 (1900).
37. G. Stein and J. Weiss, *J. Chem. Soc.*, 3265 (1951); G. R. A. Johnson, G. Stein, and J. Weiss, *ibid.*, 3275 (1951).
38. E. Warhurst, *Quart. Rev.*, **5**, 44 (1951).
39. J. E. Bennett and A. Thomas, *Nature*, **195**, 995 (1962).
40. J. L. Franklin and H. P. Broida, *Ann. Rev. Phys. Chem.*, **10**, 145 (1959).
41. G. A. Russell, *Tetrahedron*, **5**, 101 (1959).
42. D. Y. Curtin and M. J. Hurwitz, *J. Am. Chem. Soc.*, **74**, 5381 (1952).
43. A. N. Nesmeyanov, R. Kh. Freidlina, and L. I. Zakharkin, *Quart. Rev.*, **10**, 330 (1956); A. N. Nesmeyanov, R. Kh. Freidlina, and V. N. Kost, *Tetrahedron*, **1**, 241 (1957).
44. R. Gomer and G. B. Kistiakowsky, *J. Chem. Phys.*, **19**, 85 (1951); G. B. Kistiakowsky and E. K. Roberts, *ibid.*, **21**, 1637 (1953).
45. P. Alexander and D. Rosen, *Nature*, **188**, 574 (1960).
46. D. L. Bunbury, R. R. Williams, and W. H. Hamill, *J. Am. Chem. Soc.*, **78**, 6228 (1956).
47. D. Urch and R. Wolfgang, *J. Am. Chem. Soc.*, **83**, 2982 (1961).
48. B. S. Rabinowitch, S. G. Davis, and C. A. Winkler, *Can. J. Res.*, **B21**, 251 (1943).
49. F. Wöhler and J. Liebig, *Annalen*, **3**, 249 (1832).
50. J. Cuthbert, *Quart. Rev.*, **13**, 215 (1959).
51. G. N. Lewis, M. Calvin, and M. Kasha, *J. Chem. Phys.*, **17**, 804 (1949).
52. H. C. Box and H. G. Freund, *Nucleonics*, **17**, 66 (January 1959).
53. I. Miyagawa and W. Gordy, *J. Am. Chem. Soc.*, **83**, 1036 (1961).
54. R. G. W. Norrish and B. A. Thrush, *Quart. Rev.*, **10**, 149 (1956).
55. R. G. W. Norrish, *Proc. Chem. Soc.*, 247, (1958).
56. M. S. Matheson and L. M. Dorfman, *Pulse Radiolysis*, M.I.T. Press, Cambridge, Mass., 1969.
57. M. S. Matheson and L. M. Dorfman, *J. Chem. Phys.*, **32**, 1870 (1960).
58. R. L. McCarthy and A. MacLachlan, *Trans. Faraday Soc.*, **56**, 1187 (1960).
59. L. M. Dorfman, *Nucleoncs*, **19**, 54 (October 1961); *Science*, **141**, 493 (1963).
60. M. S. Matheson, *Ann. Rev. Phys. Chem.*, **13**, 77 (1962).
61. M. Ebert, J. P. Keene, A. J. Swallow, and J. H. Baxendale, *Pulse Radiolysis*, Academic, London, 1965.

CHAPTER 5

Radiolysis Kinetics

Empirical Rate Studies • Pseudo First-Order • Order of a Reaction • First-Order Half Life • Second-Order Half Life • Half Life • Molecular Kinetics • Competition Kinetics • Quantitative Effect of Radiation • Nonhomogeneous Kinetics • Effect of Solute Concentration on Molecular Yields • Relative Rate Constants • Radical Scavenging • Chain Reactions • Pulse Radiolysis • Pulse Radiation Sources • Febetron • Optical Detection • Stroboscopic Pulse Radiolysis • Picosecond Spectroscopy Using Lasers • Conductivity Detection • Electron Spin Measurement • Effect of Ionic Charge • Pulse Radiolysis Studies of Effect of Ionic Charge • References •

Empirical Rate Studies

Chemical kinetics is concerned with the rate and mechanism of chemical reactions, the rate being related to the concentration of the reacting species. In the simplest case, only one substance is changing, and the rate of change is proportional to the concentration of the substance.

$$A \rightarrow P \qquad (5.1)$$

where P is the product

$$\frac{-\mathrm{d}[A]}{\mathrm{d}t} = k[A] \qquad (5.2)$$

where [A] is the concentration at time t and k is a constant. Such a reaction is said to be *first-order*. Integrating:

$$-\log [A] = \frac{kt}{2.303} + \text{constant} \qquad (5.3)$$

The velocity constant, or specific reaction rate k, can be obtained by plotting log $[A]$ against t. The slope will be $-k/2.303$ and $k = 2.303\,(-\text{slope})$.

In a second-order reaction involving only one substance

$$\frac{-d[A]}{dt} = k[A]^2 \tag{5.4}$$

and on integration

$$k = \frac{1}{t}\frac{[A]_0 - [A]}{[A]_0[A]} \tag{5.5}$$

where $[A]_0$ is the concentration at time zero and $[A]$ at time t.

For a third-order reaction involving only one substance

$$\frac{-d[A]}{dt} = k[A]^3 \tag{5.6}$$

Where two substances are involved, say A and B, the reaction is second order if

$$\frac{-d[A]}{dt} = k[A][B]. \tag{5.7}$$

The *order* is a mathematical term. In general,

$$\frac{-d[A]}{dt} = k[A]^\alpha[B]^\beta \cdots \tag{5.8}$$

The reaction is said to be of the αth order in A, and the βth order in B etc. The overall order of the reaction is $n = \alpha + \beta + \cdots$. It is sometimes impossible to speak of an order, e.g., in the reaction between H_2 and Br_2

$$\frac{d[HBr]}{dt} = \frac{k[H_2][Br_2]^{1/2}}{1 + (k^1[HBr]/[Br_2])} \tag{5.9}$$

Pseudo First-Order

Suppose A reacts with B to form a product P

$$A + B \overset{k_2}{\to} P \tag{5.10}$$

where k_2 is the reaction rate constant. Then

$$\frac{-d[A]}{dt} = \frac{d[P]}{dt} = k_2[A][B] \tag{5.11}$$

If $[B] \gg [A]$, the concentration of B does not change significantly during the experiment and can be considered constant.

$$\frac{-\mathrm{d[A]}}{\mathrm{d}t} = k^1[A] \quad \text{where} \quad k^1 = k_2[B] \tag{5.12}$$

The rate expression is the same as for a first-order reaction. The constant $k_2 = k^1/[B]$.

Half Life

The time for half of the material A to disappear is often called its half life, $t_{1/2}$.

First-Order Half Life

$$kt = \ln_e \frac{[A]_0}{[A]_t} \tag{5.13}$$

and

$$t_{1/2} = \frac{0.693}{k} \tag{5.14}$$

Second-Order Half Life

$$kt = \frac{1}{[A]_t} - \frac{1}{[A]_0} \tag{5.15}$$

and

$$t_{1/2} = \frac{1}{k[A]_0} \tag{5.16}$$

Kinetic equations can also be developed for more complicated reactions in which there may be consecutive reactions or opposing reactions. Reactions may also take place in a flow system or may take place so rapidly that special techniques have to be developed to measure their rates [e.g., Laidler (1)].

The foregoing kinetic equations can be applied using a variety of methods to measure concentration, e.g., optical or conductometric. If the optical density (OD) has been measured at a particular wave length for which only A absorbs and for which the molar extinction coefficient is ε; $OD = \varepsilon[A]d$ where d is the optical path length. Equation 5.3 becomes

$$-\log \frac{(OD)}{\varepsilon d} = \frac{kt}{2.303} + \text{const.} \tag{5.17}$$

or

$$-\log (OD) = \frac{kt}{2.303} + \text{const.} \tag{5.18}$$

and k can easily be obtained by plotting log (OD) against t. (Note that it is not necessary to know ε in this case.)

If only the product P absorbs

$$kt = 2.303 \log \frac{OD_\infty}{OD_\infty - OD_t} \tag{5.19}$$

where OD_t is the optical density at time t and OD_∞ is the optical density when A has all disappeared.

For second-order kinetics, supposing only A absorbs

$$\frac{2kt}{\varepsilon d} = \frac{1}{OD_t} - \frac{1}{OD_0} \tag{5.20}$$

and it will be necessary to know the extinction coefficient for A. A number of extinction coefficients for transient species such as the solvated electron are now known; e.g., ε for the solvated electron at $\lambda = 720$ nm is 1.58×10^4 M^{-1} cm^{-1} (2).

In reactions involving two or more species, a wavelength is chosen for which the optical absorption for the species of interest is large compared to that of the other species present. This choice is not available for conductivity measurements where the measured conductivity will be that of all the conducting species present (e.g., p. 204 and ref. 3).

Molecular Kinetics

Reaction rates also depend on temperature; $k = Ae^{-B/T}$ where T is the absolute temperature and A and B are constants. The equation can be written in the form $k = Ae^{-E/RT}$, generally known as the Arrhenius law. This law holds for elementary reactions but does not hold for more complicated reactions in which, for example, there may be successive reactions. This leads us to another aspect of chemical kinetics, namely the elucidation of the actual reaction mechanism. This involves knowing exactly what reactive species are involved in a reaction and the nature of their participation in the reaction. This aspect has made enormous strides in the last two decades with the development of more sophisticated techniques for detecting and measuring short lived species—spectroscopic, NMR, ESR, conductivity, mass spectrography, polarography, isotopic, etc. A reaction mechanism then appears as a series of elementary reactions. For example the reaction $H_2 + Br_2 = 2HBr$, actually takes place in a number of elementary reactions such as

$$Br_2 \rightarrow 2Br \tag{5.21}$$

$$Br + H_2 \rightarrow HBr + H \tag{5.22}$$

$$H + Br_2 \rightarrow HBr + Br \text{ etc.} \tag{5.23}$$

The number of molecules entering into an elementary reaction is known as the *molecularity* of the reaction. Reaction 5.21 is a monomolecular reaction, reactions 5.22 and 5.23 are bimolecular reactions.

The Arrhenius law led to the suggestion that reactions such as $H_2 + I_2 \rightarrow 2HI$ take place as a result of collisions between molecules having excess energy known as energy of activation and in fact, a simple application of the kinetic theory of gases led to an expression of the Arrhenius type. The advent of quantum theory and quantum mechanics led to the development of the theory of absolute reaction rates in which the reacting species are said to form an *activated complex*, the rate of reaction being controlled by the rate at which these complexes travel over a free-energy barrier (see Fig. 5.1). The free energy of activation $\Delta G^{\ddagger}$ is related to the rate constant k by

$$k = \frac{\mathbf{k}t}{h} e^{\frac{-\Delta G^{\ddagger}}{RT}} \tag{5.24}$$

($\mathbf{k} = R/N_A$) where R is the gas constant and N_A is the Avogadro number.

The theoretical treatment of liquid reactions is much more difficult than that for gaseous reactions. Collisions between molecules will depend somewhat on the structure of the liquid and its electrical properties. One sometimes speaks of a "cage effect" in which the surrounding solvent molecules tend

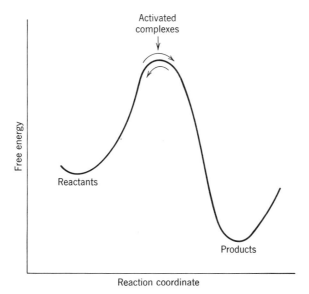

FIGURE 5.1   *The free-energy barrier, over which systems must cross to pass from the initial to the final state.*

to hold the colliding reacting molecules together and cause them to collide a number of times before they separate.

Competition Kinetics

Suppose that the radical $R\cdot$ can react with either S or X

$$R\cdot + S \overset{k_1}{\rightarrow} \text{product } P_1 \tag{5.25}$$

$$R\cdot + X \overset{k_2}{\rightarrow} \text{product } P_2 \tag{5.26}$$

Then the probability that R will react with S is given by

$$\frac{k_1[R][S]}{k_1[R][S] + k_2[R][X]} = \frac{k_1[S]}{k_1[S] + k_2[X]} \tag{5.27}$$

As an example, the production of acetone by the radiolysis of oxygen saturated aqueous solutions of 2-propanol, and methanol is thought to occur as follows (Fig. 5.2) [(2), p. 64]

$$\cdot OH + (CH_3)_2CHOH \overset{k_1}{\rightarrow} (CH_3)_2 \cdot COH + H_2O \tag{5.28}$$

$$\cdot OH + CH_3OH \overset{k_2}{\rightarrow} \cdot CH_2OH + H_2O \tag{5.29}$$

$$(CH_3)_2 \cdot COH + O_2 \rightarrow (CH_3)_2CO + HO_2 \cdot \tag{5.30}$$

$$G(CH_3COCH_3) = \frac{k_1[(CH_3)_2CHOH]}{k_1[(CH_3)_2CHOH] + k_2[CH_3OH]} \cdot G_{OH} \tag{5.31}$$

and

$$\frac{1}{G(CH_3COCH_3)} = \frac{1}{G_{OH}} \left(1 + \frac{k_2[CH_3OH]}{k_1[(CH_3)_2CHOH]}\right) \tag{5.32}$$

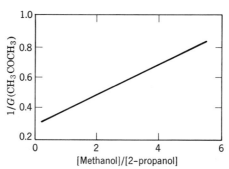

FIGURE 5.2   *Typical reciprocal plot for analysis of data for OH reactivities.*

Thus a plot of $1/G(CH_3COCH_3)$ versus $[CH_3OH]/[(CH_3)_2CHOH]$ should be linear and the intercept $= 1/G_{OH}$.

$$\text{slope} = \frac{1}{G_{OH}} \cdot \frac{k_2}{k_1}$$

$$G_{OH} = 1/\text{intercept}$$

G_{OH} is equal to 3.0 for pH $= 1.21$ (4).

Quantitative Effect of Radiation

Introducing radiation into a chemical system adds yet another variable—radiation. The effect of the radiation is taken care of in a quantitative empirical way by combining the G value of a reaction (see p. 6) with the energy absorbed by the system. From Eq. 3.28,

$$\text{molecules product formed per gram} = \frac{\text{energy absorbed (rads)}}{1.602 \times 10^{-12}} \times G\,(\text{product})$$

$$= 6.25 \times 10^{11} \text{ (energy abs in rads)} \, G\,(\text{product})$$

or pounds of product produced per kilowatt-hour of radiation energy absorbed $= G \times M \times 8.3 \times 10^{-4}$ where M is the molecular weight of the product.

A more fundamental approach to radiolysis kinetics considers the nature of the species resulting from the action of radiation on the system and the subsequent reactions of these species. These topics occupy the rest of the book. Quite obviously a good basic knowledge of classical chemical kinetics is essential to an understanding of radiolysis kinetics.

Nonhomogeneous Kinetics

Nonhomogeneous kinetics describes rates of reactions that occur in one phase, but where some of the reactants are distributed in small regions of relatively high local concentration. For example, nonhomogeneous kinetics is concerned with the reactions that occur in radiolytic spurs. Each intermediate in a spur may undergo diffusive escape from the spur. Species that escape from spurs become randomly distributed in the medium and their reactions then follow the usual ("homogeneous") kinetics. The radiolysis of a pure compound in the liquid state may be represented as follows:

$$M \xrightarrow{\quad\quad} \to [M^+ + e^-] \quad\quad\quad (5.32a)$$

$$[M^+ + e^-] \to M^* \text{ (geminate neutralization)} \quad\quad\quad (5.32b)$$

$$[M^+ + e^-] \rightarrow M^+ + e^- \text{ (free ions)} \qquad (5.32c)$$

$$M^+ + e^- \rightarrow M^* \text{ (random neutralization)} \qquad (5.32d)$$

The square brackets indicate that the species are in a spur and are exerting an appreciable coulombic force on each other. Various theoretical models have been developed by Samuel and Magee (4a), Schwarz (4b), Mozumder and Magee (4c), Hummel (4d), Freeman (4e), and others and tested against the results of scavenger experiments (see pp. 184 and 185).

Effect of Solute Concentration on the Molecular Yields from Water

Dilute solutions (generally 10^{-5} to 10^{-3} M) are normally used when making determinations of molecular yields, since more concentrated solutions may give erroneous results due to interference with the spur and track reactions forming the molecular products, particularly with high-LET radiation (chap. 7, and ref. 5).

Although, at the low concentrations normally used, solute concentration does not greatly affect the molecular yields, careful examination does show that solutes which react with hydroxyl radicals (e.g., Br^-, Cl^-) gradually lower the molecular yield of hydrogen peroxide as their concentration is increased (6), while solutes which react with hydrogen atoms or solvated electrons (e.g., O_2, H_2O_2, KNO_2, $CuSO_4$) reduce the molecular yield of hydrogen (7, 8). Sworski (6) showed that an approximately linear relationship exists between the molecular hydrogen peroxide yield produced by γ-rays and the cube root of the solute concentration. The relationship takes the form

$$G_{H_2O_2} = (G_{H_2O_2})_0 - b[S]^{1/3} \qquad (5.33)$$

where $(G_{H_2O_2})_0$ is $G_{H_2O_2}$ at infinite dilution, i.e., in the absence of solute, and b is a constant that varies slightly with different solutes; $[S]$ is the solute concentration. A similar relationship holds for G_{H_2}. The cube-root relationship found for γ-rays is fortuitous and only approximate, the exponent varying with the range of concentrations being considered. Furthermore, the value of the exponent depends upon the LET of the radiation, being smaller at higher LET (9); thus Burton and Kurien (10) give values of 0.33, 0.26, and 0.15 for cobalt-60 γ-rays, 50 kVp x-rays, and 3.4 MeV α-rays respectively.

A second relationship involving the yields of molecular products and the solute concentration was discovered by Schwarz (7), who plotted the ratios $G_{H_2}/(G_{H_2})_0$ and $G_{H_2O_2}/(G_{H_2O_2})_0$ against the logarithm of the solute concentration and obtained a curve of the shape shown in Fig. 5.3. The interesting

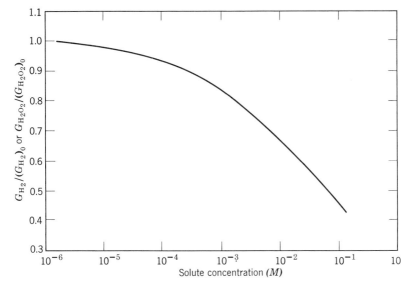

FIGURE 5.3   *The effect of solute concentration on the molecular yields of hydrogen and hydrogen peroxide.*

feature of this plot is that the shape of the curve is the same for both hydrogen and hydrogen peroxide yield and for a wide variety of solutes. Furthermore, the curves can all be brought into coincidence by multiplying the concentration by an arbitrary factor appropriate to the particular product and solute. In the case of the hydrogen yields, Schwarz was able to show that these factors were in the same ratio as the rate constants for reaction of hydrogen atoms with the solute, wherever the rate constants were known, suggesting strongly that the molecular hydrogen is formed via hydrogen atoms. It also suggests that the molecular hydrogen peroxide is formed via a radical reaction, since the hydrogen peroxide yield varies with solute concentration in the same way as the hydrogen yield.

The concentration effect in γ-irradiated solutions has been accounted for by approximate mathematical treatments based on the radical-diffusion model for the radiolysis of water (7, 9). This assumes that hydrogen atoms (or solvated electrons) and hydroxyl radicals are produced in localized regions (spurs) in the irradiated medium, that the molecular products are formed by combination of these species as they diffuse from the spurs, and that the combination reactions compete with reactions between the radicals and the solute. For the purposes of calculation it was assumed that only one radical is formed, $R\cdot$ (which represents $H\cdot$, e_{aq}^-, and $\cdot OH$), so that formation of molecular products is represented by

$$2R\cdot \xrightarrow{k_1} R_2 \qquad (5.34)$$

and the radical-solute reactions by

$$R\cdot + S \xrightarrow{k_2} RS \qquad (5.35)$$

where k_1 and k_2 are the rate constants for these reactions respectively. If the radicals and solute were homogeneously distributed in the medium, normal chemical kinetics would be applicable and the rate equation would be

$$d[R]/dt = -k_1[R]^2 - k_2[R][S] \qquad (5.36)$$

However under radiolysis conditions the radicals are not initially homogeneously distributed, and the changing radical concentration in the region of each spur must be considered. This is done by including a diffusion term, so that the rate equation becomes

$$\delta[R]/\delta t = D\nabla^2[R] - k_1[R]^2 - k_2[R][S] \qquad (5.37)$$

where D is the diffusion constant for the radicals and ∇^2 is a Laplacian operator. Calculations based on this one radical–one solute model are in quite good agreement with the experimental results, indicating that the model is probably substantially correct. Later studies of the diffusion kinetics of irradiated aqueous solutions (11, 12) have taken more complex models, which include hydrogen atoms and hydroxyl radicals separately, reactive intermediates, and more than one solute, and have eliminated some of the approximations necessary for the earlier calculations. Although choice of values for the various diffusion and rate constants and for the initial number and distribution of radicals in the spurs is still somewhat arbitrary, the success of the calculations in describing qualitatively and semiquantitatively the relationship between solute concentration and molecular yields, and isotope, LET, and dose-rate effect is strong evidence in favor of the radical-diffusion model for the radiolysis of water.

Relative Rate Constants

The fact that free radicals are very reactive entities leads to competition for them when the solution contains more than one substance with which they can react. If the solution contains two such substances, A and B, the ratio of the number of radicals reacting with A (R_A) to the number reacting with B (R_B) is

$$\frac{R_A}{R_B} = \frac{[A]k_{RA}}{[B]k_{RB}}$$

where $[A]$ and $[B]$ are the concentrations of A and B in the solution and k_{RA} and k_{RB} are the bimolecular rate constants for the reaction between radical R· and A and between R· and B respectively. Since the ratio R_A/R_B can be determined by experiment and the concentrations of the two reactants are known, the ratio of the two rate constants can be obtained. In radiation chemistry, B is generally a standard substance (e.g., ferrous ion or formic acid) for which the radiolysis reactions are established. A number of relative rate constants relevant to the radiation-induced oxidation of ferrous sulfate are given by Dainton (13), and others for the reactions of hydrogen atoms and hydroxyl radicals in acid solution are given by Swallow (13a). A more extensive survey of the relative rates of reaction of hydrogen atoms and hydroxyl radicals in aqueous solutions has been made by Ferradini (14). It should be mentioned that the determination of relative rate constants is not limited to radiation chemistry, but in fact is often more readily made using other sources of radicals.

Techniques such as intermittent irradiation with either ultraviolet light (8) or ionizing radiation (8), and pulse radiolysis (15), analogous to flash photolysis, enable absolute values for rate constants such as k_{RA} and k_{RB} to be measured. These absolute values can then be used to calculate other absolute rate constants using the rate constant ratios determined as described above (e.g., Tables 7.2, 7.10, and 7.11).

Radical Scavenging

Scavenging is the term applied to the deliberate addition to a radical reaction of a compound which will react preferentially with the radicals, at the expense of the normal radical reactions. The object is to identify the radicals taking part in the reaction and to determine what part of the overall reaction is due to (scavengeable) free radicals. In radiation-induced reactions in liquids the reaction may take place either within spurs or close to the tracks of heavy particles, where the active intermediates are formed, or in the bulk of the medium where reactions are initiated by radicals which have diffused from the track zone. Using moderate scavenger concentrations (e.g., 10^{-6} to 10^{-3} M iodine) the scavenging action will be limited to radicals which have diffused into the bulk of the sample and reactions in the track zone will be unaffected, allowing the reactions in these two regions to be differentiated. It should be pointed out that hot radicals often react within one or two collisions and therefore are not scavenged by low concentrations of scavenger. However, if the hot radicals escape reaction and are reduced to thermal energy, they will be scavenged in the normal manner.

Substances used as scavengers are either stable free radicals, compounds that eliminate radicals by electron transfer reactions, or compounds that react to give relatively stable free radicals in place of the active radicals originally present. Molecular iodine has been widely used (16–18) and reacts as follows:

$$R\cdot + I_2 \rightarrow RI + I\cdot \tag{5.38}$$

The iodine atoms that are formed are relatively stable and eventually combine, giving molecular iodine. The reaction is a favorable one energetically, since $D(R$—$I)$ is generally greater than $D(I$—$I)$. Radicals that are scavenged with iodine can be identified by identifying the iodides (RI) formed. This can be accomplished rather elegantly by using radioactive iodine-131 as a scavenger and identifying the radioactive iodides by isotope dilution (19, 20). The quantity of iodine used up gives a measure of the total number of radicals that have reacted with the scavenger. Two practical points deserve mention in connection with the use of iodine as a scavenger. First, hydrogen iodide formed by scavenging of hydrogen atoms is frequently difficult to estimate accurately, since, unless special precautions are taken, it reacts with unsaturated radiolysis products and this results in spuriously low estimates of the number of hydrogen atoms formed (18). The second point is that aromatic compounds can enter into complex reactions when irradiated with iodine, so that iodine scavenging experiments with aromatic materials must be interpreted with caution (21). Results obtained with iodine are not entirely unambiguous since iodine (and organic iodides) can react with free electrons:

$$e^- + I_2 \rightarrow I^- + I\cdot \tag{5.39}$$

This may influence the reaction by providing an alternative to the neutralization of positive ions by electrons. The alternative neutralization reaction is

$$A^+ + I^- \rightarrow A^* + I\cdot \tag{5.40}$$

which releases rather less energy than the normal neutralization process and may lead to different products.

Other radical scavengers include the stable free radicals diphenylpicrylhydrazyl (22, 23) and nitric oxide (24), ferric chloride (25, 26, 27), hydrogen iodide (28), and a variety of unsaturated compounds, such as styrene, methylmethacrylate, cyclohexene, etc., (29), including ^{14}C-labeled ethylene (30).

Charlesby, Davison, and Lloyd (31) have made a kinetic study of the effect of scavenger concentration and dose rate when anthracene is used as a scavenger in irradiated hexane, cyclohexane, and polysiloxanes. With the hydrocarbon solutions, the rate of anthracene disappearance depends on [anthracene] (dose rate)$^{-1/2}$, as expected if competition exists between a radical–radical reaction and the scavenging radical-anthracene reaction. Anthracene is a relatively inefficient scavenger, allowing the concentration and dose-rate dependence to be observed, but with many scavengers the competing radical–radical reaction is virtually eliminated, and these effects are not seen.

Chain Reactions

Reactions that were grouped earlier under the heading "radical attack on substrates" are characterized by the appearance of a radical among the products. In some situations the newly formed radical may also react with the substrate, and with favorable conditions the sequence of reactions between radical and substrate may be repeated a large number of times before the radical is destroyed; in other words, a chain reaction takes place. The sequence of reactions may be repeated anywhere from two or three to many thousand times, and the original radical may give rise to changes involving a very large number of substrate molecules.

Two basic types of chain reaction would be anticipated, corresponding to the two types of radical-substrate reaction, addition and abstraction. Successive addition reactions are typical of radical-induced polymerization, which can be represented by

$$R\cdot + AB \rightarrow R—AB\cdot \tag{5.41}$$

$$R—AB\cdot + AB \rightarrow R—AB—AB\cdot \tag{5.42}$$

Taking styrene ($PhCH=CH_2$) as an example, a typical chain propagating step is

$$\begin{array}{cccc} Ph & Ph & Ph & Ph \\ | & | & | & | \\ —CH_2—CH\cdot + CH_2=CH & \rightarrow & —CH_2—CH—CH_2—CH\cdot \end{array}$$

Chain reactions involving successive abstractions are represented by

$$\begin{array}{c} R\cdot + AB \rightarrow RA + B\cdot \\ B\cdot + RC \rightarrow BC + R\cdot \\ \hline AB + RC \rightarrow RA + BC \end{array} \tag{5.43}$$

Two different abstraction reactions must take place if the products are to be different from the reactants; in the side-chain bromination of toluene these are

$$Br\cdot + PhCH_3 \rightarrow HBr + PhCH_2\cdot \tag{5.44}$$

and

$$PhCH_2\cdot + Br_2 \rightarrow PhCH_2Br + Br\cdot \tag{5.45}$$

The repeated sequence of reactions may be more complex than the examples shown above, and it can include radical rearrangement and dissociation as well as addition and abstraction reactions. An interesting reaction between diazomethane and carbon tetrachloride to give pentaerithrytol tetrachloride involves a repeating sequence made up of eight steps; the overall chemical change is

$$4CH_2N_2 + CCl_4 \xrightarrow{h\nu} C(CH_2Cl)_4 + 4N_2 \qquad (5.46)$$

Chain reactions are very sensitive to small quantities of impurities if these are able to react with the radical intermediates. Typically, compounds that inhibit chain reactions are either stable free radicals (e.g., NO, DPPH) or compounds that react to give relatively stable free radicals, such as oxygen and benzoquinone. However, oxygen is an essential reactant in some chain reactions where it forms a peroxy radical ($RO_2\cdot$) able to react with the substrate (as in autoxidation), or else one which breaks down to give a relatively reactive radical.

Before leaving chain reactions it is worthwhile to consider the kinetics of a very simple, radiation-initiated, example. The reaction is broken down into initiation, propagation, and termination steps:

Initiation.	$A \rightsquigarrow R\cdot$	cR_M
Propagation.	$R\cdot + A \rightarrow R\cdot + P$	k_1
Termination.	$2R\cdot \rightarrow$ unreactive products	k_2

The initiation step involves the production of radicals ($R\cdot$) from the substrate (A) by the absorption of ionizing radiation. The rate of formation of radicals by this process is given by the product of the absorbed dose rate R_M (eV liter^{-1} sec^{-1}) and a constant c (moles eV^{-1}), related to the G value for the formation of radicals. Propagation is represented as reaction of the radicals with the substrate to give a product (P) and regenerate a radical ($R\cdot$; for simplicity the various radicals involved are not differentiated), the rate constant for the reaction being k_1 (liter mole^{-1} sec^{-1}). Termination is assumed to be combination of two radicals to give unreactive products, which do not include P, with a rate constant k_2 (liter mole^{-1} sec^{-1}). Under steady state conditions the radicals are formed and destroyed at the same rate:

$$\frac{d[R\cdot]}{dt} = O = cR_M - k_2[R\cdot]^2$$

and therefore

$$[R\cdot] = \left(\frac{c}{k_2}\right)^{1/2} (R_M)^{1/2}$$

The rate of formation of the product P is given by

$$\frac{d[P]}{dt} = k_1[A][R\cdot] = k_1\left(\frac{c}{k_2}\right)^{1/2}[A](R_M)^{1/2}$$

The rate of energy absorption by the system, dE/dt, is equal to the absorbed dose rate R_M, and the G value for the production of P (i.e., the yield per unit energy absorbed) is given by

$$G(P) = \frac{d[P]}{dE}$$

From this

$$G(P) = \frac{d[P]}{dt} \times \frac{dt}{dE}$$

and

$$G(P) = k_1\left(\frac{c}{k_2}\right)^{1/2}[A](R_M)^{-1/2} \tag{5.47}$$

The important conclusion here is that $G(P)$ is inversely proportional to the square root of the dose rate. Admittedly this is derived for an extremely simple chain reaction, but the conclusion arrived at, essentially that the G value will exhibit a nonlinear dependence on the dose rate, is generally true for radiation-initiated chain reactions; it is in fact a restriction on the use of chain reactions in chemical dosimetry, where their high yields would otherwise be very serviceable. When the reaction mechanism includes more than one chain-terminating step, a nonlinear dose rate dependence will be found whenever one of the competing radical-destroying reactions is second order with respect to radicals.

The dependence of G(product) on a power of the dose rate other than the first leads to an interesting result if the irradiation is intermittent rather than steady. For example, if a disc from which sectors have been cut (Fig. 5.4) is rotated between the radiation source and the sample so that the radiation is cut off from the sample for two thirds of each revolution then, when the disc is rotated slowly, the effect is essentially that of full irradiation for one third the time. For the simple chain reaction considered above $G(P)$ will still be given by Eq. 5.47, assuming R_M to be the uninterrupted dose rate. However when the sector is rotating rapidly there will be a range of speeds above which the system does not differentiate between light and dark periods, and the effect is that of continuous irradiation at one third the uninterrupted dose rate, i.e., effectively the dose rate is now $\frac{1}{3}R_M$. The change in effective dose rate means that $G(P)$ will be different at high and low speeds of rotation; a plot of the ratio of the two yields against the rate of rotation will give a

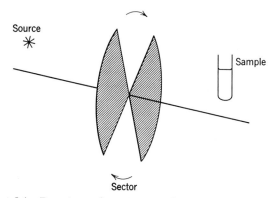

FIGURE 5.4   *Experimental arrangement for intermittent illumination.*

curve similar to Fig. 5.5. When the disc has two 60° sectors cut from it, so that the ratio of the light to dark periods is one to two, then from Eq. 5.47,

$$\frac{G(P)_{\text{fast}}}{G(P)_{\text{slow}}} = (3)^{1/2} = 1.7$$

The transition region of the curve occurs when the interval between the radiation pulses is of the same order as the average lifetime of the rate-controlling intermediates (in the present example, the average lifetime of a

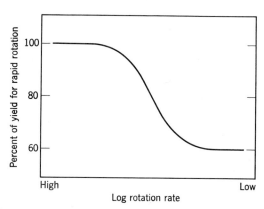

FIGURE 5.5   *Effect of intermittent illumination upon a photo or radiation-induced chain reaction.*

reaction chain between initiation and termination). Intermittent irradiation experiments can be used to determine these average lifetimes and also the individual rate constants (k_1 and k_2 in the present instance).

Experiments with intermittent radiation are common in photochemistry where it is a simple matter to interrupt the illumination, but are less common in radiation chemistry where the experimental difficulties are more formidable. Ghormley (32) has used an aluminum sector to interrupt the beam of electrons from a Van de Graaff accelerator, and found the lifetime of the shortest-lived intermediates in water whose concentration affects the steady state concentration of hydrogen peroxide to be about 10^{-3} sec under the conditions of the experiments. β-Radiation may also be stopped with relatively thin metal sectors (e.g., 33). A few experiments have been carried out with γ-rays using massive metal sectors. Thus the lifetimes of the radical chains in γ-irradiated chloral hydrate solutions (34) and in moist aerated chloroform (35) have been shown to be about 0.1 and 1 sec, respectively. The former value has been confirmed by flowing the chloral hydrate solution past a cobalt-60 source several times in rapid succession, producing the effect of intermittent irradiation (36); varying the flow rate produces similar effects to varying the speed of rotation in the rotating sector experiments.

Radiation from an accelerator may of necessity be pulsed and, if the pulse rate can be varied, can be used directly for experiments of this type. Sutton and Rotblat (37) found that G(ferric) and G(cerous) in aqueous solutions of ferrous sulfate and ceric sulfate respectively, irradiated with 15 MeV electrons from a linear accelerator, were not affected by increasing the pulse rate to an upper limit of 400 pulses $\sec^{-1}$. Though the reactions here are not chain reactions any increase in radical–radical combination at the expense of radical–solute reaction will alter the observed G values, and it could therefore be concluded that the lifetimes of the radicals involved did not exceed 2.5 msec.

Pulse Radiolysis

In pulse radiolysis, a system receives a short pulse of ionizing radiation of such intensity that the transient species produced can be observed. The pulse length is a nanosecond to a microsecond, the radiation is usually electrons, and the observation may be of the optical spectrum, electric spin resonance spectrum, or some other suitable property of the system. The development of this technique during the last decade has been quite astonishing and has led to major advances in the identification of transient species and in the measurement of fast kinetics (38, 39).

Pulsed Radiation Sources

Typical radiation sources are the Linear Electron Accelerator and the Van de Graaf Accelerator (see p. 26). A typical machine will have an electron energy of several Megaelectron volts, a pulse length somewhere between a microsecond and a nanosecond and a maximum pulse energy of about 10^{19} eV (Table 5.1). The University of Saskatchewan linac is illustrated in Figs. 5.6 and 5.7.

TABLE 5.1   *Some Linear Electron Accelerators Used for Pulse Radiolysis Work (38)*

Laboratory	Electron Energy (MeV)	Maximum Current (A)	Pulse Length (μsec)	Maximum, eV Pulse^{-1} ($\times 10^{-18}$)
1. Mount Vernon (Metropolitan Vickers Ltd.)	1.8	0.5	2	3
2. Paterson Labs (Assoc. Electrical Industries Ltd.)	4	0.1	0.2 or 2	1
3. DuPont (Varian Associates)	7	0.18	1.5–5	9
4. Argonne (Applied Radiation Corp.)	13	0.15	0.4–5	19
5. Ohio State University (Varian Associates)	4 4	0.7 0.32	0.006–0.05 1.6	11
6. University of Toronto (Vickers Eng.)	35 20	2 0.6	0.01 4	4.2 290
7. University of Saskatchewan (Varian Associates)	15–25	0.300	0.01–1.8	6

The increasing interest in radiation chemistry has prompted the introduction of machines such as the Febetron.

Febetron

The Febetron is a low cost machine. The model 706 delivers 12 J in 3 nsec with 600-keV electrons. The model 705 delivers 2-MeV electrons (40). It has been used for (*a*) emission spectroscopy, (*b*) instantaneous absorption

FIGURE 5.6 (a) *First accelerating section of linac with bending magnet to left of figure.* (b) *Irradiation cell and light system.* (*D. K. Storer, Ph.D. thesis, 1975, University of Saskatchewan.*)

FIGURE 5.7   (*a*) *Overall light system.*

spectroscopy using an internal Cerenkov light source, and (*c*) very fast kinetic studies using laser photometry for monitoring purposes (41) (Fig. 5.8).

Optical Detection

In one method, the absorption spectrum of the transient species is recorded on a spectrographic plate as a function of wavelength at a certain time after its formation (*spectrographic recording*). In the second method, the absorption of the transient at a given wavelength is recorded by a photo detector as a function of time [*spectrophotoelectric recording* (see Fig. 5.9)].

The radiation cell is commonly made of silica and may be of rectangular or circular cross section. High pressure cells of stainless steel, with windows of aluminum and silica are also used.

In a typical photographic experiment, the spectroscopic flash lamp is triggered to give a blank exposure in the absence of the transient. The linac

FIGURE 5.7 (b) Detail of light system.

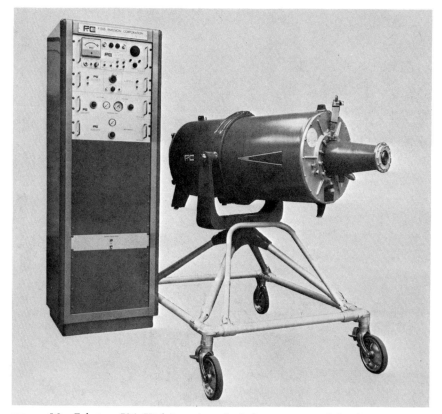

FIGURE 5.8   *Febetron 706. High intensity pulsed electron source. Pulse duration, 2.5 ns; electron beam energy per pulse, 10 J; equivalent electron energy, 500 keV.*

is then activated by a trigger pulse to produce a pulse of electrons which produces the transient. The same trigger pulse passes through a variable delay circuit and triggers off the spectroscopic flashlamp after a predetermined delay, producing a spectrum that now shows any absorption by the transient species. By varying the delay time, the change in absorption spectra of any transient or transients, can be observed. The flash will customarily be of about 1-μsec duration.

After the transient spectra have been recorded the formation and decay of the transient species can be measured using light of a suitable wavelength falling on a photodetector (Fig. 5.9). The photomultiplier signals are recorded using an oscilloscope. Figure 5.10 shows an oscilloscope trace of the formation and disappearance of e_{aq}^- in water, pH 12.6, $\lambda = 578$ nm (42).

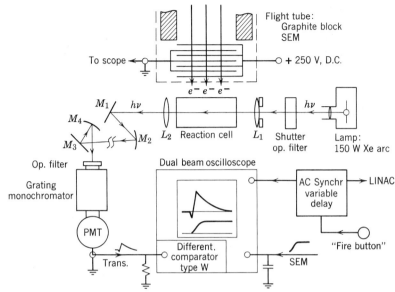

FIGURE 5.9   *Pulse Radiolysis Apparatus*; *Major Components and Their Characteristics.*
(1) Single beam, differential optical spectrophotometer. Wavelength response: ~ *200 to 700 nm (visible-near uv). Time response:* ~ *100 nsec. to minutes* (~ 10^{10}). *Absorbance range:* ~ *0.005 to 2* (~ *0.005 to 0.1). (2) Electron beam. Energy:* $15 - 20$ *MeV. Current:* ~ *300 mA. Duration:* ~ *10 nsec.* − *1.8 µsec. (3) Secondary electron emission monitor (SEM). (4) Light path 4 cm; cell diam.* ~ *2.2 cm.*

In the schematic diagram of an absorption trace (Fig. 5.11), TS represents the build up of the absorbing species, following a pulse at time T, and its decay SD. The initial intensity of the analysing light beam I_0 is measured by AC; at time t, the amount of light absorbed by the transient, I, is measured by RT. The absorbance or optical density $OD = \log \left[1/(1 - I/I_0) \right] = \varepsilon C l$ (Beers law), where ε is the extinction coefficient, C is the concentration of the absorbing species, and l is the length of the cell.

If the decay of the transient follows first-order kinetics, a plot of ln (OD) versus time should be linear, the negative of the slope being the first-order rate constant. If the decay is second order, a plot of (1/OD) versus time should be linear, the slope being $2k/\varepsilon l$, k being the second-order rate constant. More recently, computer programs have been developed which give k as a print out.

Design considerations for nanosecond pulse radiolysis studies are discussed by Hunt et al. (43).

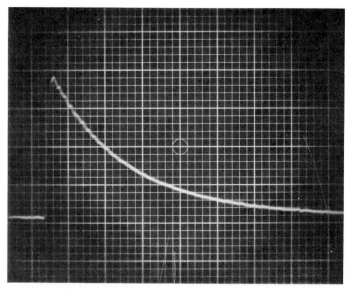

FIGURE 5.10 *Oscilloscope trace showing* e_{aq}^- *formation and disappearance in deaerated water.* 10^{-4} *M* $HClO_4$; *40-ns pulse, 35-MeV electrons,* $\sim 10^4$ *rads per pulse, horizontal scale, 40 ns per small division. Courtesy H. A. Gillis, N. V. Klassen, and G. G. Teather. N.R.C. Ottawa.*

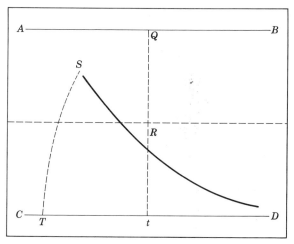

FIGURE 5.11 *Idealized schematic diagram of an absorption trace.*

Stroboscopic Pulse Radiolysis

The system just described has a time resolution limited to about 3 nsec by the length of the available radiation pulses and by the response time of optical detection systems. More recently the theoretical time resolution has been reduced to less than 20 psec by using the "fine structure" of the output from a 40-MeV linace to give extremely short (<10 psec) pulses of high energy electrons (44, 45). The pulses from a S band linac are obtained as a sequence of pulses spaced 350 psec apart (the period of the accelerating microwave field). One hundred such pulses are contained in a single 35-nsec linac output pulse.

Figure 5.12 (ref. 44, Fig. 2, p. 334) indicates the underlying principle of the *stroboscopic pulse radiolysis system.* When pulses of high energy electrons traverse a medium in which they exceed the speed of light in the medium, simultaneous Cerenkov light flashes are produced. The spectrum of the light flashes is continuous, and the flashes have the same time spread as the fine structure electron pulses. The Cerenkov light flashes are delayed for a period of time by a set of movable mirrors and are then transmitted through the irradiated sample where they serve to measure the concentration of the absorbing product (Fig. 5.13). After transmission through the sample cell these light flashes pass through the usual monochromator-photomultiplier detector system. The electron pulse also produces Cerenkov radiation from the cell. Three types of light pulse are used to derive an absorption signal.

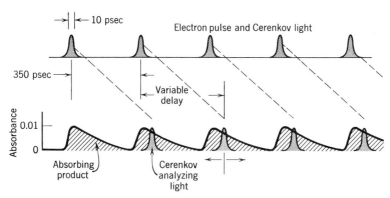

FIGURE 5.12   *Principle of the stroboscopic pulse radiolysis technique. The upper line shows the fine structure electron pulses and the simultaneous flashes of Cerenkov light. The lower line shows the production and rapid decay of an absorbing product following each fine structure pulse. The Cerenkov light flashes are delayed for a variable period and used as analyzing light flashes to measure the concentration of the absorbing product.*

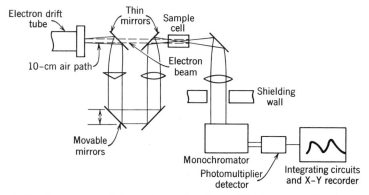

FIGURE 5.13   *Components of the stroboscopic pulse radiolysis system. The electron beam (broken lines) emerges from the drift tube, passes through a 10-cm air path and two thin mirrors, and irradiates the sample cell. The Cerenkov analyzing light (solid lines) is transmitted through a variable length optical system before passing through the irradiated sample. This light is focused through a monochromator and is detected by a photomultiplier. The subsequent electronic circuits derive a dc signal which is displayed on an X-Y recorder.*

They are produced sequentially by a synchronously rotating wheel on which six beam stoppers and six light choppers are suitably mounted (46). If the Cerenkov analyzing light signal is I_0, the Cerenkov signal due to the sample is I_1 and the absorption signal is S, the three signals A, B and C, in Fig. 5.14 will be A $= I_0 + I_1 +$ S, B $= I_0$, and C $= I_1$. Then S $=$ A $-$ (B $+$ C), and this can be obtained using a differential amplifier (44).

Summation over the train of delayed light flashes produces a light signal corresponding to a fixed point in time relative to the electron fine structure pulse. The detection system adds up the total number of photons received to provide a usable signal. By varying the delay of the light flashes, a sweep over the time interval between fine structure pulses may be obtained. This variable time delay is about 0.6 nsec, corresponding to 18-cm path length in air. The stroboscopic effect is thus produced by varying the phase difference between the Cerenkov light flashes and the fine structure electron pulses. Figure 5.15 gives an example of the decay of the hydrated electron in acidic solution, obtained by this method (47).

A recent paper (48) describes a modification of the stroboscopic pulse radiolysis system to allow determination of the *spectrum* of the transient species after 30 psec or several nanoseconds. In this modification, scattered Cerenkov radiation produced by the electron beam in the sample cell provides the light beam.

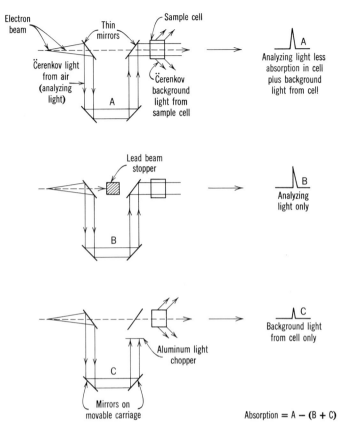

Three reference light signals from system

Signal observed

Electron beam

Thin mirrors

Sample cell

Čerenkov light from air (analyzing light)

A

Čerenkov background light from sample cell

A

Analyzing light less absorption in cell plus background light from cell

Lead beam stopper

B

B

Analyzing light only

Aluminum light chopper

C

Mirrors on movable carriage

C

Background light from cell only

Absorption = A − (B + C)

FIGURE 5.14   *The three types of light pulse used to derive an absorption signal. Both light and electron paths are open for signal A. Signal B consists of only Čerenkov analyzing light and signal C consists of the background Čerenkov light produced in the detection cell and neighbouring air path. These three signals are produced sequentially at a 60-Hz rate by a synchronously rotating wheel on which six beam stoppers and six light choppers are suitably mounted.*

350 psec

FIGURE 5.15   *Decay of the hydrated electron in perchloric acid solution as measured by stroboscopic detection system.*

201

Picosecond Spectroscopy Using Lasers

Lasers producing nanosecond pulses have been used for some time to study the lifetime of short-lived intermediates in organic reactions. Recently, the appearance of mode-locked lasers, with their picosecond resolution, has facilitated the study of processes occuring in the 10^{-10} to 10^{-12} sec range (49). It has been used in particular to measure the ultrafast relaxation process of an excess electron from the quasi-free state to the localized state in water. Quasi-free electrons generated by photo-ionization were localized within 2 psec. The "normal" absorption of the hydrated electron was observed in ~ 4 psec (50). It was suggested that the corresponding experiment using pulse radiolysis to generate the quasi-free electrons could be done (50).

The ordinary Fabry-Perot laser-cavity resonator permits oscillation in any one of a discrete number of longitudinal modes. In the case of a Nd^{3+} glass laser, mode locking is accomplished by inserting a dye absorber cell in the laser cavity. The dye bleaches in proportion to the intensity of the laser oscillation traversing it so that preferential transmission of high intensity oscillations occurs. After several passes, the lower intensity out of phase components are rejected, and oscillations build up in phase for the longitudinal modes. Eastman Kodak 9860 dye is a suitable absorber for a Nd^{3+} glass laser, producing picosecond pulses (see Fig. 5.16). The optimum resolution is obtained when a single pulse is used. This is accomplished by placing a Pockels cell energized by a spark gap charged to ~ 25 kV between crossed polarizers downstream from the oscillator (Fig. 5.17). High resolution time spectroscopy is accomplished using a stepped-delay echelon, fabricated from

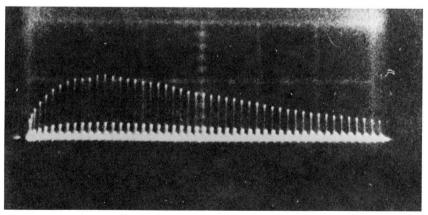

FIGURE 5.16   *Oscilloscope trace showing train of picosecond pulses from a mode-locked laser.*

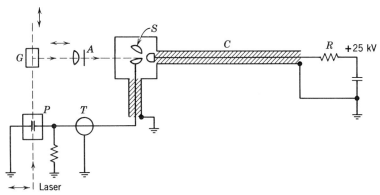

FIGURE 5.17 *Apparatus for selection of a single pulse from a mode-locked train. Initial pulses in the train are rejected by the cross polarizers and focused into a spark gap S. When the pulses become sufficiently intense, breakdown across the gap occurs, dumping a high voltage into the Pockels cell P. The polarization of the next pulse is rotated by 90°, thereby allowing its transmission through the second polarizer G.*

a stack of ordinary glass slides (Fig. 5.18). The delay between adjacent parts of the echelon is $t = d/c[(n^2 - \sin^2 \theta)^{1/2} - \cos \theta)]$; $n =$ refractive index, $d =$ slide thickness, and angle of incidence $= \theta$. For glass, 1 mm thick and $\theta = 45°$, $t = 3$ psec.

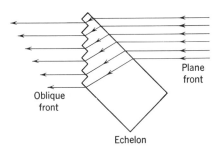

FIGURE 5.18 *Transmission echelon for obtaining stepped delays.*

The experimental arrangement for picosecond absorption spectroscopy applied to transient absorption of electrons produced from a solution of $K_4Fe(CN)_6$ in water is shown in Fig. 5.19. (For details, see ref. 49, p. 488). In this case the echelon divided the single continuum pulse into a set of pulses of identical duration and band width but separated by 2.1 psec. The

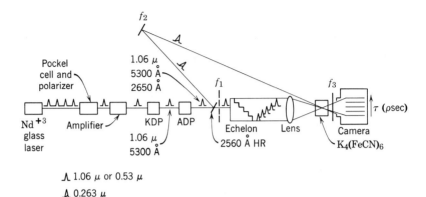

FIGURE 5.19   *Experimental arrangement for picosecond continuum absorption spectroscopy applied to transient absorption of electrons in water.*

results of the experiment indicated (*a*) that the infrared optical absorption spectrum of the excess electrons in water at 1.06 μ is observed within 2 psec after the appearance of the quasi-free electrons and (*b*) the "normal" absorption spectrum of the hydrated electron is developed after 4 psec. It is clear that the techniques of ultrafast picosecond spectroscopy utilizing mode-locked lasers show great promise for the study of electron localization in polar solvents and similar problems (see also refs. 51, 52, and 53).

A new "double beam" picosecond spectrometer has recently been described which improves the sensitivity of picosecond spectroscopy (54).

Laser magnetic resonance (LMR) offers very high sensitivity, for example, 2×10^8 molecules OH cm^{-3} as compared with 2×10^{12} molecules cm^{-3} with ESR. It has been used to detect rotational transitions of gaseous HO_2 at far infrared wavelengths. Approximate values of rotational constants have been derived from the spectra (54a). The transient free radicals OH, CH, and HCO have also been observed using LMR techniques. The technique has also been used in kinetic radical studies (54b) and could no doubt be readily applied in radiation kinetics.

Conductivity Detection

Since irradiation produces charged species in solution, measurement of changes in conductivity should provide information on the behavior of charged transient species. The conductivity

$$\sigma = \frac{F}{1000} \sum_i [i] Z_i \mu_i \qquad (5.48)$$

where F is the Faraday, $[i]$ is the molar concentration of the charged species, Z_i the charge on each ion i, and μ_i is the mobility of the ion i. A pulse will produce a change in conductivity $\Delta\sigma_t$

$$\Delta\sigma_t = \frac{F}{1000} \sum_i [\Delta i]_t Z_i \mu_i \qquad (5.49)$$

where $[\Delta i]_t$ is the change in concentration at time t (47).

The method suffers from the disadvantage that the observed conductivity is the summation for all the ions present in solution but under certain circumstances it can produce very useful information (3).

The ionization of dielectric liquids by high energy radiation, including pulsed radiation, has been widely studied. Mobilities, diffusion, and recombination coefficients have been calculated (55–57).

Electron Spin Measurement

Free radicals, which form an important species in radiolysis studies, are characterized by their EPR spectrum. Measurements of EPR absorption are thus of considerable importance. In a typical setup, the electron beam from a pulsed 3-MeV Van de Graaf accelerator was directed through an axial hole in the magnet on to the ESR cavity. A time resolution of about 2 μsec has been achieved (58–60).

Among a number of studies have been those of the pulse radiolysis of cyclopentene, cyclohexane, various alcohols, and the effect of scavengers (58). The rates of reaction of hydrogen and deuterium atoms, produced in water, with a variety of substances such as alcohols, phenol, and benzoic acid have also been measured (59). The unique ESR spectrum gives a direct method for making such measurements. Its use is particularly indicated where a prior identification of a radical is a problem or where several species with similar optical spectra are present (59).

There are two types of experiment, those in which the time-averaged radical concentration is examined and those in which both growth and decay are directly observed. The former is the radiolytic analog of the photochemical rotating sector experiment. In such an experiment the intermittent production of radicals is achieved by pulsing the electron beam. The on to off ratio is maintained constant, and the variation of average radical concentration with pulse repetition frequency is compared with calculated curves (61). As an example, the mean lifetime of ethyl radicals in liquid ethane at $-177°C$ is 7.8 msec. The rate constant for disappearance of ethyl radicals was $1.7 \times 10^8\ M^{-1}\ \text{sec}^{-1}$. (Fig. 5.20)

Direct observation of the growth and decay of radicals requires the averaging of a number of decay curves (to reduce the noise to signal ratio

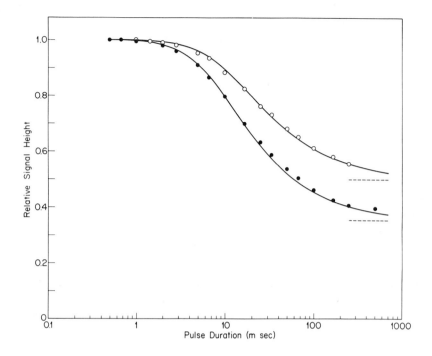

FIGURE 5.20   *Dependence of relative amplitude of the ethyl-radical signal on the pulse length during intermittent irradiation of liquid ethane. Data are given for duty cycles of 1 in 4($\circ$) and 1 in 8($\bullet$). Solid curves are calculated from rotating-sector theory on the assumption that radical disappearance is via a second-order process and that the mean radical lifetime under the conditions of the experiments is 7.8 msec.*

to a reasonable level). The averaging can be done in various ways, with a computer or electronically (see for example Fig. 5.21, from ref. 62 showing the growth and decay of the ethyl radical in liquid ethane).

Effect of Ionic Charge

The rate of reactions between ions will depend on both the ionic charge and on the dielectric constant; it is found, for example, that log k varies as the reciprocal of the dielectric constant (1). The effect of ionic charge has been widely investigated for aqueous solutions at 25°C.

According to the Brønsted-Bjerrum equation

$$\log \frac{k}{k_0} = 1.02 Z_A Z_B \frac{\mu^{1/2}}{1 + \mu^{1/2}}$$

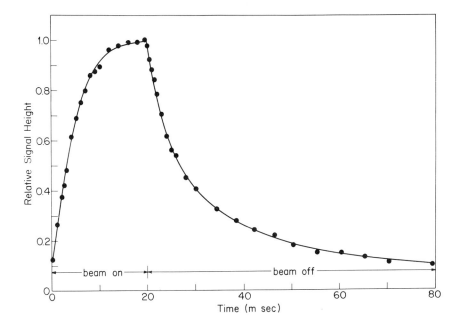

FIGURE 5.21 *Growth and decay of the ethyl-radical signal in liquid ethane, as observed by sampling the spectrometer output at various periods during the cycle of a repetitive experiment. The solid curves correspond to the calculated time dependence of the concentration on the assumption of a constant production rate during the period the beam is on and second-order disappearance during the entire cycle.*

where k and k_0 are the rate coefficients at ionic strengths μ and 0, respectively, and Z_A, Z_B are the algebraic numbers of the charges on the ions.

Pulse Radiolysis Studies of Effect of Ionic Charge

Studies on the effect of ionic strength using pulse radiolysis have been carried out on the reactions

$$e_{aq}^- + Fe(CN)_6^{3-} \rightarrow Fe(CN)_6^{4-} \tag{5.50}$$

and

$$e_{aq}^- + NO_3^- \rightarrow NO_3^{2-} \tag{5.51}$$

When $\log k/k_0$ is plotted against $\mu^{1/2}/(1 + \alpha\mu^{1/2})$, the expected slopes of 3 and 1 are obtained, consistent with the unit charge of the hydrated electron (63) (Fig. 5.22).

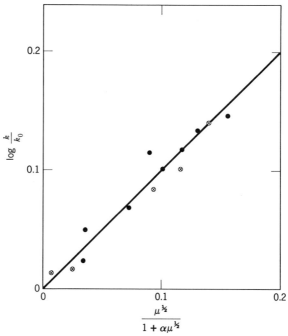

FIGURE 5.22   *Effect of ionic strength on the reactivity of NO_3^- and*

$[Co(CN)_5O_2Co(CN)_5]^{5-}$. •, NO_3^-: $\alpha = 1.33$; $k_0 = 8.5 \times 10^9\ M^{-1}sec^{-1}$;

$\otimes$, $[Co(CN)_5O_2Co(CN)_5]^{5-}$: $\alpha = 2.5$; $k_0 = 2.8 \times 10^9\ M^{-1}sec^{-1}$.

REFERENCES

1. K. J. Laidler, *Reaction Kinetics*, Pergamon Press, Oxford, 1963.
2. G. Hughes, *Radiation Chemistry*, Oxford Chemical Series, Clarendon Press, Oxford, 1973.
3. K. D. Asmus, *Int. J. Rad. Phys. Chem.*, **4**, 417 (1972).
4. G. Hughes and H. A. Makada, *Trans. Far. Soc.*, **64**, 6276 (1968).
4a. A. H. Samuel and J. L. Magee, *J. Chem. Phys.*, **21**, 1085 (1953).
4b. H. A. Schwarz, *J. Phys. Chem.*, **73**, 1928 (1969).
4c. A. Mozumder and J. L. Magee, *J. Chem. Phys.*, **47**, 939 (1967).
4d. A. Hummel, *J. Chem. Phys.*, **48**, 3268 (1968).
4e. G. R. Freeman, *Int. J. Radiat. Phys. Chem.*, **4**, 237 (1972).
5. C. B. Senvar and E. J. Hart, *Proc. 2nd Int. Conf. Peaceful Uses Atomic Energy*, United Nations, Geneva, **29**, 19 (1958).
6. T. J. Sworski, *J. Am. Chem. Soc.*, **76**, 4687 (1954); *Radiat. Res.*, **2**, 26 (1955).
7. H. A. Schwarz, *J. Am. Chem. Soc.*, **77**, 4960 (1955).
8. J. A. Ghormley and C. J. Hochanadel, *Radiat. Res.*, **3**, 227 (1955).

9. J. L. Magee, *J. Chim. Phys.*, **52**, 528 (1955); A. K. Ganguly and J. L. Magee, *J. Chem. Phys.*, **25**, 129 (1956).
10. M. Burton and K. C. Kurien, *J. Phys. Chem.*, **63**, 899 (1959).
11. A. Kuppermann, *Nucleonics*, **19**, 38 (1961 October); in *Actions Chimiques et Biologiques des Radiations* (ed. M. Haissinsky), Masson et Cie, Paris, 1961, Vol. 5, p. 85.
12. A. Kuppermann and G. G. Belford, *J. Chem. Phys.*, **36**, 1412 (1962).
13. F. S. Dainton, *Radiat. Res.*, *Suppl. 1.*, **1** (1959).
13a. A. J. Swallow, *Radiation Chemistry of Organic Compounds*, Pergamon Press, Oxford, 1960, pp. 55, 56.
14. C. Ferradini, in *Advances in Inorganic and Radiochemistry* (eds. H. J. Emeleus and A. G. Sharpe), Academic, New York, Vol. 3, 1961.
15. L. M. Dorfman, R. E. Bühler, and I. A. Taub, *J. Chem. Phys.*, **36**, 549, 3051 (1962).
16. R. W. Fessenden and R. H. Schuler, *J. Am. Chem. Soc.*, **79**, 273 (1957).
17. R. H. Schuler, *J. Phys. Chem.*, **62**, 37 (1958).
18. G. Meshitsuka and M. Burton, *Radiat. Res.*, **10**, 499 (1959).
19. R. R. Williams and W. H. Hamill, *J. Am. Chem. Soc.*, **72**, 1857 (1950).
20. L. H. Gevantman and R. R. Williams, *J. Phys. Chem.*, **56**, 569 (1952).
21. A. T. Fellows and R. H. Schuler, *J. Phys. Chem.*, **65**, 1451 (1961).
22. A. Chapiro, J. W. Boag, M. Ebert, and L. H. Gray, *J. Chim. Phys.*, **50**, 468 (1953).
23. L. Bouby and A. Chapiro, *J. Chim. Phys.*, **52**, 645 (1955); L. Bouby, A. Chapiro, and E. Chapiro, *ibid.*, **58**, 442 (1961).
24. K. Yang, *J. Phys. Chem.*, **65**, 42 (1961).
25. E. A. Cherniak, E. Collinson, F. S. Dainton, and G. M. Meaburn, *Proc. Chem. Soc.*, **54**, (1958).
26. G. E. Adams, J. H. Baxendale, and R. D. Sedgwick, *J. Phys. Chem.*, **63**, 854 (1959).
27. J. Kumamoto, H. E. De La Mare, and F. F. Rust, *J. Am. Chem. Soc.*, **82**, 1935 (1960).
28. R. H. Schuler, *J. Phys. Chem.*, **61**, 1472 (1957).
29. A. Chapiro, *J. Phys. Chem.*, **63**, 801 (1959).
30. R. A. Holroyd and G. W. Klein, *J. Am. Chem. Soc.*, **84**, 4000 (1962); *Int. J. Appl. Radiat. Isotop.*, **13**, 493 (1962).
31. A. Charlesby and D. G. Lloyd, *Proc. Roy. Soc. (London)*, *Ser. A*, **249**, 51 (1958); A. Charlesby, W. H. T. Davison, and D. G. Lloyd, *J. Phys. Chem.*, **63**, 970 (1959).
32. J. A. Ghormley, *Radiat. Res.*, **5**, 247 (1956).
33. R. F. Platford and J. W. T. Spinks, *Can. J. Chem.*, **37**, 1022 (1959).
34. G. R. Freeman, A. B. Van Cleave, and J. W. T. Spinks, *Can. J. Chem.*, **32**, 322 (1954).
35. R. W. Hummel, A. B. Van Cleave, and J. W. T. Spinks, *Can. J. Chem.*, **32**, 522 (1954).
36. R. G. McIntosh, R. L. Eager, and J. W. T. Spinks, *Science*, **131**, 992 (1960).
37. H. C. Sutton and J. Rotblat, *Nature*, **180**, 1332 (1957).
38. Max S. Matheson and Leon M. Dorfman, *Pulse Radiolysis*, MIT Press, Mass., 1969.
39. M. Ebert, J. P. Keene, A. J. Swallow, and J. H. Baxendale, *Pulse Radiolysis*, Academic, London, 1965.
40. Field Emission Corporation, McMinnville, Oregon, U.S.A.
41. G. A. Kenney-Wallace, E. A. Shaede, and D. C. Walker, *Int. J. Radiat. Phys. Chem.*, **4**, 209 (1972).
42. H. A. Gillis, personal communication.
43. J. W. Hunt, C. L. Greenstock, and M. J. Bronskil, *Int. J. Radiat. Phys. Chem.*, **4**, 87 (1972).
44. M. J. Bronskil, W. B. Taylor, R. K. Wolff, and J. W. Hunt, *Rev. Sci. Inst.*, **41** (3), 333 (1970).
45. R. K. Wolff, Thesis, University of Toronto, 1972.

46. J. Bronskil and J. W. Hunt, *J. Phys. Chem.*, **72**, 3762 (1968).
47. L. Dorfman, Pulse Radiation, in *Techniques of Chemistry* (ed. G. G. Hammes), Vol. 6, Wiley-Interscience, 1974, Chap. 2.
48. J. E. Aldrich, P. Foldvary, J. W. Hunt, W. B. Taylor, and R. K. Wolff, *Rev. Sci. Inst.*, **43** (7), 991 (1972).
49. T. L. Netzel, W. S. Struve, and P. M. Rentzepis, *Ann. Rev. Phys. Chem.*, **24**, 473 (1973).
50. P. M. Rentzepis, R. P. Jones, and J. Jortner, *J. Chem. Phys.*, **59**, 766 (1973).
51. P. M. Rentzepis, R. P. Jones, and J. Jortner, *Chem. Phys. Lett.*, **15**, 480 (1972).
52. P. M. Rentzepis, *Electrons in fluids*, (eds. J. Jortner and N. R. Kestner) Springer Verlag, Berlin, 1973, p. 103.
53. P. M. Rentzepis, Picosecond spectroscopy and molecular relaxation, *Advances in Chemical Physics*, Interscience-Wiley, New York, 1973, Vol. 23, p. 189.
54. T. L. Netzel and P. M. Rentzepis, Picosecond kinetics of photoionized tetracene dianions (to appear in *Chem. Phys. Lett.*, 1975).
54a. H. E. Radford, K. M. Evenson, and C. J. Howard, *J. Chem. Phys.*, **60**, 3178 (1974).
54b. C. J. Howard and K. M. Evenson, *J. Chem. Phys.*, **61**, 1943 (1974).
55. A. Hummel and W. F. Schmid, *Radiat. Res. Rev.*, **5**, 199 (1974).
56. J. P. Dodelet and G. R. Freeman, *Can. J. Chem.*, **50**, 2667 (1972).
57. K. H. Schmidt, *Int. J. Rad. Phys. Chem.*, **4**, 439 (1972).
58. R. W. Fessenden and R. H. Schuler, in *Advances in Radiation Chemistry* (eds. M. Burton and J. L. Magee), Vol. 2, Wiley-Interscience, New York, 1970, Chap. 1.
59. B. Smaller, E. C. Avery, and J. R. Remko, *J. Chem. Phys.*, **55**, 2414 (1971).
60. B. Smaller, J. R. Remko, and E. C. Avery, *J. Chem. Phys.*, **48**, 5174 (1968).
61. H. W. Meville and G. M. Burnett, *Technique of Organic Chemistry*, Vol. 8, Wiley-Interscience, New York, 1953, p. 138.
62. R. W. Fessenden, *J. Phys. Chem.*, **68**, 1508 (1964).
63. E. J. Hart and M. Anbar, *The Hydrated Electron*, Wiley-Interscience, New York, 1970, p. 55.

CHAPTER 6

Gases[1]

Values of W for the Noble Gases • Para-Ortho Hydrogen Conversion • Hydrogen–Deuterium Mixtures • Tritium Labeling • Oxygen • Oxygen–Nitrogen Mixtures • Nitrous Oxide • Ammonia • Carbon Dioxide • Hydrogen Bromide • Methane • Ethane • Ethylene • Acetylene • Addition to Unsaturated Hydrocarbons • Pulse Radiolysis of Gases • Attachment of Electrons to Gas Molecules • References •

GASES

From the point of view of radiation chemistry, gaseous systems are both simpler and more amendable to diverse experimental investigation than are liquids or solids. They are simpler because their lower density greatly reduces the effect of LET, e.g., α-particles and γ-rays produce practically the same product yields, and because the active species are not confined in the particle tracks where they are formed. Thus the overall effect of irradiation is to produce these species, which are initially positive ions, electrons, and excited atoms or molecules, distributed more or less homogeneously throughout the gas. This is in marked contrast to the situation in liquids or solids, where the active species are formed close together in spurs and are hindered in moving apart by the close-packed surrounding molecules. Experimentally,

[1] Vapors of compounds that are normally liquid at room temperature and pressure are considered later with the parent liquid.

211

gases can be investigated by techniques which often are not applicable to liquids or solids, or else are restricted in application with these condensed systems. Typical experimental techniques are: measurement of the ionization produced in the gas; mass spectrometry, which gives information about the ions formed and ion-molecule reactions; and the study of the photochemical or photosensitized decomposition of the gas which provides information about excited species that may be present, their reactions, and the reactions of free radicals formed from them.

One of the first effects of high-energy radiation to be observed and measured was the production of ionization in air. Similar relatively simple ionization measurements are still valuable and enable the number of ion pairs formed in a gas by irradiation to be determined and also, when the energy absorbed is known, the important quantity W, the mean energy absorbed per ion pair formed. Neither the number of ion pairs produced nor W can be measured directly for liquids or solids. Yields of products formed on irradiation are related to the number of ion pairs formed by the *ion-pair* or *ionic yield*, represented by M/N where M is the number of molecules of product formed and N the number of ion pairs (1). The use of M/N to express yields in gaseous systems was widespread during the earlier years of radiation chemistry, but fell into disuse when the emphasis in the subject shifted from gaseous to condensed systems, since with these N could not readily be determined. In condensed media, the G value has become the standard means of recording radiation yields, and this practice has spread to include gaseous systems (in fact most of the yields given in this chapter are in the form of G values). However, as Back and his colleagues (2) have pointed out, available methods for determining the energy absorbed in a gas, which is necessary if a G value is to be calculated, are far from satisfactory. Consequently they have made the reasonable suggestion that for gases, yields again be expressed in terms of M/N, and they describe a simple radiolysis vessel that may also be used to measure saturation ion currents. Where, or when, reliable values of W are available, ionic yields and G values are readily interconvertible (Eq. 1.1; footnote 4, p. 215).

The early radiation chemistry of gases was strongly influenced by recognition of the close relationship between ionization and product formation. Radiolysis mechanisms at this time were based largely on the formation and neutralization of ions, and the ion-cluster theory (cf. Chapter 1)[2] was advanced to account for ion-pair yields greater than unity. Then, in 1936, in two classical papers on the α-particle induced *para-ortho* hydrogen conversion and the synthesis and decomposition of hydrogen bromide respectively,

---

[2] A reappraisal of the clustering of ions in irradiated gases has been made by Magee and Funabashi (3).

Eyring, Hirschfelder, and Taylor (4, 5) proposed that excitation, as in photo-chemistry, was an important first step in radiation-initiated reactions and, furthermore, that both ionization and excitation might lead to free radicals and, in some cases, free-radical chain reactions. Thereafter free-radical mechanisms dominated radiation chemistry. This situation continued until fairly recently, when extensive mass spectrometric investigation of the properties and reactions of ions showed that ions also are of considerable significance in radiolysis reactions. In the following pages the radiolysis of a number of gaseous systems is described, and it will be apparent that all three species, ions, excited molecules and atoms, and free radicals are important intermediates.

It should perhaps be mentioned here that electric discharges in gases produce chemical effects similar to those produced by high-energy radiation (cf. 6). Furthermore, reactions taking place in the upper atmosphere are often similar to those studied in the radiation chemistry of gaseous systems, though the latter are normally at appreciably greater pressures. As would be anticipated from the composition of the atmosphere the reactions taking place in its upper levels generally involve oxygen and nitrogen, the nitrogen oxides, and, to a lesser extent, water vapor, carbon dioxide, etc.

Values of W for the Noble Gases

Working with ^{210}Po α-particles,[3] Jesse and Sadauskis (7) found that values of W for the noble gases were very sensitive to traces of impurities. For example W for very pure helium was found to be 41.3 eV/ion pair while for helium containing 0.13% argon W dropped to 29.7 eV/ion pair. Similarly W for pure neon (36.3 eV/ion pair) is reduced to 26.1 eV/ion pair by the addition of 0.12% of argon. The authors attributed these phenomena to the transfer of excitation energy from long-lived (metastable) excited states of helium and neon to argon, followed by ionization of the excited argon, e.g.,

$$\text{He}^* + \text{Ar} \rightarrow \text{He} + \text{Ar}^* \qquad (6.1)$$

$$\text{Ar}^* \rightarrow \text{Ar}^+ + e^- \qquad (6.2)$$

Thus in the presence of argon the energy associated with excited helium atoms, which normally is dissipated without producing ionization, can give

---

[3] α-Particles were, and still are, frequently used in studies on gaseous systems because the radiation energy can be absorbed within a reasonably small volume of gas. However, the yields obtained with different types of radiation and with radiations with differing LET's are generally very nearly the same in gaseous systems. In this chapter we have generally omitted details of the radiation involved and assumed that in gases all radiations produce identical effects.

rise to ions. Consequently W, the ratio of the energy absorbed to the number of ion pairs formed, is lowered. These systems afford an interesting example of energy transfer in gas-phase radiolysis.

The process represented by Eqs. 6.1 and 6.2 is only possible because the energy of an excited helium atom in its lowest excited state (19.8 V) is greater than the ionization potential of argon (15.8 V), though a favorable energy balance is no guarantee that energy transfer will in fact take place, since the transfer also requires that appropriate energy levels be present in the two reactants. Neon has no effect on W_{He}, since the ionization potential of neon (21.6 V) is greater than the energy available from the excited helium atom.

Similar effects have been found with other gas mixtures (8, 9), and in some cases it is necessary to postulate energy transfer from higher electronically excited states than the lowest to account for the results.

Para-Ortho *Hydrogen Conversion*

The conversion of *para*-hydrogen to *ortho*-hydrogen under the influence of α-particles was studied by Capron (10) who found it to be a chain reaction with a chain length between 700 and 1100. Eyring, Hirschfelder, and Taylor (4), combining mass spectral, thermal, and kinetic data and treating the reaction theoretically, showed that Capron's results could be accounted for quantitatively by the following mechanism.

INITIATION (the formation of hydrogen atoms)

$$H_2 \rightsquigarrow H_2^+ + e^- \tag{6.3}$$

$$H_2 \rightsquigarrow H_2^* \tag{6.4}$$

$$H_2^* \rightarrow 2H\cdot \tag{6.5}$$

$$H_2^+ + H_2 \rightarrow H_3^+ + H\cdot \tag{6.6}$$

$$H_3^+ + e^- \rightarrow 3H\cdot \tag{6.7}$$

and, to a smaller extent,

$$H_2^+ + e^- \rightarrow 2H\cdot \tag{6.8}$$

CHAIN PROPAGATION

$$H\cdot + p\text{-}H_2 \rightarrow o\text{-}H_2 + H\cdot \tag{6.9}$$

TERMINATION

$$H\cdot \rightarrow \tfrac{1}{2}H_2 \text{ (mainly at a surface)} \tag{6.10}$$

Hydrogen-Deuterium Mixtures

Very similar to the *para-* to *ortho*-hydrogen conversion is the exchange reaction between hydrogen and deuterium to give hydrogen deuteride,

$$H_2 + D_2 \rightarrow 2HD \qquad (6.11)$$

The α-particle induced exchange was studied by Mund and his collaborators (11) and shown to be a chain reaction with an ion-pair yield[4] of about 1000 at atmospheric pressure and rather less at lower pressures. A free radical mechanism was postulated with the chain propagating steps:

$$H\cdot + D_2 \rightarrow D\cdot + HD \qquad (6.12)$$

$$D\cdot + H_2 \rightarrow H\cdot + HD \qquad (6.13)$$

The chain reaction is inhibited by traces of oxygen, which reacts with hydrogen atoms:

$$H\cdot + O_2 \rightarrow HO_2\cdot \qquad (6.14)$$

By carefully removing both oxygen and other inhibitors (e.g., stopcock grease) Dorfman and Shipko (12) were able to increase the exchange yield considerably and found, for example, an ion-pair yield of about 6×10^4 at 276 mm Hg pressure. Physical processes such as charge transfer can also inhibit the exchange reaction. Thompson and Schaeffer (13, 14) showed that the addition of small amounts of krypton and particularly xenon, both of which have ionization potentials lower than those of possible ionic chain carriers (H^+, H_2^+, H_3^+) markedly reduced the yield of hydrogen deuteride. Moderate amounts (less than 2%) of helium, neon, or argon have little effect on the reaction, for which Thompson and Schaeffer (13) found ion-pair yields of 18,000 at 100 mm Hg pressure. These authors also showed that none of the noble gases has any effect on the thermal or mercury photosensitized exchange reactions, which are known to be free radical in character, and interpreted these results to mean that the high yields obtained in the radiation-initiated process are due to an ionic rather than a free radical chain reaction. The mechanism proposed by Schaeffer and Thompson (14) is as follows.

INITIATION

$$H_2 \rightsquigarrow H_2^+ + e^- \qquad (6.3)$$

$$H_2^+ + H_2 \rightarrow H_3^+ + H\cdot \qquad (6.6)$$

---

[4] Both ion-pair yields and G values are used in this chapter. They are related by the expression: $G = M/N \times 100/W$ or, roughly, $G = 3M/N$.

PROPAGATION

$$H_3^+ + D_2 \rightarrow HD_2^+ + H_2 \qquad (6.15)$$

$$HD_2^+ + H_2 \rightarrow H_2D^+ + HD \qquad (6.16)$$

$$H_2D^+ + D_2 \rightarrow HD_2^+ + HD \quad \text{etc.} \qquad (6.17)$$

Termination is by neutralization of the positive ions, either by free electrons or at a surface, while inhibition by krypton or xenon involves proton transfer from the chain-propagating positive ions, e.g.,

$$H_3^+ + Xe \rightarrow XeH^+ + H_2 \qquad (6.18)$$

Lind (15) has suggested that the very high exchange yields (M/N of the order of 10^4), which are reduced in the presence of krypton and xenon, are due to the ionic chain reaction and the smaller yields (M/N of the order 100 to 1000), found with less pure gases or in the presence of the inhibiting noble gases, are due to a radical chain reaction (Eqs. 6.12 and 6.13), as are the thermal and photosensitized exchange reactions.

Tritium Labeling

Exposure of organic compounds to tritium gas often leads to replacement of part of the hydrogen present in the compound by tritium, so that the procedure can frequently be used to obtain compounds labeled with tritium (16–18). The tritium is not necessarily uniformly distributed in the labeled compound and the susceptibility of compounds to labeling in this way varies markedly, and not always predictably, with their chemical structure. Incorporation of tritium into a compound can often be accelerated by simultaneous exposure to ultraviolet light, ionizing radiation, or an electric discharge; but a competing process is decomposition of the compound by the tritium β-radiation.

The mechanism of the tritium labeling process is not yet certain, but it may involve recoil tritium ions, T^+, or the ion $^3HeT^+$, formed by the β-decay of tritium,

$$T_2 \rightarrow T^+ + {}^3He + e^- \qquad (\text{or } {}^3HeT^+ + e^-) \qquad (6.19)$$

and ion-molecule reactions, e.g.,

$$T^+ + \text{compound} \rightarrow \text{tritiated products} \qquad (6.20)$$

Alternatively tritiation of the compound may be induced by the β-radiation emitted,

$$\text{compound} \xrightarrow{\beta} \text{ions, excited molecules,} \xrightarrow{T_2} \text{tritiated products}$$
$$\text{and (?) radicals} \qquad (6.21)$$

(The mechanism is discussed in refs. 17, 19, and 20, among others.)

Riesz and Wilzbach (17) showed that when *n*-hexane, cyclohexane, and benzene are irradiated by mixing with tritium gas, the radiolysis products are labeled, and they suggested that exposure to tritium offered both a source of radiation and a means (by isotope dilution) of identifying the radiolysis products.

Oxygen

Undoubtedly the earliest recorded reference to a reaction typical of radiation chemistry is that of Homer to the sulfurous smell accompanying lightning. A similar observation was made many years later of air exposed to salts of the then recently discovered element, radium. In both cases the substance chiefly responsible is ozone, whose characteristic odor has more recently come to be associated with high-voltage electrical equipment; it is normally prepared by passing oxygen through a silent electric discharge.

Experimental work on the irradiation of oxygen was first carried out by Lind (21). Lind measured the yield of ozone formed in a slow stream of oxygen exposed to α-radiation and found ion-pair yields of about 0.5, though these were poorly reproducible. Subsequent work by Lind and Bardwell (22) and D'Olieslager (23) showed that in a static system the yield of ozone is very small and that in a flowing system it depends upon the rate of flow and, inversely, on the dose rate; ion-pair yields as high as 2 to 2.5 were obtained by Lind and Bardwell at high flow rates.[5] These observations are explained by the existence of an efficient back reaction which destroys ozone, and which therefore competes with the process producing ozone. The back reaction is in fact a chain reaction with a high ion-pair yield (e.g., 15,000 at 300 mm Hg pressure of pure ozone (25)), though the forward process forming ozone is not. Magee and Burton (26) have suggested a mechanism for the formation of ozone in irradiated oxygen:

$$O_2 \rightsquigarrow O_2^+ + e^- \qquad (6.22)$$

$$\rightsquigarrow O_2^* \qquad (6.23)$$

$$O_2 + e^- \rightarrow O_2^- \qquad (6.24)$$

followed by one or more of the reactions,

$$O_2^+ + O_2^- \rightarrow O + O_3 \qquad (6.25)$$

or

$$\rightarrow O_2^* + O_2 \qquad (6.26)$$

[5] A summary of radiation-induced yields of ozone for various types of radiation is given by Kircher et al. (24). For gaseous oxygen, ion-pair yields for the formation of ozone are mainly in the range 1 to 3.

or

$$\rightarrow 2O + O_2 \qquad (6.27)$$

(one or both oxygen atoms will be excited)

and

$$O + O_2 + M \rightarrow O_3 + M \qquad (6.28)$$

$$O_2{}^* \rightarrow 2O \qquad (6.29)$$

$$O_2{}^* + O_2 \rightarrow O + O_3 \qquad (6.30)$$

$$O + O_3 \rightarrow 2O_2 \qquad (6.31)$$

(in pure oxygen $M = O_2$). The decomposition of ozone becomes a chain reaction if reaction 6.31 is replaced in part by

$$O + O_3 \rightarrow 2O + O_2 \qquad (6.32)$$

Ozone is also formed when liquid oxygen or liquid mixtures of oxygen with nitrogen or the noble gases are irradiated (24, 27); $G(O_3) = 6$ for cobalt-60 γ-irradiation of liquid oxygen. This may present a hazard when samples are simultaneously cooled in liquid nitrogen and irradiated if any oxygen (b.p. $-183°C$) has condensed into the nitrogen (b.p. $-196°C$). Ozone has a higher boiling point than nitrogen (ozone boils at $-112°C$) and any formed on irradiation remains as the liquid evaporates. The liquid ozone can cause a dangerous explosion if it comes into contact with organic matter.

Oxygen–Nitrogen Mixtures

When mixtures of oxygen and nitrogen are irradiated, ozone will be formed as described above, but oxides of nitrogen will also be produced. The formation of nitrogen oxides and, in the presence of moisture, nitric acid in this way can create a corrosion and a health problem near nuclear reactors and particle accelerators where air is exposed to high radiation doses. However, the same reaction also presents the possibility of converting atmospheric nitrogen into a commercially useful form (e.g., nitrate fertilizers) and this possibility has been explored in a number of laboratories (28).

The radiolysis of oxygen–nitrogen systems is complex and changes as the concentration of products builds up. However Harteck and Dondes (28, 29) believe that initially the dominant reactions are

$$N_2 \rightsquigarrow 2N \qquad (G \text{ value } 4-5) \qquad (6.33)$$

$$O_2 \rightsquigarrow 2O \qquad (G \text{ value } 8) \qquad (6.34)$$

$$N + O_2 \rightarrow O + NO \qquad (6.35)$$

$$2NO + O_2 \rightarrow 2NO_2 \qquad (6.36)$$

$$N + NO_2 \rightarrow 2NO \qquad (6.37)$$

or

$$\rightarrow O + N_2O \qquad (6.38)$$

or

$$\rightarrow 2O + N_2 \qquad (6.39)$$

$$N + NO \rightarrow O + N_2 \qquad (6.40)$$

and, illustrating the many possible ionic reactions,

$$NO + M^+ \rightarrow NO^+ + M \qquad (6.41)$$

$$NO^+ + e^- \rightarrow N + O \qquad (6.42)$$

or

$$\rightarrow NO + h\nu \qquad (6.43)$$

(reactions involving ozone are omitted). Pshezhetsky and Dmitriev (30) have given a rather different group of reactions which they consider predominate:

$$N_2^+ + O_2 \rightarrow NO^+ + NO \qquad (6.44)$$

$$(N_2^+)^* + O_2 \rightarrow NO_2^+ + N \qquad (6.45)$$

$$N + O_2 \rightarrow O + NO \qquad (6.46)$$

$$N + O_2 + M \rightarrow NO_2 + M \qquad (6.47)$$

Regardless of the detailed mechanism, an equilibrium is finally reached in dry gas mixtures between the formation and decomposition of the nitrogen oxides. When moisture is present the main product is nitric acid, which is formed until the water vapor is exhausted (31).

Nitrous Oxide

Harteck and Dondes (32; cf. 33) have suggested that the radiation-induced decomposition of nitrous oxide be used as a gas-phase chemical dosimeter. The decomposition products are nitrogen, oxygen, and nitrogen dioxide, and in the range 5×10^4 to 10^7 R the combined yield of nitrogen and oxygen is linear with dose; the dosimeter can be used to measure doses up to almost 10^{10} R by means of calibration curves. In the upper part of the dose range (3×10^7 to 3×10^9 R), sufficient nitrogen dioxide is formed to measure colorimetrically. This may be done without opening the irradiation vessel and offers a convenient alternative to measuring the amount of nitrogen and oxygen formed. Advantages of this dosimeter include the high-dose range which it covers, the stability of both unirradiated and irradiated dosimeter systems, independence of temperature [the dosimeter can be used at temperatures ranging from -80 to $+200°C$), dose rate, and LET (β and γ-radiation and fission products (thermal neutrons + uranium-235 oxide) were tested by Harteck and Dondes]. The authors list the main reactions taking place as

$$\left.\begin{array}{l} N_2O \rightsquigarrow N_2O^+ + e^- \\[6pt] \rightsquigarrow N_2O^* \end{array}\right\} \rightarrow \begin{array}{ll} N_2 + O & \text{(about 80\%)} \\[6pt] N + NO & \text{(about 20\%)} \end{array} \tag{6.48}$$

or

followed by reactions 6.36 to 6.39 and

$$N_2O^+ + NO \rightarrow N_2O + NO^+ \tag{6.49}$$

$$NO^+ + e^- \rightarrow N + O \tag{6.42}$$

or

$$\rightarrow NO + h\nu \tag{6.43}$$

A mechanism based largely upon ionic reactions has been proposed by Burtt and Kircher (34).

With nitrous oxide pressures above 500 mm Hg Harteck and Dondes found the ion-pair yield to be 4.2 ± 0.1 molecules N_2O decomposed per ion pair formed, corresponding to $G(-N_2O) = 12$; the ratio of the products was $N_2:O_2:NO_2 = 1:0.14:0.48$. Burtt and Kircher (34) found an ion-pair yield ($-N_2O$) of 4.85 at both 200 and 555 mm Hg pressure which increased to 6.87 as the pressure was decreased to 50 mm Hg. At lower pressures, therefore, the experimental results appear inconsistent and, for dosimetric purposes, pressures of 500 mm Hg or more should be used; the dosimeter is independent of pressure above 500 mm Hg (32).

A review paper by Johnson (35) suggests a $G(N_2)$ value of 10.0 at room temperature (290 to 300°K) and pressures in the region of 760 mm (see also ref. 36 p. 312, refs. 36a and 36b).

Ammonia

Both the synthesis of ammonia from its elements and its decomposition under the influence of radiation have been studied extensively since early work on these systems by Cameron and Ramsay (37) and Usher (38).

The radiation-induced decomposition of ammonia produces nitrogen, hydrogen, and hydrazine (N_2H_4) with $G(-NH_3)$ about 3 to 4; the same products are formed by photochemical decomposition (39) and by subjecting ammonia to an electric discharge (40). The mechanism of the photochemical decomposition is probably as follows, though the reaction producing nitrogen is uncertain:

$$NH_3 + hv \rightarrow NH_3* \tag{6.50}$$

$$NH_3* + M \rightarrow NH_3 + M \tag{6.51}$$

$$NH_3* \rightleftharpoons \cdot NH_2 + H\cdot \tag{6.52}$$

$$2\cdot NH_2 + M \rightarrow N_2H_4 + M \tag{6.53}$$

$$2\cdot NH_2 \rightarrow N_2 + 2H_2 \tag{6.54}$$

$$H\cdot \rightarrow \tfrac{1}{2}H_2 \quad \text{(at wall)} \tag{6.10}$$

Reactions involving hydrazine will occur once this has reached a sufficient concentration; these may be

$$H\cdot + N_2H_4 \rightarrow \cdot NH_2 + NH_3 \tag{6.55}$$

or

$$\cdot NH_2 + N_2H_4 \rightarrow N_2H_3\cdot + NH_3 \tag{6.56}$$

and

$$2N_2H_3\cdot \rightarrow N_2 + 2NH_3 \tag{6.57}$$

In addition to these reactions the radiolysis mechanism (41) will include steps involving ions. The mass spectrum of ammonia (42) suggests that these will probably be mainly NH_3^+ and NH_2^+. Dorfman and Noble (43), also by mass spectrometry, have shown that NH_3^+ undergoes an efficient ion-molecule reaction,

$$NH_3^+ + NH_3 \rightarrow \cdot NH_2 + NH_4^+ \tag{6.58}$$

The NH_2^+ ion probably reacts by charge transfer,

$$NH_2^+ + NH_3 \rightarrow \cdot NH_2 + NH_3^+ \tag{6.59}$$

while the NH_4^+ ions, and possibly some of the other positive ions, are neutralized by electrons, partly in the gas phase and partly at the wall of the irradiation vessel,

$$NH_4^+ + e^- \rightarrow H\cdot + NH_3 \tag{6.60}$$

$$NH_3^+ + e^- \rightarrow NH_3 \quad \text{(at wall)} \tag{6.61}$$

or

$$\rightarrow \cdot NH_2 + H\cdot \tag{6.62}$$

Essex and his colleagues (44, 45) have irradiated ammonia in the presence of an electric field to estimate to what extent ion-recombination in the gas contributes to the overall reaction. The apparatus used is represented diagrammatically in Fig. 6.1. Two metal disc electrodes are sealed at either end

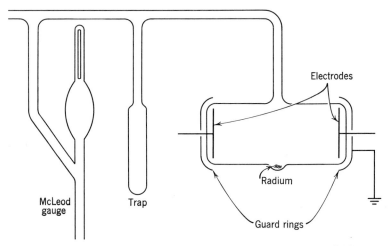

FIGURE 6.1   *Irradiation of ammonia in presence of an electric field.*

of a glass cylinder containing the gas under examination, and the gas is irradiated with α-particles from a radium preparation placed in a depression in the wall of the cylinder. The cylinder is connected to a trap, which can be cooled in liquid nitrogen to condense out the original gas after irradiation, and to a McLeod gauge where the (noncondensable) product gases can be measured. When irradiations are carried out with a potential applied between the electrodes, some or all of the ions formed are collected on the electrodes and neutralized there rather than in the gas itself, the proportion of ions collected depending on the potential applied. By using a sufficiently high potential all the ions formed can be collected and hence their number determined; the current passing between the electrodes under these conditions is known as the *saturation current*. Once the number of ions formed with a given gas and α-particle source is known, ion-pair yields can be determined without, and for various values of, the applied field and hence the effect of the electric field on the radiolysis determined. For ammonia it was estimated that at both 30 and 100°C the ion-pair yield for the decomposition process is reduced by about 30% by applying a field high enough to produce saturation. This suggests that when gaseous ammonia is irradiated, about 30% of the reaction results from an ion-recombination reaction (e.g., 6.62) which does not occur when the ions are neutralized at an electrode. Other gases that have been irradiated in the presence of an electric field by this technique include nitrous oxide, ethane, and azomethane (45). Woodward and Back (46), in similar experiments, have irradiated ethane, propane, and the butanes with γ-radiation. These authors also observed a dose-rate effect in the irradiated gases (in the absence of the electric field) which they attribute to competition between neutralization of the ions in the gas phase and diffusion of the ions to the vessel wall; high dose rates favoring the former process, and low dose rates the latter.

The radiation-induced decomposition of liquid ammonia has also been studied. For cobalt-60 γ-irradiation, Cleaver, Collinson, and Dainton (47) found the yields of nitrogen, hydrogen, and hydrazine formed in the early stages of the irradiation to be: $G(N_2) = 0.22$, $G(H_2) = 0.81$, and $G(N_2H_4) = 0.13$. The mechanism is probably similar to that postulated for gaseous ammonia with the exception that excited ammonia is unimportant in the liquid phase because of the Franck-Rabinowitch effect (recombination of the dissociation products from an excited molecule as in Eq. 6.52 and deactivation of the excited product, Eq. 6.51). The radiolysis of liquid ammonia is of special interest, since the liquid has many properties in common with water. Theories concerning the radiolysis of water can therefore be tested to some extent by applying them to the radiolysis of liquid ammonia. For a detailed review of the radiolysis of NH_3, see Peterson (48).

Carbon Dioxide

Gaseous carbon dioxide is practically unaffected by ionizing radiation (49) whereas liquid or solid carbon dioxide decomposes to give carbon monoxide and oxygen with a $G(-CO_2)$ value of about 5 (28, 50). Hirschfelder and Taylor (51) attributed the stability of carbon dioxide in the gas phase (below pressures of about 10 atm) to a back reaction which they postulated to be

$$CO + O_3 \rightarrow CO_2 + O_2 \qquad (6.63)$$

However, although this reaction is exothermic, it has an activation energy of over 28 kcal (52) so that it would be very slow at room temperature. Harteck and Dondes (28, 52) proposed instead the following mechanism for the radiolysis of pure carbon dioxide:

$$CO_2 \rightsquigarrow CO + O \quad (G \text{ value } 8)^6 \qquad (6.64)$$

and, to a minor extent,

$$CO_2 \rightsquigarrow C + 2O \qquad (6.65)$$

then,

$$CO + C + M \rightarrow C_2O + M \qquad (6.66)$$

$$C_2O + CO + M \rightarrow C_3O_2 + M \qquad (6.67)$$

$$O + O_2 + M \rightarrow O_3 + M \qquad (6.68)$$

$$C_3O_2 + O \rightarrow C_2O + CO_2 \qquad (6.69)$$

$$C_3O_2 + O_3 \rightarrow C_2O + CO_2 + O_2 \qquad (6.70)$$

$$CO + O + M \rightarrow CO_2 + M \qquad (6.71)$$

The back reaction between carbon monoxide and oxygen atoms or ozone is catalyzed here by C_2O. Part of the C_2O and C_3O_2 diffuse to the walls of the vessel and form polymerization products; as a result a small surplus of oxygen remains after irradiation. In the liquid or solid phase, only reactions 6.64 and 6.65 occur, the oxygen atoms combining in pairs to form molecular oxygen or with carbon atoms to form carbon monoxide,

$$2O \rightarrow O_2 \qquad (6.72)$$

$$C + O \rightarrow CO \qquad (6.73)$$

Small amounts of nitrogen dioxide inhibit the back reaction to reform carbon dioxide by reacting rapidly with both oxygen and carbon atoms (28, 52), and $G(-CO_2)$ in the gas in this case is about 8[6]. The reactions of the inhibitor are

$$O + NO_2 \rightarrow NO + O_2 \qquad (6.74)$$

$$C + NO_2 \rightarrow NO + CO \qquad (6.75)$$

and

$$2NO + O_2 \rightarrow 2NO_2 \qquad (6.36)$$

Irradiation of carbon monoxide gives carbon dioxide, carbon suboxide polymers $(C_3O_2)_n$, and carbon, which can be accounted for by a mechanism which is consistent with the reactions (6.64 to 6.71) postulated for the radiolysis of carbon dioxide (52, 54),

$$CO \rightsquigarrow C + O \qquad (6.76)$$

$$CO + C + M \rightarrow C_2O + M \qquad (6.66)$$

$$C_2O + CO + M \rightarrow C_3O_2 + M \qquad (6.67)$$

$$C_3O_2 + O \rightarrow C_2O + CO_2 \qquad (6.68)$$

$$C \rightarrow graphite \qquad (at\ wall) \qquad (6.77)$$

$$C_2O, C_3O_2 \rightarrow polymerization\ products \qquad (at\ wall) \qquad (6.78)$$

The radiation-induced decomposition of carbon dioxide is of practical significance in reactor technology, since carbon dioxide is used as a heat-transfer agent in a number of power-producing reactors. In a graphite

[6] Another determination of the carbon dioxide yield in the presence of a radical scavenger (nitrogen dioxide or sulfur dioxide) has given a value of $G(CO) = 3.51 \pm 0.23$ for γ-radiation, and the same value, within the limits of the experimental accuracy, for reactor radiation (53). The steady state concentration of carbon monoxide produced by γ-radiation when no scavenger is present is given as less than 0.01% at a dose rate of about 500 rads min^{-1}.

moderated reactor cooled by circulating carbon dioxide the reaction

$$CO_2 + \text{graphite} \rightleftharpoons 2CO \qquad (6.79)$$

is also relevant, though apparently not a major problem with present types of reactors (55). Under reactor conditions this reaction appears to be initiated by active intermediates (possibly O or O_3) formed by irradiation of the gas.

Hydrogen Bromide

The synthesis and decomposition of hydrogen bromide,

$$H_2 + Br_2 \rightleftharpoons 2HBr \qquad (6.80)$$

are reactions that have been studied since the early days of radiation chemistry. Lind (1, 56) observed the combination of hydrogen and bromine under the influence of α-radiation ($M_{\text{HBr}}/N = 0.54$) but failed to find any decomposition of gaseous hydrogen bromide, though liquid hydrogen bromide was decomposed with a yield calculated later to be about $M_{-\text{HBr}}/N = 2.2$. Later studies, including further work by Lind (57), showed that gaseous hydrogen bromide can be decomposed by radiation, and the data for both synthesis and decomposition were used by Eyring, Hirschfelder, and Taylor (5) to test their theory of radiation-induced action. The mechanism proposed by these authors for the synthesis of hydrogen bromide gave results in satisfactory agreement with the experimental results of Lind and Livingston (57); hydrogen was assumed to give hydrogen atoms by the series of reactions given earlier (6.3 to 6.8) and bromine to give bromine atoms by a similar series of reactions and also by steps which are made possible by the greater electron affinity of bromine, i.e.,

$$Br_2 \cdot + e^- \rightarrow Br \cdot + Br^- \qquad (6.81)$$

$$(Br^+ + Br_2 \rightarrow Br_3^+) \qquad (6.82)$$

and

$$Br^- + Br_3^+ \rightarrow 4Br \cdot \qquad (6.83)$$

Bromine atoms will not react appreciably with hydrogen or hydrogen bromide at room temperature because of the high energy requirements and the subsequent reactions of the hydrogen and bromine atoms will be

$$H \cdot + Br_2 \rightarrow Br \cdot + HBr \qquad (6.84)$$

$$2Br \cdot + M \rightarrow Br_2 + M \qquad (6.85)$$

The mechanism postulated by Eyring, Hirschfelder, and Taylor (5) for the decomposition of hydrogen bromide was

$$HBr \rightsquigarrow HBr^+ + e^- \qquad (6.86)$$

$$\rightsquigarrow HBr^* \qquad (6.87)$$

$$HBr^* \rightarrow H\cdot + Br\cdot \qquad (6.88)$$

$$HBr + e^- \rightarrow H\cdot + Br^- \qquad (6.89)$$

$$HBr^+ + Br^- \rightarrow H\cdot + 2Br\cdot \qquad (6.90)$$

followed by

$$H\cdot + HBr \rightarrow Br\cdot + H_2 \qquad (6.91)$$

$$2Br\cdot + M \rightarrow Br_2 + M \qquad (6.85)$$

They also considered the possibility of formation of the ion H_2Br^+,

$$HBr^+ + HBr \rightarrow H_2Br^+ + Br\cdot \qquad (6.92)$$

and its neutralization,

$$H_2Br^+ + Br^- \rightarrow 2H\cdot + 2Br\cdot \qquad (6.93)$$

However, Zubler, Hamill, and Williams (58) consider the two latter reactions unimportant, since with reactions 6.86, 6.89, 6.91, and 6.85 they lead to an ion-pair yield for HBr decomposition of 6.0; the authors find ion-pair yields between 4.0 and 5, and therefore favor reactions 6.89 and 6.90 with a smaller contribution from excited hydrogen bromide (Eq. 6.88).[7]

Zubler, Hamill, and Williams (58) also irradiated mixtures of hydrogen bromide with a much larger amount of a noble gas (argon, krypton, or xenon) and found ion-pair yields, M_{-HBr}/N, between 4.0 and 4.7.

In contrast with the synthesis of hydrogen bromide, which gives low ion-pair yields, the yields for the synthesis of hydrogen chloride are high (e.g., 10^3 to 10^5); indicative of a chain reaction. The difference arises from the greater reactivity of chlorine atoms as compared to bromine atoms. Thus atomic chlorine reacts with hydrogen at room temperature whereas atomic bromine will only react at relatively high temperatures (e.g., 200 to 300°C), and a chlorine-hydrogen chain reaction is possible with the propagation steps,

$$Cl\cdot + H_2 \rightarrow H\cdot + HCl \qquad (6.94)$$

$$H\cdot + Cl_2 \rightarrow Cl\cdot + HCl \qquad (6.95)$$

---

[7] Nevertheless, the ion-molecule reaction depicted by Eq. 6.92 has been observed experimentally (59), and it seems possible that it occurs and is followed by the neutralization reaction

$$H_2Br^+ + Br^- \rightarrow H_2 + 2Br\cdot$$

rather than reaction 6.93. The sequence of reactions proposed by Eyring, Hirschfelder, and Taylor then gives an ion-pair yield of 4, neglecting the contribution by excited molecules, which apparently do not all decompose.

Methane

The radiolysis of methane has been studied extensively, and the mechanism is reasonably well understood, though not yet established beyond contention. Information derived from the radiolysis of methane and its mixtures (cf. Table 6.1) is well supported by pertinent data from mass spectroscopy and photochemistry.

T A B L E 6.1   *Radiolysis of Mixtures Containing Methane (Fast Electron or γ-Irradiation)*

	G(product)		Relative Yields		
Product	Pure CH_4 (ref. 60)	CH_4 + NO (ref. 60)	CH_4 + Ar (ref. 61)	CH_4 + Ar + I_2 (ref. 61)	CH_4 + I_2 (ref. 62)
H_2	6.4	3.6	100	100	
C_2H_2			0	1.6	
C_2H_4	0.13	0.64	0.5	14	
C_2H_6	2.1	0.32	45	0.6	
C_3H_6	0	0.03	0.2	0	
C_3H_8	0.26	0.01	7	0	
n-C_4H_{10}	0.13	0	3.5	0	
i-C_4H_{10}	0.06	0			
i-C_5H_{10}	0.05	0			
Liquid polymer, $(CH_2)_n$					
HI					8.5
CH_3I				34	70
C_2H_5I				18	4.5
C_3H_7I				0.5	
C_4H_9I				0.1	
CH_2I_2				9	8.5

In the mass spectrometer the relative abundances of the ions formed from methane are (63): CH_4^+, 48%; CH_3^+, 40%; CH_2^+, 8%; CH^+, 4%; C^+, 1%. These values are for electrons with energies in the range 50 to 70 eV, but they are probably reasonably typical of high-energy radiation also. Other mass-spectrometric experiments with methane, at rather higher pressures, have demonstrated the existence of two rapid ion-molecule reactions involving the major primary ions (61, 64–66),

$$CH_4^+ + CH_4 \rightarrow CH_5^+ + \cdot CH_3 \qquad (6.96)$$

$$CH_3^+ + CH_4 \rightarrow C_2H_5^+ + H_2 \qquad (6.97)$$

At very much higher ionization chamber pressures, the secondary ions produced in this way predominate in the mass spectrum of methane. At

0.2 mm Hg, for example, Wexler and Jesse (67) find that CH_5^+ and $C_2H_5^+$ account for about 70% of the ions collected. The parent ion and its dissociation products (i.e., CH_4^+ and CH_3^+, CH_2^+, etc.) make a relatively small contribution, and most of the remaining ions collected contain two or three carbon atoms, though ions with up to seven carbon atoms are detected. The manner in which the ion yields vary with increasing pressure over the range 10^{-6} to 0.2 mm Hg shows, as expected, that ion-molecule reactions are favored by increasing pressure, and also that as the pressure is increased the secondary ions tend to react further, giving C_3 and higher ions by successive ion-molecule reactions. However, neglecting for the moment these secondary ion-molecule reactions, which nevertheless will contribute to methane radiolysis, the main ionic reactions taking place in irradiated methane can be written as

$$CH_4 \rightsquigarrow CH_4^+ + e^- \qquad (6.98)$$
$$\rightsquigarrow CH_3^+ + H\cdot + e^- \qquad (6.99)$$
$$\rightsquigarrow CH_2^+ + H_2 \text{ (or 2H·)} + e^- \qquad (6.100)$$
$$CH_4^+ + CH_4 \rightarrow CH_5^+ + \cdot CH_3 \qquad (6.96)$$
$$CH_3^+ + CH_4 \rightarrow C_2H_5^+ + H_2 \qquad (6.97)$$
$$CH_2^+ + e^- \rightarrow \dot{C}H_2 \qquad (6.101)$$
$$CH_5^+ + e^- \rightarrow \cdot CH_3 + H_2 \qquad (6.102)$$
$$C_2H_5^+ + e^- \rightarrow \cdot C_2H_5 \qquad (6.103)$$

or

$$\rightarrow C_2H_4 + H\cdot \qquad (6.104)$$

These will be followed by various free radical reactions, including

$$H\cdot + CH_4 \rightarrow \cdot CH_3 + H_2 \qquad (6.105)$$
$$2\cdot CH_3 \rightarrow C_2H_6 \qquad (6.106)$$
$$2\cdot C_2H_5 \rightarrow C_4H_{10} \qquad (6.107)$$

or

$$\rightarrow C_2H_6 + C_2H_4 \qquad (6.108)$$
$$\cdot CH_3 + \cdot C_2H_5 \rightarrow C_3H_8 \qquad (6.109)$$

or

$$\rightarrow CH_4 + C_2H_4 \qquad (6.110)$$
$$2CH_2 \rightarrow C_2H_4 \qquad (6.111)$$
$$\cdot CH_2 + CH_4 \rightarrow C_2H_6 \qquad (6.112)$$
$$R\cdot + C_2H_4 \rightarrow R{-}CH_2{-}CH_2\cdot \xrightarrow{C_2H_4} \text{polymer} \qquad (6.113)$$

where R· is a hydrogen atom or an organic radical. (Third bodies, which may be necessary in some of these neutralization and radical reactions, have been omitted.) Three carbon and higher hydrocarbons may be formed by radical reactions, as shown, or by further ion-molecule reactions, e.g.,

$$C_2H_5^+ + CH_4 \rightarrow C_3H_7^+ + H_2 \qquad (6.114)$$

Several authors have attempted to correlate radiation chemical yields with mass spectral data. Dorfman (68) suggested that molecular detachment yields (the yield of product formed directly in molecular form without the agency of scavengeable free radicals) rather than overall product yields be compared with these data, since the overall yields often change as products build up in the irradiated system. Furthermore, the molecular detachment yields, or "molecular" yields, are generally derived from a relatively small number of processes, in contrast to the overall yields, and there is less temptation to adjust the calculated yields by selecting suitable reactions. Dorfman and Sauer (68) estimated the (experimental) molecular yield of hydrogen from methane ($G_{H_2} = 3.3 \pm 0.2$)[8] from the overall yield of hydrogen and the contribution "molecular" hydrogen makes to the overall yield, as determined by scavenger and isotopic techniques. If the reactions taking place in irradiated methane are those already given, molecular hydrogen arises from reactions 6.96 + 6.102, 6.97, and 6.100. Assuming W_{CH_4} to be 27.3 eV/ion-pair, then the G value for molecular hydrogen, estimated from the relative ion abundance for CH_4^+, CH_3^+, and CH_2^+ and the reactions cited, is $G_{H_2} = 3.5$, in satisfactory agreement with the experimental value of 3.3 ± 0.2. The process can be taken further and free radical yields (determined by scavenger experiments) (69) and overall product yields (61) correlated with the mass spectral data. The agreement between calculated and experimental yields is surprisingly good, and probably fortuitous.

Radical scavengers have been used to demonstrate the presence of free radicals in irradiated methane. Gevantman and Williams (62) and Meisels, Hamill, and Williams (61) used iodine to identify and estimate the radicals formed in pure methane and in mixtures of methane with a noble gas respectively (Table 6.1). Principal radical products detected by the appearance of the corresponding iodide were H·, ·CH₃, ·C₂H₅, and CH₂. The increase in the yield of ethylene in the presence of scavenger is interesting and is presumably due to the removal of radicals which would otherwise add to ethylene (Eq. 6.113), leading to polymerization.

Though radiolysis yields from methane can be explained almost completely without assuming any contribution from excited molecules it is probable that these also contribute to the overall chemical change. Indeed it has been claimed that ions, though undoubtedly formed, play no important part in

---

[8] The use of subscripts to denote "molecule" yields is described in Chapter 7.

the formation of products, which can be accounted for satisfactorily solely by reactions involving excited molecules and the free radicals derived from them. It is worthwhile to reflect that these two extreme views (ionization versus excitation) can be held about a system as thoroughly studied as the radiolysis of methane; very few of the mechanisms given in this book are established beyond the possibility of correction, or even drastic change, at some future date.

The mercury-photosensitized decomposition of methane produces identical products to the radiolysis (70). In this instance the first step is reaction between an excited mercury atom and methane to give a hydrogen atom and a methyl radical (71, 72), which must be responsible for the products formed in the early stages of the reaction,

$$Hg^* + CH_4 \rightarrow Hg + \cdot CH_3 + H \cdot \qquad (6.115)$$

As the concentration of initial products builds up, however, these also react with the excited mercury atoms, e.g.,

$$Hg^* + C_2H_6 \rightarrow Hg + \cdot C_2H_5 + H \cdot \qquad (6.116)$$

and the concentration of initial products finally reaches a steady state, though the higher hydrocarbon products continue to build up. This is not the case with radiolysis experiments, where all the products continue to build up as the irradiation proceeds. When methane is irradiated with 1236 Å light, the major primary process is elimination of molecular hydrogen (73)

$$CH_4 + h\nu \rightarrow CH_2 + H_2 \qquad (6.117)$$

which is accompanied by a smaller amount of breakdown to methyl radicals and hydrogen atoms, as in the mercury photosensitized reaction.

Mains and Newton have irradiated methane in the presence of a fivefold excess of mercury vapor (70). The mercury acts here as an ion scavenger since the ionization potential of mercury (10.43 V) is lower than that of methane (13.0 V):

$$CH_4^+ + Hg \rightarrow Hg^+ + CH_4 \qquad (6.118)$$

Subsequently the mercury ion is neutralized and the excited mercury atom formed reacts with methane, as in the mercury sensitized photolysis,

$$Hg^+ + e^- \rightarrow Hg^* \qquad (6.119)$$

$$Hg^* + CH_4 \rightarrow Hg + \cdot CH_3 + H \cdot \qquad (6.115)$$

The authors found that at 260°C the presence of the mercury did not alter the distribution of products in the radiolysis of methane, and they concluded that ion-molecule reactions contribute little to the radiolysis at this temperature. They noted however that the ion CH_3^+ would not be scavenged by mercury, since the ionization potential of the methyl radical is only 10.0 V. However, the authors concluded that the CH_4^+ ion, from which CH_3^+ was presumed to be formed by unimolecular dissociation, would not have a sufficient lifetime to dissociate in their experiments; the estimated lifetime of the ion in the radiolysis experiments was about 10^{-9} sec while the ion collection time in a mass spectrometer is about 10^{-6} sec.

The reverse of the situation when methane is irradiated with excess mercury occurs when the diluting gas is argon or krypton (61). Here charge transfer takes place in the opposite direction:

$$Ar^+ + CH_4 \rightarrow Ar + CH_3^+ + H\cdot \qquad (6.120)$$

$$Kr^+ + CH_4 \rightarrow Kr + CH_4^+ \qquad (6.121)$$

$$(Kr^+)^* + CH_4 \rightarrow Kr + CH_3^+ + H\cdot \qquad (6.122)$$

Charge transfer from argon releases sufficient energy to cause dissociation of the ion CH_4^+ (Eq. 6.120) whereas charge transfer from the krypton ion in its ground state does not (Eq. 6.121), though dissociation is possible if the krypton ion is excited (74). Meisels, Hamill, and Williams (61) attributed differences in the radiolysis of the mixtures methane-argon and methane-krypton to the different results of charge transfer with the two noble gases, as illustrated by reactions 6.120 and 6.121. Melton (75) has suggested that the differences may arise through the formation of the ion CH_2^+ in the mixtures containing argon but not in those containing krypton,

$$Ar^+ + CH_4 \rightarrow Ar + CH_2^+ + H_2 \qquad (6.123)$$

The corresponding charge-transfer reaction with krypton is endothermic and not probable. More recently direct experimental evidence of the abundance of secondary ions in these mixtures has been provided by Cermak and Herman (76). Using a mass spectrometer modified so that only ions produced by secondary reactions were collected, they showed that in methane–argon mixtures the relative abundance was $CH_4^+:CH_3^+:CH_2^+ = 2.0:79.0:19.0$. In methane–krypton mixtures the relative abundance of these ions was 64.5:35.0:0.5. The results show therefore that with argon the principal charge transfer reaction is 6.120, with a smaller contribution from 6.123. With krypton the principal reaction is 6.121, but there is an important contribution by 6.122. Xenon releases less energy by charge transfer with methane than either argon or krypton. When xenon is the sensitizing gas the major secondary ion is CH_4^+, the ratio of $CH_4^+:CH_3^+:CH_2^+$ being 97.0:2.5:0.5.

The liquid products formed when methane is subjected to large doses of radiation have been examined by Hummel (77), who found that the liquid contains saturated hydrocarbons with carbon contents ranging from C_6 to at least C_{27}. It was suggested that these hydrocarbons, together with molecular hydrogen, result from a series of ion-molecule reactions of the type (cf. ref. 67)

$$CH_3^+ + CH_4 \rightarrow C_2H_5^+ + H_2 \qquad (6.97)$$

$$C_2H_5^+ + CH_4 \rightarrow C_3H_7^+ + H_2 \qquad \text{etc.} \qquad (6.114)$$

A certain amount of saturated hydrocarbon product will be formed by polymerization reactions, for example, reaction 6.113, which consume unsaturated products produced by irradiation.

When methane is irradiated in the presence of oxygen, oxidation products are formed. With excess oxygen α-particle irradiation eventually converts methane to carbon dioxide and water (78). Irradiation of mixtures containing a smaller proportion of oxygen also produces carbon dioxide and water, but a variety of other gaseous and liquid products are also formed. Products which have been identified include hydrogen, carbon monoxide, formaldehyde, formic acid, alcohols, and peroxides (79), including methyl hydroperoxide (80), CH_3OOH. The irradiation of hydrocarbons in the presence of oxygen shows some similarities to the slow gas-phase combustion of these compounds.

Ethane

The radiolysis of ethane and higher members of the paraffin series is similar to that of methane; hydrogen, dimeric products, and unsaturated compounds are major products, though compounds resulting from carbon—carbon bond fission are also formed. The products obtained from ethane by γ-irradiation are shown in Table 6.2, which also includes yields found in the presence of nitric oxide, a radical scavenger. Yields shown in parentheses were obtained at low doses with very low conversion of ethane to products (e.g., $10^{-3}\%$ conversion); they illustrate the influence that buildup of radiolysis products can have on the observed product yields.

As in the case of methane, mass spectrometric evidence is helpful when postulating a radiolysis mechanism. Ions formed from ethane in the mass spectrometer are, in order of decreasing abundance, $C_2H_4^+$, $C_2H_3^+$, $C_2H_6^+$, $C_2H_5^+$, $C_2H_2^+$, C_2H^+.

Irradiation of ethane mixed with iodine vapor has demonstrated the presence of the radicals $H\cdot$, $\cdot C_2H_5$, $\cdot CH_3$ and, in smaller amounts, $n\text{-}C_3H_7\cdot$, $n\text{-}C_4H_9\cdot$, and CH_2, by the appearance of the corresponding iodides (62). The radicals will react with ethane and with one another in a similar manner to the radicals formed from methane (Eqs. 6.105 to 6.113). Ethyl radicals will be more important here than in the radiolysis of methane.

TABLE 6.2   *Radiolysis of Ethane[a]*

Product	G(product) Pure C_2H_6	C_2H_6 + NO
H_2	6.8 (8.8)	2.4[b]
CH_4	0.61 (0.39)	0.34
C_2H_2	0 ($\sim$0.2)	0.27
C_2H_4	0.05 ($\sim$2)	1.4
C_3H_6	0	0.03
C_3H_8	0.54 ($\sim$0.8)	0.14
C_4H_8	0	0.01
$n\text{-}C_4H_{10}$	1.0 (2.5)	0.27
$i\text{-}C_4H_{10}$	0.03	0
$i\text{-}C_5H_{10}$	0.54	0
Liquid polymer		

[a] γ-Irradiation; ref. 60. Values in parentheses are for very low conversions and are from ref. 81.
[b] Ref. 82.

Table 6.2 illustrates the fact that yields of some products, which presumably arise through free radical reactions, are reduced to zero by the presence of a radical scavenger. Yields of other products may be unchanged or merely lowered by the scavenger and these must be formed at least in part by nonradical reactions. Molecular products, for example, are not affected by scavengers or by the buildup of unsaturated radiolysis products, which can also act as radical scavengers. That the "molecular" hydrogen does in fact come from a single molecule of ethane has been demonstrated by irradiating mixtures of C_2H_6, C_2D_6, and a scavenger, when the hydrogen formed consists entirely of H_2 and D_2, apart from a small amount of HD from isotopic impurities (83, 84).[9] It is also clear from Table 6.2 that yields of unsaturated products are raised in the presence of a scavenger. When no scavenger is present these unsaturated compounds are probably converted to polymer via radical-initiated reactions.

The presence of ethyl radicals in liquid ethane irradiated with fast electrons has been demonstrated by observing the epr spectrum of the liquid *during irradiation* (86). The only other paramagnetic species detected were methyl radicals, whose concentration was estimated to be 4% of that of the ethyl radical.

[9] Okabe and McNesby (85) found for ethane irradiated with 1470 and 1295 Å light that hydrogen is not only formed from a single molecule, but comes preferentially from a single carbon atom of the molecule.

Ethylene

The radiolysis products from ethane, ethylene, and acetylene are compared in Table 6.3. Ethylene and acetylene are more sensitive to radiation than the saturated hydrocarbon and are converted to products (mainly polymeric) with moderately high G values. At the same time the yield of hydrogen is less from the unsaturated hydrocarbons than from ethane. Ethylene occupies a position between acetylene and ethane both with respect to the amount of polymer produced and the yield of hydrogen.

T A B L E　6.3　*Radiolysis of Ethane, Ethylene, and Acetylene*

	G(product)		
Product	Ethane (γ-Irradiation, ref. 60)	Ethylene (Fast Electron Irradiation, ref. 87)	Acetylene (Tritium β-Particle, ref. 88)
—(Original gas)		15.5	71.9
H_2	6.8	1.28	
CH_4	0.61	0.12	
C_2H_2	0	1.46	
C_2H_4	0.05	—	
C_2H_6	—	0.27	
C_3H_6	0	0.23	
C_3H_8	0.54	0.11	
Butenes	0	0.40	
Butanes	1.1	0.48	
Pentenes	0.54		
Pentanes		0.06	
Hexanes		0.13	
Polymer		Approx. 11[a]	Benzene 5.1 Cuprene 57[a]

[a] Molecules of parent gas converted to polymer per 100 eV energy absorbed.

It might be noted here that since unsaturated hydrocarbons are attacked very readily by free radicals, they are effective radical scavengers and will act in this manner if produced during radiolysis. Because radiolysis products may act as radical scavengers, or otherwise intrude into the radiolysis mechanism, it is always advisable to use as small a radiation dose as possible when determining product yields. This point was not always realized in the past, and radiolysis yields determined after the conversion of more than 1 or

2% (possibly much less; cf. 81, 89, and Table 6.2) of the original material to products should be suspect.

The radiation-induced polymerization of ethylene is of particular interest, since ethylene polymerized by other means constitutes the commercially important polymer polyethylene. At atmospheric pressure and room temperature the polymeric material produced by irradiating gaseous ethylene is a liquid. At rather higher pressures (about 21 atm) γ-irradiation produces a waxy solid with $G(-C_2H_4)$ about 160, and the yield of polymer is increased still further by increasing the temperature (e.g., $G(-C_2H_4)$ is about 10^4 at 460°F and 21.3 atm) though the polymer produced at higher temperatures is liquid (90).

The radiolysis mechanism for ethylene is complex, though a considerable amount of data is available about the reactions likely to be concerned. Mass spectrometry shows that the ions formed under the conditions prevailing in the mass spectrometer are, in order of decreasing importance, $C_2H_4^+$, $C_2H_3^+$, $C_2H_2^+$, C_2H^+, and C_2^+, and that many rapid ion-molecule reactions are possible (65, 66, 91, 92).

When ethylene is irradiated in the presence of 5% of nitric oxide, ethane formation is suppressed but the yields of hydrogen, acetylene, and *n*-butane are unchanged, suggesting that ethane, but not the other products, arises by a free radical process (93). Ethylene itself is, of course, also a radical scavenger. Irradiation of C_2H_4—C_2D_4 mixtures gives hydrogen composed almost entirely of H_2 and D_2 (94), indicating that the hydrogen is split from a single ethylene molecule by a "molecular" process. The two hydrogen atoms of the molecule may come from the same carbon atom or one may come from each carbon, since H_2, HD, and D_2 are all formed when either CH_2CD_2 or CHDCHD is irradiated (94). The proportions of the three hydrogen isotopes are not the same from the two isomeric ethanes, showing that the elimination process does not involve an intermediate in which all four hydrogen atoms in the molecule are equivalent.

Mixtures of ethylene and argon have been irradiated (87), and it appears that energy absorbed by the argon is transferred very efficiently to the ethylene, either by charge transfer[10] or the transfer of excitation energy,[11]

$$Ar^+ + C_2H_4 \rightarrow Ar + C_2H_4^+ \qquad (6.124)$$

or

$$Ar^* + C_2H_4 \rightarrow Ar + C_2H_4^+ + e^- \qquad (6.125)$$

[10] Cermak and Herman (76), using a modified mass spectrometer, found the relative abundance of secondary ions formed by charge transfer between argon ions and ethylene to be $C_2H_4^+:C_2H_3^+:C_2H_2^+ = 3 \cdot 0:76 \cdot 0:21 \cdot 0$. Charge exchange reactions between noble gas ions and ethylene have also been studied by Franklin and Field (95).

[11] The excitation energies of the lowest excited states of argon are greater than the ionization potential of ethylene.

The relative yields of the products are the same whether energy is absorbed directly by the ethylene or indirectly through argon, suggesting that the mechanism of ethylene radiolysis is the same in both cases. However, the distribution of products changes when sufficient hydrogen is added to the ethylene-argon mixture that ethylene is a minor component. Energy is absorbed mainly by either hydrogen or argon and the only significant products found are acetylene, ethane, *n*-butane and, at high partial pressures of argon and with high ethylene conversion, propane. These four gases account for 70 to 80% of the ethylene reacting, in contrast with a total of about 30% gaseous products when pure ethylene or ethylene-argon mixtures are irradiated. Lampe (87) postulates that energy absorbed by either hydrogen or argon eventually yields hydrogen atoms which, reacting with ethylene, can give both ethyl and methyl radicals,

$$H\cdot + C_2H_4 \rightarrow \cdot C_2H_5 \qquad (6.126)$$

$$H\cdot + \cdot C_2H_5 \rightarrow C_2H_6{}^* \qquad (6.127)$$

$$C_2H_6{}^* \rightarrow 2\cdot CH_3 \qquad (6.128)$$

Radical-radical reactions involving the two organic radicals can then give the products ethane, propane, and *n*-butane. Acetylene is presumably formed via the normal ethylene radiolysis mechanism, upon which these extra reactions are superimposed.

Polymerization of ethylene under the influence of radiation is probably a free radical process represented by

$$R\cdot + CH_2{=}CH_2 \rightarrow R{-}CH_2{-}CH_2\cdot \xrightarrow{C_2H_4} \text{polymer} \qquad (6.113)$$

where $R\cdot$ is a hydrogen atom or an organic radical. When solid ethylene is irradiated at $-196°C$, low molecular weight branched chain polymers are formed by a chain reaction represented (96) by

$$C_nH_{2n}^+ + C_2H_4 \rightarrow C_{n+2}H_{2n+4}^+ \qquad (6.129)$$

and it is possible that this process may also contribute in the gas phase radiolysis. However, in a number of instances where both free radicals and ions can initiate polymerization, it appears that radical polymerization predominates at normal temperatures and the ionic process at low temperatures.

The yield of hydrogen from irradiated ethylene or propylene is independent of the absorbed dose, and it has been suggested (81) that measurement of the hydrogen produced be used as a means of dosimetry in gas-phase systems; $G(H_2)_{\text{ethylene}} = 1.2$ and $G(H_2)_{\text{propylene}} = 1.1$.

Acetylene

Acetylene forms benzene and a polymer, cuprene, upon irradiation, in relative proportions of about one to five (based on the acetylene consumed); little or no gaseous product is formed except where this may be produced by irradiation of the initial products. The polymerization of acetylene, initiated both by radiation and by other means, has been studied extensively, but the radiolysis mechanism is still uncertain. The polymer produced is generally referred to as cuprene, though its properties depend on the method of formation. As prepared by irradiation, cuprene is a yellowish powder, insoluble in all common solvents, which neither melts nor sublimes. Little is known of its chemical properties beyond the facts that it is inflammable and will absorb oxygen (up to 25% by weight) when exposed to air.

An interesting feature of the radiolysis of acetylene is the ion-pair yield, $M_{-C_2H_2}/N$, which under a variety of experimental conditions is remarkably constant at about 20 (97, 98). The yields from polymerization reactions are generally influenced very markedly by any change in the reaction conditions, and the polymerization of acetylene is therefore not a typical chain reaction; the ion-pair yield is reproducible, and largely independent of dose rate, pressure, and temperature; it is not easily changed by inhibitors. The first radiolysis mechanism postulated for acetylene accounted for these facts on the basis of the ion-cluster theory (98, 99). Ions formed in the gas were believed to attract neutral acetylene molecules and hold them as a group of about twenty molecules by polarization forces. Energy released when the cluster is neutralized initiates polymerization of the molecules present to form cuprene,

$$C_2H_2 \rightsquigarrow C_2H_2^+ + e^- \tag{6.130}$$

$$C_2H_2^+ + 19C_2H_2 \rightarrow (19C_2H_2, C_2H_2^+) \tag{6.131}$$

$$(19C_2H_2, C_2H_2^+) + e^- \rightarrow \underset{\text{cuprene}}{C_{40}H_{40}} \tag{6.132}$$

This accounted successfully for the constant ion-pair yield, even in the presence of an inert gas (nitrogen or one of the noble gases) where only part of the ions formed would be produced from acetylene. However, later mechanisms, while maintaining the importance of ions and ionic reactions, have removed the emphasis from cluster formation as an intermediate step, though without offering an alternative explanation of the constancy of the ion-pair yield.

Dorfman (88, 100) has suggested that two independent mechanisms occur in irradiated acetylene, one leading to benzene and the other to cuprene. The benzene precursor is postulated to be an excited acetylene molecule (probably in a triplet state) which for steric reasons is able to give benzene by the sequence of reactions,

$$C_2H_2 \rightsquigarrow C_2H_2^* \qquad (6.133)$$

$$C_2H_2^* + C_2H_2 \rightarrow (C_2H_2)_2^* \xrightarrow{C_2H_2} C_6H_6 \qquad (6.134)$$

while cuprene is formed by a chain mechanism for which the initiation steps may be

$$C_2H_2 \begin{cases} \rightsquigarrow C_2H_2^+ + e^- \\ \rightarrow \cdot C_2H + H\cdot \\ \rightsquigarrow C_2H_2'^* \end{cases} \qquad (6.135)$$

The excited acetylene shown here is presumably in a different excited state to that which leads to benzene. Both radicals, $\cdot C_2H$ and $H\cdot$, add to acetylene to initiate a chain reaction leading to a polymeric cuprene,

$$H\cdot + C_2H_2 \rightarrow CH_2{=}CH\cdot \rightarrow \text{cuprene} \qquad (6.136)$$

$$\cdot C_2H + C_2H_2 \rightarrow HC{\equiv}C{-}CH{=}CH\cdot \rightarrow \text{cuprene} \qquad (6.137)$$

Reaction 6.136 explains the absence of molecular hydrogen in the radiolysis products from acetylene, even though rupture of a carbon—hydrogen bond does occur, as is shown by isotopic exchange when a mixture of C_2H_2 and C_2D_2 is irradiated (88),

$$C_2H_2 + C_2D_2 \rightarrow 2C_2HD \qquad (6.138)$$

Dorfman and Wahl (100) investigated the importance of excitation in the radiolysis of acetylene by irradiating mixtures of acetylene with a preponderance of helium. Under these conditions acetylene radiolysis is initiated by energy transfer to acetylene from an ionized or an excited helium atom, which in both cases leads to ionization of the acetylene,

$$He^+ + C_2H_2 \rightarrow He + C_2H_2^+ \qquad (6.139)$$

$$He^* + C_2H_2 \rightarrow He + C_2H_2^+ + e^- \qquad (6.140)$$

Excited states of acetylene are not formed and, if an excited state is the necessary precursor of benzene, no benzene should be produced. None was found. Cyclization to benzene is also eliminated when the radiolysis is sensitized by argon, krypton, or xenon. In all the noble gas-sensitized radiolyses, the polymerization yield is apparently unchanged.

Dorfman and Wahl also showed that the fraction of reacted acetylene converted to benzene, which in direct radiolysis at moderate pressures is constant at one fifth, decreases sharply at very low pressures. This is due to deactivation of the transient precursors of benzene at the wall of the irradiation vessel, diffusion to the wall being more rapid at low than at high pressures. From a consideration of the experimental conditions it was shown that the transient intermediates have a lifetime of at least 10^{-4} sec, which is consistent with the postulate that they are molecules excited to a triplet state. Benzene and cuprene are also formed from acetylene by photopolymerization (101) and mercury photosensitization (102). In both cases the yield of benzene falls at low pressures but, unlike the radiation-induced reaction, the yield of cuprene also falls, so that the ratio of benzene to cuprene is constant even at very low pressures. This is nevertheless compatible with the radiolysis mechanism since under these circumstances both benzene and cuprene must be formed via excited molecules (neither excitation process transfers sufficient energy to ionize acetylene).

More recently, Mains, Niki, and Wijnen (103) have obtained evidence that radicals are involved in benzene formation, and that the process cannot be initiated exclusively by excited molecules. Their experiments include the irradiation of equimolar mixtures of C_2H_2 and C_2D_2 and analysis of the mixture of isotopically substituted benzenes formed. The product was found to contain about 25% of the compounds C_6H_5D, $C_6H_3D_3$, and C_6HD_5, which can only arise by a mechanism involving C—H bond rupture at some stage during the radiolysis. Other experiments showed that benzene formation is completely suppressed by oxygen or iodine. Both experiments suggest a free radical mechanism, and the authors postulate that benzene is formed by a side reaction

$$C_6H_7 \cdot \rightarrow C_6H_6 + H \cdot \tag{6.141}$$

of the hydrogen atom-initiated chain polymerization given above (Eq. 6.136). Since the yield of benzene falls with increasing dose rate, this reaction is believed to compete with

$$C_6H_7 \cdot + R \cdot \rightarrow \text{products} \tag{6.142}$$

as increasing dose rate raises the radical concentration.

An alternative process which may accompany radical-initiated polymerization of acetylene is polymerization through a series of ion-molecule reactions (cf. 104–106). Ions formed from acetylene in the mass spectrometer, with their relative abundances (107), are (*a*) (ionization by 75 V electrons), $C_2H_2^+$ (100), C_2H^+ (20.7), C_2^+ (5.7), CH^+ (4.4), C^+ (1); and (*b*) (ionization by 5.1 MeV α-particles), $C_2H_2^+$ (100), C_2H^+ (7.3). Feasible ion–molecule reactions include

$$C_2H_2^+ + C_2H_2 \rightarrow C_4H_4^+ \tag{6.143}$$

or

$$\rightarrow C_4H_3^+ + H \cdot \tag{6.144}$$

$$CH^+ + C_2H_2 \rightarrow C_3H_3^+ \tag{6.145}$$

and these may be followed by chain reactions of the type,

$$C_mH_n^+ + C_2H_2 \rightarrow C_{m+2}H_{n+2}^+ \tag{6.146}$$

Unsaturated hydrocarbons appear rather prolific in the matter of ion-molecule reactions, and other such reactions given in the references cited might equally contribute to the radiolysis mechanism. A second, rather less probable, alternative to free radical polymerization is polymerization via excited molecules alone (cf. 101), e.g.,

$$C_2H_2^* + C_2H_2 \rightarrow C_4H_4^* \xrightarrow{C_2H_2} \text{cuprene} \tag{6.147}$$

When benzene vapor is mixed with acetylene prior to irradiation, the α-particle-induced polymerization is reduced by about a third (108). Benzene will often inhibit the radiation decomposition of other organic liquids, but examples of similar protective action in the gas phase are rare. Rudolph and Melton (109) have shown by means of the α-ray mass spectrometer that a very efficient charge transfer occurs between acetylene and benzene,

$$C_2H_2^+ + C_6H_6 \rightarrow C_6H_6^+ + C_2H_2 \tag{6.148}$$

which will adequately account for the observed decrease in polymer yield. Benzene itself is relatively stable towards radiation decomposition.

Addition to Unsaturated Hydrocarbons

Free radical attack on olefinic and acetylenic systems generally produces either a polymer or an addition compound. Examples of radical-polymerization induced by radiation have already been met when discussing ethylene and acetylene. When these gases are mixed with a second gas and irradiated, a chain reaction may follow giving an addition product. For example, ethyl bromide is formed with an ion-pair yield of the order of 10^5 when a mixture of ethylene and hydrogen bromide is irradiated with γ-radiation (110), the overall reaction being

$$CH_2{=}CH_2 + HBr \rightarrow CH_3CH_2Br \tag{6.149}$$

The reaction is very similar to the photo-initiated bromination of ethylene, and a similar radical-initiated chain mechanism is proposed with the chain-propagating steps,

$$Br\cdot + CH_2{=}CH_2 \rightleftharpoons \cdot CH_2{-}CH_2Br \tag{6.150}$$

$$\cdot CH_2{-}CH_2Br + HBr \rightarrow Br\cdot + CH_3CH_2Br \tag{6.151}$$

Reaction is initiated by free radicals formed from hydrogen bromide or ethylene in the manner already described. When ethylene is in excess the reaction yield is inversely proportional to the square root of the dose rate.

Addition reactions are not limited to halogens or compounds containing halogen. Lampe (111) has described a radiation-induced alkylation reaction between ethylene and propane which takes place under conditions of relatively high temperature and pressure. The major product is isopentane $[G = 42.8$ at 650°F, 500 psig, 8.5 mole percent ethylene; $G(-C_2H_4) = 93.6]$ which is probably formed as follows:

$$C_3H_8 \rightsquigarrow \text{free radicals (R·)} \tag{6.152}$$

$$R· + CH_3CH_2CH_3 \rightarrow CH_3\dot{C}HCH_3 + RH \tag{6.153}$$

$$CH_3\dot{C}HCH_3 + CH_2{=}CH_2 \rightarrow (CH_3)_2CH{-}CH_2{-}CH_2· \tag{6.154}$$

$$(CH_3)_2CHCH_2CH_2· + CH_3CH_2CH_3 \rightarrow$$
$$CH_3\dot{C}HCH_3 + (CH_3)_2CHCH_2CH_3 \tag{6.155}$$

Reactions 6.154 and 6.155 comprise a chain reaction, and this accounts for the relatively high yield of isopentane. Chain termination is by combination of two of the free radicals involved and, as is to be expected for a chain reaction of this sort, G(isopentane) is inversely proportional to the square root of the dose rate. It is also a function of the temperature and pressure. Isopentane makes up less than half the total radiolysis product and the detailed radiolysis mechanism is obviously very much more complex than shown above.

A similar radiation-induced chain alkylation occurs between propene and isobutane (112). At 55 atm pressure and 350 to 400°C the chain length is of the order 20 to 100, and it increases with increasing temperature and pressure and is inversely proportional to the square root of the dose rate. A complex mixture of products is obtained and it was noted that under these conditions purely thermal alkylation is negligibly slow.

Radiation-induced addition may also take place with acetylenic compounds. Bartok and Lucchesi (113) have reported the chain alkylation of acetylene by propane to give predominantly 3-methylbut-1-ene with an ion-pair yield between 7 and 12, depending on the dose rate, at 10–15 atm and 250 to 400°C (mixed reactor radiation):

$$C_2H_2 + CH_3CH_2CH_3 \rightarrow CH_2{=}CH{-}\overset{\displaystyle CH_3}{\underset{\displaystyle |}{CH}}{-}CH_3 \tag{6.156}$$

A point of interest about this reaction is that there is no thermal counterpart.

Pulse Radiolysis of Gases

A relatively small number of gaseous systems have been investigated by the pulse radiolysis technique due to the limitation of detecting transient species by optical absorption spectroscopy in a gaseous system. Multiple reflection systems and high pressure have been used to overcome these limitations (114, 115). Using a relatively low pressure of oxygen and up to 100 atm argon, the rate curves for ozone formation have been found to be first order (116, 117)

$$O + O_2 \rightleftharpoons O_3^* \qquad (6.157)$$

$$O_3^* + Ar \rightarrow O_3 + Ar \qquad (6.158)$$

Pulse radiolysis experiments with ammonia gas yielded NH radicals which decayed by a second-order process (118).

$$NH + NH \rightarrow N_2 + H_2 \qquad (6.159)$$

A rate constant estimated at 10^{12} M^{-1} sec^{-1} indicates reaction at every collision in pure ammonia.

The radiolysis of N_2O-O_2 mixtures by high intensity electron pulses has been studied (119, 119a). For pure N_2O, $G(N_2) = 12.4$ and $G(NO) = 5.6$. The major excitation reaction in N_2O is thought to be formation of $N_2(X^1\Sigma)$ and $O(^1D)$. Other excitation reactions may give $O(^3P)$, $N_2(A^3\Sigma)$, and $N_2(B^3\pi)$. NO^+ and N_2O^+ are also formed;

$$N_2O \rightsquigarrow N_2O^+ + e \qquad (6.160)$$

$$\rightsquigarrow NO^+ + N + e \qquad (6.161)$$

$$\rightsquigarrow N^+ + NO + e \qquad (6.162)$$

$$\rightsquigarrow N_2^+O + e \qquad (6.163)$$

$$\rightsquigarrow O^+ + N_2 + e \qquad (6.164)$$

Other gaseous investigations include the addition of H atoms to benzene and toluene to form the cyclohexadienyl radicals. The radicals have a convenient absorption

$$H + C_6H_6 \rightarrow \dot{C}_6H_7 \qquad (6.165)$$

$$H + CH_3C_6H_5 \rightarrow CH_3\dot{C}_6H_6 \qquad (6.166)$$

maximum in the 300-nm region (120).

H-Atom kinetics have been studied directly by utilizing the absorption of the Lyman α line (121, 122).

Attachment of Electrons to Gas Molecules

Rate curves for the disappearance of electrons by attachment to molecular oxygen have been obtained by microwave detection.

$$e^- + O_2 + M \rightarrow O_2^- + M$$

An electron pulse from a Van der Graaf accelerator was used. The three-body rate constant for oxygen in air was 2.6×10^{-30} cm^2 sec^{-1} (123, 124). Similar curves have been obtained for N_2O (124) and some organic compounds (115, p. 29).

REFERENCES

A comprehensive and authoritative account of the radiation chemistry of gases is given by S. C. Lind in *Radiation Chemistry of Gases*, Reinhold Publishing, New York, 1961. W. Mund has given a briefer review in *Actions Chimiques et Biologiques des Radiations* (ed. M. Haissinsky), Masson et Cie, Paris, Vol. 2, 1956 (see also refs. 115 and 125).

1. S. C. Lind, *J. Phys. Chem.*, **16**, 564 (1912).
2. R. A. Back, T. W. Woodward, and K. A. McLauchlan, *Can. J. Chem.*, **40**, 1380 (1962).
3. J. L. Magee and K. Funabashi, *Radiat. Res.*, **10**, 622 (1959).
4. H. Eyring, J. O. Hirschfelder, and H. S. Taylor, *J. Chem. Phys.*, **4**, 479 (1936).
5. H. Eyring, J. O. Hirschfelder, and H. S. Taylor, *J. Chem. Phys.*, **4**, 570 (1936).
6. G. Glockler and S. C. Lind, *The Electrochemistry of Gases and Other Dielectrics*, Wiley, New York, 1939.
7. W. P. Jesse and J. Sadauskis, *Phys. Rev.*, **88**, 417 (1952).
8. G. Bertolini, M. Bettoni, and A. Bisi, *Phys. Rev.*, **92**, 1586 (1953).
9. C. E. Melton, G. S. Hurst, and T. E. Bortner, *Phys. Rev.*, **96**, 643 (1954).
10. P. C. Capron, *Ann. Soc. Sci. Bruxelles*, **55**, 222 (1935).
11. W. Mund, L. Kaertkemeyer, M. Vanpee, and A. van Tiggelen, *Bull. Soc. Chim. Belges*, **49**, 187 (1940); W. Mund, T. de M. de Hornes, and M. van Meersche, *ibid.*, **56**, 386 (1947); W. Mund and M. van Meersche, *ibid.*, **57**, 88 (1948).
12. L. M. Dorfman and F. J. Shipko, *J. Phys. Chem.*, **59**, 1110 (1955).
13. S. O. Thompson and O. A. Schaeffer, *J. Am. Chem. Soc.*, **80**, 553 (1958).
14. O. A. Schaeffer and S. O. Thompson, *Radiat. Res.*, **10**, 671 (1959).
15. S. C. Lind, *Radiation Chemistry of Gases*, Reinhold, New York, 1961, p. 91.
16. K. E. Wilzbach, *J. Am. Chem. Soc.*, **79**, 1013 (1957).
17. P. Riesz and K. E. Wilzbach, *J. Phys. Chem.*, **62**, 6 (1958).
18. L. M. Dorfman and K. E. Wilzbach, *J. Phys. Chem.*, **63**, 799 (1959).
19. K. Yang and P. L. Gant, *J. Chem. Phys.*, **31**, 1589 (1959).
20. T. H. Pratt and R. Wolfgang, *J. Am. Chem. Soc.*, **83**, 10 (1961).
21. S. C. Lind, *Am. Chem. J.*, **47**, 397 (1912).
22. S. C. Lind and D. C. Bardwell, *J. Am. Chem. Soc.*, **51**, 2751 (1929).
23. J. D'Olieslager, *Bull. Sci. Acad. Roy. Belg.*, **11**, 711 (1925); W. Mund and J. D'Olieslager, *ibid.*, **12**, 309 (1926); W. Mund and J. D'Olieslager, *Bull. Soc. Chim. Belges*, **36**, 399 (1927).

24. J. F. Kircher, J. S. McNulty, J. L. McFarling, and A. Levy, *Radiat. Res.*, **13**, 452 (1960).
25. B. Lewis, *J. Phys. Chem.*, **37**, 533 (1933).
26. J. L. Magee and M. Burton, *J. Am. Chem. Soc.*, **73**, 523 (1951).
27. D. W. Brown and L. A. Wall, *J. Phys. Chem.*, **65**, 915 (1961).
28. P. Harteck and S. Dondes, *Proc. 2nd Int. Conf. Peaceful Uses Atomic Energy*, United Nations, Geneva, **29**, 415 (1958).
29. P. Harteck and S. Dondes, *J. Chem. Phys.*, **27**, 546 (1957); **28**, 975 (1958); *J. Phys. Chem.*, **63**, 956 (1959).
30. S. Y. Pshezhetsky and M. T. Dmitriev, *Soviet J. At. Energy*, **3**, 350 (1957); *Int. J. Appl. Radiat. Isotop.*, **5**, 67 (1959).
30a. M. T. Dmitriev, *Russ. J. Phys. Chem.*, **40**, 819 and 939 (1966).
31. A. R. Jones, *Radiat. Res.*, **10**, 655 (1959).
32. P. Harteck and S. Dondes, *Nucleonics*, **14**, 66 (March 1956).
33. R. E. Simpson, *Health Phys.*, **8**, 143 (1962).
34. B. P. Burtt and J. F. Kircher, *Radiat. Res.*, **9**, 1 (1958).
35. G. R. A. Johnson, U.S. Dept. of Commerce, NSRDS-NBS 45 (1973).
36. P. Auloos (Ed.), *Fundamental Processes in Radiation Chemistry*, Wiley-Interscience, New York, 1968.
36a. T. J. Sworski, *J. Phys. Chem.*, **70**, 1546 (1966).
36b. K. G. McLaren, *Int. J. Appl. Radiat. Isotopes*, **25**, 87 (1974).
37. A. T. Cameron and W. Ramsay, *J. Chem. Soc.*, **93**, 966 (1908).
38. F. L. Usher, *J. Chem. Soc.*, **97**, 389 (1910).
39. C. C. McDonald, A. Kahn, and H. E. Gunning, *J. Chem. Phys.*, **22**, 908 (1954); C. C. McDonald and H. E. Gunning, *ibid.*, **23**, 532 (1955).
40. J. C. Devins and M. Burton, *J. Am. Chem. Soc.*, **76**, 2618 (1954).
41. B. P. Burtt and A. B. Zahlan, *J. Chem. Phys.*, **26**, 846 (1957).
42. M. M. Mann, A. Hustrulid, and J. T. Tate, *Phys. Rev.*, **58**, 340 (1940).
43. L. M. Dorfman and P. C. Noble, *J. Phys. Chem.*, **63**, 980 (1959).
44. H. Essex and D. FitzGerald, *J. Am. Chem. Soc.*, **56**, 65 (1934); C. Smith and H. Essex, *J. Chem. Phys.*, **6**, 188 (1938); M. J. McGuinness and H. Essex, *J. Am. Chem. Soc.*, **64**, 1908 (1942).
45. H. Essex, *J. Phys. Chem.*, **58**, 42 (1954).
46. T. W. Woodward and R. A. Back, *Can. J. Chem.*, **41**, 1463 (1963).
47. D. Cleaver, E. Collinson, and F. S. Dainton, *Trans. Faraday Soc.*, **56**, 1640 (1960).
48. D. B. Peterson, U.S. Dept. of Commerce, NSRDS-NBS 44 (1974).
49. S. C. Lind and D. C. Bardwell, *J. Am. Chem. Soc.*, **47**, 2675 (1925).
50. P. Harteck and S. Dondes, *J. Chem. Phys.*, **26**, 1727 (1957).
51. J. O. Hirschfelder and H. S. Taylor, *J. Chem. Phys.*, **6**, 783 (1938).
52. P. Harteck and S. Dondes, *J. Chem. Phys.*, **23**, 902 (1955).
53. A. R. Anderson, J. V. Best, and D. A. Dominey, *J. Chem. Soc.*, 3498 (1962).
54. A. C. Stewart and H. J. Bowlden, *J. Phys. Chem.*, **64**, 212 (1960).
55. A. R. Anderson, H. W. Davidson, R. Lind, D. R. Stranks, C. Tyzack, and J. Wright, *Proc. 2nd Int. Conf. Peaceful Uses Atomic Energy*, United Nations, Geneva, **7**, 335 (1958).
56. S. C. Lind, *Le Radium*, **8**, 289 (1911).
57. S. C. Lind and R. Livingston, *J. Am. Chem. Soc.*, **58**, 612 (1936).
58. E. G. Zubler, W. H. Hamill, and R. R. Williams, *J. Chem. Phys.*, **23**, 1263 (1955).
59. D. O. Schissler and D. P. Stevenson, *J. Chem. Phys.*, **24**, 926 (1956).
60. K. Yang and P. J. Manno, *J. Am. Chem. Soc.*, **81**, 3507 (1959).
61. G. G. Meisels, W. H. Hamill, and R. R. Williams, *J. Phys. Chem.*, **61**, 1456 (1957).
62. L. H. Gevantman and R. R. Williams, *J. Phys. Chem.*, **56**, 569 (1952).

63. American Petroleum Institute, *Catalog of Mass Spectral Data*, Carnegie Institute of Technology, Pittsburgh, Pa., 1947–1956.
64. V. L. Tal'roze and A. K. Lyubimova, *Dokl. Akad. Nauk SSSR*, **86**, 909 (1952).
65. D. O. Schissler and D. P. Stevenson, *J. Chem. Phys.*, **24**, 926 (1956).
66. F. H. Field, J. L. Franklin, and F. W. Lampe, *J. Am. Chem. Soc.*, **79**, 2419 (1957).
67. S. Wexler and N. Jesse, *J. Am. Chem. Soc.*, **84**, 3425 (1962).
68. L. M. Dorfman and M. C. Sauer, *J. Chem. Phys.*, **32**, 1886 (1960).
69. L. Reinisch, *J. Chim. Phys.*, **57**, 1064 (1960).
70. G. J. Mains and A. S. Newton, *J. Phys. Chem.*, **65**, 212 (1961).
71. K. Morikawa, W. S. Benedict, and H. S. Taylor, *J. Chem. Phys.*, **5**, 212 (1937).
72. R. A. Back and D. van der Auwera, *Can. J. Chem.*, **40**, 2339 (1962).
73. B. H. Mahan and R. Mandal, *J. Chem. Phys.*, **37**, 207 (1962).
74. G. G. Meisels, *J. Chem. Phys.*, **31**, 284 (1959).
75. C. E. Melton, *J. Chem. Phys.*, **33**, 647 (1960).
76. V. Cermak and Z. Herman, *Nucleonics*, **19**, 106 (September 1961).
77. R. W. Hummel, *Nature*, **192**, 1178 (1961).
78. S. C. Lind and D. C. Bardwell, *J. Am. Chem. Soc.*, **48**, 2335 (1926).
79. B. M. Mikhailov, M. E. Kuimova, and V. S. Bogdanov, *Trans. First All Union Conf. Radiat. Chem., 1957*, Moscow, 1958, p. 234.
80. G. R. A. Johnson and G. A. Salmon, *J. Phys. Chem.*, **65**, 177 (1961).
81. K. Yang and P. L. Gant, *J. Phys. Chem.*, **65**, 1861 (1961).
82. K. Yang, *Can. J. Chem.*, **38**, 1234 (1960).
83. L. M. Dorfman, *J. Phys. Chem.*, **62**, 29 (1958).
84. L. J. Stief and P. Ausloos, *J. Chem. Phys.*, **36**, 2904 (1962).
85. H. Okabe and J. R. McNesby, *J. Chem. Phys.*, **34**, 668 (1961).
86. R. W. Fessenden and R. H. Schuler, *J. Chem. Phys.*, **33**, 935 (1960).
87. F. W. Lampe, *Radiat. Res.*, **10**, 691 (1959).
88. L. M. Dorfman and F. J. Shipko, *J. Am. Chem. Soc.*, **77**, 4723 (1955).
89. R. A. Back, *J. Phys. Chem.*, **64**, 124 (1960).
90. J. C. Hayward and R. H. Bretton, *Chem. Eng. Progr. Symp. Ser.*, **50** (13), 73 (1954).
91. C. E. Melton and P. S. Rudolph, *J. Chem. Phys.*, **32**, 1128 (1960).
92. F. H. Field, *J. Am. Chem. Soc.*, **83**, 1523 (1961).
93. K. Yang and P. J. Manno, *J. Phys. Chem.*, **63**, 752 (1959).
94. P. Ausloos and R. Gorden, *J. Chem. Phys.*, **36**, 5 (1962).
95. J. L. Franklin and F. H. Field, *J. Am. Chem. Soc.*, **83**, 3555 (1961).
96. C. D. Wagner, *J. Phys. Chem.*, **65**, 2276 (1961); **66**, 1158 (1962).
97. W. Mund and W. Koch, *Bull. Soc. Chim. Belges*, **34**, 119, 241 (1925); *J. Phys. Chem.*, **30**, 289 (1926).
98. S. C. Lind, D. C. Bardwell, and J. H. Perry, *J. Am. Chem. Soc.*, **48**, 1556 (1926).
99. W. Mund, C. Velghe, C. de Vos, and M. Vanpee, *Bull. Soc. Chim. Belges*, **48**, 269 (1939).
100. L. M. Dorfman and A. C. Wahl, *Radiat. Res.*, **10**, 680 (1959).
101. M. Zelikoff and L. M. Aschenbrand, *J. Chem. Phys.*, **24**, 1034 (1956).
102. S. Shida, Z. Kuri, and T. Furuoya, *J. Chem. Phys.*, **28**, 131 (1958).
103. G. J. Mains, H. Niki, and M. H. J. Wijnen, *J. Phys. Chem.*, **67**, 11 (1963).
104. F. H. Field, J. L. Franklin, and F. W. Lampe, *J. Am. Chem. Soc.*, **79**, 2665 (1957).
105. R. Barker, W. H. Hamill, and R. R. Williams, *J. Phys. Chem.*, **63**, 825 (1959).
106. P. S. Rudolph and C. E. Melton, *J. Phys. Chem.*, **63**, 916 (1959).
107. C. E. Melton and P. S. Rudolph, *J. Chem. Phys.*, **30**, 847 (1959).
108. S. C. Lind and P. S. Rudolph, *J. Chem. Phys.*, **26**, 1768 (1957).
109. P. S. Rudolph and C. E. Melton, *J. Chem. Phys.*, **32**, 586 (1960).

110. D. A. Armstrong and J. W. T. Spinks, *Can. J. Chem.*, **37**, 1210 (1959).
111. F. W. Lampe, *Nucleonics*, **18**, 60 (April 1960).
112. P. J. Lucchesi and C. E. Heath, *J. Am. Chem. Soc.*, **81**, 4770 (1959).
113. W. Bartok and P. J. Lucchesi, *J. Am. Chem. Soc.*, **81**, 5918 (1959).
114. M. S. Matheson and L. M. Dorfman, *Pulse Radiolysis*, M.I.T. Press, Cambridge, Mass., 1969.
115. R. F. Firestone and L. M. Dorfman, in *Actions Chimiques et Biologiques des Radiations*, (ed. M. Haissinsky), Vol. 15, Masson et Cie, Paris, 1971, Chap. 1.
116. M. C. Sauer and L. M. Dorfman, *J. Am. Chem. Soc.*, **87**, 3801 (1965).
117. J. F. Riley and R. W. Cahill, *J. Phys. Chem.*, **52**, 3297 (1970).
118. S. Gordon and G. M. A. C. Meaburn, *J. Phys. Chem.*, **7**, 1592 (1968).
119. C. Willis, A. W. Boyd, and A. O. Miller, *Radiat. Res.*, **46**, 428 (1971).
119a. C. Willis, A. W. Boyd, and P. E. Binder, *Can. J. Chem.*, **50**, 1557 (1972).
120. M. C. Sauer and B. Ward, *J. Phys. Chem.*, **71**, 3971 (1967).
121. W. P. Bishop and L. M. Dorfman, *J. Chem. Phys.*, **52**, 3210 (1970).
122. T. Hikida, J. Eyre, and L. M. Dorfman, *J. Chem. Phys.*, **54**, 3422 (1971).
123. R. W. Fessenden and J. M. Warman, *Adv. Chem. Series* **82**, 222 (1968).
124. J. M. Warman and R. W. Fessenden, *J. Chem. Phys.*, **49**, 4718 (1968).
125. Ion-Molecule Reactions in the Gas Phase, *Adv. Chem. Ser.*, *Amer. Chem. Soc.*, **58** (1966).

CHAPTER 7

Water and Aqueous Solutions

Water • Water Vapor • Liquid Water • Ice • Aqueous Solutions • The Reducing Radicals (e^-_{aq} and H) • The Oxidizing Species (OH, HO$_2$, and Their Anions, H$_2$O$_2$) • Gaseous Solutes • Ferrous Sulfate (Fricke Dosimeter) • Ferrous Sulfate-Cupric Sulfate • Ceric Sulfate • Determination of Molecular and Radical Yields • Ferrocyanide-Ferricyanide • Halide and Pseudohalide Ions • Aliphatic Acids • Alcohols and Carbonyl Compounds • Carbohydrates • Organic Halogen Compounds • Organic Nitrogen Compounds • Organic Sulfur Compounds • Aromatic and Unsaturated Compounds • Other Solutes • References •

WATER

The radiolysis of water and aqueous solutions is described in this chapter and organic systems in the next. This division is convenient, since many of the systems of most interest to chemists and biologists are aqueous solutions, and it can be justified on the grounds that organic substances are both more complex and less polar than water. Thus organic compounds tend to form many more products upon radiolysis than water and to do so via processes that are not strongly based on ion solvation; as will be seen, a key step in the radiolysis of water and aqueous solutions is the stabilization by hydration of the electrons released. Water is an excellent solvent, and another difference between this chapter and the next is that this deals mainly with dilute solutions while the next is generally concerned with the radiolysis of pure compounds. The two chapters deal largely, therefore, with "indirect" and "direct" effects, the term indirect signifying that the radiant energy is absorbed by the major component of the system (the solvent) and that it only alters the minor component (the solute) indirectly. Both chapters deal predominantly with liquid systems, since the greater part of the published material deals with these, and they are the systems most readily studied by kinetic methods.

Historically, the evolution of gas from aqueous solutions containing radium salts was one of the earliest radiation-induced chemical reactions to be observed and then studied. Early work showed that α-particles decompose water into hydrogen and oxygen and that part of the oxygen remains in solution in the form of hydrogen peroxide. In aerated solutions the yield of hydrogen peroxide is increased by the presence of oxygen. x-Rays and γ-rays give much smaller yields of hydrogen peroxide from aerated solutions and produce very little change at all when pure oxygen-free water is irradiated in a closed system. Both oxidation and reduction of dissolved substances were observed, and as far back as 1914 Debierne (1) suggested that free radicals formed from the water might be responsible for the chemical action of radiation. Similar suggestions were made by Risse (2) and later developed by Weiss (3) and the group working on radiation chemistry in the United States Atomic Energy Project (cf. 4, 5), though in the meantime other, non-radical, mechanisms were also considered [e.g., the activated-water hypothesis (6)]. Initially the radicals were believed to be hydrogen atoms and hydroxyl radicals (OH), although it was later shown that the major reducing species in neutral and basic solutions is the hydrated electron (e_{aq}^-), which in acid solutions is rapidly converted into a hydrogen atom by reaction with H_3O^+. The possibility that hydrated electrons might be formed was first suggested by Platzman (7) who, following Lea (8) and Gray (9), calculated that the secondary electrons produced by ionization in the spurs will travel an average of at least 5 nm from the parent positive ion before being reduced to thermal energy. At this distance the electron would be essentially free of the electrostatic attraction of the positive ion and could become hydrated in a similar manner to the solvation of other ions in polar solvents. Alternative calculations by Samuel and Magee (10) suggested that the secondary electron would travel a shorter distance from the parent ion while reaching thermal energy and would be attracted back to the ion, neutralizing it and forming a geminate radical pair. Although it is now known that this is not the case with water and other highly polar solvents, ion-pair recombination as proposed by Samuel and Magee does occur with weakly polar and nonpolar solvents in the absence of ion scavengers. The development of these concepts in so far as they concern water have been described by Hart (11), Allen (12, 13), Draganić and Draganić (14), and in the earlier edition of this book (15).

Water Vapor

The photochemistry of water vapor has been studied extensively at wavelengths ranging from 186 to below 120 nm (16). Primary processes are

$$H_2O + hv \rightarrow H + OH \qquad (7.1)$$

and, at the shorter wavelengths and with a lower probability,

$$H_2O + hv \rightarrow H_2 + O \qquad (7.2)$$

$$\rightarrow H_2O^+ + e^- \qquad (7.3)$$

The latter reaction, photoionization, is only possible at wavelengths below 98.3 nm (threshold energies for processes of importance in the decomposition of water vapor are given in Table 7.1). In closed systems the photolysis products, hydrogen and oxygen, build up to a small equilibrium concentration, at which point the rate of formation is balanced by the rate at which

T A B L E 7.1 *Threshold Energies for Processes in Water Vapor*

Process	Minimum Energy Requirement			Ref.
	kcal mole^{-1}	eV molecule^{-1}	nm	
$H_2O \rightarrow H_2O^*$ (lowest triplet)	97–104	4.2–4.5[a]	295–275	17
				18
$H_2O \rightarrow H_2O^*$ (lowest singlet)	152	6.6	187[b]	19
				20
$H_2O \rightarrow H_2(^1\Sigma_g^+) + O(^3P)$	116	5.02	246.8[b]	16
				21
$H_2O \rightarrow H(^2S) + OH(X^2\Pi)$	118	5.12	242.0[b]	16
				21
$H_2O \rightarrow H_2(^1\Sigma_g^+) + O(^1D)$	162	7.03	176.3[b]	16
				21
$H_2O \rightarrow H_2(^1\Sigma_g^+) + O(^1S)$	217	9.39	132.0[b]	16
				21
$H_2O \rightarrow H_2O^+(^2B_1) + e^-$	291	12.61 ± 0.05[c]	98.3	22
				23
$H_2O \rightarrow H^+ + OH + e^-$	369	16.0 ± 0.3[c]	77.5	22
$H_2O \rightarrow OH^+(^3\Sigma) + H + e^-$	417	18.1 ± 0.1[c]	68.5	22

[a] Energy of the triplet state above the ground state.
[b] From the absorption spectrum.
[c] Appearance potential of positive ion in mass spectrometer.

the products are removed by back reactions. The presence of back reactions is confirmed by flow experiments in which the vapor is analyzed after a very brief irradiation. These show much higher yields of products corresponding to initial quantum yields for water decomposition rising from about 0.4 for 186 to 143 nm light to values greater than unity for light in the 143 to 125 nm region. Hydrogen peroxide is also observed as a product in the flow experiments. Reactions that have been proposed to account for the products include

$$H + OH + M \rightarrow H_2O + M \qquad (7.4)$$

$$H + H + M \rightarrow H_2 + M \qquad (7.5)$$

$$OH + OH + M \rightarrow H_2O_2 + M \qquad (7.6)$$

$$H + OH \rightarrow H_2 + O \qquad (7.7)$$

$$OH + OH \rightarrow H_2O + O \qquad (7.8)$$

$$O + OH \rightarrow O_2 + H \qquad (7.9)$$

where M is a water molecule, or the wall of the container, and serves to remove excess energy from the products. The relative importance of reactions such as 7.4 and 7.7 or 7.6 and 7.8, which involve the same pairs of reactive species, depends on the excited state of the reactants and the water vapor pressure, the termolecular reactions being favored by high vapor pressures and the bimolecular reactions by short wavelength radiation and low vapor pressures. Energy-rich (i.e., "hot") hydrogen atoms formed when short wavelength radiation is used may react directly with water to give molecular hydrogen (21, 24)

$$H^* + H_2O \rightarrow H_2 + OH \qquad (7.10)$$

Other reactions, e.g.,

$$OH + H_2 \rightarrow H_2O + H \qquad (7.11)$$

become significant as the concentration of products builds up and are responsible for the low steady state yields observed.

The most abundant ions formed when water vapor is bombarded with electrons or α-particles in the mass spectrometer are listed in Table 7.2 with their appearance potentials and the reactions that have been proposed to account for their formation; more detailed information on the mass spectroscopy of water vapor is given by Dixon (16). The same ions are produced by electron and α-particle irradiation but the ion abundances differ, the α-particles producing rather less fragmentation than the electrons, which have a comparable velocity. The ion abundances also change as the water

TABLE 7.2 *Principal Ions Formed in Water Vapor in the Mass Spectrometer*

Ion	Appearance Potential (V) (16, 22)	Relative Intensity			Probable Formation Process (16, 22, 27)	
		100-eV Electron (25)	50-eV Electron (26)	2.2-MeV α-Particle (26)		
H^-	4.8	0.6			$H_2O + e^- \rightarrow H^- + OH$	(7.12)
OH^-	4.7				$H^- + H_2O \rightarrow OH^- + H_2$	(7.13)
	6.0				$H_2O + e^- \rightarrow OH^- + H$	(7.14)
O^-	7.4	0.15			$H_2O + e^- \rightarrow O^- + H_2$	(7.15)
					$\rightarrow O^- + 2H$	(7.16)
H_2O^+	12.61	100	100	100	$H_2O + e^- \rightarrow H_2O^+ + 2e^-$	(7.17)
H_3O^+	12.67		0.32^a	0.54^a	$H_2O^+ + H_2O \rightarrow H_3O^+ + OH$	(7.18)
OH^+	18.1	23	23.1	17.95	$H_2O + e^- \rightarrow OH^+ + H + 2e^-$	(7.19)
H^+	19.6	5	4.3	1.11	$H_2O + e^- \rightarrow H^+ + OH + 2e^-$	(7.20)
O^+	18.8	2	2.0	1.11	$H_2O + e^- \rightarrow O^+ + H_2 + 2e^-$	(7.21)
	29.15				$\rightarrow O^+ + 2H + 2e^-$	(7.22)

^{a} Pressure dependent.

vapor pressure is increased, the abundances of OH^+, H^+, and O^+ falling with increasing pressure and those of H_2O^+ and H_3O^+ increasing. The abundance of H_3O^+ increases in proportion to the square of the pressure, confirming that this ion is the product of an ion-molecule reaction (Eq. 7.18). It has been suggested that at high pressures H_3O^+ may also be formed via excited water (28),

$$H_2O^* + H_2O \rightarrow H_3O^+ + OH + e^- \qquad (7.23)$$

Negative ions are less abundant than positive ions (29) and, because their formation involves incorporation of the incident electron in the products, require relatively low energy electrons (5 to 15 eV); formation of negative ions may reach a maximum at certain electron energies that correspond to preferred energy states of the products of the reaction (i.e., the reaction may be a resonance process). At low pressures, H^- is the most abundant negative ion and very little OH^- is formed, but at higher pressures the opposite is true and OH^- is the more abundant ion. This tendency is seen when the negative ion abundancies given in Table 7.2 are compared with the following relative abundances found by Muschlitz and Bailey (30) at 4.6 mm Hg pressure (a high pressure in the context of mass spectrometry), namely $H^-:O^-:OH^- = 1.2:23:31$. The changes are consistent with the suggestion that OH^- is formed, in part, by an ion-molecule reaction of H^- (Eq. 7.13). However the total yield of negative ions is always much smaller than the total yield of positive ions [Melton (29) has estimated that in water vapor at 6×10^{-6} mm Hg pressure bombarded with 100-eV electrons less than 1% of the products result from the formation of negative ions], and the electron and hydrated electron remain the negative species of greatest interest. The negative ion H_2O^- is not observed in the mass spectrometer and must therefore have a very brief existence.

Ion-molecule reactions observed in the mass spectrometer with water vapor are listed in Table 7.3. Reactions 7.18 and 7.25 are most significant

T A B L E 7.3   *Ion-Molecule and Neutralization Reactions in Water Vapor (16)*

Reaction		Rate Constant^a $(M^{-1}\,sec^{-1})$	$-\Delta H$ (kcal mole^{-1})
$H^- + H_2O \rightarrow OH^- + H_2$	(7.13)	$\sim 3 \times 10^{12}$	
$O^- + H_2O \rightarrow OH^- + OH$	(7.24)		
$H_2O^+ + H_2O \rightarrow H_3O^+ + OH$	(7.18)	9×10^{11}	25–30
$OH^+ + H_2O \rightarrow H_3O^+ + O$	(7.25)	1×10^{12}	58
$H_3O^+ + (n-1)H_2O \rightarrow H^+(H_2O)_n$	(7.26)	($n = 1$ to 8; refs. 31, 32)	
$H_3O^+ + e^- \rightarrow H_2O + H$	(7.27)	1.4×10^{14}	~ 145
$\rightarrow 2H + OH$	(7.28)	1.4×10^{14}	~ 25
$H^+(H_2O)_n + e^- \rightarrow H + nH_2O$	(7.29)		~ 45

^{a} Rate constants for gas phase reactions are generally given in units of cm^3 molecule^{-1} sec^{-1}. To facilitate comparison with other rate constants given in this chapter, these units have been converted to liter mole^{-1} sec^{-1} (M^{-1} sec^{-1}) by multiplying rate constants given in units of cm^3 molecule^{-1} sec^{-1} by 6.022×10^{20}.

in the radiation chemistry of water vapor with the former making the greater contribution; the higher pressures used in radiolysis experiments favor reaction or collisional quenching of excited H_2O^+ ions rather than fragmentation to give OH^+. Oxygen atoms formed by dissociation of excited molecules (Eq. 7.2), in reactions 7.7 and 7.8, or in reaction 7.25 probably react with water molecules

$$O + H_2O + M \rightarrow H_2O_2 + M \qquad (7.30)$$

$$O + H_2O \rightarrow 2OH \qquad (7.31)$$

Reaction 7.24 is an intermediate step in the conversion of electrons to hydroxyl radicals by reaction with nitrous oxide (p. 294)

$$e^- \ (e_{aq}^- \text{ in aqueous soln.}) + N_2O \rightarrow N_2 + O^- \ (\xrightarrow{H_2O} OH) \qquad (7.32)$$

Cluster formation between a positive or negative ion and water molecules (e.g. Eq. 7.26) has been observed relatively recently (16, 31, 32), confirming theoretical predictions by Magee and Funabashi (33) and earlier suggestions that this might occur by Lind (34). Formation of clusters by successive addition of water molecules to an ion is significant because neutralization of the hydrated ion (e.g., Eq. 7.29) releases less energy than neutralization of the bare ion (e.g., Eq. 7.27 or 7.28) so that the products formed may differ in the two cases.

The overall features of the radiation-induced decomposition of water vapor are similar to the photochemical decomposition, namely that breakdown occurs to give hydrogen and oxygen or hydrogen peroxide which, because of back reactions, reach only small steady state concentrations in pure water vapor (16, 35). The equilibrium yields obtained depend in part on the experimental conditions (e.g., purity of the water used, nature and condition of the vessel walls, radiation type, and dose rate) and published results are not entirely consistent. However, representative results are those of Anderson and Best (reported in 35) who found 10^{-2} mole % hydrogen and 3.8×10^{-3} mole % oxygen for proton irradiation at 1 atm pressure, 125°C, and 5×10^{20} eV g^{-1} sec^{-1}. The information available from mass spectrometry and photochemistry suggests that the following reactions will contribute when water vapor is irradiated under the conditions normally employed (i.e., pressures of 500 to 1000 mm Hg or 0.65 to 1.3 atm, and temperatures below 150°C),

$$H_2O \rightsquigarrow H_2O^+, e^-, H_2O^* \tag{7.33}$$

$$H_2O^+ + H_2O \rightarrow H_3O + OH \tag{7.18}$$

$$H_3O^+ + (n-1)H_2O \rightarrow H^+(H_2O)_n \tag{7.26}$$

$$H^+(H_2O)_n + e^- \rightarrow H + nH_2O \tag{7.29}$$

$$H_2O^* \rightarrow H + OH \text{ (the products may be excited,}$$
$$\text{i.e., } H^* \text{ and } OH^*) \tag{7.1}$$

$$\rightarrow H_2 + O \tag{7.2}$$

and possibly

$$H^* + H_2O \rightarrow H_2 + OH \tag{7.10}$$

$$H_2O^* + H_2O \rightarrow H_3O^+ + OH + e^- \tag{7.23}$$

Hydration of the proton will be reduced at lower pressures and higher temperatures and reaction 7.29 may be replaced by 7.27 or 7.28 under these conditions. In any event, the reactions shown will be followed by others (e.g., Eq. 7.4 to 7.9, 7.30, 7.31) in which the radicals formed are converted to chemically stable products (H_2, O_2, H_2O_2, and H_2O). Back reactions such as reaction 7.11 and

$$H + H_2O_2 \rightarrow H_2O + OH \tag{7.34}$$

assume importance as the concentration of molecular products builds up and are responsible for the low steady state concentrations of products observed.

The W value for water vapor, 30.1 ± 0.3 eV per ion-pair for electrons and 37.6 eV per ion-pair for α-particles (36), is two to three times the ionization potential for water (12.61 eV) so that between 55 and 65% of the absorbed energy will be in the form of electronic or vibrational excitation of H_2O and H_2O^+. The decrease in abundance of fragment ions with increasing pressure in the mass spectrometer suggests that excited ions produced by irradiation at even higher pressures will dissipate their energy without fragmentation. However, to account for radical yields measured in the presence of scavengers, a substantial part of the excitation energy transferred to neutral water molecules must result in dissociation of the molecules. The extent of water decomposition in vapor phase irradiation can be estimated theoretically using the so-called optical approximation introduced by Platzman (37, 38) in which the abundance of a particular ionized or excited state is taken to be proportional to f/E, where f is the dipole oscillator strength for the transition from the ground state to the ionized or excited state and E is the energy of this state above the ground state. Applying this approach, Santar and Bednar (39) have estimated $G_{-H_2O}{}^1$ to be in the range 6.8 to 7.8 for water vapor with $G_{ion} = 3.32\,(= 100/W_{H_2O})$ and $G_{excited\,states} = 3.5$ to 4.5; the range in yields results in part from uncertainties regarding yields from collisions of slow (<100 eV) electrons. Platzman also pointed out that molecules may be raised to superexcited states in which they have energies in excess of the ionization energy, and that these states may subsequently ionize or dissociate giving excited radicals (H^* and OH^* from water) capable of reactions (e.g., Eq. 7.10) which are energetically not possible with ground state radicals. Inclusion of reactions of this type with a G value of about 1 raises the estimated G_{-H_2O} to 7.8 ± 1.0, which compares well with measured yields. Since the bond dissociation energy in water, D (HO–H), is 5.16 eV, this G value corresponds to utilization of 35 to 45% of the absorbed energy in bond scission; the remainder is dissipated in the form of thermal energy or, to a very limited extent, light.

Experimental G values for the primary species are determined by adding ion or radical scavengers to intercept the reactive intermediates. The yield of electrons, for example, can be estimated by irradiating water vapor con-

---

[1] The symbols G_{ion}, $G_{excited\,states}$, G_H, G_{OH}, etc. are used to represent the primary yields of the species indicated, i.e., the earliest detectable yields and before any subsequent reaction of the species. Thus G_{-H_2O} refers to the total number of water molecules dissociated per 100 eV energy absorbed while $G(-H_2O)$ refers to the number of water molecules found to be transformed into stable products other than H_2O at the end of the reaction; $G(-H_2O)$ is generally smaller than G_{-H_2O} because some of the molecules dissociated are reformed, e.g., by combination of H and OH. Several other symbols are, or have been, used for the primary yields. The most common is the use of a lower case g, as in $g(-H_2O)$ and $g(ion)$ ($= G_{-H_2O}$ and G_{ion}) (16, 40); the alternative symbols are only used for the primary yields; final yields are always given in the form $G(product)$.

taining nitrous oxide (41, 42), when $G_{e^-} = G(N_2) = 3.0 \pm 0.4$ due to

$$e^- + N_2O \rightarrow N_2O^- \rightarrow N_2 + O^- \tag{7.32}$$

While organic compounds (e.g., alcohols and alkanes) scavenge hydrogen atoms

$$H + CH_3OH \rightarrow H_2 + \cdot CH_2OH \tag{7.35}$$

so that $G(H_2) = G_{H_2} + G_H$, hydroxyl radicals are also scavenged but do not form H_2. Mixtures containing deuterium have also been used to estimate the yield of hydrogen atoms (43, 44), since

$$H + D_2 \rightarrow HD + D \tag{7.36}$$

$$2D \rightarrow D_2 \tag{7.37}$$

and $G(HD)$ should be equal to G_H (hydroxyl radicals are scavenged but do not form HD). However hydrogen atom yields estimated using deuterium ($G_H = 13.1$) (44) are significantly higher than yields ($G_H = 7.2 \pm 0.4$) estimated using organic scavengers (16) or hydrogen halides (HCl, and HBr also form H_2 with hydrogen atoms) (45), and it appears that HD can be formed in other processes in addition to reaction 7.36. These might involve reaction of triplet excited water molecules with D_2 or the formation of HD when deuterium atoms combine in the presence of water (16, 45).

Irradiation of water vapor in the presence of both an electron scavenger (N_2O or SF_6) and a hydrogen atom scavenger (an organic compound or hydrogen halide) gives the yield of hydrogen formed by molecular processes (e.g., Eq. 7.2 or 7.10) and via excited molecules (Eq. 7.1), since the neutralization process (Eq. 7.29) is inhibited. Such experiments give $G(H_2) = G_{H_2} + G_{H(exc)} = 5.0 \pm 0.5$. The yield of hydrogen formed by molecular processes, i.e., those that cannot be inhibited by adding a radical scavenger, can be estimated by measuring the yield of hydrogen in the presence of unsaturated organic compounds or the yield of H_2 from solutions containing C_6D_6 (46), since under these conditions H and OH are scavenged but do not form H_2. Experiments of this type give $G_{H_2} = 0.5 \pm 0.1$. The yield of hydroxyl radicals, G_{OH}, is generally not measured but estimated from the stoichiometric relationship $G_{OH} = 2G_{H_2} + G_H$, which assumes that no hydrogen peroxide is formed by molecular processes. The yield of water molecules dissociated is given by $G_{-H_2O} = 2G_{H_2} + G_H$. In summary, then, in irradiated water vapor,

$$G_{e^-} = 3.0 \pm 0.4 \qquad G_{H_2} = 0.5 \pm 0.1 \qquad G_H = 7.2 \pm 0.4$$

$$G_{OH} = 8.2 \pm 0.6 \qquad G_{-H_2O} = 8.2 \pm 0.6$$

The primary yields listed above apply particularly to electron and gamma irradiation but are expected to be relatively independent of LET, since the low density of the vapor allows the active species to diffuse away from the region of the particle track before reaction. Boyd and Miller (47) found the same steady state and molecular yields of hydrogen from water irradiated with low-LET gamma radiation and very high-LET fission fragments, although in the presence of an organic scavenger lower hydrogen atom yields were observed with the fission fragment irradiation due to competition between radical–radical and radical–scavenger reactions. Radical yields increase at temperatures above about $150°C$, $G(H_2)$ measured in the presence of organic scavengers increasing by about 2 to 3.5 G units. The increase is approximately equal to G_{e^-}, and it has been suggested that at the higher temperatures the neutralization reaction is 7.28 rather than 7.29, although not all the experimental evidence is consistent with this proposal (16).

$$H_3O^+ + e^- \rightarrow 2H + OH \qquad (7.28)$$

$$H^+(H_2O)_n + e^- \rightarrow H + nH_2O \qquad (7.29)$$

Radical and molecular yields are not significantly different when D_2O, rather than H_2O, is irradiated (16).

Liquid Water

The radiation chemistry of liquid water differs from that of water vapor in the following respects: (1) in the liquid all charged species become hydrated[2] within about 10^{-11} sec if their kinetic energy is not greater than thermal energy, (2) excitational energy can be dissipated by collisional processes more rapidly in liquids than in gases, and in water this is facilitated by the highly hydrogen-bonded structure of the system; and (3) diffusion is slower

---

[2] Hydration resembles the clustering of water molecules about H_3O^+ observed in irradiated water vapor, but in the liquid all ions become hydrated, including the electron once it has slower to thermal energy. The hydration (or solvation) process involves water molecules becoming orientated about the charged species and it therefore takes a finite time, of the order of 10^{-11} sec, the relaxation time for dipoles in water. Various models have been proposed for the hydrated electron, e_{aq}^-, but for the present purposes it is sufficient to regard it as an electron smeared over a group of orientated water molecules. Other hydrated ions can be considered as ions surrounded by shells of water molecules, one or two molecules thick, held more or less loosely by electrostatic attraction between the ion and solvent dipoles. Formation of the ion-dipole bonds releases energy of the order of 1 eV ion^{-1}, which may lead to significant differences in ΔH for reactions of the bare and hydrated ion (cf. reactions 7.27 and 7.29 in Table 7.3).

The hydrogen ion, H^+, can form a definite hydrate, H_3O^+, the oxonium ion. In aqueous solution (as opposed to the gas phase) and after sufficient time has elapsed for the ions to become solvated, the symbols H^+, H_3O^+, and $H^+(aq)$ all represent the same thing, namely the hydrated hydrogen ion; both H^+ and H_3O^+ are used in this chapter to represent $H^+(aq)$ as proves convenient. All other ionic species discussed in this chapter are also hydrated in solution, although the symbol "aq" is generally omitted from their formulae.

in a liquid than in gas so that species formed close together, particularly those formed with only thermal energy, will be constrained to remain close to one another for a longer period in the liquid phase. The latter leads to track and cage effects in liquids that are largely absent in gas-phase radiolysis, and to the use in liquid-phase studies of the concepts of spurs and columnar particle tracks and of diffusion kinetics to describe the nonhomogeneous distribution of the primary species. Differences between the radiolysis of liquid water and water vapor are most apparent when radical and molecular yields of primary species in the two systems are compared, so that this will be done before describing the mechanism responsible for the formation of the species in liquid water.

The determination of radical and molecular yields in irradiated water by the use of low concentrations of scavengers is described later in this chapter (p. 304), but a selection of reported values is given in Table 7.4. The hydrated electron and hydrogen atom are often described collectively as the "reducing radicals," since they frequently bring about reduction of dissolved substances, while the hydroxyl and perhydroxyl (HO_2) radicals and hydrogen peroxide are described as the "oxidizing products," since they tend to bring about oxidation. Hydrogen and hydrogen peroxide are "molecular" products and the hydrated electron, hydrogen atom, hydroxyl, and perhydroxyl radicals the "radical" products, while the molecular and radical products are often referred to collectively as the "primary products" or "primary species" and their yields as the "primary yields." This is somewhat anomalous as ions and excited molecules are precursors of the "primary products"; however we have continued to use the term. The primary radical and molecular products are chiefly responsible for the chemical changes observed in irradiated aqueous solutions, and the numerical values given in Table 7.4 are their yields about 10^{-8} sec after passage of a charged particle deposits energy in the system. To maintain a material balance, the radical and molecular yields must be related by

$$G_{-H_2O} = 2G_{H_2} + G_H + G_{e_{aq}^-} = 2G_{H_2O_2} + G_{OH} \qquad (7.38)$$

or, including HO_2, by

$$G_{-H_2O} = 2G_{H_2} + G_H + G_{e_{aq}^-} - G_{HO_2} = 2G_{H_2O_2} + G_{OH} + 2G_{HO_2} \qquad (7.39)$$

In these equations G_{-H_2O} is the net yield of water molecules converted to primary products and not the total number of water molecules originally dissociated, since some water is reformed by radical combination (e.g., Eq. 7.4) (this comment does not apply to water vapor where it is possible to scavenge the radicals before such combination and $G_{-H_2O} = G_{-H_2O}^{total}$). To a first approximation in liquid water, assuming that a collision between two radicals has an equal probability of forming water as of forming one of the molecular

TABLE 7.4 Radical and Molecular Product Yields in Irradiated Water and Water Vapor

Radiation	pH	G_{-H_2O}	G_{H_2}	$G_{H_2O_2}$	$G_{e_{aq}^-}$	G_H	G_{OH}	G_{HO_2}	Ref.
Water vapor									
x or γ-Rays, electrons		8.2	0.5	0	(G_{e^-} = 3.0)	7.2	8.2	0	Previous section
Liquid water									
γ-Rays and fast electrons with energies in the range 0.1 to 20 MeV	0.46	4.45	0.40	0.78	0	3.65	2.90	0.008[a]	48
									14
	3–13	4.08	0.45	0.68	2.63	0.55	2.72	0.026[b]	49
									14
(D$_2$O solutions)[c]	1.3	4.32	0.32	0.66	3.67		3.00		14
(D$_2$O solutions)[c]	4–13	4.12	0.36	0.66	2.96	0.43	2.84		14
Tritium β-particles (E$_{av}$ 5.7 keV)	1	3.97	0.53	0.97	0	2.91	2.0		50
32 MeV He^{2+}	~7	3.01	0.96	1.00	0.72	0.42	0.91	0.05	51
12 MeV He^{2+}	~7	2.84	1.11	1.08	0.42	0.27	0.54	0.07	51
Polonium α-particles (5.3 MeV)	0.46	3.62	1.57	1.45	0	0.60	0.50	0.11	52
^{10}B(n, α)^{7}Li recoil nuclei	0.46	3.55	1.65	1.55	0	0.25	0.45		53
Particle with infinitely high LET, extrapolated from results with accelerated ^{12}C and ^{14}N ions	0.46	~2.9	~1.45	~1.45	0	0	0		54

[a] Ref. 55.
[b] Ref. 51.
[c] For D$_2$O, D should replace H where the latter appears in each of the column headings.

products,

$$G_{-H_2O}^{total} = G_{-H_2O}^{net} + G_{H_2} + G_{H_2O_2} \qquad (7.40)$$

Perhydroxyl radicals are formed in regions of high radical concentration by reaction of hydroxyl radicals with hydrogen peroxide

$$OH + H_2O_2 \rightarrow H_2O + HO_2 \qquad (7.41)$$

However, with low-LET radiation the yields are small and may not make a significant contribution to the reactions taking place. Of the values listed in Table 7.4 for γ-radiation and fast electrons, there is general agreement on the molecular yields of H_2 and H_2O_2 and the radical yields at low pH (a large number of determinations have been made at the pH of the Fricke dosimeter, 0.46, which is the pH of 0.4 M H_2SO_4) but some uncertainty, attributable to incomplete knowledge of the radiolysis mechanisms, of radical yields in neutral and basic solutions.

At low pH the hydrated electron is rapidly scavenged by hydrogen ions, and it is consequently reported as H in Table 7.4 at pH below 1;

$$e_{aq}^- + H^+ \rightarrow H \qquad k_{42} = (2.3 - 2.4) \times 10^{10} \, M^{-1} \, sec^{-1} \qquad (7.42)$$

(Unless specific references are given, the rate constants quoted in this chapter are mean values derived from the tables compiled by Anbar and Neta (56), Anbar, Bambenek, and Ross (57), and Dorfman and Adams (58). Most of the rate constants referred to are collected in Tables 7.7, 7.10, and 7.11.) At pH above about 4, the perhydroxyl radical dissociates

$$HO_2 \rightleftharpoons O_2^- + H^+ \qquad pK \ 4.88 \qquad (7.43)$$

while in very basic solutions (pH > 11) the hydroxyl radical and hydrogen peroxide also dissociate

$$OH \rightleftharpoons O^- + H^+ \qquad pK \ 11.9 \qquad (7.44)$$

$$H_2O_2 \rightleftharpoons HO_2^- + H^+ \qquad pK \ 11.6 \qquad (7.45)$$

The effect of these acid-base equilibria on the primary yields from water is illustrated in Fig. 7.1, which shows the nature and yields of the primary species that would be expected to react with a dilute ($< 10^{-3}$ M) solution of scavenger at various pH in the range 0.5 to 13. It was assumed in preparing the graph that acid-base equilibria for OH and H_2O_2 would be established before reaction with the solute, but that the proportions of H and e_{aq}^- reacting with the solute would be governed by competition between the various possible reactions, i.e., between reactions 7.42 and 7.46

$$H + OH^- \rightarrow e_{aq}^- + H_2O \qquad k_{46} = 2.3 \times 10^7 \, M^{-1} \, sec^{-1} \qquad (7.46)$$

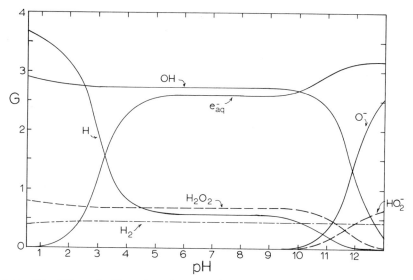

FIGURE 7.1   *Effect of pH on the primary products of water irradiation Yields are for γ-rays and fast electrons with energies of the order of 0.1 to 20 MeV and assume that OH and H_2O_2 achieve acid-base equilibrium. The yields of H and e_{aq}^- are the amounts of these species which would react with 10^{-3} M scavenger (S) if $k(e_{aq}^- + S) = 2 \times 10^{10}$ M^{-1} sec^{-1} and $k(H + S) = 10^7$ M^{-1} sec^{-1}.*

and reaction of H and e_{aq}^- with 10^{-3} M solute for which rate constants of 10^7 and 2×10^{10} M^{-1} sec^{-1}, respectively, were assumed. This procedure was adopted because hydrated electron reactions often have extremely high rate constants while hydroxyl radicals react more slowly (though still very rapidly), so that there is some possibility of establishing an acid-base equilibrium with the latter, although not with the former, before reaction with the solute. Figure 7.1 does not include the possibility that in strongly acid solution H will form H_2^+.

Values reported in Table 7.4 are for low concentrations of scavenger (less than about 10^{-3} M), room temperature, and moderate dose rates. Higher concentrations of scavenger (p. 263) and higher temperatures (59) give rather higher radical yields and somewhat lower molecular yields. Very high dose rates (above 10^9 rad sec^{-1}) (60 to 63), on the other hand, give lower radical and higher molecular yields, as does use of higher-LET radiation. G_{-H_2O} is lower when the molecular yields are lower and, by extrapolating results obtained with accelerated ^{12}C and ^{14}N ions which have very high LET's, Imamura, Matsui, and Karasawa (54) have estimated that G_{-H_2O} has a lower limit of about 2.9 for radiation with an infinitely high LET. Substitution of

D_2O for H_2O has little effect on $G_{-\text{water}}$ but the yield of radical products is a little higher and the yield of molecular products a little lower with the D_2O.

Comparison of the primary yields from water vapor and liquid water shows several differences that must be accounted for by any radiolysis mechanism. Most obvious are the higher yields G_{-H_2O}, G_H, and G_{OH} for water vapor compared with liquid water and the effects of pH and LET upon yields for the liquid. Less obvious are the greater effects of solute concentration, dose rate, and substitution of D_2O for H_2O on yields from liquid water. One significant difference that cannot be deduced from Table 7.4 is a rather higher initial yield of hydrated electrons in liquid water than of electrons in the vapor; this result, derived from picosecond pulse radiolysis experiments, is discussed a little later.

The radiolysis mechanism for liquid water is similar to that for water vapor but includes several additional terms necessitated by the greater density of the liquid phase as outlined at the beginning of this section. Energy deposition by passage of a charged particle through a liquid results in the formation of spurs, blobs, and short tracks as described earlier (Fig. 2.22). These are groups of molecules within the liquid containing the initial products of energy deposition, namely ionized and excited molecules and electrons, in proportion to the energy deposited; spurs representing an energy loss of about 6 to 100 eV, blobs about 100 to 500 eV, and short tracks about 500 to 5 keV. The concentration of reactive species increases in the sequence spur, blob, short track and determines the proportions in which these species escape into the bulk of the liquid or react with one another, i.e., the relative proportions of "radical" and "molecular" products. The fraction of the total absorbed energy deposited in spurs, blobs, and short tracks has been estimated for various types of radiation by Mozumder and Magee (64, 65), who find that for ^{60}Co γ-rays 64% of the energy is deposited in isolated spurs, 11% in blobs, and 25% in short tracks. For tritium β-particles, which have an average energy of 5.7 keV, the proportions are spurs 18%, blobs 7.6%, and short tracks 74.4%, illustrating a general tendency toward more energy deposition in short tracks as the LET of the radiation increases. Mozumder and Magee (64, 66) have also estimated the distribution of spur sizes for fast electrons and γ-rays stopped in water, finding that 56% of the absorbed energy is deposited in spurs with one to two radical pairs, 20.6% in spurs with three to six radical pairs, 10.6% in blobs with more than seven radical pairs, and 12.8% in short tracks. The spurs and blobs are pictured as more or less spherical collections of molecules and the short tracks as columnar zones containing a high proportion of ionized and excited molecules along the track of a fast secondary electron. Spurs, blobs, and short tracks produced by the absorption of low-LET radiation are widely separated from one another along the tracks of the ionizing particles, but those produced by high-LET radiation will be closer together and may coalesce to give a cylindrical "track" (65, 67). Formation of the spurs and tracks (from this point the term spur will be taken to include both spurs and the larger blobs) can be represented by

$$H_2O \rightsquigarrow [H_2O^+, H_2O^*, e^-] \qquad (7.47)$$

where the brackets signify that, for the moment, the species are located close together, and H_2O^* represents both singlet and triplet excited molecules.

A number of very rapid processes can occur before the spurs and tracks expand and dissipate by the normal processes of diffusion. These include the ion-molecule reaction

$$H_2O^+ + H_2O \rightarrow H_3O^+ + OH \qquad k_{18} = 9 \times 10^{11} \ M^{-1} \ sec^{-1} \quad (7.18)$$

dissociation of molecules excited to repulsive states

$$H_2O^* \rightarrow H + OH \qquad (7.1)$$

$$\rightarrow H_2 + O \qquad (7.2)$$

and transfer of excitation energy from some of the excited ions and molecules to adjacent water molecules without producing any additional chemical change. Electron hydration precedes spur expansion and is accompanied by neutralization of a fraction of the ions present,

$$H_3O+ \ (or \ \ H_3O^+_{aq}) + e^- \ (or \ \ e^-_{aq}) \rightarrow H + H_2O \qquad (7.48)$$

The net effect of these processes is to produce a relatively high concentration of radicals (of the order of 1 M) in the spurs and particle tracks soon after they are formed. The radical concentration falls as the spurs and tracks expand, but in the early stages the conditions favor such interradical reactions as

$$H + OH \rightarrow H_2O \qquad k_4 = 3.2 \times 10^{10} \ M^{-1} \ sec^{-1} \quad (7.4)$$

$$H + H \rightarrow H_2 \qquad k_5 = 1.3 \times 10^{10} \ M^{-1} \ sec^{-1} \quad (7.5)$$

$$OH + OH \rightarrow H_2O_2 \qquad k_6 = 5.3 \times 10^9 \ M^{-1} \ sec^{-1} \quad (7.6)$$

$$e^-_{aq} + OH \rightarrow OH^- \qquad k_{49} = 3.0 \times 10^{10} \ M^{-1} \ sec^{-1} \quad (7.49)$$

$$e^-_{aq} + e^-_{aq} \rightarrow H_2 + 2OH^- \qquad k_{50} = 5.4 \times 10^9 \ M^{-1} \ sec^{-1} \quad (7.50)$$

$$e^-_{aq} + H \rightarrow H_2 \qquad k_{51} = 2.5 \times 10^{10} \ M^{-1} \ sec^{-1} \quad (7.51)$$

which reform water and produce the molecular products hydrogen and hydrogen peroxide. Radicals that do not react while the spurs and tracks are expanding escape into the bulk of the solution and represent the "radical" products listed in Table 7.4; at this later stage they are usually assumed to be homogeneously distributed in the solution. Events to this point are summarized by the equation

$$H_2O \rightsquigarrow H_2, H_2O_2, e_{aq}^-, H, OH, H_3O^+, {}^3H_2O^*(?) \qquad (7.49)$$

and the yields are those listed in Table 7.4; $G_{H_3O^+} = G_{e_{aq}^-}$. The inclusion of triplet excited water is speculative.

The mechanism described accounts for the main features of the radiolysis of liquid water. For instance, the shift from higher radical yields with low-LET radiation to higher molecular yields with high-LET radiation parallels the shift in energy deposition from spurs to short tracks with increasing LET; radical concentrations are higher in the tracks and these therefore favor interradical reactions forming the molecular products. Very high dose rates will also favor these reactions by increasing the possibility of spur overlap, and an increase in molecular yields is observed at dose rates above 10^9 rad sec^{-1}. High concentrations of radical scavengers can compete with the interradical reactions, reducing the yield of molecular products and the number of radicals that combine to reform water. The pH effect is an example of this, though it is a special case because the hydrogen ion is a specific scavenger for hydrated electrons (Eqs. 7.42 and 7.48) and the reaction produces hydrogen atoms and does not consume hydroxyl radicals. Increasing the hydrogen ion concentration, therefore, has little effect upon the molecular yields, although it does increase the radical yield and G_{-H_2O}. Raising the temperature above room temperature leads to some increase in the radical yield at the expense of the molecular products and can be attributed to rather more rapid diffusion of radicals from the spurs at the higher temperatures. Slightly higher radical yields in D_2O than in H_2O have been attributed (68) to the longer dielectric relaxation time in D_2O, which might allow the electron to diffuse further from the track. Quantitative calculations based on the expanding spur model have been carried out by a kinetic treatment (diffusion kinetics) which takes into account the nonhomogeneous distribution of the radicals, with results that are generally in good agreement with experiment (14, 55, 69).

To this point, the most striking difference between the radiolysis of water vapor and liquid water, namely the significantly higher water-decomposition and radical yields in the vapor, has not been explained. The usual explanation is that formation of radicals by the dissociation of excited molecules (reactions 7.1 and 7.2) is much less efficient in the liquid than in the vapor both because the excited molecules tend to be quenched in the liquid without dissociation and because geminate radical pairs that are formed cannot easily diffuse apart (i.e., they are caged) and will tend to recombine. Certainly the quantum yields for photochemical dissociation of pure water are low (e.g., 0.02 for

H_2 using 184.9-nm light) but increase very considerably in the presence of an organic scavenger (reported quantum yields for hydrogen include 0.33 at 184.9 nm, 0.7 at 147 nm, and 1.03 at 123.6 nm in the presence of a low concentration of organic scavenger, e.g., 5×10^{-3} M methanol) (70 to 72) due to

$$H_2O + h\nu \rightleftarrows [H + OH] \xrightarrow{\text{organic solute}} H_2 + \text{other products} \qquad (7.52)$$

The radicals in brackets represent a geminate radical pair confined, briefly, in a cage of solvent molecules. While the low quantum yield with pure water is consistent with the radiolysis results, the higher quantum yields when a low concentration of organic scavenger is present are not, since they would imply a much larger increase in radical yield with scavenger concentration than is observed in the radiolysis experiments. Anbar (73) has suggested that the anomaly can be explained if photolysis of liquid water brings about excitation of isolated (monomeric) water molecules that exist in small concentration in liquid water and that possess a higher extinction coefficient than the predominant water aggregates. Irradiation, on the other hand, would bring about excitation in monomeric and polymeric water molecules in proportion to their concentration, that is predominantly in the aggregates whose structure favors energy dissipation without dissociation. It is also possible that some of the excited states that dissociate to radical pairs in the vapor give rise to ionization in liquid water.

$$H_2O^* \rightarrow e_{aq}^- + OH + H_3O^+ \qquad (7.53)$$

since Boyle and his colleagues (74) find that the threshold energy for the formation of hydrated electrons by photoionization in liquid water is about half the ionization energy (12.6 eV) in water vapor. This is consistent with the observation by picosecond pulse radiolysis of hydrated electron yields of $G \geqslant 4$ at very short times (3×10^{-11} sec) after the pulse (75 to 77). An initial $G_{e_{aq}^-}$ of 4.8 at 10^{-11} sec has been estimated on the basis of diffusion kinetic calculations (69). Photolysis of water produces predominantly singlet excited states, and these are the states presumed to be involved when similar reactions are initiated radiolytically. Such states will have a lifetime of the order of 10^{-8} sec and are unlikely to survive long enough to escape from the spurs. Relatively little has been published about the triplet excited states of water (e.g., 17, 18) though they would be expected to have a long enough lifetime to escape from the spurs. However, any role assigned to excited molecules that escape from the spurs appears, at the present time, to be largely speculative.

In the case of liquid water it is not possible to correlate the yield of ions formed upon irradiation with the energy necessary to form an ion-pair in the liquid (W), since the latter is not known. Very limited evidence suggests that W is about the same in a liquid and its vapor (36), in which case G_{ion}

by direct ionization would be the same as in water vapor, i.e., about 3.3. It has been suggested that it may be possible to scavenge the "dry" electron and the corresponding "dry hole" (H_2O^+) before they have a chance to react or become hydrated by using sufficiently high concentrations (>1 M) of suitable scavengers (e.g., 78–81).

The formation of H_3O^+ in reaction 7.18 renders the spur more acid than the surrounding medium. Evidence for this effect has been obtained by irradiating solutions of diethyl acetal buffered to pH 7 and showing that hydrolysis occurs although it normally requires a pH much lower than 7 (82); the authors estimate that the pH within the spur must be about 1.4 to account for the observed hydrolysis.

$$CH_3CH(OC_2H_5) + H_2O \xrightarrow{H^+} CH_3CHO + 2C_2H_5OH \qquad (7.54)$$

A second experiment indicative of an acid spur is the observation of absorption attributed to Cl_2^- when concentrated (>1 M), neutral, solutions of inorganic chloride are irradiated; the formation of Cl_2^- normally requires an acid medium (83).

Hydrogen and hydrogen peroxide, the molecular products, are believed to be formed largely by the inter-radical reactions given above. Small amounts of hydrogen may be formed by the ion-molecule reaction

$$H^- + H_2O \rightarrow OH^- + H_2 \qquad (7.13)$$

though the hydride ion becomes less abundant in the mass spectrometer as the water vapor pressure is increased. Hydrogen peroxide may be formed in part by reaction of oxygen atoms with water

$$O + H_2O \rightarrow H_2O_2 \qquad (7.30)$$

The oxygen atoms arise from dissociation of excited water molecules (Eq. 7.2) or the ion-molecule reaction

$$OH^+ + H_2O \rightarrow H_3O^+ + O \qquad (7.25)$$

Cyclopentene reacts with oxygen atoms, forming ethylene and acrolein, and, using this as a scavenger, Brown and Hart (84) have detected a small yield (G 0.008 to 0.02) of a species believed to be the oxygen atom in γ-irradiated water.

Irradiation of pure water in a closed system by low-LET radiation leads to the establishment of a steady state in which low concentrations of hydrogen, oxygen, and hydrogen peroxide are present. The following reactions serve to maintain the concentrations of the molecular products at their low values,

$$OH + H_2 \rightarrow H_2O + H \qquad k_{11} = 4.9 \times 10^7 \ M^{-1} \ sec^{-1} \qquad (7.11)$$

$$H + H_2O_2 \rightarrow H_2O + OH \qquad k_{34} = 9 \times 10^7 \ M^{-1} \ sec^{-1} \qquad (7.34)$$

$$OH + H_2O_2 \rightarrow H_2O + HO_2 \qquad k_{41} = 2.7 \times 10^7 \, M^{-1} \, \text{sec}^{-1} \qquad (7.41)$$

$$e_{aq}^- + H_2O_2 \rightarrow OH^- + OH \qquad k_{55} = 1.2 \times 10^{10} \, M^{-1} \, \text{sec}^{-1} \qquad (7.55)$$

$$e_{aq}^- + O_2 \rightarrow O_2^- \qquad k_{56} = 1.9 \times 10^{10} \, M^{-1} \, \text{sec}^{-1} \qquad (7.56)$$

$$H + O_2 \rightarrow HO_2 \qquad k_{57} = 1.9 \times 10^{10} \, M^{-1} \, \text{sec}^{-1} \qquad (7.57)$$

$$HO_2 \rightleftharpoons H^+ + O_2^- \qquad pK = 4.88 \qquad (7.43)$$

$$2O_2^- + 2H^+ \rightarrow H_2O_2 + O_2 \qquad k_{58} = 5 \times 10^7 \, M^{-1} \, \text{sec}^{-1} \qquad (7.58)$$

The steady state concentrations depend on the LET of the radiation, increasing with increasing LET until with high-LET radiation such as the recoil nuclei from $^{10}B(n, \alpha)^7Li$ the water is decomposed continuously and a steady state is not reached. Irradiation of pure water in a system that is not closed leads to the buildup of a steady state concentration of hydrogen peroxide in the solution and the continual escape of hydrogen and oxygen from the system; i.e., the irradiation effectively decomposes the water to hydrogen and oxygen.

In some cases it is possible to follow the changes taking place in irradiated water and aqueous solutions directly. The hydrated electron, for example, absorbs light over a broad range of wavelengths centered at 715 nm and has been observed as early as 3×10^{-11} sec after an electron pulse and 4×10^{-12} sec after a light pulse by the techniques of picosecond pulse radiolysis and picosecond spectroscopy respectively (see Chapter 5). The decay of the hydrated electron in irradiated solutions is readily followed after times of the order of 10^{-8} sec by conventional pulse radiolysis techniques. The hydroxyl radical also absorbs light, although the absorption maximum is in the ultraviolet and is weaker than that of e_{aq}^-, and has been observed and its decay measured by pulse radiolysis using high dose rates. When the products of radical reactions absorb light, these too may be studied directly using the same techniques, so that a large number of rate constants are available for reactions that occur in irradiated aqueous systems. Reaction rates and pathways are also often inferred from the effect produced by adding a radical scavenger. By combining data obtained in these and other ways, it is possible to draw up a time scale for the events that follow absorption of radiation energy, as in Table 7.5 (see also refs. 65, 85 to 87). Times given in the table are those after passage of an ionizing particle through the system and are expressed using a logarithmic scale in which $pt = -\log t$, so that 10^{-16} sec becomes pt 16 and 10^{-5} sec pt 5. The time scale refers specifically to liquid water but a very similar time scale of events will apply to other liquids. The table is divided into three stages, physical, physiochemical, and chemical, as suggested by Platzman (88) and Kuppermann (55), corresponding with the initial dissipation of energy in the system, the establishment of thermal equilibrium, and the establishment of chemical equilibrium, respectively.

Ice

The primary effects of radiation on ice are probably similar to those produced in liquid water, though the solid structure appears to favor energy dissipation without reaction, and caging and subsequent recombination of radicals, so that yields are considerably lower than in liquid water and water vapor; G_{-H_2O} falls from 4.1 in neutral water at 20°C to 3.4 in ice at −10°C, 1.0 at −78°C, and 0.5 at −200°C (89, 90).

Radicals trapped in irradiated ice can be detected and identified by means of their electron paramagnetic resonance spectra, and additional information can often be obtained by following the spectrum as the frozen sample warms up. For example, irradiation of pure ice at −269°C produces hydrogen atoms which are stable at this temperature and can be identified by their epr spectrum. As the temperature is raised the hydrogen atoms begin to disappear and are completely gone at −196°C (91). However, pure ice irradiated at −196°C shows absorption characteristic of the hydroxyl radical (Fig. 7.2) which in turn disappears in the temperature range −170 to −140°C (92) (an optical absorption peak at 280 nm disappears in the same temperature range). Hydrogen atoms are stable in ice containing sulfuric, phosphoric, or perchloric acid up to −180°C but at higher temperatures disappear by a second-order process, suggesting that they combine to form molecular hydrogen (93); both hydrogen and hydrogen peroxide have been identified in melted irradiated ice. Trapped electrons are not observed when neutral or acid ices are irradiated but are formed with alkaline ices, the irradiated alkaline ice turning blue with an absorption peak near 600 nm in the optical spectrum and a single line near the free electron position in the epr spectrum (94, 95).

Radiation-induced chemical reactions can take place in frozen solutions. Ghormley and Stewart (90) showed that hydrogen peroxide can be decomposed when ice containing dissolved hydrogen and hydrogen peroxide is irradiated at low temperatures. The amount of decomposition is increased when the irradiation is carried out at temperatures at which the hydroxyl radical is believed to be mobile (i.e., above about −50°C). Under these conditions both H and OH are able to diffuse through the solid and the mechanism is probably similar to that occurring in the liquid phase.

Irradiated ice also exhibits fluorescence and thermoluminescence, both of which are influenced by substances present in solution (96).

TABLE 7.5 *Approximate Time Scale for the Radiolysis of Liquid Water*

Stage	pt	Events	Reactions	Species Present
Physical stage (Interaction of the primary radiation with the water; energy transferred is in the form of electronic motion.)	18	Electron with energy in the megaelectron volt range traverses a distance of the order of a molecular diameter.		H_2O^+
	17	α-Particle with energy in the megaelectron volt range traverses a molecule.	$H_2O \rightsquigarrow H_2O^+ + e^-$	e^-
	16	Loss of energy by secondary electrons.	$H_2O \rightsquigarrow {}^1H_2O^*, {}^3H_2O^*$	$^1H_2O^*$
	15	Time between successive ionizations produced by a megaelectron volt energy electron.		$^3H_2O^*$ (Localized in spurs or tracks)
		Time for "vertical" excitation to an electronic excited state.		
	14	Ion-molecule reactions.	$H_2O^+ + H_2O \rightarrow [H_3O^+ + OH]$	$[H_3O^+ + OH]$ In solvent cage
		Period of molecular vibration. Part of the electronic energy transformed into vibrational energy.		
		Dissociation of molecules excited to repulsive states.	$^1H_2O^* \rightarrow [H + OH] \rightarrow [H_2 + O]$	$[H + OH]$ $[H_2 + O]$ In solvent cage
Physiochemical stage (Part of the transferred energy degraded to vibrational and rotational motion; energy transfer, dissociation, ion-molecule reactions)	13	Secondary electrons reduced to thermal energy (about 0.025 eV).		
		Some electrons captured by positive ions. Geminate recombination of caged radicals.	$H_3O^+ + e^- \rightarrow H_2O + H$ $[H + OH] \rightarrow H_2O$	H
		Internal conversion from higher to lowest electronically excited state.		
	12	Radical moves one molecular diameter by diffusion.		
		Spur temperature reaches a maximum estimated to be 50°C above ambient for a		

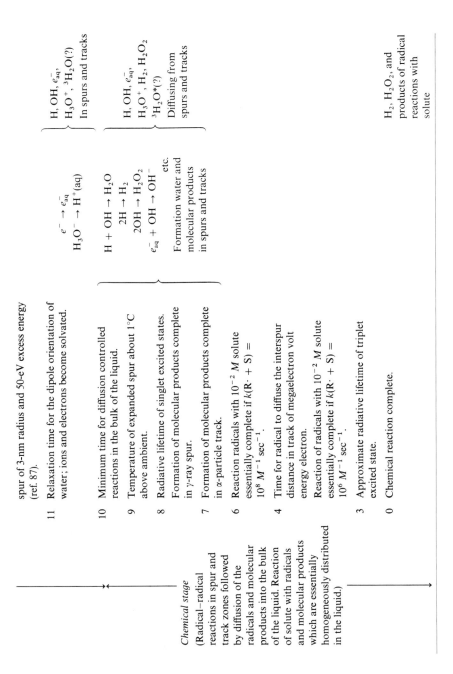

spur of 3-nm radius and 50-eV excess energy (ref. 87).

11 Relaxation time for the dipole orientation of water; ions and electrons become solvated.

$$e^- \rightarrow e^-_{aq}$$
$$H_3O^- \rightarrow H^+(aq)$$

10 Minimum time for diffusion controlled reactions in the bulk of the liquid.

9 Temperature of expanded spur about 1°C above ambient.

8 Radiative lifetime of singlet excited states. Formation of molecular products complete in γ-ray spur.

7 Formation of molecular products complete in α-particle track.

6 Reaction radicals with 10^{-2} M solute essentially complete if $k(R\cdot + S) = 10^8\ M^{-1}\ sec^{-1}$.

$$H + OH \rightarrow H_2O$$
$$2H \rightarrow H_2$$
$$2OH \rightarrow H_2O_2$$
$$e^-_{aq} + OH \rightarrow OH^-$$
etc.

Formation water and molecular products in spurs and tracks

4 Time for radical to diffuse the interspur distance in track of megaelectron volt energy electron. Reaction of radicals with 10^{-2} M solute essentially complete if $k(R\cdot + S) = 10^6\ M^{-1}\ sec^{-1}$.

3 Approximate radiative lifetime of triplet excited state.

0 Chemical reaction complete.

Chemical stage
(Radical–radical reactions in spur and track zones followed by diffusion of the radicals and molecular products into the bulk of the liquid. Reaction of solute with radicals and molecular products which are essentially homogeneously distributed in the liquid.)

H, OH, e^-_{aq}, H$_3$O$^+$, ^{3}H$_2$O(?)
In spurs and tracks

H, OH, e^-_{aq}, H$_3$O$^+$, H$_2$, H$_2$O$_2$, ^{3}H$_2$O*(?)
Diffusing from spurs and tracks

H$_2$, H$_2$O$_2$, and products of radical reactions with solute

269

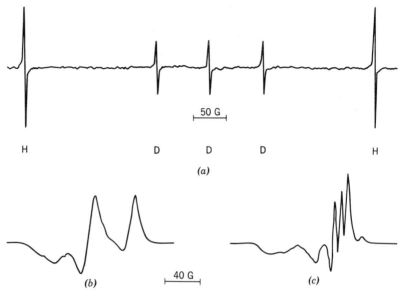

FIGURE 7.2   *Electron paramagnetic resonance spectrum of ice irradiated with tritium β-particles at − 196° C (92). (a) Acid H_2O—D_2O ice, H doublet and D triplet, (b) H_2O ice, OH doublet, (c) D_2O ice, OD triplet.*

AQUEOUS SOLUTIONS

In the following pages a rather arbitrary selection of inorganic and organic systems is described which, it is felt, illustrates the main features of the reactions induced by radiation in dilute aqueous solutions. Very often the initial products from these reactions will themselves react with the primary radicals, particularly if the irradiation is prolonged, but this experimental complication has been ignored in the following account.

When dilute aqueous solutions are irradiated practically all the energy absorbed is deposited in water molecules and the observed chemical changes are brought about *indirectly* via the molecular and, particularly, the radical products. *Direct* action due to energy deposited directly in the solute is generally unimportant in dilute solutions (i.e., at solute concentrations below about 0.1 *M*). At higher solute concentrations direct action may be significant, and there is some evidence that solutes may react directly with the initial products of energy deposition, the "dry" electron and the "positive hole" (i.e., H_2O^+) (cf. p. 310).

An estimate of the upper limit for the radical concentration in the bulk of an irradiated solution can be made by equating the rate of radical formation with that for radical removal, assuming that the radicals are initially homogeneously distributed (which they are not). In the following calculation radicals formed with a G value of 10 and at a dose rate of 10^2 rads sec^{-1} are assumed to disappear by reacting with one another with a bimolecular rate constant of 10^{10} M^{-1} sec^{-1}. Then, assuming a steady state radical concentration of $[R \cdot]$ mole $liter^{-1}$,

rate of radical reaction = rate of radical formation

$$[R \cdot]^2 \times 10^{10} \frac{(liter)}{(mole\ sec)} = \frac{10}{100} \frac{(radicals)}{(eV)} \times \frac{1}{6.022 \times 10^{23}}$$

$$\frac{(mole)}{(molecule\ or\ radical)} \times 10^2 \frac{(rad)}{(sec)} \times 6.241 \times 10^{13} \frac{(eV)}{(g\ rad)} \times 10^3 \frac{(g)}{(liter)}$$

and

$$[R \cdot] = 1.0 \times 10^{-8}\ M$$

At 10^5 and 10^8 rads sec^{-1}, the estimated radical concentrations are 3×10^{-7} and 1×10^{-5} M, respectively. The radicals are not, in fact, homogeneously distributed, and if allowance is made for radical combination in the spurs, the radical concentration in the bulk of the solution is even lower than estimated above. The importance of these estimates in regard to the radiolysis of aqueous solutions is that they show that even though radical–radical reactions have very high rate constants, they will rarely compete with radical–solute reactions unless the solute is also present at an unusually low concentration. In air-saturated solutions, for example, the approximately 2.5×10^{-4} M concentration of oxygen is more than sufficient to scavenge all H-atoms in competition with the dimerization reaction to give molecular hydrogen,

$$H + O_2 \rightarrow HO_2 \qquad k_{57} = 1.9 \times 10^{10}\ M^{-1}\ sec^{-1} \qquad (7.57)$$

$$2H \rightarrow H_2 \qquad k_5 = 1.3 \times 10^{10}\ M^{-1}\ sec^{-1} \qquad (7.5)$$

Assuming a dose rate of 10^2 rads sec^{-1}, which gives a steady state radical concentration of less than 10^{-8} M, the ratio of the number of hydrogen atoms scavenged by oxygen to the number forming H_2 is given by

$$\frac{H + O_2}{H + H} = \frac{k_{57}[O_2]}{k_5[H]} > \frac{1.9 \times 10^{10} \times 2.5 \times 10^{-4}}{1.3 \times 10^{10} \times 10^{-8}}$$

$$> \frac{37,000}{1}$$

The properties of the radicals produced by the radiolysis of water are now well established and are described in the next section, since they provide a rational basis for discussing radiation-induced changes in the aqueous systems described afterwards. It might be noted here that most of the product yields given in the remainder of the chapter are for fast electron or γ-irradiation. The same products are formed by higher-LET radiation but, generally, with some variation in yield due to the different radical and molecular yields from water with the higher-LET radiation.

The Reducing Radicals (e_{aq}^- and H)

Up until about 1960, the reducing species in irradiated water was assumed to be the hydrogen atom produced by the dissociation of excited water molecules formed directly or by ion recombination. However, evidence was accumulating that rate constants for reactions of the reducing species varied with the pH of the system, and that the reducing species formed in neutral solutions reacted at a different rate to hydrogen atoms formed by

$$OH + H_2 \rightarrow H + H_2O \qquad (7.11)$$

or produced externally in a discharge and introduced into the solution (97–102). Soon afterwards, relative reaction rate determinations made in the presence of chemically inert salts (103, 104) showed that the reducing species in neutral solutions behaved as though it had a unit negative charge. The method depends on the fact that rate constants for reactions between ions of similar charge increase with increasing ionic strength, while rate constants for reactions between ions of opposite charge decrease, and there is relatively little effect if one of the reactants is uncharged. Czapski and Schwarz (103) studied the effect of ionic strength upon the following rate constant ratios, $k(R\cdot + O_2)/k(R\cdot + H_2O_2)$, $k(R\cdot + H^+)/k(R\cdot + H_2O_2)$, and $k(R\cdot + NO_2^-)/k(R\cdot + H_2O_2)$, finding that the first was unaffected, the second decreased, and the third increased by increasing the ionic strength, consistent with a negative charge on the radical $R\cdot$. Further analysis of the results showed that the charge corresponded to a single negative charge. Collinson and Dainton and their colleagues (104) looked at the effect of ionic strength on the reactions of the reducing radical with Ag^+ and acrylamide, concluding that at pH 4 the radical had a unit negative charge but that at pH 2 it was uncharged.[3] Subsequently, Hart and Boag (106) irradiated deaerated water using a high intensity electron pulse and observed a broad transient absorption band at about 700 nm which was similar to the known absorption of the solvated electron in liquid ammonia. The intensity of the transient absorption was reduced by low concentrations of electron scavengers such as oxygen, carbon dioxide, and nitrous oxide, and was weak, if present at all, in acid solutions. The conclusion drawn from these, and many other, experiments was that the

---

[3] Similar experiments intended to establish the charge, if any, on the oxidizing radical produced by irradiation of water showed this to be uncharged (105), supporting the belief that this is the OH radical.

major reducing species in neutral and basic solutions is the hydrated electron and that in acid solutions this is rapidly converted to the hydrogen atom. Further work has fully confirmed these conclusions and, at the present time, the hydrated electron is probably the most extensively studied of any radical species (e.g., 73, 107–113).

The absorption spectrum of the hydrated electron is shown in Fig. 7.3 and some of its properties are listed in Table 7.6. For many purposes it is sufficient to regard the hydrated electron as an electron trapped by a small group of water molecules which have become suitably orientated as a consequence of the electron's presence, though more detailed models are available (Refs. 113, 116, 117). Orientation of the water molecules that constitute the trap takes about 10^{-11} sec once a free thermal (or "dry") electron is introduced into the system. Hydrated electrons are believed to be formed as transient intermediates when alkali metals dissolve in water (109, 118) and at the cathode when dilute aqueous solutions of salts are electrolyzed (109). They are also produced by the photolysis of a variety of inorganic ions and organic compounds in polar solvents such as water (109, 110, 119, 120), e.g.,

$$I^-(aq) + h\nu \rightleftharpoons \left[I + e_{aq}^-\right] \qquad (7.59)$$

The products are formed within a solvent cage and normally recombine, but may be scavenged by electron or radical scavengers; using a scavenger, the quantum yield of reaction 7.59 is 0.29 at 25°C with 253.7-nm light (119).

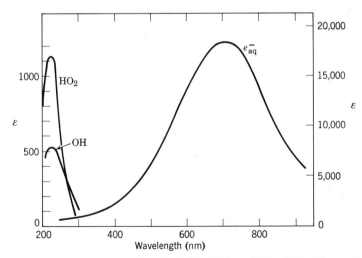

FIGURE 7.3   *Absorption spectra of* e_{aq}^- *(113), OH (114), and* HO_2 *(115). The vertical axes give the extinction coefficients* $(M^{-1}\ cm^{-1})$*, the lefthand scale for OH and* HO_2 *and the righthand for* e_{aq}^-.

TABLE 7.6   *Properties of the Hydrated Electron[a]*

Radius of charge distribution	0.25 to 0.3 nm
Diffusion constant	$4.90 \times 10^{-5} \, cm^2 \, sec^{-1}$
Absorption maximum	
Wavelength	715 nm
Energy equivalent	1.73 eV
Extinction coefficient at 715 nm	$1.85 \times 10^4 \, M^{-1} \, cm^{-1}$
Enthalpy of hydration (ΔH)	$-38.1 \, kcal \, mole^{-1}$
Entropy of hydration (ΔS)	$-1.9 \, cal \, mole^{-1} \, K^{-1}$
$E^\circ \, (e_{aq}^- + H^+ \to \frac{1}{2}H_2)$	2.77 V
Half-life	
In neutral water (pH 7.0)	$2.1 \times 10^{-4} \, sec$
In basic solution	$7.8 \times 10^{-4} \, sec$

[a] Data are for 25°C and are from tables prepared by Thomas (112) and Hart and Anbar (113).

Addition of sodium to liquid ammonia produces a relatively stable solution of solvated electrons which, in this form, have been used as reducing agents in organic chemistry for many years (121–123).

The standard redox potential (E°) provides a measure of the strength of the hydrated electron as a reducing agent and is derived using the following cycle (124),

$$e_{aq}^- + H_2O \to H(aq) + OH^- \qquad \Delta G^\circ = +8.4 \, kcal \, mole^{-1} \qquad (7.60)$$

$$H(aq) \to H(g) \qquad \Delta G^\circ = -4.6 \, kcal \, mole^{-1} \qquad (7.61)$$

$$H(g) \to \tfrac{1}{2}H_2 \qquad \Delta G^\circ = -48.6 \, kcal \, mole^{-1} \qquad (7.62)$$

$$H_3O^+ + OH^- \to H_2O \qquad \Delta G^\circ = -19.1 \, kcal \, mole^{-1} \qquad (7.63)$$

$$\overline{e_{aq}^- + H_3O^+ \to \tfrac{1}{2}H_2 \qquad \Delta G^\circ = -63.9 \, kcal \, mole^{-1}} \qquad (7.64)$$

For reaction 7.60, ΔG° is calculated from the equilibrium constant by substituting in $\Delta G^\circ = -RT \ln K$ and taking $K = k_{forward}/k_{back} = 16/2.3 \times 10^7$ or 7.0×10^{-7}, while ΔG° for reaction 7.61 is assumed to be the same as the free energy of hydration of the helium atom, $-4.6 \, kcal \, mole^{-1}$; the remaining free energies are established values (125). The quantity E° for reaction 7.64 is found by substituting in $\Delta G^\circ = -nFE^\circ$ where the number of moles of electrons transferred between the species is one ($= n$) and F is the Faraday (23.05 kcal equiv^{-1} V^{-1}), giving $E^\circ = 2.77$ V. This is also the standard redox potential for the half reaction

$$e_{aq}^- \rightleftharpoons nH_2O + e^- \qquad (7.65)$$

since $E°$ for the second half reaction contained in reaction 7.64,

$$H_3O^+ + e^- \rightleftharpoons \tfrac{1}{2}H_2 \qquad (7.66)$$

is, by definition, zero. The corresponding value for the hydrogen atom ($E° = 2.31$ V) is found by summing Eqs. 7.61 and 7.62.

Rate constants for typical reactions of the hydrated electron are collected in Table 7.7 and compared with the corresponding reactions of the hydrogen atom. To facilitate comparison of the rate constants given in this table and in Tables 7.10 and 7.11, all are given in units of 10^7 M^{-1} sec^{-1}; i.e., the actual values are divided by 10^7 to give numbers that can be compared more readily. Many of the hydrated electron reactions listed have very low activation energies so that most, if not all, of the collisions between the reactants lead to reaction, and the rate of reaction is governed by the rate at which the reactants diffuse together; such "diffusion-controlled" reactions have rate constants of the order of 10^{10} M^{-1} sec^{-1} (i.e., ~ 1000 on the scale used in the tables). The expected rate of a diffusion controlled reaction can be calculated (126) using

$$k_{AB} = \frac{4\pi r_{AB} D_{AB} N_A}{1000} \left\{ \frac{Z_A Z_B e^2}{r_{AB}\varepsilon kT} \bigg/ \left(\exp\left[\frac{Z_A Z_B e^2}{r_{AB}\varepsilon kT} \right] - 1 \right) \right\} \qquad (7.67)$$

where r_{AB} (cm) is the sum of the radii of the reacting species A and B, D_{AB} (cm^2 sec^{-1}) is the sum of their diffusion constants, N_A is Avogadro's number, Z_A and Z_B are the charges on A and B, respectively, e (esu) is the charge on the electron, k (ergs molecule^{-1} K^{-1}) the Boltzman constant, and ε the dielectric constant of the solution. If either A or B is uncharged, the expression reduces to

$$k_{AB} = \frac{4\pi r_{AB} D_{AB} N_A}{1000} \; M^{-1} \; \text{sec}^{-1} \qquad (7.68)$$

which is the encounter rate of A and B; the additional terms in Eq. 7.67 make allowance for electrostatic forces if both A and B are charged. Some hydrated electron reactions are even faster than calculated for diffusion-controlled processes, and it has been suggested (e.g., 113) that in these instances the electron migrates by a tunneling mechanism, moving from trap to solute without displacing the intervening molecules. Tunneling can also explain the mobility of the hydrated electron, which is much higher than expected for a singly charged ion with a radius of 0.3 nm.

Hydrated electron reactions are single electron transfer processes represented by

$$e_{aq}^- + S^n \rightarrow S^{n-1} \qquad (7.69)$$

where n is the charge ($+$ve, 0, or $-$ve) on the solute. Cations having oxidation potentials (Table 7.8) less than that of the hydrated electron are reduced,

TABLE 7.7 Rate Constants for Reactions of the Hydrated Electron and Hydrogen Atom[a]

Solute	Reaction with e_{aq}^- Products	Reaction with e_{aq}^- Rate Constant ($10^7\ M^{-1}\ sec^{-1}$)	Reaction with H Products	Reaction with H Rate Constant ($10^7\ M^{-1}\ sec^{-1}$)
Species present in irradiated water				
	$(H_3O^+ + OH^- \to 2H_2O\ \ 14{,}300)$			
e_{aq}^-	$H_2 + 2OH^-$	540[b]	$H_2 + OH^-$	2500
H	$H_2 + OH^-$	2500	H_2	1300[b]
OH	OH^-	3000	H_2O	3200
O^- (pH 13)	$2OH^-$	2200	OH^-	2
HO_2	HO_2^-	1300	H_2O_2	2000
O_2^- (pH 11)	O_2^{2-}	2350	HO_2^-	
H^+ (aq)	H		H_2^+	0.00026
OH^-	—	<1	e_{aq}^-	2.3
H_2			—	
O_2	O_2^-	1900	HO_2	1900
H_2O	$H + OH^-$ (16 $M^{-1}\ sec^{-1}$)		—	
H_2O_2	$OH + OH^-$	1200	$OH + H_2O$	9
Inorganic solutes				
Br_2^-	$2Br^-$	1300		
Cd^{2+}	Cd^+	5300		<0.01
Ce^{3+}		<100		
CO	$CO^-(\to H\dot{C}O)$	100		
CO_2	CO_2^-	770		
HCO_3^-	HCO_3^{2-}	<0.1		<0.1
$Cr_2O_7^{2-}$		5000		~0.003
Cu^{2+}	Cu^+	3500	$Cu^+ + H^+$	1600
Fe^{2+}	Fe^+	20	$Fe^+ + H^+$	60
Fe^{3+}	Fe^{2+}	~5000	$Fe^{2+} + H^+$	1.6
$Fe(CN)_6^{4-}$		<0.01		5
				4
				200

Species	Reaction / Product	Value	Note	Value
I_3^-	$I_2 + I^-$	2000		0.01
N_2O	$N_2 + O^- \; (\to OH)$	870		100
NO	$NO^- \; (\to HNO)$	2700		1.4
NO_2^-	NO_2^{2-}	450		
NO_3^-	$NO_3^{2-} \; (\to NO_2 + 2OH^-)$	1100		
Na^+		<0.01		<0.0002
NH_4^+	$H + NH_3$	0.2		
$H_2PO_4^-$	$H + HPO_4^{2-}$	0.42		
HS^-	$H + S^{2-}$	0.06		
H_2S	$H + HS^- \; (65\%),$ $H_2 + S^- \; (35\%)$	1350		
SF_6	$6F^- + SO_4^{2-} + 7H^+$	1650		
SO_4^{2-}		<0.1		

Organic solutes

Species	Reaction / Product	Value	Note	Value
Acetaldehyde		350		3.4
Acetic acid	$CH_3COO^- + H$	8.4 (ref. 128)	(pH 1)	0.02
	$CH_3CO\cdot + OH^-$	9.6 (ref. 128)		
Acetate ion (pH 10)		<0.1	(pH 7)	0.027
Acetone	$(CH_3)_2\dot{C}\!-\!O^-$	590	H addition	0.04
			H abstraction	0.19
Acetonitrile	$CH_3CH\!=\!\dot{N} + OH^-$	2.5	$CH_3CH\!=\!\dot{N}$	0.35
N-Acetylalanine (pH ~7)		1.0	(pH 1)	0.8
Acetylene		3500		
Acrylamide	$CH_3\dot{C}HCONH_2$	1800	$CH_3\dot{C}HCONH_2$	1800
Allyl alcohol		<0.1		230
Aniline		<2		180
Benzene	$C_6H_7 + OH^-$	1.3		53
Benzoic acid (pH 5.4)		3300		100
Benzoate ion (pH 5 to 14)		330	(pH 7)	87
Benzophenone		3000		
Benzyl alcohol	$(C_6H_5)_2\dot{C}\!-\!O^-$	13		65

TABLE 7.7 *(Cont'd)*

Solute	Reaction with e_{aq}^-		Reaction with H	
	Products	Rate Constant $(10^7\ M^{-1}\ sec^{-1})$	Products	Rate Constant $(10^7\ M^{-1}\ sec^{-1})$
Benzyl chloride		550		
Bromoacetate ion (pH 10)	$Br^- + \cdot CH_2COO^-$	620 (pH 9)	$HBr + \cdot CH_2COO^-$ $H_2 + Br\dot{C}HCOO^-$	35 <0.2
Bromobenzene		430		
p-Bromophenol		1200		
Chloroacetic acid (pH 1 to 1.5)	$Cl^- + \cdot CH_2COOH$	690	$HCl + \cdot CH_2COOH$ $H_2 + Cl\dot{C}HCOOH$	0.008 0.018
Chloroacetate ion (pH 7 to 11)	$Cl^- + \cdot CH_2COO^-$	120	$HCl + \cdot CH_2COO^-$ $H_2 + Cl\dot{C}HCOO^-$	0.026 0.26
Chlorobenzene		50		
Chloroform	$Cl^- + \cdot CHCl_2$	3000		1.2
Cystamine (RSSR) (pH 7.3)	$RSSR^-$	4000		
Cysteine				
+ve Ion (pH 1)		3000	$H_2 + \cdot SCH_2CH(NH_3^+)COOH$	250
Zwitterion (pH 5.5 to 7)	$HS^- + \cdot CH_2CH(NH_3^+)COO^-$	820	$H_2 + \cdot SCH_2CH(NH_3^+)COO^-$ $H_2S + \cdot CH_2CH(NH_3^+)COO^-$	100 12
−ve Ion (pH 11.6)		7.5	H abstraction from C—H	5
Cystine (RSSR)				
+ve Ion (pH 1)	$RSSR^-$	1300	$RSH + RS\cdot$	500
Zwitterion (pH 6.1)		300		>150
−ve Ion (pH 10.7 to 12)		<1		
Diethyl ether		<0.01		4.7
Ethanol	$C_2H_5O^- + H$	5.9		1.7
Ethyl acetate		0.1		0.06
Ethylamine (pH > 11)	$CH_3\dot{C}H_2 + NH_3 + OH^-$	0.25		
+ve Ion (pH < 10)	$H + CH_3CH_2NH_2$			

Compound (pH)	Products		Products	
Fluorobenzene				0.5
Formaldehyde	HCOO⁻ + H	6.5		0.11
Formic acid	HĊO + OH⁻	<1	H₂ + ·COOH	
	HĊO + OH⁻	15 (ref. 128)		
		19 (ref. 128)		22
Formate ion (pH 9 to 11)		~0.001		4
D-Glucose		~0.03		
Glycine				1.7
+ve Ion (pH 3)		47	(pH 2)	0.008
Zwitterion (pH 6.4 to 8.5)	NH₃ + ·CH₂COO⁻	0.7	H₂ + NH₃⁺ĊHCOO⁻	
−ve Ion (pH 11)		0.18		15
Glycylglycine zwitterion (pH 6.4)		25		
Iodoacetate ion (pH 10)	I⁻ + ·CH₂COO⁻	1200		
Iodobenzene		1200		
Iodoethane	I⁻ + ·C₂H₅	1500		
Methacrylate ion (pH 10)		840		
Methane		<1		0.16
Methane thiol	SH⁻ + ·CH₃	1800		
Methanol	CH₃O⁻ + H	<0.001		
Methanol radical	·CH₂OH → CH₂OH⁻	<10		170
Nitrobenzene	C₆H₅NO₂⁻ (→ C₆H₅ṄO₂H)	3000	·C₆H₆NO₂	4.4
Nitromethane (pH 0 to 6)	CH₃NO₂⁻	2500	(pH 1)	
−ve Ion (pH 12)	CH₃NO₂⁻	660		0.04
Oxalic acid (pH 1.3)	OĊCOOH + OH⁻	2500		
− Monoanion (pH 2.8 to 4)		330		
− Dianion (pH 7 to 10)		3		420
Phenol (pH 6.3 to 6.8)		1.8	(pH 7)	12
−ve Ion (pH ~11)		0.4		
DL-Phenylalanine zwitterion (pH ~7)		1300		5
2-Propanol			CH₃ĊOHCH₃	

279

TABLE 7.7 (Cont'd)

Solute	Reaction with e_{aq}^-		Reaction with H	
	Products	Rate Constant ($10^7\ M^{-1}\ sec^{-1}$)	Products	Rate Constant ($10^7\ M^{-1}\ sec^{-1}$)
Purine (pH 7.2)		1700		
Pyridine (pH 5.5 to 7.3)		300	(pH 7)	6.4
Pyrolle (pH 10.3)		0.06		
Ribose		<1		
Styrene		1300		
Tetrachloroethylene		1300		
Tetracyanoethylene		1500		
Tetranitromethane (pH 5.5 to 7)	$C(NO_2)_3^- + NO_2$	5000	(pH 7)	260
Thiourea		300		
Thiophene		6.5		
Thymine		1800		
Trichloroacetate ion (pH ~10)		1200		
Uracil		1100		
Urea		0.028		300

[a] Hydrated electron rate constants are for neutral or basic solutions and are taken from ref. 57; rate constants for hydrogen atom reactions are from refs. 56 or 127; values are generally known to $\pm 25\%$.
[b] The quantity k rather than $2k$.

T A B L E 7.8   *Standard Electrode Potentials*[a]

Reaction	$E°$ (V)	
Species present in irradiated water		
$e_{aq}^- \rightleftharpoons H_2O + e^-$	2.77	Reduction by e_{aq}^-
$H(aq) \rightleftharpoons H^+ + e^-$	2.31	Reduction by H
$OH + H^+ + e^- \rightleftharpoons H_2O$	2.8	Oxidation by OH
$OH + e^- \rightleftharpoons OH^-$	1.4	Oxidation by OH (basic solns.)
$HO_2 + H^+ + e^- \rightleftharpoons H_2O_2$	1.5	Oxidation by HO_2
$HO_2 + 2H^+ + 2e^- \rightleftharpoons H_2O + OH$	1.35[b]	Oxidation by HO_2
$O_2^- + H_2O + e^- \rightleftharpoons HO_2^- + OH^-$	0.4	Oxidation by HO_2 (basic solns.)
$HO_2 \rightleftharpoons O_2 + H^+ + e^-$	0.3[b]	Reduction by HO_2
$O_2^- \rightleftharpoons O_2 + e^-$	0.56	Reduction by HO_2 (basic solns.)
$H_2O_2 + 2H^+ + 2e^- \rightleftharpoons 2H_2O$	1.78	Oxidation by H_2O_2
$H_2O_2 + H^+ + e^- \rightleftharpoons H_2O + OH$	0.72	Oxidation by H_2O_2
$HO_2^- + H_2O + 2e^- \rightleftharpoons 3OH^-$	0.87	Oxidation by H_2O_2 (basic solns.)
$HO_2^- + H_2O + e^- \rightleftharpoons 2OH^- + OH$	−0.24	Oxidation by H_2O_2 (basic solns.)
$H_2O_2 \rightleftharpoons O_2 + 2H^+ + 2e^-$	−0.68	Reduction by H_2O_2
$H_2O_2 + 2OH^- \rightleftharpoons O_2 + 2H_2O + 2e^-$	0.15	Reduction by H_2O_2 (basic solns.)
$HO_2^- + OH^- \rightleftharpoons O_2 + H_2O + 2e^-$	0.08	Reduction by H_2O_2 (basic solns.)
Other solutes		
$Li \rightleftharpoons Li^+ + e^-$	3.045	
$Na \rightleftharpoons Na^+ + e^-$	2.711	
$H_2(g) \rightleftharpoons 2H^+ + 2e^-$	0.000	
$Cu^+ \rightleftharpoons Cu^{2+} + e^-$	−0.158	
$H_2SO_3 + H_2O \rightleftharpoons SO_4^{2-} + 4H^+ + 2e^-$	−0.17	
$2I^- \rightleftharpoons I_2 + 2e^-$	−0.535	
$Fe(CN)_6^{4-} \rightleftharpoons Fe(CN)_6^{3-} + e^-$	−0.69	(−0.46 in basic soln.)
$Fe^{2+} \rightleftharpoons Fe^{3+} + e^-$	−0.770	
$2Br^- \rightleftharpoons Br_2(aq) + 2e^-$	−1.087	
$Tl^+ \rightleftharpoons Tl^{3+} + 2e^-$	−1.247	
$2Cl^- \rightleftharpoons Cl_2(g) + 2e^-$	−1.358	
$Ce^{3+} \rightleftharpoons Ce^{4+} + e^-$	−1.443	

[a] Unless otherwise stated, values are for acid solutions and are from ref. 129, where more comprehensive tables are available.
[b] Ref. 130.

for example

$$e_{aq}^- + Cu^{2+} \rightarrow Cu^+ \tag{7.70}$$

Negative ions formed very often dissociate or react with water. With proton donors, the negative ion dissociates giving a hydrogen atom, particularly important cases in water radiolysis being the reaction with water itself and with the hydrated proton,

$$e_{aq}^- + H_2O \rightarrow H + OH^- \quad k_{71} = 16 \ M^{-1} \sec^{-1} \quad (7.71)$$

$$e_{aq}^- + H_3O^+ \rightarrow H + H_2O \quad k_{42} = 2.35 \times 10^{10} \ M^{-1} \sec^{-1} \quad (7.42)$$

The former sets an upper limit to the lifetime of the hydrated electron in water, the half-life being given by $\tau_{1/2} = 0.693/k_{71}[H_2O] = 7.8 \times 10^{-4}$ sec, since the reaction is pseudo first order. The rate constant for reaction 7.71 was determined by following the decay of the hydrated electron absorption in slightly basic solutions (pH 8.3 to 9) containing exceptionally low concentrations ($<5 \times 10^{-9} \ M$) of impurities and given low absorbed doses of 1 to 2 rads to give a concentration of e_{aq}^- of about $5 \times 10^{-9} \ M$ (131); under these conditions no other reaction of e_{aq}^- need be considered. In neutral solutions (pH 7), the low concentration of hydrogen ions present is sufficient to scavenge a significant proportion of the hydrated electrons and the half-life of e_{aq}^- is reduced to 2.1×10^{-4} sec. Reaction 7.42, however, is of most importance at low pH (see Fig. 7.1) where it may lead to changes in mechanism and yield by scavenging hydrated electrons before they can react with other solutes; radical yields increase at low pH (Table 7.4) as a consequence of electron scavenging within the spurs and tracks. The intermediate in this reaction, H_3O, is usually assumed to have too short a lifetime to react directly with solutes, but is sometimes postulated as an independent species (e.g., 73, cf. 132). Negative ions formed by reaction of e_{aq}^- with a solute frequently fragment so as to form a small stable negative ion such as OH^-, HS^-, or Cl^- as in

$$e_{aq}^- + H_2O_2 \rightarrow OH^- + OH \quad (7.55)$$

$$e_{aq}^- + H_2S \rightarrow HS^- + H \quad (7.72)$$

$$\rightarrow S^- + H_2 \quad (7.73)$$

or react with water with the formation of OH^-, e.g.,

$$e_{aq}^- + CO \rightarrow CO^- \xrightarrow{H_2O} H\dot{C}O + OH^- \quad (7.74)$$

$$e_{aq}^- + N_2O \rightarrow N_2O^- \xrightarrow{H_2O} N_2 + OH + OH^- \quad (7.75)$$

Inorganic solutes containing unpaired electrons generally react at diffusion-controlled rates and give relatively stable negative ions,

$$e_{aq}^- + OH \rightarrow OH^- \quad k_{49} = 3.0 \times 10^{10} \ M^{-1} \sec^{-1} \quad (7.49)$$

$$e_{aq}^- + O_2 \rightarrow O_2^- \quad k_{56} = 1.9 \times 10^{10} \ M^{-1} \sec^{-1} \quad (7.56)$$

though organic radicals are generally less reactive.

Reactions of the hydrated electron with organic compounds have been reviewed by Anbar and Hart (111, 113) and shown to be those expected of a nucleophilic reagent, attack on alkene double bonds, for example, being enhanced by adjacent electron-withdrawing substituents. Neighboring substituents also affect the reactivity of other multiple bonds ($C\equiv C$, $C=O$, $N=O$) in aliphatic compounds toward the hydrated electron. Singly bonded compounds containing only C, H, O, N, or F are unreactive, but compounds containing halogen (other than fluorine), disulfide (—S—S—), or un-ionized thiol (—SH) groups react rapidly. Aliphatic halogen compounds (RX where X = Cl, Br, or I) are quantitatively dehalogenated by reaction with e_{aq}^-,

$$e_{aq}^- + RX \rightarrow RX^- \rightarrow R\cdot + X^- \tag{7.76}$$

the rate of reaction increasing in the series Cl < Br < I. Nitro compounds are also highly reactive and the rates of reaction with the polychloro- and polynitro-methanes appear to be higher than calculated assuming the reaction to be diffusion controlled, suggesting that a tunnelling mechanism may be concerned (113). Among the aromatic compounds, benzene, phenol, and aniline react relatively slowly with hydrated electrons but benzoic acid and the halogen and nitro-substituted benzenes react at rates that are, or are close to, diffusion controlled.

The reaction with tetranitromethane,

$$e_{aq}^- + C(NO_2)_4 \rightarrow C(NO_2)_3^- + NO_2 \tag{7.77}$$

was used to establish the molar extinction coefficient of the hydrated electron as the product (nitroform) anion is stable and its extinction coefficient known. The experiment involved simultaneous observation of the decrease in the hydrated electron absorption at 578 nm and the increase in the anion absorption at 366 nm, where the anion has a molar extinction coefficient of $10,200$ M^{-1} cm^{-1}, and led to a value of $\varepsilon_{578} = 10,600$ M^{-1} cm^{-1} for the hydrated electron (133); the extinction coefficient at other wavelengths could then be calculated knowing the absorption spectrum of e_{aq}^-.

The properties and reactions of the hydrogen atom have not been documented as extensively as those of the hydrated electron, primarily because the hydrogen atom does not absorb in a readily accessible region of the spectrum (in the gas phase, ground state hydrogen atoms do not absorb at wavelengths longer than 122 nm). However, a substantial number of rate constants for hydrogen atom reactions have been determined indirectly by pulse radiolysis and competition experiments (56; Table 7.7) and the behavior of hydrogen atoms in aqueous systems is well understood. The hydrogen atom ($E°$ 2.31 V) is a slightly less powerful reducing agent than the hydrated electron ($E°$ 2.77 V), but reduces cations with lower oxidation potentials, e.g.,

$$H + Cu^{2+} \rightarrow H^+ + Cu^+ \tag{7.78}$$

Addition reactions occur with species containing unpaired electrons

$$H + OH \rightarrow H_2O \qquad (7.4)$$

$$H + O_2 \rightarrow HO_2 \qquad (7.57)$$

and with unsaturated and aromatic organic compounds

$$H + CH_3C\equiv N \rightarrow CH_3CH=\dot{N} \quad (or\ CH_3\dot{C}=NH) \qquad (7.79)$$

$$H + \text{⬡} \longrightarrow \text{⬡}\langle{}^H_H \qquad (7.80)$$

The usual reaction with saturated organic compounds is hydrogen abstraction to give molecular hydrogen and an organic radical,

$$H + CH_3OH \rightarrow H_2 + \cdot CH_2OH \qquad (7.81)$$

In strongly basic solutions (pH > 10), the hydrogen atom may react with OH^- to form a hydrated electron,

$$H + OH^- \rightarrow e_{aq}^- + H_2O \qquad k_{46} = 2.3 \times 10^7\ M^{-1}\ sec^{-1} \qquad (7.46)$$

The reaction has a fairly high rate constant and will often compete with hydrogen atom-solute reactions. Reaction of hydrogen atoms with hydrogen ions can occur in strongly acid solutions to give H_2^+,

$$H + H^+ \rightarrow H_2^+ \qquad k_{82} = 2.6 \times 10^3\ M^{-1}\ sec^{-1} \qquad (7.82)$$

but this reaction has a relatively low rate constant and is generally not significant.

One of the main differences between the hydrogen atom and the hydrated electron, apart from reaction rates, is that the former often abstracts hydrogen from organic solutes giving molecular hydrogen while the latter reacts to give other products. Advantage is taken of this difference to measure G_H in neutral solutions by irradiating solutions containing an electron scavenger (e.g., acetone, ferricyanide, and bicarbonate) and an organic compound that will give H_2 with H atoms (e.g., 2-propanol, formate, and methanol), when $G_{H_2} + G_H = G(H_2)$. Hydrogen atoms can be produced by passing hydrogen through an electrodeless high-frequency electric discharge, and this method has been used in studying hydrogen atom reactions in several systems (e.g., 99, 101, 134).

Molecular hydrogen, though a reducing species and a product of water radiolysis, plays a minor role in the radiolysis of aqueous solutions and is not included in this discussion of the reducing species. This is so both because the solubility of hydrogen in water is low, so that most of it escapes from the solution, and because the rate constants for molecular hydrogen reactions are rather low. The reaction of most direct interest

$$H_2 + OH \rightarrow H + H_2O \qquad k_{11} = 4.9 \times 10^7 \, M^{-1} \, sec^{-1} \qquad (7.11)$$

cannot normally compete with other hydroxyl radical reactions unless the pressure of hydrogen is increased above atmospheric (a solution saturated with H_2 at atmospheric pressure contains about $7.8 \times 10^{-4} \, M \, H_2$). However this reaction is significant if pure water is irradiated in a closed system, since it contributes to the back reactions keeping the net decomposition low, or if aqueous solutions are irradiated under high pressures of hydrogen (135).

The Oxidizing Species (OH, HO_2, *and Their Anions*, H_2O_2)

The hydroxyl radical, OH, is the main oxidizing radical formed when aqueous solutions are irradiated. With high-LET radiation, considerable interradical reaction occurs in the particle track producing H_2O, H_2, and H_2O_2 and reaction of OH with the latter creates a small yield of perhydroxyl radicals, HO_2, before the track disperses.

$$OH + H_2O_2 \rightarrow H_2O + HO_2 \qquad k_{41} = 2.7 \times 10^7 \, M^{-1} \, sec^{-1} \qquad (7.41)$$

Larger yields of HO_2 are formed in aerated solutions by reaction of e_{aq}^- and H with oxygen

$$e_{aq}^- + O_2 \rightarrow O_2^- \qquad k_{56} = 1.9 \times 10^{10} \, M^{-1} \, sec^{-1} \qquad (7.56)$$

$$H + O_2 \rightarrow HO_2 \qquad k_{57} = 1.9 \times 10^{10} \, M^{-1} \, sec^{-1} \qquad (7.57)$$

Both the hydroxyl and the perhydroxyl radical behave as weak acids and in solution are in equilibrium with their anions, O^- (the oxide ion) and O_2^- (the superoxide or perhydroxide ion), the relative proportions of acid and anion depending on the pH of the solution. Properties of the hydroxyl and perhydroxyl radicals are summarized in Table 7.9.

The pK of the hydroxyl radical is rather high and the undissociated, acid form predominates at pH up to about 12 (Fig. 7.1). The radical is also reactive so that reaction with a solute may precede establishment of the acid-base equilibrium, and, in many cases, it is more convenient to regard hydroxyl ions and solute as competing for hydroxyl radicals in strongly basic solution than to treat the OH as a weak acid; the hydroxyl ion-hydroxyl radical reaction is

$$OH + OH^- \rightarrow O^- + H_2O \qquad k_{83} = 1.2 \times 10^{10} \, M^{-1} \, sec^{-1} \qquad (7.83)$$

Rate constants for hydroxyl radical reactions are collected in Table 7.10. The hydroxyl radical has a rather weak absorption and the rate constants are normally measured by following the formation of a product, if this has

TABLE 7.9 *Properties of Hydroxyl and Perhydroxyl Radicals*

Hydroxyl radical[a]

Diffusion constant	2.3×10^{-5} cm^2 sec^{-1}
Absorption maximum	
OH	230 nm, $\varepsilon_{230} = 530 \ M^{-1}$ cm^{-1}
	$\varepsilon_{260} = 370 \ M^{-1}$ cm^{-1}
O$^-$	240 nm, $\varepsilon_{240} = 240 \ M^{-1}$ cm^{-1}
Enthalpy of hydration (ΔH)	-88 kcal mole^{-1}
$E°$ (OH + $e^- \rightleftharpoons$ OH$^-$)	2.8 V
pK (OH $\rightleftharpoons$ O$^-$ + H$^+$)	11.9 ± 0.2

Perhydroxyl radical[b]

Absorption maximum	
HO$_2$	240 nm, $\varepsilon_{240} = 1150 \ M^{-1}$ cm^{-1}
O$_2^-$	245 nm, $\varepsilon_{245} = 1970 \ M^{-1}$ cm^{-1}
pK (HO$_2$ $\rightleftharpoons$ O$_2^-$ + H$^+$)	4.88 ± 0.10
pK (H$_2$O$_2^+$ $\rightleftharpoons$ HO$_2$ + H$^+$)	1.0 to 1.2

[a] Ref. 58.
[b] Ref. 136 to 138. See also Table 7.8.

TABLE 7.10 *Rate Constants for Reactions of the Hydroxyl Radical and Oxide Ion[a]*

Reactant	Products or Reaction Type	Rate Constant ($10^7 \ M^{-1}$ sec^{-1})
Species present in irradiated water		
e_{aq}^-	OH$^-$	3000 (2200)
H	H$_2$O	2000
OH	H$_2$O$_2$ (HO$_2^-$)	530[b] ($<$2600)
(2 O$^-$ → O$_2^{2-}$)		($<$80)[b]
HO$_2$	H$_2$O$_3$ (→ H$_2$O + O$_2$)	900
H$_2$O$_2^+$ (pH $<$1.51)	H$_3$O$^+$ + O$_2$	1270
O$_2^-$ (pH $>$2.74)	OH$^-$ + O$_2$	1010
OH$^-$	O$^-$ + H$_2$O	1200 (ref. 139)
H$_2$	H + H$_2$O (H + OH$^-$)	4.9 (8)
O$_2$	$-$ (O$_3^-$)	$-$ (290)
H$_2$O	$-$ (OH + OH$^-$)	$-$ (0.17 to 2)
H$_2$O$_2$	HO$_2$ + H$_2$O (O$_2^-$ + OH$^-$)	2.7 (53)
Inorganic solutes		
BH$_4^-$	BH$_4$ + OH$^-$	1200
Br$^-$ (pH 6 to 9)	Br + OH$^-$	110
CO	H + CO$_2$	45
CO$_2$		$<$0.1
HCO$_3^-$	CO$_3^-$ + H$_2$O	1.25
CO$_3^{2-}$	CO$_3^-$ + OH$^-$	41 ($<$1)

Reactant	Products or Reaction Type	Rate Constant $(10^7 \ M^{-1} \ sec^{-1})$
Ce^{3+}	$Ce^{4+} + OH^-$	7.2
CNS^-	$CNS + OH^-$	1100 (100)
$C(NO_2)_3^-$	$C(NO_2)_3 + OH^-$	300
Cu^{2+}	$Cu^{3+} + OH^-$	35
Fe^{2+}	$Fe^{3+} + OH^-$	35
$Fe(CN)_6^{4-}$	$Fe(CN)_6^{3-} + OH^-$	1200 (<7)
I^-	$I + OH^-$	1550 (96)
NO_2^-	$NO_2 + OH^-$	620 (28)
H_2S	$HS \cdot + H_2O$	1830
$KHSO_4$ (pH 7)		0.15
H_2SO_4 (pH 0.1 to 2)	$HSO_4 + OH^-$	0.07
Tl^+	$Tl^{2+} + OH^-$	7.6
Organic solutes		
Acetic acid (pH 1 to 2)	$\cdot CH_2COOH + H_2O$	1.65
Acetate ion (pH >6)	$\cdot CH_2COO^- + H_2O$	7.5
Acetone	$\cdot CH_2COCH_3 + H_2O$	8.5
Acetonitrile	$CH_3C(OH)\!\!=\!\!\dot{N}$ or $CH_3\dot{C}\!\!=\!\!NOH$	0.55
N-Acetylalanine	Abstraction	46
Acrylamide	Addition	450
Aniline	Addition	790
Benzene	Addition	530
Benzoic acid (pH 3)	Addition	400
Benzoate ion (pH >6)	Addition	540 (<0.6)
Benzyl alcohol	Addition	840
Bromoacetate ion	$\cdot CHBrCOO^- + H_2O$	4.4
1-Butanol	Abstraction	390
2-Butanol	Abstraction	260
Isobutanol	Abstraction	340
t-Butanol	Abstraction	51
Chloroacetic acid	$\cdot CHClCOOH + H_2O$	5.6
p-Chlorobenzoate ion	Addition	500
Chlorobenzene	Addition	620
Chloroform	$\cdot CCl_3 + H_2O$	1.4
Cysteamine (pH 1.4 to 9)		1470
Cysteine (pH 1 to 2)	$\cdot SCH_2CH(NH_3^+)COOH + H_2O$	~900
Cystine (pH 2) (RSSR)	$RSOH + RS\cdot$	480
Diethyl ether	$\cdot C_2H_4OC_2H_5 + H_2O$	235
Ethanol	$\cdot C_2H_4OH + H_2O$	180 (95)
Ethyl acetate	Abstraction	32
Ethylamine (pH >11)	$H_2O + CH_3CH_2\dot{N}H$ and $CH_3\dot{C}HNH_2$	630
+ve Ion (pH <10)	$H_2O + \cdot CH_2CH_2NH_2$	50

TABLE 7.10   (*Cont'd*)

Reactant	Products or Reaction Type	Rate Constant $(10^7 \ M^{-1} \ \text{sec}^{-1})$
Ethylene	$HOCH_2CH_2\cdot$	180
Formaldehyde	$\cdot CHO + H_2O$	~200
Formic acid (pH 1)	$\cdot COOH + H_2O$	13
Formate ion (pH 6 to 11)	$\cdot COO^- + H_2O$	280
Glycine		
+ve Ion (pH 3)	$NH_3^+\dot{C}HCOOH + H_2O$	1.1
Zwitterion (pH ~7)	$NH_3^+\dot{C}HCOO^- + H_2O$	0.9
−ve Ion (pH >9.5)	Abstraction	260
Glycylglycine (pH ~6)	Abstraction	25
D-Glucose	Abstraction	190
Methane	$\cdot CH_3 + H_2O$	24
Methanol	$\cdot CH_2OH + H_2O$	84 (55)
Nitrobenzene	Addition	340
Nitromethane	Addition	<0.9
−ve Ion	Addition	850
p-Nitroso-N,N-dimethyl aniline (PNDA)	Addition	1250
Oxalic acid (pH 2)	$\cdot OOCCOOH + H_2O$	0.8
Dianion	$\cdot OOCCOO^- + OH^-$	0.95 (1.6)
Phenol (pH 6 to 9)	Addition	1200
Phenylalanine (pH ~7)		630
1-Propanol	$\cdot C_3H_6OH + H_2O$	265
Isopropanol	$CH_3\dot{C}(OH)CH_3 + H_2O$	200
Pyridine (pH 7)	Addition	250
Ribose	Abstraction	210
Thymine	Addition	530
Toluene	Addition	300
Uracil	Addition	630

[a] Rate constants are mean values calculated from the data collected by Dorfman and Adams (58); values are generally known to ±25%. Reaction products and rate constants in parentheses are for the oxide radical ion, O^-, and were determined at pH 11, or above; unless pH ranges are specified, the remaining rate constants (for OH) should be applicable at all pH below 11. Abstraction is H-atom abstraction.
[b] The quantity k rather than $2k$.

a conveniently large optical absorption, or by a competition method. In pulse radiolysis determinations, the competition is between a solute (S) whose rate of reaction (k_S) with OH is required and a reference solute (R) for which the rate of reaction with OH is known and which either absorbs strongly itself or forms a product having a strong optical absorption. The two reactions in competition are then

$$OH + S \rightarrow P_S \qquad \text{rate constant, } k_S \qquad (7.84)$$

and

$$OH + R \rightarrow P_R \qquad \text{rate constant, } k_R \qquad (7.85)$$

Assuming the product from the reference solute (P_R) to be the species that absorbs and that neither S nor P_S absorb, the absorbance (A_0) is measured for a solution that contains only the reference solute and also, under as nearly identical irradiation conditions as possible, for a solution that contains both the same concentration of R as before ([R]) and a known concentration of the second solute S ([S]). If the absorbance of the solution containing both solutes is A_s, then the extent of reaction 7.85 in the two solutions, one containing only R the other both R and S, is proportional to A_0 and A_s, respectively, and, applying competition kinetics (p. 180),

$$\frac{1}{A_s} = \frac{1}{A_0} + \frac{k_S[S]}{A_0 k_R[R]} \qquad (7.86)$$

The only unknown in this expression is k_S which can therefore be determined, generally by carrying out a series of experiments at different solute ratios and plotting the results in the form shown in Fig. 7.4. Reference solutes that have been used in this way include (58) the carbonate ion, which must be used in basic solution and forms the absorbing ion CO_3^- (λ_{max} 600 nm, ε_{600} 1.9×10^3 M^{-1} cm^{-1}), the thiocyanate ion, CNS^-, which forms $(CNS)_2^-$

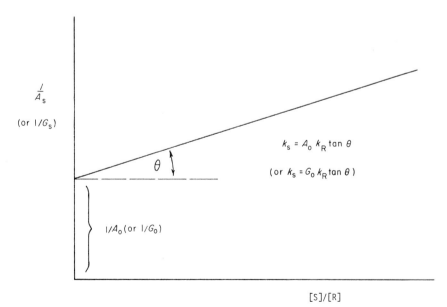

FIGURE 7.4 *Competition plot for determination of a rate constant.*

λ_{max} 475 nm, ε_{475} 7.6 $\times$ 10^3 M^{-1} cm^{-1}) in a two step process, and the ferro-cyanide ion, $Fe(CN)_6^{4-}$, which is oxidized by OH to the ferricyanide ion, $Fe(CN)_6^{3-}$ (λ_{max} 420, ε_{420} 1.0 $\times$ 10^3 M^{-1} cm^{-1}).

Equation 7.86 also applies to competition experiments carried out using continuous, rather than pulsed, irradiation if A_0 and A_s are changed to G_0 and G_s, the yields of product from the reference solute when irradiated alone and in the presence of S, respectively. The product yields can be determined by any convenient method of analysis and do not require use of an electron accelerator, although they do require a complete knowledge of the radiolysis mechanism and of possible competing processes; the mechanism can sometimes be simplified by saturating the solution with N_2O to convert e_{aq}^- to OH, giving a ratio of OH/H of about 10. Reference solutes that have been used in such steady state competition experiments include paranitroso-N,N-dimethylaniline (PNDA) and thymine. Both solutes absorb strongly, PNDA at 440 nm (ε_{440} 3.42 $\times$ 10^4 M^{-1} sec^{-1}) and thymine at 264 nm (ε_{264} 7.95 $\times$ 10^3 M^{-1} sec^{-1}), and the reaction is followed by the loss of solute absorption (58). The competition technique is not limited to hydroxyl radicals but may be applied to other transient species such as the hydrated electron. However, it is always necessary to be sure that competing reactions do not invalidate the kinetics.

With inorganic solutes the hydroxyl radical generally behaves as an oxidizing agent (in keeping with its high oxidation potential) and gives the hydroxide ion by simple electron transfer,

$$OH + S^n \rightarrow OH^- + S^{n+1} \tag{7.87}$$

The solute may be neutral or a positive or negative ion, as in

$$OH + Fe^{2+} \rightarrow OH^- + Fe^{3+} \tag{7.88}$$

$$OH + CO_3^{2-} \rightarrow OH^- + CO_3^- \tag{7.89}$$

Rapid addition reactions occur with free radicals (but not with oxygen unless the OH is ionized to the oxide ion), e.g.,

$$OH + H \rightarrow H_2O \tag{7.4}$$

$$OH + HO_2 \underset{\longrightarrow}{\overset{[H_2O_3 \rightarrow}{\rule{0pt}{0pt}}} \Big\} H_2O + O_2 \tag{7.90}$$

The intermediate, H_2O_3 (hydrogen sesquioxide or hydrogen trioxide), in reaction 7.90 has a significant, pH dependent, half life which is of the order of seconds at pH 1; it is apparently not formed at every OH–HO_2 collision (115, 138).

Organic compounds containing aromatic systems or carbon–carbon multiple bonds also undergo addition reactions with OH, in a similar manner to the addition of H to these compounds,

$$OH + CH_2\!\!=\!\!CH_2 \rightarrow HOCH_2\text{---}CH_2\cdot \qquad (7.91)$$

$$OH + \text{⬡} \rightarrow \text{⬡} \genfrac{}{}{0pt}{}{OH}{H} \qquad (7.92)$$

while hydrogen abstraction is the usual reaction with saturated organic compounds, including those containing carbonyl groups,

$$OH + CH_3COCH_3 \rightarrow H_2O + \cdot CH_2COCH_3 \qquad (7.93)$$

Weaker C—H bonds are broken in preference to stronger bonds; with 2-propanol, for example, 95% of the hydrogen is abstracted from the secondary (α) carbon atom

$$OH + CH_3CH(OH)CH_3 \rightarrow H_2O + CH_3\dot{C}(OH)CH_3 \qquad (7.94)$$

compared with only 37% from the α-carbon atom with 1-butanol, where the C—H bonds are more nearly equal in strength. Inductive effects also influence the site and rate of hydroxyl radical attack, since the radical is a strongly electrophilic reagent, reacting preferentially at sites of high electron density. This is illustrated by the higher rate of reaction with carboxylate anions than with the undissociated acid and the increase in rate of reaction in the series of alcohols methanol, ethanol, 1-propanol, and 1-butanol as larger alkyl groups are substituted for H in CH_3OH (Table 7.10).

Relatively few rate constants are available for the oxide radical ion, O^-, which must be studied in strongly basic solution because of the high pK of the hydroxyl radical. Rate constants which have been measured show that O^- is a little less active in hydrogen abstraction reactions and very much less active in addition to aromatic compounds than OH. The oxide ion is also less effective than OH in oxidizing inorganic anions ($E°$ is lower in basic solutions, Table 7.8), but reacts readily with oxygen to form the ozonide ion, O_3^-, whereas OH is unreactive toward oxygen.

$$O^- + O_2 \rightleftharpoons O_3^- \qquad (7.95)$$

Hydroxyl radicals can be produced without the use of ionizing radiation by the photolysis of hydrogen peroxide,

$$H_2O_2 + h\nu \rightarrow 2OH \qquad (7.96)$$

[253.7 nm light gives a primary quantum yield of 0.5 (140)], and by reaction of certain transition metal ions (e.g., Fe^{2+} and Ti^{3+}) with hydrogen peroxide (e.g., 141, 142).

$$H_2O_2 + Fe^{2+} \rightarrow Fe^{3+} + OH^- + OH \qquad (7.97)$$

However the latter reaction is probably more complex than the equation implies and may form a complex of radical and metal ion rather than free OH.

The perhydroxyl radical (HO_2) and its anion (O_2^-) play a minor role in the radiolysis of oxygen-free solutions but become important when oxygenated systems are irradiated because oxygen is an efficient scavenger for e_{aq}^- and H (Eqs. 7.56 and 7.57). The pK of HO_2 (4.88) is such that HO_2 predominates below about pH 4.5 and O_2^- above about pH 5. A protonated form, $H_2O_2^+$, is stable in strongly acid solutions. Properties of the perhydroxyl radical are given in Table 7.8 ($E°$), Table 7.9 (light absorption and pK), and Table 7.11 (rate constants) and have been extensively reviewed by Bielski and Gebicki (115) and Czapski (138, 143).

The perhydroxyl radical may react as an oxidizing agent or reducing agent depending on the solute. Ferrous ions, which are readily oxidized,

T A B L E 7.11   *Rate Constants for Reactions of the Perhydroxyl Radical and Its Ions*[a]

Reactant	Products	Rate Constant (10^7 M^{-1} sec^{-1})
Reactions of $H_2O_2^+$ (pH < 0.7)		
$H_2O_2^+$	$H_2O_2 + O_2 + 2H^+$	0.11^b
OH	$H_2O + O_2 + H^+$	1270
Reactions of HO_2 (pH 1 to 4)		
HO_2	$H_2O_2 + O_2$	0.25^b
O_2^-	$HO_2^- + O_2$	4.4
H	H_2O_2	2000
OH	$H_2O_3 (\rightarrow H_2O + O_2)$	900
Ce^{3+}	$Ce^{4+} + HO_2^-$	0.021
Cu^{2+}	$Cu^+ + O_2 + H^+$	1.5
Fe^{2+}	$Fe^{3+} + HO_2^-$	0.073
$Fe(CN)_6^{4-}$	$Fe(CN)_6^{3-} + H_2O_2$	0.0164
Te(IV)	Te(VI) + OH	2.5×10^{-5}
Br_2	$Br + Br^- + O_2 + H^+$	15
Br_2^-	$Br_2 + HO_2^-$	360
$C(NO_2)_4$	$C(NO_2)_3^- + NO_2 + O_2 + H^+$	0.056
Reactions of O_2^- (pH > 5)		
O_2^-	$O_2^{2-} + O_2$	5^b
OH	$OH^- + O_2$	1010
$C(NO_2)_4$	$C(NO_2)_3^- + NO_2 + O_2$	200
CO_3^-		~40

[a] Rate constants are mean values calculated from the data given by Czapski (143) and Bielski and Gebicki (115).

[b] The quantity k rather than $2k$.

are oxidized to ferric ion

$$HO_2 + Fe^{2+} \rightarrow Fe^{3+} + HO_2^- \ (+H^+ \rightarrow H_2O_2) \qquad (7.98)$$

while the ceric ion, which is a strong oxidizing agent, is reduced to cerous ion

$$HO_2 + Ce^{4+} \rightarrow Ce^{3+} + O_2 + H^+ \qquad (7.99)$$

Electrode potentials for the perhydroxyl radical in acid and basic solutions (Table 7.8) suggest that HO_2 is a considerably stronger oxidizing agent than O_2^-, while the anion is the slightly stronger reducing agent. Both forms of the radical are inert toward organic compounds unless these contain a particularly weakly bonded hydrogen atom (e.g., ascorbic acid, cysteine, and hydroquinone) when hydrogen abstraction can occur, for example, with cysteine (RSH),

$$HO_2 + RSH \rightarrow H_2O_2 + RS \cdot \qquad (7.100)$$

In the absence of other reaction, perhydroxyl radicals react together forming hydrogen peroxide,

$$2HO_2 \rightarrow H_2O_2 + O_2 \qquad (7.101)$$

$$2O_2^- \ (+2H_2O) \rightarrow H_2O_2 + O_2 + 2OH^- \qquad (7.102)$$

Reactions analogous to reactions 7.100 to 7.102 occur with organic peroxy radicals.

Perhydroxyl radicals can be produced photochemically by photolysis of hydrogen peroxide solutions under conditions that allow the hydroxyl radicals formed to react with H_2O_2 (144), and chemically by the reaction of ions such as Ce^{4+} and Ti^{4+} with excess hydrogen peroxide (138, 145).

$$H_2O_2 + Ce^{4+} \rightarrow Ce^{3+} + HO_2 + H^+ \qquad (7.103)$$

Salts of the radical such as potassium superoxide, KO_2, are stable solids.

Hydrogen peroxide, like the perhydroxyl radical, can act as an oxidizing agent (Eq. 7.97) or a reducing agent (Eq. 7.103) depending on the material present and as a weak acid, dissociating in strongly basic solution

$$H_2O_2 \rightleftharpoons HO_2^- + H^+ \qquad pK = 11.6 \qquad (7.45)$$

and giving rise to a series of salts such as potassium peroxide, K_2O_2. Hydrogen peroxide is inert toward most organic compounds but may react with some organic radicals, e.g.,

$$H_2O_2 + \cdot COOH \rightarrow CO_2 + H_2O + OH \qquad (7.104)$$

Gaseous Solutes

Aqueous systems are generally irradiated in the presence of air or of some other gas used to displace air from the solution. In solutions saturated with air or oxygen, which contain about 2.5×10^{-4} and $1.3 \times 10^{-3} M\ O_2$ respectively at room temperature and atmospheric pressure, the oxygen rapidly scavenges hydrated electrons and hydrogen atoms

$$e_{aq}^- + O_2 \rightarrow O_2^- \qquad k_{56} = 1.9 \times 10^{10}\ M^{-1}\ sec^{-1} \qquad (7.56)$$

$$H + O_2 \rightarrow HO_2 \qquad k_{57} = 1.9 \times 10^{10}\ M^{-1}\ sec^{-1} \qquad (7.57)$$

unless other solutes capable of reacting rapidly with the radicals are present in greater concentration. At low pH, for example, hydrogen ions can compete with the oxygen for hydrated electrons

$$e_{aq}^- + H^+ \rightarrow H \qquad k_{42} = 2.35 \times 10^{10}\ M^{-1}\ sec^{-1} \qquad (7.42)$$

although the competition will have little effect upon product yields if reaction 7.42 is followed reaction 7.57. In pure water, the perhydroxyl radicals dismutate giving hydrogen peroxide (Eqs. 7.101 or 7.102), but they may also react with dissolved substances to bring about distinctly different changes to their parent reducing radicals. Since organic radicals also react rapidly with oxygen, the presence of air or oxygen can lead to quite different products to those formed by irradiation under oxygen-free conditions. Oxygen can be removed from aerated solutions by repeated freeze-pump-thaw cycles or, generally more conveniently, by passing a stream of inert gas through the solution for at least several minutes. Argon, helium, and nitrogen are frequently used for this purpose.

In solutions saturated with hydrogen at atmospheric pressure (about $7.8 \times 10^{-4}\ M\ H_2$ at 25°C) hydroxyl radicals may be converted to hydrogen atoms by

$$OH + H_2 \rightarrow H_2O + H \qquad k_{11} = 4.9 \times 10^7\ M^{-1}\ sec^{-1} \qquad (7.11)$$

but substantially higher hydrogen pressures are necessary if this reaction is to compete with other hydroxyl radical reactions in the systems described in this chapter.

Aqueous solutions are very often purged with nitrous oxide so that hydrated electrons are converted to hydroxyl radicals by the sequence of reactions represented by

$$e_{aq}^- + N_2O \rightarrow N_2O^- \xrightarrow{H_2O} N_2 + OH^- + OH$$
$$k_{75} = 8.7 \times 10^9\ M^{-1}\ sec^{-1} \qquad (7.75)$$

The rate constant and solubility of nitrous oxide in water (a saturated solution contains about $0.02\ M\ N_2O$ at 1 atm, 25°C) are sufficiently high that this reaction can compete successfully with other e_{aq}^- reactions provided the other reactants are present at concentrations of about $10^{-3}\ M$ or less. Estimates of the lifetime of the intermediate N_2O^- ion vary from about

10^{-13} sec to greater than 10^{-9} sec (e.g., 113, 146, 147); if the longer estimates are correct, the intermediate may be able to react directly with some solutes.

Carbon dioxide, which is often adventitiously present in air-saturated solutions, reacts rapidly with hydrated electrons but only slowly with hydroxyl radicals; the hydrated electron reaction gives CO_2^-, which is described later as an intermediate in the radiolysis of formate solutions.

Ferrous Sulfate (Fricke Dosimeter)

The radiation-induced oxidation of ferrous sulfate solutions is one of the most thoroughly studied reactions in radiation chemistry and is the basis for the widely used Fricke dosimeter (see 93, ref. 148). Since the mechanism is well established, the reaction is also used to derive information about the primary radical and molecular yields in irradiated water.

The oxidation is invariably studied in acid solution to avoid precipitation of basic iron compounds, the most frequently used solvent being 0.4 *M* sulfuric acid (pH 0.46). Solutions may be saturated with air or oxygen or purged with an inert gas to remove oxygen, but yields of ferric ion are appreciably lower when oxygen is removed, ^{60}Co γ-radiation, for example, giving $G(Fe^{3+})$ values of 15.5 and 8.2, respectively, in the presence and absence of oxygen. Oxygen present in the solution is consumed during the course of the irradiation, but the yield of ferric ion is not affected significantly by the changing oxygen concentration until the oxygen is virtually exhausted, when the yield rapidly falls to the value characteristic of the oxygen-free solution (Fig. 7.5). The consumption of oxygen is rarely measured precisely but assuming an air-saturated solution to contain about 2.5×10^{-4} *M* O_2, the break in Fig. 7.5 at about 55 krad corresponds to $G(-O_2)$ about 3.7 for ^{60}Co γ-rays. Curves of the same general shape as Fig. 7.5 are typical of many other radiation-induced reactions in aqueous solution where the yield of product is dependent on the presence of oxygen. The break in the curve is often quite sharp, showing that oxygen is effective down to very low concentrations.

Reactions that are likely to contribute to the radiolysis of acid ferrous sulfate solutions can be found in Tables 7.7, 7.8, 7.10, and 7.11. For example, assuming that primary radicals (e_{aq}^-, H, and OH) present in relatively low concentration will react preferentially with the stable solutes (Fe^{2+}, H^+, HSO_4^-, SO_4^{2-}, O_2) present in much higher concentration, the following reactions can be derived by combining pairs of half reactions from Table 7.8

$$e_{aq}^- + Fe^{2+} \rightarrow Fe^+ \qquad E° = \text{about } -0.3 \text{ V} \quad (7.105)$$

$$e_{aq}^- + H^+ \rightarrow H \qquad E° = 0.46 \text{ V} \quad (7.42)$$

$$e_{aq}^- + O_2 \rightarrow O_2^- \qquad E° = 2.21 \text{ V} \quad (7.56)$$

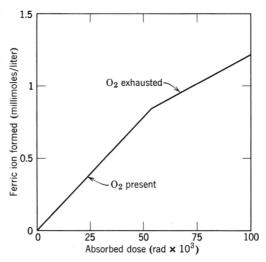

FIGURE 7.5   *Radiation-induced oxidation of air-saturated ferrous sulfate solutions* (^{60}Co *γ-radiation*).

$$OH + Fe^{2+} + H^+ \rightarrow H_2O + Fe^{3+} \qquad E^\circ = 2.03 \text{ V} \qquad (7.88)$$

$$H + Fe^{2+} \rightarrow H^+ + Fe^+ \qquad E^\circ = \text{about } -0.8 \text{ V} \quad (7.106)$$

$$H + O_2 \rightarrow HO_2 \qquad E^\circ = 2.01 \text{ V} \qquad (7.57)$$

Under the standard conditions (25°C, unit activity of reactants and products) reactions having positive values of E° are thermodynamically possible, since this corresponds to a negative free energy change. In fact, the reactions will not be occurring under conditions where each species is at unit activity, and the actual activities should be substituted in the Nernst equation to determine the actual electrode potential, E, for the reaction. The Nernst equation is

$$E = E^\circ - \frac{RT}{nF} \ln \frac{a_C^c a_D^d}{a_A^a a_B^b} \qquad (7.107)$$

for the reaction $aA + bB \rightarrow cC + dD$ where a is the activity of the species specified, R the ideal gas constant, T the absolute temperature, n the number of electrons transferred from the species oxidized to the species reduced in the equation given, and F the Faraday; for practical purposes, the activity of a species present in dilute solution is the same as its concentration (cf. 129). The Nernst equation shows that the reaction will have a positive value of E (i.e., will be spontaneous) if E° is positive and the concentrations of reactants are higher than the concentrations of products, or if the magnitude of a negative E° is smaller than the positive effect due to high reactant and low product concentrations. In practice, positive $E°$ values greater than about 0.5 V are strong presumptive evidence that the reaction will be spontaneous under normal radiolysis conditions.

Although tables of electrode potentials can be used to determine which reactions are thermodynamically possible, they do not give any information about the rates of these reactions: whether, for example, a particular reaction will be fast enough to contribute significantly, or which of several competing reactions will predominate. This information is obtained from the tables of rate constants. The reactions listed above include three in which the hydrated electron takes part and which are, or may be, spontaneous based on their $E°$ values. The rates of the three reactions with Fe^{2+}, H^+, and O_2 are (k_{105}) 2×10^8, (k_{42}) 2.35×10^{10}, and (k_{56}) 1.9×10^{10} M^{-1} sec^{-1}, respectively and, in the Fricke dosimeter solution, the concentrations of these three solutes are 10^{-3} M Fe^{2+}, 0.35 M H^+ (at pH 0.46), and about 2.5×10^{-4} M O_2. The relative contribution of the three reactions is therefore,

$$e_{aq}^- + Fe^{2+} : e_{aq}^- + H^+ : e_{aq}^- + O_2 = k_{105}[Fe^{2+}] : k_{42}[H^+] : k_{56}[O_2]$$
$$= 1 : 4 \times 10^4 : 24$$

In other words, about 99.94% of the hydrated electrons that escape from the spurs and tracks will react with H^+, so that this is the only reaction of e_{aq}^- that need be considered. Proceeding along these lines it is found that reactions 7.88 and 7.57, which are both spontaneous and rapid, account for the behaviour of OH and H in this system. With the selection of HO_2 and H_2O_2 reactions on the same basis, the following mechanism can be developed for the oxidation of ferrous ion in the Fricke dosimeter solution:

$$e_{aq}^- + H^+ \rightarrow H \qquad\qquad k_{42} = 2.35 \times 10^{10}\ M^{-1}\ sec^{-1} \quad (7.42)$$

$$H + O_2 \rightarrow HO_2 \qquad\qquad k_{57} = 1.9 \times 10^{10}\ M^{-1}\ sec^{-1} \quad (7.57)$$

$$OH + Fe^{2+} \rightarrow OH^- + Fe^{3+} \qquad k_{88} = 3.5 \times 10^8\ M^{-1}\ sec^{-1} \quad (7.88)$$

$$HO_2 + Fe^{2+} \rightarrow HO_2^- + Fe^{3+} \qquad k_{98} = 7.3 \times 10^5\ M^{-1}\ sec^{-1} \quad (7.98)$$

$$HO_2^- + H^+ \rightarrow H_2O_2 \qquad\qquad\qquad\qquad\qquad (7.108)$$

$$H_2O_2 + Fe^{2+} \rightarrow Fe^{3+} + OH^- + OH \quad k_{97} = 50\ M^{-1}\ sec^{-1} \quad (7.97)$$

At normal dose rates the concentrations of radicals are very much lower than the solute concentrations and the reactions can be treated as pseudo first order, so that the half life and the time for the reaction to reach 99% completion for each step are given by $\ln 2/(k[S])$ $(= 0.693/k[S])$ and $\ln 100/(k[S])$ $(= 4.605/k[S])$, respectively. Substitution of k_{88} and $[Fe^{2+}] = 10^{-3}$ M in the second of these expressions shows that reaction 7.88 will be 99% complete within about 1.5×10^{-5} sec (15 μsec). Reactions 7.42 and

7.57 will also be complete within this time, but the oxidation of Fe^{2+} by HO_2 is slower and requires 6 to 7×10^{-3} sec (6 to 7 msec) for completion. Reaction of hydrogen peroxide (both the molecular product yield and that formed by reaction 7.108) with ferrous ion (Eq. 7.97) is even slower and requires about 90 sec for 99% reaction with 10^{-3} M Fe^{2+}; the reaction is faster with higher concentrations of Fe^{2+}, but slower if $[Fe^{2+}]$ is less than 10^{-3} M. As a result of these differences in reaction rate, $G(Fe^{3+})$ is time-dependent if the radiation is delivered as a pulse which is short compared with the lifetimes of the various reactions. For 10^{-3} M Fe^{2+}, for example, the rate constants suggest that the following stages will be observed following a short (nanosecond or microsecond) radiation pulse (the values in parentheses are G values for fast electron and γ-radiation obtained by substituting primary yields from Table 7.4 in the expressions given below for $G(Fe^{3+})$; G_{HO_2} was taken to be 0.026 but the other primary yields are those given for pH 0.46):

At 20 μsec

$$G(Fe^{3+}) = G_{OH} \qquad\qquad (G = 2.90) \quad (7.109)$$

At 10 msec

$$G(Fe^{3+}) = G_{e_{aq}^-} + G_H + G_{OH} + G_{HO_2} \qquad\qquad (G = 6.58) \quad (7.110)$$

At times > 100 sec

$$G(Fe^{3+}) = 2G_{H_2O_2} + 3(G_{e_{aq}^-} + G_H + G_{HO_2}) + G_{OH} \quad (G = 15.5) \quad (7.111)$$

$$G(-O_2) = G_{e_{aq}^-} + G_H \qquad\qquad (G = 3.65)$$

Keene (149) has indeed observed these stages in the pulse radiolysis of ferrous sulfate solutions and has also shown that the absorption spectrum of the product changes with time from that of uncomplexed Fe^{3+} (stage 1) to that of a mixture containing a hydroxyl complex of Fe^{3+} formed by reaction 7.98 (stage 2) and, finally, to the spectrum of a sulfate complex of Fe^{3+} ($FeSO_4^+$) which is stable in 0.4 M H_2SO_4 (stage 3). The intermediate complexes formed in the oxidation of ferrous ion have been described in some detail by Jayson, Parsons, and Swallow (150) but, for simplicity, they are omitted from the equations given in this section. Also in the interests of clarity, ferrous and ferric ions are represented as Fe^{2+} and Fe^{3+} rather than as complexes with water and anions, e.g., $Fe^{2+}(H_2O)_5HSO_4^-$, that would more properly represent the ions in sulfuric acid solutions. Thus the equations used give the correct stoichiometric relationships but do not describe the detailed mechanism of the reactions. Dainton and his colleagues (151, 152) have pointed out that reaction 7.97 can be neglected if the ferric ion formed is measured immediately following a short irradiation of a dilute

($<10^{-5}$ M) ferrous ion solution, since at these concentrations the half life of the reaction is of the order of 20 min or longer; $G(Fe^{3+})$ is then given by Eq. 7.110. The normal yield of ferric ion is obtained if the solution is allowed to stand for several hours.

Values of $G(Fe^{3+})$ are affected by factors that alter the primary yields, e.g., pH (Fig. 7.1) and LET (Table 7.4), dose rate at extremely high dose rates, and the concentration of acid and ferrous ion. Increasing LET (e.g., by using α-particles instead of γ-radiation) has the effect of reducing the radical yields but increasing the molecular product yields from water (Table 7.4). Substitution of typical yields in Eqs. 7.109 to 7.111 shows that this leads to a decrease in $G(Fe^{3+})$ with increasing LET (see also Table 3.9). At the very high dose rates attainable with electron accelerators the G value is reduced by the occurrence of inter radical reactions such as

$$2HO_2 \rightarrow H_2O_2 + O_2 \qquad (7.101)$$

which compete with the radical-solute reactions (e.g., Eq. 7.98) that predominate at lower dose rates. The effect of higher dose rates can be compensated to some extent by using higher concentrations of ferrous ion and oxygen; the "super Fricke" dosimeter consists of a 10^{-2} M $FeSO_4$ solution in 0.4 M H_2SO_4 saturated with oxygen and is useable at dose rates up to 7×10^9 rad sec^{-1} (cf., 153). Lower than normal concentrations of acid and ferrous ion increase the possibility that other reactions will compete with those listed above, in addition to the effect that changing pH has upon the primary yields. Low ferrous ion concentrations also increase the possibility of competition between Fe^{3+} and Fe^{2+} for reactive species as the concentration of Fe^{3+} builds up, for example, competition between 7.98 and

$$HO_2 + Fe^{3+} \rightarrow Fe^{2+} + H^+ + O_2 \qquad (7.112)$$

(In spite of the latter possibility, the response of the normal, 10^{-3} M $FeSO_4$, dosimeter solution is linear to an absorbed dose of at least 40,000 rads, which corresponds to the conversion of 65% of the ferrous ion originally present to ferric ion.)

Organic impurities (RH) can increase the yield of ferric ion in aerated solutions (148, 154) by the sequence of reactions:

$$OH + RH \rightarrow R\cdot + H_2O \qquad (7.113)$$

$$R\cdot + O_2 \rightarrow RO_2\cdot \qquad (7.114)$$

$$RO_2\cdot + H^+ + Fe^{2+} \rightarrow Fe^{3+} + RO_2H \qquad (7.115)$$

$$RO_2H + Fe^{2+} \rightarrow Fe^{3+} + RO\cdot + OH^- \qquad (7.116)$$

$$RO\cdot + H^+ + Fe^{2+} \rightarrow Fe^{3+} + ROH \qquad (7.117)$$

Thus each hydroxyl radical brings about the oxidation of three ferrous ions instead of only one, and the yield of ferric ion is spuriously high. Deliberate addition of organic compounds sometimes leads to even higher yields of ferric ion than would be predicted from reactions 7.113 to 7.117 (148). For example, $G(Fe^{3+})$ values up to 75 in the presence of ethanol (154) and about 250 in solutions containing formic acid (155) have been reported. In these examples a chain reaction must be taking place and reaction 7.117 is probably replaced by

$$RO\cdot + RH \rightarrow R\cdot + ROH \qquad (7.118)$$

(The chain is made up of reactions 7.114 to 7.116 and 7.118.) Addition of a small quantity of sodium chloride (about $10^{-3}\ M$) reduces the effect of organic impurities by converting the hydroxyl radicals to chlorine atoms in a reaction whose overall stoichiometry is represented by

$$OH + Cl^- + H^+ \rightarrow Cl + H_2O \qquad (7.119)$$

The chlorine atom may react with either ferrous ion or the organic material

$$Cl + Fe^{2+} \rightarrow Cl^- + Fe^{3+} \qquad (7.120)$$

$$Cl + RH \rightarrow R\cdot + Cl^- + H^+ \qquad (7.121)$$

but, in fact, reacts predominantly with Fe^{2+} with the net result that each OH leads to the oxidation of one ferrous ion regardless of the presence of the organic material.

In the absence of oxygen, reaction 7.57 is replaced by

$$H + H^+ + Fe^{2+} \rightarrow H_2 + Fe^{3+} \qquad (7.122)$$

in which the hydrogen atom takes the unusual role of an oxidizing agent, and the yield of ferric ion is given by

$$G(Fe^{3+}) = 2G_{H_2O_2} + G_{e_{aq}^-} + G_H + G_{OH} + 3G_{HO_2} \qquad (7.123)$$

$$G(H_2) = G_{H_2} + G_{e_{aq}^-} + G_H \qquad (7.124)$$

[Substituting primary yields from Table 7.4 for γ-radiation and pH 0.46 with $G_{HO_2} = 0.026$, these expressions give $G(Fe^{3+}) = 8.19$ and $G(H_2) = 4.05$ for fast electron and γ-radiation, which agree with experimentally determined values.] Although the stoichiometry of reaction 7.122 is well established, the mechanism is less certain. Alternative mechanisms that have been proposed include the association of H and H^+ to form H_2^+, which might readily accept an electron from Fe^{2+} to form H_2 (156, 157) although the association reaction itself is relatively slow (Table 7.7), abstraction by the hydrogen atom of hydrogen from a polarized water molecule in the solvation shell of the ferrous ion (158, 159)

$$H + Fe^{2+}H_2O \rightarrow Fe^{3+}OH^- + H_2 \qquad (7.125)$$

and the formation of an intermediate hydride complex, FeH^{2+} (160–162),

$$H + Fe^{2+} \rightleftharpoons FeH^{2+} \xrightarrow{H^+} Fe^{3+} + H_2 \qquad (7.126)$$

The part played by sulfuric acid and its anions is generally ignored when the ferrous sulfate system is described, although with the usual dosimeter solution (0.4 M H_2SO_4) about 40% of the hydroxyl radicals react with HSO_4^-

$$OH + HSO_4^- \rightarrow OH^- + HSO_4 \qquad k_{127} = 7 \times 10^5 \, M^{-1} \, sec^{-1} \qquad (7.127)$$

This does not affect Eqs. 7.109 to 7.111 or 7.123 and 7.124, which relate $G(Fe^{3+})$ and $G(H_2)$ to the primary yields from water, since the HSO_4 radical behaves like OH and oxidizes one ferrous ion to ferric

$$HSO_4 + Fe^{2+} \rightarrow HSO_4^- + Fe^{3+} \qquad (7.128)$$

or dimerizes to a product which, like H_2O_2, oxidizes two ferrous ions (163–165). Lesigne, Ferradini, and Pucheault (166) have shown that even in solutions as dilute as 0.4 M, a small yield ($G = 0.18$) of HSO_4 is also formed by the direct action of the radiation on HSO_4^-

$$HSO_4^- \rightsquigarrow HSO_4 + e^- \qquad (7.129)$$

Perchloric acid may be substituted for sulfuric acid without change in the yield of ferric ion but according to Schwarz (167) hydrochloric acid cannot be used because back reactions occur between H and the ferric complexes $FeCl_2^+$ and $FeCl^{2+}$.

Ferrous Sulfate-Cupric Sulfate

A number of modified ferrous sulfate systems have been suggested for use as chemical dosimeters or investigated to establish radiolysis mechanisms and yields (148). One of the earliest studied and best known is a mixture of ferrous and cupric sulfates dissolved in dilute sulfuric acid which was suggested as a dosimeter by Hart (168–170); the solution used for this purpose contains $10^{-3} \, M$ $FeSO_4$, $10^{-2} \, M$ $CuSO_4$, $5 \times 10^{-3} \, M$ H_2SO_4, and is air-saturated. As a dosimeter, the solution has the advantage over the normal Fricke dosimeter that the yield of ferric ion does not depend on the presence of oxygen so that the range of the dosimeter is much greater (10^5 to 10^7 or 10^8 rad).

Examination of Tables 7.7, 7.8, 7.10, and 7.11 suggests that the following reactions of the cupric ion may contribute to the radiolysis mechanism:

$$e_{aq}^- + Cu^{2+} \rightarrow Cu^+ \qquad E^\circ = 2.93 \, V \quad k = 3.5 \times 10^{10} \, M^{-1} \, sec^{-1} \qquad (7.70)$$

$$H + Cu^{2+} \rightarrow Cu^+ + H^+ \qquad E^\circ = 2.47 \, V \quad k = 6 \times 10^8 \, M^{-1} \, sec^{-1} \qquad (7.78)$$

$$OH + Cu^{2+} \rightarrow Cu^{3+} + OH^- \qquad\qquad k = 3.5 \times 10^8 \ M^{-1} \ sec^{-1} \tag{7.130}$$

$$HO_2 + Cu^{2+} \rightarrow Cu^+ + O_2 + H^+ \quad E^\circ = 0.46 \ V \quad k = 1.5 \times 10^7 \ M^{-1} \ sec^{-1} \tag{7.131}$$

$$Cu^+ + Fe^{3+} \rightarrow Cu^{2+} + Fe^{2+} \qquad E^\circ = 0.61 \ V \tag{7.132}$$

Cuprous ion may also react with HO_2 and H_2O_2 although these reactions are likely to be slower than the electron-transfer reaction 7.132, which will also be favored by the increasing concentration of ferric ion as the irradiation progresses. Cupric ion is present in relatively high concentration in this system and scavenges over 75% of the hydrated electrons and about 55% of the hydrogen atoms in competition with the hydrogen ion (Eq. 7.42) and oxygen (Eq. 7.57) scavenging reactions respectively. Virtually all HO_2 radicals formed by the latter reactions react with cupric ion (Eq. 7.131) so that effectively each hydrated electron or hydrogen atom formed eventually reduces one cupric ion to a cuprous ion. About 10% of the hydroxyl radicals formed oxidize ferrous ion to ferric (Eq. 7.88) while about 90% react via reaction 7.130. However experiment shows that the Cu^{3+} produced oxidize Cu^+ to Cu^{2+} or Fe^{2+} to Fe^{3+}, so that effectively each OH oxidizes one Fe^{2+} to Fe^{3+}. Hydrogen peroxide oxidizes two ferrous ions to ferric as in the normal Fricke dosimeter, while cuprous ions reduce ferric ions back to ferrous ion (Eq. 7.132), so that the overall yield of ferric ion is given by

$$G(Fe^{3+}) = 2G_{H_2O_2} - G_{e_{aq}^-} - G_H + G_{OH} - G_{HO_2} \tag{7.133}$$

Oxygen enters into the sequence of reactions but is not consumed since each HO_2, formed with the consumption of one molecule of oxygen, releases a molecule of oxygen when it reacts with cupric ion. At high dose rates $G(Fe^{3+})$ increases because the higher radical concentration favors dismutation of HO_2 to give H_2O_2 (Eq. 7.101) rather than reaction with cupric ion (Eq. 7.131) (171).

When oxygen is not present all hydrogen atoms react directly with Cu^{2+} (Eq. 7.78), but the yield of ferric ion is unchanged, since, stoichiometrically, it is immaterial whether H or HO_2 reduces Cu^{2+}. However oxygen formed when deaerated solutions are irradiated can be used to measure the primary yield of perhydroxyl radicals since

$$G(O_2) = G_{HO_2} \tag{7.134}$$

under these conditions.

Ceric Sulfate

Solutions of ceric sulfate in sulfuric acid are also used in dosimetry and, like the ferrous sulfate system, played an important part in the development of an understanding of the radiation chemistry of water and aqueous solutions. In the ceric sulfate dosimeter, solutions containing 10^{-5} to 0.4 M Ce^{4+} in air-saturated 0.4 M H_2SO_4 are reduced to Ce^{3+} by a process which does not consume oxygen and which is only slightly affected by the oxygen concentration (172). Several of the rate constants necessary to evaluate the extent of competition between various possible reactions in the system have not yet been measured, although the main features of the radiolysis mechanism are well established (e.g., 173). Ceric ion is a strong oxidizing agent, though weaker than the hydroxyl radical, and oxidizes all the species with which it reacts except OH, leading to the following mechanism:

$$e_{aq}^- + H^+ \rightarrow H \qquad (7.42)$$

$$H + O_2 \rightarrow HO_2 \qquad (7.57)$$

$$HO_2 + Ce^{4+} \rightarrow Ce^{3+} + H^+ + O_2 \qquad (7.99)$$

$$OH + Ce^{3+} \rightarrow Ce^{4+} + OH^- \qquad (7.135)$$

$$H_2O_2 + Ce^{4+} \rightarrow Ce^{3+} + HO_2 + H^+ \qquad (7.103)$$

which gives

$$G(Ce^{3+}) = 2G_{H_2O_2} + G_{e_{aq}^-} + G_H - G_{OH} + G_{HO_2} \qquad (7.136)$$

a relationship that has been confirmed by experiment and that predicts an increase in $G(Ce^{3+})$ with increasing LET, as observed (Table 3.10). In the absence of oxygen, reactions 7.57 and 7.99 are replaced by

$$H + Ce^{4+} \rightarrow Ce^{3+} + H^+ \qquad (7.137)$$

which does not affect the stoichiometry or Eq. 7.136. When no oxygen is present initially,

$$G(O_2) = G_{H_2O_2} + G_{HO_2} \qquad (7.138)$$

The mechanism given above has been simplified by describing the cerium ions as simple ions rather than ions complexed with H_2O, HSO_4^-, and SO_4^{2-}, and by ignoring any direct reduction of Ce^{4+} by e_{aq}^-, although this is to be expected at high ceric concentrations, and any interaction of the primary radicals with H_2SO_4 and its anions. However, these omissions should not affect the stoichiometry of the reaction. At high absorbed dose, back reactions, for example,

$$HO_2 + Ce^{3+} \rightarrow HO_2^- + Ce^{4+} \qquad (7.139)$$

become more probable as the concentration of cerous ion increases, and $G(Ce^{3+})$ falls.

When sufficient concentrations of thallous ion (174) or formic acid (173, 175) are added to the system, these scavenge hydroxyl radicals and replace them with species that are weaker oxidizing agents than ceric ion; under these condition all radical and molecular products except hydrogen are oxidized by ceric ion and

$$G(Ce^{3+}) = 2G_{H_2O_2} + G_{e_{aq}^-} + G_H + G_{OH} + G_{HO_2} \qquad (7.140)$$

The scavenging and subsequent oxidation reactions are

$$OH + Tl^+ \rightarrow OH^- + Tl^{2+} \qquad (7.141)$$

$$Tl^{2+} + Ce^{4+} \rightarrow Tl^{3+} + Ce^{3+} \qquad (7.142)$$

and

$$OH + HCOOH \rightarrow H_2O + \cdot COOH \qquad (7.143)$$

$$\cdot COOH + Ce^{4+} \rightarrow Ce^{3+} + CO_2 + H^+ \qquad (7.144)$$

Determination of Molecular and Radical Yields

Molecular and radical yields can be determined from the extent of radiation-induced changes produced in aqueous systems if the detailed radiolysis mechanism is known. In acid solutions the systems described above and based on the oxidation of ferrous ion or the reduction of ceric ion can be used, as described below taking numerical G values obtained with ^{60}Co γ-radiation as an illustration. These are, for 0.4 M H_2SO_4 solutions:

$$G(Fe^{3+})_{+O_2}, 15.5 \qquad G(H_2)_{Fe^{2+}, -O_2}, 4.05 \qquad G(Ce^{3+}), 2.35$$

$$G(Fe^{3+})_{-O_2}, 8.2 \qquad G(O_2)_{Fe^{2+}, Cu^{2+}}, 0.026 \qquad G(Ce^{3+})_{Tl^+}, 8.15$$

$$G(Fe^{3+})_{Cu^{2+}}, 0.78$$

The subscripts identify the system (if necessary) and any special conditions. The G values are related to the primary yields by the expressions derived in the previous pages and collected in Table 7.12. Individual molecular and radical yields are derived algebraically, for example:

$$G_{e_{aq}^-} + G_H = 0.5[G(Fe^{3+})_{+O_2} - G(Fe^{3+})_{-O_2}] = 3.65$$

$$G_{H_2} = G(H_2)_{Fe^{2+}, -O_2} - (G_{e_{aq}^-} + G_H) \qquad = 0.40$$

$$G_{OH} = 0.5[G(Ce^{3+})_{Tl^+} - G(Ce^{3+})] \qquad = 2.90$$

$$G_{H_2O_2} = 0.25[G(Fe^{3+})_{Cu^{2+}} + G(Ce^{3+})] \qquad = 0.78$$

$$G_{HO_2} = G(O_2)_{Fe^{2+}, Cu^{2+}} \qquad = 0.026$$

The experimental data can obviously be combined in other ways to derive the primary yields, while the number of experimental determinations necessary can be reduced by including the material balance equations given at the head of Table 7.12. With low-LET radiation G_{HO_2} is small and is often ignored, leading to simpler relationships, for example,

$$G(Fe^{3+})_{+O_2} = 2G_{H_2O_2} + 3(G_{e_{aq}^-} + G_H) + G_{OH} \qquad (7.145)$$

$$G(Fe^{3+})_{-O_2} = 2G_{H_2O_2} + G_{e_{aq}^-} + G_H + G_{OH} \qquad (7.146)$$

$$G(Ce^{3+}) = 2G_{H_2O_2} + G_{e_{aq}^-} + G_H - G_{OH} \qquad (7.147)$$

and

$$G_{e_{aq}^-} + G_H = 0.5[G(Fe^{3+})_{+O_2} - G(Fe^{3+})_{-O_2}]$$

$$G_{OH} = 0.5[G(Fe^{3+})_{-O_2} - G(Ce^{3+})]$$

$G_{H_2O_2}$ can be found by substituting the radical yields in one of the Eq. 7.145 to 7.147, and G_{H_2} by substitution in the material balance equation (Table 7.12). The quantities $G_{e_{aq}^-}$ and G_H cannot be determined separately using the strongly acid ferrous and ceric sulfate systems, although they are readily distinguished in neutral and basic solutions by choice of suitable scavengers (Table 7.12; see also refs. 12, 14, 176, 177). In choosing a scavenger, it is not essential that each primary radical react completely with a single solute if the rate constants for competing reactions are known so that corrections can be applied; it is essential, however, that the radiolysis mechanism be thoroughly understood. In practice, results from a wide variety of systems are considered when attempting to obtain definitive values of the primary yields so that experimental and interpretative errors are minimized.

Scavenger concentrations used in estimating primary yields are generally in the millimolar range, since they need to be large enough to scavenge all reactive species and yet not so high that significant scavenging occurs within the spurs and tracks. The optimum range of concentration will depend on the rate constants for solute-radical reactions and will vary from solute to solute; Buxton (48) has suggested that intraspur scavenging becomes significant when k(radical + solute) [solute] is greater than 10^7 sec^{-1}. Such scavenging is unavoidable at low pH and is the reason for the higher radical yields found in acid solutions.

Ferrocyanide-Ferricyanide

The ferrous sulfate system played an important role in the development of the radiation chemistry of aqueous solutions but is limited to acid solutions. In order to extend the range of experiment to higher pH a variety of other inorganic redox systems have been investigated, one of the earliest being the ferrocyanide-ferricyanide system in which ferrous and ferric ions are stabilized by formation of a complex ion with cyanide; the complex ions are stable over the pH range from acid to strongly basic solution.

TABLE 7.12 Systems for Measuring Molecular and Radical Yields in Aqueous Solutions

System	pH Range	Yield Equation[a]
Material balance equations		$\begin{cases} 2G_{H_2} + G_{e_{aq}^-} + G_H = 2G_{H_2O_2} + G_{OH} + 3G_{HO_2} \\ G_{-H_2O} = 2G_{H_2} + G_{e_{aq}^-} + G_H - G_{HO_2} = 2G_{H_2O_2} + G_{OH} + 2G_{HO_2} \end{cases}$
Boiling water	Acid to basic	$G(H_2) = 2G(O_2) = G_{H_2}$
Bromide (0.001 to 10 mM)	Acid or neutral	
Oxygen-free		$G(H_2) = G_{H_2}$
+ Oxygen		$G(H_2O_2) = G_{H_2O_2} + 0.5(G_{e_{aq}^-} + G_H - G_{OH} + G_{HO_2})$
Carbon monoxide (0.15 to 0.65 mM) + Oxygen (0.15 to 0.35 mM)	Acid to basic	$\begin{cases} G(H_2) = G_{H_2} \\ G(CO_2) = G_{OH} \\ G(H_2O_2) = G_{H_2O_2} + 0.5(G_{e_{aq}^-} + G_H + G_{OH} + G_{HO_2}) \end{cases}$
Ceric sulfate (0.01 to 10 mM)	Acid	
With or without oxygen		$G(Ce^{3+}) = 2G_{H_2O_2} + G_{e_{aq}^-} + G_H - G_{OH} + G_{HO_2}$
Oxygen-free		$G(O_2) = G_{H_2O_2} + G_{HO_2}$
Ceric sulfate + thallous sulfate or formic acid + oxygen	Acid	$G(Ce^{3+}) = 2G_{H_2O_2} + G_{e_{aq}^-} + G_H + G_{OH} + G_{HO_2}$
Deuterium, oxygen-free	Neutral	$G(HD) = G_H$
Ferrous sulfate (1 mM) + oxygen	Acid	$\begin{cases} G(Fe^{3+}) = 2G_{H_2O_2} + 3(G_{e_{aq}^-} + G_H + G_{HO_2}) + G_{OH} \\ G(H_2) = G_{H_2} \end{cases}$
Ferrous sulfate (0.01 mM), with or without oxygen; rapid analysis	Acid	$G(Fe^{3+}) = G_{e_{aq}^-} + G_H + G_{OH} + G_{HO_2}$
Ferrous sulfate (1 mM), oxygen-free	Acid	$\begin{cases} G(Fe^{3+}) = 2G_{H_2O_2} + G_{e_{aq}^-} + G_H + G_{OH} + 3G_{HO_2} \\ G(H_2) = G_{H_2} + G_{e_{aq}^-} + G_H \end{cases}$

System	Condition	Equation
Ferrous sulfate (1 mM) + cupric sulfate (10 mM)		
With or without oxygen	Acid	$G(\text{Fe}^{3+}) = 2G_{\text{H}_2\text{O}_2} - G_{e^-_{aq}} - G_\text{H} + G_\text{OH} - G_{\text{HO}_2}$
Oxygen-free		$G(\text{O}_2) = G_{\text{HO}_2}$
Ferrocyanide (5 mM) + oxygen[b]	Acid or basic	$G(\text{Fe(CN)}_6^{3-})_{\mu s} = G_\text{OH}$
Water + hydrogen, oxygen-free[b]	Neutral or basic	$G(e^-_{aq})_{\mu s} = G_{e^-_{aq}}(+ G_\text{OH} + G_\text{H} \quad \text{if strongly basic})$
Water + ethanol (10 mM), oxygen-free[b]	Neutral or basic	$G(e^-_{aq})_{\mu s} = G_{e^-_{aq}}(+ G_\text{H} \quad \text{if strongly basic})$
Acrylamide (0.1 mM), oxygen-free	Acid to basic	$G(\text{H}_2\text{O}_2) = G_{\text{H}_2\text{O}_2}$
Formic acid or formate (1 mM), or oxalic acid or oxalate (50 mM) + oxygen	Acid to basic	$\begin{cases} G(\text{H}_2) = G_{\text{H}_2} \\ G(\text{H}_2\text{O}_2) = G_{\text{H}_2\text{O}_2} + 0.5(G_{e^-_{aq}} + G_\text{H} + G_\text{OH} + G_{\text{HO}_2}) \\ G(-\text{O}_2) = 0.5(G_{e^-_{aq}} + G_\text{H} + G_\text{OH} - G_{\text{HO}_2}) \\ G(\text{CO}_2)_{\text{formic}} = G_\text{OH} \\ G(\text{CO}_2)_{\text{oxalic}} = 2G_\text{OH} \\ G(-\text{oxalic acid}) = G_\text{OH} \end{cases}$
Formic acid (1 mM), oxygen-free	Acid	$G(\text{H}_2) = G_{\text{H}_2} + G_{e^-_{aq}} + G_\text{H}$
2-Propanol (25 mM) + H$_2$O$_2$ (1 mM), oxygen-free	Neutral	$G(\text{H}_2) = G_{\text{H}_2} + G_\text{H}$
2-Propanol (25 mM) + N$_2$O (0.01 to 0.1 mM), oxygen-free	Neutral	$\begin{cases} G(\text{H}_2) = G_{\text{H}_2} + G_\text{H} \\ G(\text{N}_2) = G_{e^-_{aq}} \end{cases}$

[a] The equations refer to initial yields, since accumulations of radiolysis products may alter the stoichiometry.

[b] Product yield measured by pulse radiolysis at microsecond times after the pulse; hydrated electron yields corrected for any decay prior to the measurement.

307

On the basis of the standard electrode potentials for the reactions (Table 7.8), the following can be suggested as thermodynamically possible steps when solutions containing ferrocyanide, $Fe(CN)_6^{4-}$, and ferricyanide, $Fe(CN)_6^{3-}$, are irradiated; in several cases it has been shown that the reactions are not only possible but fast, by measurement of rate constants.

$$OH + Fe(CN)_6^{4-} \rightarrow Fe(CN)_6^{3-} + OH^-$$
$$E^\circ = 2.11 \text{ V} \qquad k_{148} = 1.2 \times 10^{10} \ M^{-1} \text{ sec}^{-1} \quad (7.148)$$

$$e_{aq}^- + Fe(CN)_6^{3-} \rightarrow Fe(CN)_6^{4-}$$
$$E^\circ = 3.46 \text{V} \qquad k_{149} = 3.0 \times 10^9 \ M^{-1} \text{ sec}^{-1} \quad (7.149)$$

$$H + Fe(CN)_6^{3-} \rightarrow Fe(CN)_6^{4-} + H^+$$
$$E^\circ = 3.00 \text{ V} \qquad k_{150} = 3.9 \times 10^9 \ M^{-1} \text{ sec}^{-1} \quad (7.150)$$

$$HO_2 + Fe(CN)_6^{4-} \rightarrow Fe(CN)_6^{3-} + HO_2^-$$
$$E^\circ = 0.81 \text{ V} \qquad k_{151} = 1.6 \times 10^5 \ M^{-1} \text{ sec}^{-1} \quad (7.151)$$

$$HO_2^- + H^+ \rightarrow H_2O_2 \qquad\qquad\qquad\qquad\qquad\qquad (7.108)$$

$$HO_2 + Fe(CN)_6^{3-} \rightarrow Fe(CN)_6^{4-} + H^+ + O_2 \qquad E^\circ \text{ about 1.0 V} \quad (7.152)$$

$$H_2O_2 + 2H^+ + 2Fe(CN)_6^{4-} \rightarrow 2Fe(CN)_6^{3-} + 2H_2O$$
$$E^\circ = 1.09 \text{ V} \quad (7.153)$$

The E° values are for acid or neutral solutions and change in magnitude and even sign in basic solutions; in strongly basic solution OH is replaced by O^- which reacts with ferrocyanide with a rate constant of less than $7 \times 10^7 \ M^{-1} \text{ sec}^{-1}$, i.e., at less than 1% of the rate at which OH reacts with ferrocyanide. In neutral or acid solutions containing concentrations of the order of $10^{-2} \ M$ ferrocyanide and saturated with air or oxygen, e_{aq}^- and H are scavenged by oxygen, and ferrocyanide is oxidized by OH, HO_2, and H_2O_2, the HO_2 and H_2O_2 oxidizing three and two equivalents of ferrocyanide respectively. Then,

$$G[Fe(CN)_6^{3-}] = 2G_{H_2O_2} + 3(G_{e_{aq}^-} + G_H + G_{HO_2}) + G_{OH} \quad (7.154)$$

The reactions with HO_2 and H_2O_2 are relatively slow so that if the system is observed within a few microseconds of a brief radiation pulse, only OH has brought about oxidation (Eq. 7.148), and

$$G[Fe(CN)_6^{3-}]_{\mu s} = G_{OH} \quad (7.155)$$

This is the basis for the use of ferrocyanide solutions as dosimeters in pulse radiolysis. The yield of ferricyanide can be approximately doubled by saturating the solution with nitrous oxide to convert e_{aq}^- to OH, when

$$G[Fe(CN)_6^{3-}]_{N_2O, \mu s} \simeq G_{OH} + G_{e_{aq}^-} \quad (7.156)$$

In the N_2O saturated solutions, hydrogen atoms will be scavenged by both ferrocyanide and the radiation-produced ferricyanide, so that Eq. 7.156 may not be strictly true; the extent of any back reaction will depend on the ferrocyanide concentration and the absorbed dose.

The radiolysis of ferrocyanide and ferricyanide solutions (178) is appreciably more complex than the radiolysis of ferrous and ferric solutions both because of the greater pH range covered and because additional reactions become possible as the solute becomes more complex. Reaction of hydrated electrons with ferricyanide, for example, gives not only ferrocyanide (Eq. 7.149) but also about 10% of a pentacyano complex formed by loss of a CN^- ion,

$$e_{aq}^- + Fe(CN)_6^{3-} \xrightarrow{\sim 90\%} Fe(CN)_6^{4-} \qquad (7.149)$$

$$\xrightarrow{\sim 10\%} Fe(CN)_5H_2O^{3-} + CN^- \qquad (7.157)$$

Radical and molecular yields determined using oxygen-free ferrocyanide and ferricyanide systems should, therefore, be treated with some caution until it is certain that the radiolysis mechanisms are fully understood. Aqueous solutions of many other transition metal complexes (e.g., of Ag, Cd, Co, Cr, Cu, Pt, Rh, are Ru) have been irradiated (57, and the references cited therein), both for their intrinsic interest and for the light they shed on mechanisms in biological processes.

Halide and Pseudohalide Ions

The halide ions (X^-) are inert toward e_{aq}^- and H but undergo rapid electron-transfer type reactions with OH to give products $(Cl_2^-, Br_2^-,$ and $I_2^-)$ which incorporate a second halide ion and are stoichiometrically equivalent to $X + X^-$.

$$OH + X^- \xrightarrow{X^-} X_2^- + OH^- \qquad (7.158)$$

With chloride and bromide the reaction is more rapid in the presence of hydrogen ions, but with iodide it is independent of pH. Pulse radiolysis experiments have shown that the reaction with chloride ion takes place in the following steps (179):

$$OH + Cl^- \rightarrow ClOH^- \qquad k_{159} = 4.3 \times 10^9 \ M^{-1} \ sec^{-1} \qquad (7.159)$$

$$ClOH^- + H^+ \rightarrow Cl + H_2O \qquad k_{160} = 2.1 \times 10^{10} \ M^{-1} \ sec^{-1} \qquad (7.160)$$

$$Cl + Cl^- \rightarrow Cl_2^- \qquad k_{161} = 2.1 \times 10^{10} \ M^{-1} \ sec^{-1} \qquad (7.161)$$

each of which is reversible. The authors estimate that in the Fricke dosimeter solution containing $10^{-3} \ M \ Cl^-$ all OH, including those formed by reaction of H_2O_2 with Fe^{2+} (Eq. 7.97), enter into reactions 7.159 and 7.160 and form Cl, and that 81% of the chlorine atoms formed react directly with ferrous ion (Eq. 7.120) while 19% form Cl_2^-, which also oxidizes ferrous ion,

$$Cl_2^- + Fe^{2+} \rightarrow 2Cl^- + Fe^{3+} \qquad (7.162)$$

The Cl_2^- ion absorbs strongly at 340 nm ($\varepsilon_{340} = 8800\ M^{-1}\ cm^{-1}$) and is readily detected when acid chloride solutions are pulse irradiated. The characteristic absorption is also detected, though less strongly, immediately following pulse irradiation of more concentrated (above 1 M) neutral chloride solutions where the higher pH precludes formation of Cl by reaction 7.160, and it has been suggested that this demonstrates the presence of H_3O^+ in the spurs where the rapid formation of Cl_2^- is presumed to take place (83). Hamill and his colleagues (79, 81, 180, 181) have offered an alternative explanation, suggesting that formation of Cl_2^- in concentrated neutral chloride solutions is not the result of OH scavenging in an acid spur but rather the result of H_2O^+ (i.e., positive hole) scavenging by Cl^-,

$$H_2O^+ + Cl^- \rightarrow H_2O + Cl \qquad (7.163)$$

which competes with

$$H_2O^+ + H_2O \rightarrow H_3O^+ + OH \qquad (7.18)$$

The competition is described by the empirical expression

$$P = \frac{G(Cl_2^-)}{G^\circ} = \frac{\sigma_1[Cl^-]}{\sigma_1[Cl^-] + \sigma_2} \qquad (7.164)$$

where P is the probability of electron transfer from Cl^- to H_2O^+, $G(Cl_2^-)$ the observed yield, and G° the yield of H_2O^+ available to react with Cl^-; $\sigma_1[Cl^-]$ and σ_2 are the relative probabilities that H_2O^+ will react with Cl^- or undergo an alternative first-order or pseudo first-order reaction (i.e., reaction 7.18); rate constants are not used, since the processes described take place within an extremely short time following energy deposition and far too rapidly for the occurrence of diffusion-related processes. Applied to the data of Anbar and Thomas (83), this expression gives $\sigma_1/\sigma_2 = 0.38$ and $G^\circ = 3.8$; the latter may be compared with G values of about 4 for the earliest yields of e_{aq}^- detectable by picosecond pulse radiolysis. This approach has given consistent results when applied by Hamill to hole trapping by anions other than Cl^-, including anions such as acetate and oxalate which do not react rapidly with e_{aq}^-, H, or OH and which must, therefore, be reacting with some other reactive species.

Dimerization of the species produced by reaction of OH with halide ions produces the corresponding halogen or an ion derived from it, i.e.,

$$2X_2^- \rightarrow X_2 + 2X^- \qquad (or\ X_3^- + X^-) \qquad (7.165)$$

However, low steady state concentrations of the halogens are produced when halide solutions are irradiated, and it is clear that back reactions with e_{aq}^-

and H are occurring which reform halide ion, e.g.,

$$e_{aq}^- + X_2 \rightarrow X_2^- \tag{7.166}$$

$$e_{aq}^- + X_2^- \rightarrow 2X^- \tag{7.167}$$

$$H + X_2 \rightarrow H^+ + X^- + X \tag{7.168}$$

The rate constants for these reactions are high and, with reaction 7.158, they serve to protect the molecular hydrogen formed from radical attack. Thus G_{H_2} can be determined by measuring the yield of hydrogen when dilute (10^{-6} to 10^{-2} M) solutions of Br^- or I^- in deaerated acid or neutral solution are irradiated (182, 183); under these conditions $G(H_2) = G_{H_2}$. In the presence of air or oxygen, e_{aq}^- and H form HO_2 which reacts

$$2HO_2 \rightarrow H_2O_2 + O_2 \tag{7.101}$$

or

$$HO_2 + Br_2 \rightarrow H^+ + Br^- + Br + O_2 \tag{7.169}$$

so that when dilute aerated bromide solutions are irradiated

$$G(H_2O_2) = G_{H_2O_2} + 0.5(G_{e_{aq}^-} + G_H - G_{OH} + G_{HO_2}) \tag{7.170}$$

Higher concentrations of hydrogen peroxide are broken down by a chain reaction in the presence of bromide in which chain propagating steps are reaction 7.169 and

$$Br + H_2O_2 \rightarrow HO_2 + H^+ + Br^- \tag{7.171}$$

The thiocyanate ion behaves like a halide ion in forming a dimeric radical ion following reaction with OH,

$$CNS^- + OH \rightarrow CNS + OH^- \tag{7.172}$$

$$CNS + CNS^- \rightarrow (CNS)_2^- \quad k_{173} = 1.1 \times 10^{10} \, M^{-1} \, \text{sec}^{-1} \tag{7.173}$$

The dimer absorbs strongly at 475 nm ($\varepsilon_{475} = 7.6 \times 10^3 \, M^{-1} \, cm^{-1}$) and use is made of this fact in pulse radiolysis experiments for dosimetry and for the determination of hydroxyl radical rate constants by applying competition kinetics (58). As with the chloride ion, the reaction involves the intermediate formation of a hydroxyl radical adduct, $(CNSOH)^-$, which has been observed by pulse radiolysis (184). The dimer disappears by a second-order reaction that may be

$$2(CNS)_2^- \rightarrow 2CNS^- + (CNS)_2 \tag{7.174}$$

although the complete radiolysis mechanism has not been established. Final products from the radiolysis of relatively concentrated thiocyanate solutions (>0.1 M) include CN^-, SO_4^{2-}, and sulfur in aerated solution and CN^-, H_2S, and sulfur in the absence of oxygen; cyanide and sulfur are formed by a direct effect at higher thiocyanate concentrations (185).

Reaction of OH with CN^- gives an adduct that does not break down to CN and OH^- but instead reacts with water and rearranges to give the form-amide radical, $\cdot CONH_2$, which is also formed by reaction of hydroxyl radicals with formamide, $HCONH_2$ (186).

$$OH + (C\equiv N)^- \rightarrow (HO-\dot{C}=N)^- \xrightarrow{+H_2O}$$

$$HO-\dot{C}=NH + OH^- \; (\rightarrow \cdot CONH_2) \quad (7.175)$$

Radical ions similar to Cl_2^- and $(SCN)_2^-$ are formed when HS^- and thiols react with hydroxyl radicals (cf. p. 343).

Aliphatic Acids

If ferrous sulfate is the classic inorganic solute studied during the development of the radiation chemistry of water, formic acid is the classic organic solute. Formic acid solutions have been studied extensively, particularly by Hart and his colleagues (187 to 194, cf. 49), the early work showing that hydrogen peroxide and carbon dioxide are the major products formed in the presence of oxygen and that hydrogen is an important product when oxygen-free solutions are used.

Possible reactions of the primary radicals in formic acid solutions containing oxygen are

$$e_{aq}^- + H^+ \rightarrow H \qquad\qquad k_{42} = 2.35 \times 10^{10} \; M^{-1} \, sec^{-1} \quad (7.42)$$

$$e_{aq}^- + O_2 \rightarrow O_2^- \qquad\qquad k_{56} = 1.9 \times 10^{10} \; M^{-1} \, sec^{-1} \quad (7.56)$$

$$e_{aq}^- + HCOOH \rightarrow HCOO^- + H \quad k_{176} = 1.5 \times 10^8 \; M^{-1} \, sec^{-1} \quad (7.176)$$

$$\rightarrow H\dot{C}O + OH^- \quad k_{177} = 1.9 \times 10^8 \; M^{-1} \, sec^{-1} \quad (7.177)$$

$$H + O_2 \rightarrow HO_2 \qquad\qquad k_{57} = 1.9 \times 10^{10} \; M^{-1} \, sec^{-1} \quad (7.57)$$

$$H + HCOOH \rightarrow H_2 + \cdot COOH \quad k_{178} = 1.1 \times 10^6 \; M^{-1} \, sec^{-1} \quad (7.178)$$

$$OH + HCOOH \rightarrow H_2O + \cdot COOH \quad k_{179} = 1.3 \times 10^8 \; M^{-1} \, sec^{-1} \quad (7.179)$$

Application of competition kinetics to the first three reactions shows that for a 0.1 M formic acid solution (pH 2.4) saturated with air ($\sim 2.5 \times 10^{-4}$ M O_2) about 71% of the hydrated electrons react with H^+, about 4% with O_2, and about 25% with HCOOH; at lower formic acid concentrations the proportion reacting with HCOOH and H^+ falls and a larger proportion of the hydrated electrons react with oxygen. In the same air-saturated 0.1 M solution, 2% of the hydrogen atoms formed directly and in reactions 7.42 and 7.176 react with formic acid and the remainder with oxygen; at higher formic acid concentrations a larger proportion of the hydrogen atoms react with HCOOH and $G(H_2)$ is higher, at lower solute concentrations virtually all hydrogen atoms are scavenged by oxygen. Hydroxyl radicals abstract hydrogen from HCOOH (Eq. 7.179) and both the carboxyl radicals formed in this reaction and the formyl radicals formed in reaction 7.177 add oxygen and subsequently dissociate,

$$\cdot COOH + O_2 \rightarrow CO_2 + HO_2 \qquad k_{180} = 2.4 \times 10^9 \, M^{-1} \, sec^{-1} \quad (7.180)$$

$$H\dot{C}O + O_2 \rightarrow CO_2 + OH \qquad\qquad\qquad\qquad\qquad\qquad (7.181)$$

The perhydroxyl radicals dismutate forming hydrogen peroxide

$$2HO_2 \rightarrow H_2O_2 + O_2 \qquad k_{101} = 2.5 \times 10^6 \, M^{-1} \, sec^{-1} \quad (7.101)$$

At formic acid concentrations of about $10^{-3} \, M$ all e_{aq}^- and H are scavenged by oxygen and the following relationships exist between experimental and primary yields:

$$G(H_2) = G_{H_2} \qquad\qquad\qquad\qquad\qquad\qquad\qquad (7.182)$$

$$G(H_2O_2) = G_{H_2O_2} + 0.5(G_{e_{aq}^-} + G_H + G_{OH} + G_{HO_2}) \qquad (7.183)$$

$$G(CO_2) = G_{OH} \qquad\qquad\qquad\qquad\qquad\qquad\qquad (7.184)$$

$$G(-O_2) = 0.5(G_{e_{aq}^-} + G_H + G_{OH} - G_{HO_2}) \qquad\qquad (7.185)$$

In neutral and basic solutions the active solute is the formate ion rather than the undissociated acid, and reactions 7.176, 7.177, 7.178, and 7.179 are replaced by

$$e_{aq}^- + HCOO^- \xrightarrow{H_2O} H\dot{C}O + 2OH^- \quad k_{186} = \sim 10^4 \, M^{-1} \, sec^{-1} \quad (7.186)$$

$$H + HCOO^- \rightarrow H_2 + \cdot COO^- \quad k_{196} = 2.2 \times 10^8 \, M^{-1} \, sec^{-1} \quad (7.187)$$

$$OH + HCOO^- \rightarrow H_2O + \cdot COO^- \quad k_{188} = 2.8 \times 10^9 \, M^{-1} \, sec^{-1} \quad (7.188)$$

The reactions are stoichiometrically equivalent to reactions occurring under acid conditions, but the reaction rates are much different. In neutral and basic solutions saturated with air or oxygen, hydrated electrons react exclusively with oxygen, but hydrogen atoms only react predominantly ($> 95\%$) with oxygen if the $[O_2]/[HCOO^-]$ ratio is greater than 0.22, i.e., if the concentration of formate is not greater than $10^{-3} \, M$ in air-saturated solutions; provided this restriction is observed, the relationships described in Eq. 7.182 to 7.185 will apply to at least pH 11. At pH greater than about 5, the stable form of the perhydroxyl radical is the anion O_2^-, although substitution of the anion for HO_2 in the preceding mechanism does not affect the stoichiometry.

The radiolysis mechanism is more complex when oxygen is absent. In dilute ($\sim 10^{-3} \, M$), oxygen-free, formic acid solutions, e_{aq}^- reacts predominantly with H^+ and the hydrogen atoms formed abstract hydrogen from the acid, so that both reducing radicals and OH react with HCOOH with the eventual formation of one ·COOH radical. This radical is described as ·COOH rather than HCOO· because irradiation of DCOOH in H_2O (194) and HCOOD in D_2O (195) gives hydrogen composed largely of HD. Probable reactions of the carboxyl radical are

$$·COOH + H_2O_2 \rightarrow CO_2 + OH + H_2O \qquad (7.104)$$

and

$$2·COOH \rightarrow HCOOH + CO_2 \qquad (7.189)$$

since approximately equimolar amounts of H_2 and CO_2 are formed with a G value of about 4 (187, 192). Under these conditions,

$$G(H_2) = G_{H_2} + G_{e^-_{aq}} + G_H \qquad (7.190)$$

Reactions 7.104 and 7.179 constitute the propagation steps of a chain reaction in which hydrogen peroxide oxidizes formic acid to carbon dioxide and water. A reaction of this type has been observed when formic acid solutions containing hydrogen peroxide are irradiated (193, 196), although oxygen inhibits the chain by scavenging the carboxyl radicals (Eq. 7.180). At pH above 3 the yield of CO_2 falls and oxalic acid is formed in its place as reaction 7.189 is replaced by

$$2·COO^- \rightarrow {}^-OOC—COO^- \qquad (7.191)$$

Formaldehyde is a product when more concentrated formic acid solutions are irradiated with γ-radiation (189) and, with glyoxal, glyoxylic acid, and other products, when dilute, oxygen-free, formic acid solutions are irradiated with accelerated helium nuclei (197, 198). The sources of these products are probably reactions of the formyl radical, for example,

$$H\dot{C}O + HCOOH \rightarrow HCHO + ·COOH \qquad (7.192)$$

$$H\dot{C}O + ·COOH \rightarrow OHC—COOH \qquad \text{(glyoxylic acid)} \quad (7.193)$$

$$2H\dot{C}O \rightarrow OHC—CHO \qquad \text{(glyoxal)} \quad (7.194)$$

$$\xrightarrow{H_2O} HCOOH + HCHO \qquad (7.195)$$

Formyl radicals (which form the hydrate, $H\dot{C}(OH)_2$, in aqueous solution) will be formed by reaction 7.177 at formic acid concentrations of the order of 0.1 M, or higher.

At very high concentrations of formic acid, from 1 M to essentially pure formic acid, the yield of CO_2 increases to G values as high as 12. Hart (199) has postulated that under these conditions a chain reaction occurs made up of reaction 7.192 and

$$·COOH + HCOOH \rightarrow H\dot{C}O + CO_2 + H_2O \qquad (7.196)$$

with additional reactions such as

$$\cdot COOH + HCHO \rightarrow \cdot CH_2OH + CO_2 \qquad (7.197)$$

and

$$\cdot CH_2OH + HCOOH \rightarrow CH_3OH + \cdot COOH \qquad (7.198)$$

to remove formaldehyde, which only builds up to a very low steady-state concentration. Carbon monoxide is also formed at formic acid concentrations greater than about 10^{-3} M (199, 200), and it has been suggested that this is the result of excitation of the formic acid by subexcitation electrons at the lower concentrations and of direct action (direct excitation of the solute by fast electrons) at the higher formic acid concentrations.

The carboxyl and formyl radicals that are intermediates in the radiolysis of formic acid and formate solutions are also formed by the irradiation of oxygen-free CO_2 and CO solutions, respectively,

$$e_{aq}^- + CO_2 \rightarrow \cdot COO^- \qquad (7.199)$$

$$e_{aq}^- + CO \rightarrow CO^- \xrightarrow{H_2O} H\dot{C}O + OH^- \qquad (7.74)$$

and lead to similar products. Irradiation of oxygen-free carbon dioxide solutions (201), for example, yields numerous products that include formaldehyde, formic acid, and oxalic acid, while irradiation of oxygen-free carbon monoxide solutions (202) gives carbon dioxide, formaldehyde, formic acid, and glyoxal. Solutions containing carbon monoxide and oxygen have been employed by Dainton and his colleagues (203) to estimate primary yields over a wide range of pH from 0.4 to 12 (Table 7.12).

The radiolysis of acetic acid appears to be similar to that of formic acid in many respects, hydrogen atoms and hydroxyl radicals abstracting hydrogen from both acetic acid and the acetate ion and hydrated electrons reacting relatively rapidly with the acid but much more slowly with the anion:

$$H \text{ (or OH)} + CH_3COOH \rightarrow H_2 \text{ (or } H_2O) + \cdot CH_2COOH \quad (7.200)$$

$$e_{aq}^- + CH_3COOH \rightarrow CH_3COO^- + H \qquad (7.201)$$

$$\rightarrow CH_3\dot{C}O + OH^- \qquad (7.202)$$

Hydrogen, hydrogen peroxide, and succinic acid are formed when dilute oxygen-free solutions are irradiated (204, 205), and it appears that hydrogen peroxide does not react with the intermediate organic radicals, which dimerize,

$$2\cdot CH_2COOH \rightarrow (CH_2COOH)_2 \quad \text{(succinic acid)} \qquad (7.203)$$

Higher carboxylic acids such as malic, citric, and tricarballylic

$$HOOCCH(CH_2COOH)_2$$

are produced when the irradiation is prolonged, probably by secondary reactions involving the succinic acid (206). Carbon dioxide, methane, and smaller amounts of other products formed at high acetic acid concentrations ($>1\ M$) (204, 207) are also products of the radiolysis of pure acetic acid and may be the result of excitation by subexcitation electrons or direct action when the aqueous solutions are irradiated. Succinic acid formation is suppressed when oxygen is present, and carbon dioxide, formaldehyde, glycollic acid, glyoxylic acid, and oxalic acid are produced (205, 208) via the peroxy radical $\cdot O_2CH_2COOH$ formed by addition of oxygen to $\cdot CH_2COOH$.

The radiation chemistry of aqueous oxalic acid and oxalate solutions has been studied fairly extensively because of the interest in these systems as chemical dosimeters and as a means of measuring primary yields over a wide range of pH (209). When oxygen is present, hydrogen, hydrogen peroxide, and carbon dioxide are the only stable products formed over the pH range 1.3 to 13 with oxalate concentrations from 10^{-3} to 0.2 M. The mechanism under these conditions can be represented by:

$$\text{H (or }\ e_{aq}^-) + O_2 \rightarrow HO_2 \text{ (or }\ O_2^-) \tag{7.204}$$

$$OH + HOOC\text{—}COOH \rightarrow H_2O + HOOC\text{—}COO\cdot \tag{7.205}$$

$$OH + {}^-OOC\text{—}COO^- \rightarrow OH^- + {}^-OOC\text{—}COO\cdot \tag{7.206}$$

$$HOOC\text{—}COO\cdot \text{ (or }\ {}^-OOC\text{—}COO\cdot) + O_2 \rightarrow$$
$$2CO_2 + HO_2\cdot \text{ (or }\ O_2^-) \tag{7.207}$$

$$2HO_2 \text{ (or }\ O_2^-) \rightarrow O_2 + H_2O_2 \text{ (or }\ O_2^{2-}) \tag{7.208}$$

($HOOC\text{—}COO^-$ reacts in a similar manner to ${}^-OOC\text{—}COO^-$.) Stoichiometrically, it is immaterial whether oxalic acid reacts in the acid form or an ionic form, or whether H and e_{aq}^- react with oxygen or with oxalic acid (Eq. 7.209 to 7.212), since equivalent products are formed in each case.

$$H + HOOC\text{—}COOH \rightarrow HOOC\text{—}\dot{C}(OH)_2 \tag{7.209}$$

$$HOOC\text{—}\dot{C}(OH)_2 + O_2 \rightarrow HOOC\text{—}COOH + HO_2 \tag{7.210}$$

$$e_{aq}^- + HOOC\text{—}COOH \rightarrow HOOC\text{—}\dot{C}(OH)O^- \tag{7.211}$$

$$HOOC\text{—}\dot{C}(OH)O^- + O_2 \rightarrow HOOC\text{—}COOH + O_2^- \tag{7.212}$$

If the absorbed dose is kept low (<4 krad) so that the concentration of hydrogen peroxide remains low, and it does not take part in the reaction mechanism, the primary yields can be determined from the product yields and the consumption of oxygen and oxalic acid (Table 7.12). As with formic acid, the reaction is more complex in the absence of oxygen, although the formation of the main products in acid solutions (pH 1 to 3), carbon dioxide, glyoxylic acid, and dihydroxytartaric acid, can be explained qualitatively by reactions of the radicals formed when H and OH react with oxalic acid (210).

$$2HOOC-\dot{C}(OH)_2 \rightarrow HOOC-C(OH)_2-C(OH)_2-COOH$$
$$\text{(dihydroxytartaric acid)} \qquad (7.213)$$

$$HOOC-\dot{C}(OH)_2 + HOOC-COO\cdot \rightarrow$$
$$OHC-COOH + 2CO_2 + H_2O \quad (7.214)$$

The quantity $G(-\text{oxalic acid})$ falls as the products accumulate in the solution and as the pH is increased above 2, but the precise mechanism of these changes has not been established. Use of oxygen-free oxalic acid solutions to estimate absorbed dose in the megarad to 100-Mrad range by measuring the loss of oxalic acid is based on empirical relationships rather than an established mechanism (209).

Alcohols and Carbonyl Compounds

Alcohols are frequently added to aqueous systems to scavenge H and OH, methanol, for example, reacting

$$\left.\begin{array}{l} H + CH_3OH \rightarrow H_2 + \cdot CH_2OH \\ \qquad\qquad \rightarrow H_2 + CH_3O\cdot \end{array}\right\} \quad k = 1.6 \times 10^6 \, M^{-1} \sec^{-1} \quad \begin{array}{l}(7.81)\\(7.215)\end{array}$$

$$\left.\begin{array}{l} OH + CH_3OH \rightarrow H_2O + \cdot CH_2OH \\ \qquad\qquad\;\; \rightarrow H_2O + CH_3O\cdot \end{array}\right\} \quad k = 8.4 \times 10^8 \, M^{-1} \sec^{-1} \quad \begin{array}{l}(7.216)\\(7.217)\end{array}$$

The reaction with e_{aq}^- is very much slower

$$e_{aq}^- + CH_3OH \rightarrow H + CH_3O^- \qquad k_{218} < 10^4 \, M^{-1} \sec^{-1} \quad (7.218)$$

and, except possibly at high methanol concentrations, will not compete with other reactions of e_{aq}^-; at concentrations less than about $10^{-2} \, M$ CH_3OH, hydrated electrons react with water

$$e_{aq}^- + H_2O \rightarrow H + OH^- \qquad k_{71} = 16 \, M^{-1} \sec^{-1} \quad (7.71)$$

rather than with the methanol. The ratio of hydrogen abstraction from oxygen to abstraction from carbon (e.g., the ratio of $CH_3O\cdot/\cdot CH_2OH$ in the products of reactions 7.215 and 7.81) are 0.15 for H atoms (211) and 0.075 for OH (212). Hydrogen, formaldehyde, and ethylene glycol are major products in the absence of oxygen (192, 213, 214), and it appears that the methoxy radicals abstract hydrogen from methanol

$$CH_3O\cdot + CH_3OH \rightarrow CH_3OH + \cdot CH_2OH \qquad (7.219)$$

and that the methanol radicals both disproportionate and dimerize,

$$2 \cdot CH_2OH \rightarrow CH_3OH + HCHO \qquad 2k = 2.4 \times 10^9 \ M^{-1} \ sec^{-1} \quad (7.220)$$
$$\rightarrow (CH_2OH)_2 \qquad \qquad (ref. \ 215) \qquad \qquad (7.221)$$

though they may also react with H_2O_2

$$\cdot CH_2OH + H_2O_2 \rightarrow HCHO + H_2O + OH \qquad (7.222)$$

In acid, oxygen-free, solutions

$$G(H_2) = G_{H_2} + G_{e_{aq}^-} + G_H \qquad (7.223)$$

since all e_{aq}^- are scavenged by hydrogen ions. This expression should also apply to neutral solutions, but the requisite conditions are more difficult to achieve, since all traces of oxygen, carbon dioxide, and other electron scavengers must be removed (216).

In solutions containing dissolved air or oxygen, all e_{aq}^- are scavenged by oxygen (assuming the pH is not too low) and H will be scavenged (>95%) by oxygen if the ratio $[O_2]/[CH_3OH]$ is higher than 625, i.e., if the concentration of methanol in an air-saturated solution is less than about 0.15 M. The methanol radicals also add oxygen to form peroxy radicals

$$\cdot CH_2OH + O_2 \rightarrow \cdot O_2CH_2OH \qquad k_{224} = 4.2 \times 10^9 \ M^{-1} \ sec^{-1} \quad (7.224)$$

which generally disappear by a second-order process in acid or neutral solutions (217). The reactions responsible have not been identified unequivocally but may be

$$2 \cdot O_2CH_2OH \rightarrow 2HCHO + H_2O_2 + O_2 \qquad (7.225)$$

$$\cdot O_2CH_2OH + HO_2 \rightarrow HCHO + H_2O_2 + O_2 \qquad (7.226)$$

since hydrogen peroxide and carbonyl compounds are the main products when alcohols are irradiated in the presence of oxygen. No distinctive products due to reaction of $CH_3O \cdot$ with O_2 have been reported, and it is probable that this reaction also leads eventually to H_2O_2 and HCHO; certainly both radicals formed by reaction of OH with ethanol give acetaldehyde in the presence of oxygen since $G(CH_3CHO) = G_{OH}$.

The radiation chemistry of other simple alcohols in aqueous solution is similar to that of methanol. Hydrogen atoms and hydroxyl radicals react with the alcohol by abstracting hydrogen predominantly (212) from the α-carbon atom, while hydrated electrons are rather inert toward the alcohols and react preferentially with other solutes or even with water. The organic radicals formed by hydrogen abstraction both dimerize and disproportionate, giving glycols and carbonyl compounds, respectively, when oxygen is not present but rapidly add oxygen forming peroxy radicals when it is available. The peroxy radicals subsequently react forming hydrogen peroxide and carbonyl compounds. Acid ethanol solutions, for example, give hydrogen, hydrogen peroxide ($=G_{H_2O_2}$), acetaldehyde, and 2,3-butanediol when oxygen-free solutions are irradiated, and hydrogen ($=G_{H_2}$), hydrogen peroxide, and acetaldehyde, but no butanediol, when oxygen-saturated solutions are irradiated (Table 7.13). The G values fall if the irradiation is prolonged due to radical attack on the products, but at moderate doses the yields from oxygen-saturated solutions are relatively constant from pH 1 to 11, and the system has been used to estimate primary yields in irradiated water (218, 219), when

$$G(H_2O_2) = G_{H_2O_2} + 0.5(G_{e_{aq}^-} + G_H + G_{OH} + G_{HO_2}) \quad (7.227)$$

$$G(CH_3CHO) = G_{OH} \quad (7.228)$$

Allan, Hayon, and Weiss (220) showed that the same products are formed when deaerated ethanol solutions are irradiated at room temperature or are frozen and irradiated at temperatures down to $-196°C$, though the yields are lower at the lower irradiation temperatures. At $-196°C$, for example, a 0.1 M solution (pH 1.2) gave: $G(H_2)$ 0.40, $G(H_2O_2)$ 0.05, $G(CH_3CHO) = 0.27$.

T A B L E 7.13 *Radiolysis of Aqueous Ethanol Solutions*[a]

| Product | G(Product) | |
	Oxygen-Free Solution	Oxygen-Saturated Solution
H$_2$	4.2	About 0.6
H$_2$O$_2$	About 0.6	4.15
CH$_3$CHO	1.90	2.6
Glycol	1.65	0

[a] Ref. 213; 200 kVp x-rays, 3.4×10^{-2} M ethanol, pH 1.2.

Carbonyl compounds are readily attacked by the radicals from irradiated water but, with the possible exception of acetone, have not been studied in great detail. In the case of formaldehyde, the simplest aldehyde, radical attack produces the methanol and formyl radicals already described as intermediates in the radiolysis of CO, HCOOH, and CH$_3$OH solutions (221):

$$e_{aq}^- + HCHO \rightarrow H_2\dot{C}{-}O^- \xrightarrow{H_2O} \cdot CH_2OH + OH^-$$
$$k_{229} < 10^7 \ M^{-1} \ sec^{-1} \quad (7.229)$$

$$H + HCHO \rightarrow H_2 + H\dot{C}O \quad k_{230} = 5 \times 10^6 \ M^{-1} \ sec^{-1} \quad (7.230)$$

$$OH + HCHO \rightarrow H_2O + H\dot{C}O \quad k_{231} = {\sim}2 \times 10^9 \ M^{-1} \ sec^{-1} \quad (7.231)$$

Products expected from the reaction of these radicals with each other and with H_2O_2 include CH_3OH, $HCOOH$, $OHCCOOH$, and $OHCCHO$. When oxygen is present, both e_{aq}^- and H will be scavenged forming perhydroxyl radicals and oxidation of the strongly reducing formyl radical occurs,

$$H\dot{C}O + O_2 \rightarrow \cdot O_2CHO \qquad k_{232} = 7.7 \times 10^8 \ M^{-1} \ \text{sec}^{-1} \quad (7.232)$$

$$\cdot O_2CHO + HO_2 \xrightarrow{H_2O} HCOOH + H_2O_2 + O_2 \qquad (7.233)$$

Allan (222) has identified H_2O_2 and two organic hydroperoxides (probably HO_2CH_2CHO and CH_3COO_2H), formaldehyde, and acetic acid as products of the radiolysis of acetaldehyde solutions containing oxygen.

Acetone is a good scavenger for hydrated electrons and reacts also with H and OH:

$$e_{aq}^- + CH_3COCH_3 \rightarrow (CH_3)_2\dot{C}-O^- \xrightarrow{H_2O} CH_3\dot{C}OHCH_3 + OH^-$$
$$k_{234} = 5.9 \times 10^9 \ M^{-1} \ \text{sec}^{-1} \quad (7.234)$$

$$OH + CH_3COCH_3 \rightarrow H_2O + \cdot CH_2COCH_3$$
$$k = 8.5 \times 10^7 \ M^{-1} \ \text{sec}^{-1} \quad (7.235)$$
$$\rightarrow (CH_3)_2C(OH)O \cdot \qquad (7.236)$$

$$H + CH_3COCH_3 \rightarrow H_2 + \cdot CH_2COCH_3$$
$$k_{237} = 1.9 \times 10^6 \ M^{-1} \ \text{sec}^{-1} \quad (7.237)$$
$$\rightarrow CH_3\dot{C}OHCH_3$$
$$k_{238} = 4 \times 10^5 \ M^{-1} \ \text{sec}^{-1} \quad (7.238)$$

The main products from oxygen-free 0.01 M acetone solutions are H_2, H_2O_2, 2-propanol, hydroxyacetone, and 2,5-hexadione (223), formed by

$$2CH_3\dot{C}OHCH_3 \rightarrow CH_3CHOHCH_3 + CH_3COCH_3 \quad (7.239)$$

$$2 \cdot CH_2COCH_3 \rightarrow (CH_2COCH_3)_2 \quad \text{(2,5-hexadione)} \quad (7.240)$$

$$(CH_3)_2C(OH)O \cdot + R \cdot \rightarrow CH_3COCH_2OH + RH \qquad (7.241)$$

where $R \cdot$ is an organic radical. At higher (0.5 M) acetone concentrations, 2-propanol is also formed by

$$CH_3\dot{C}OHCH_3 + CH_3COCH_3 \rightarrow$$
$$CH_3CHOHCH_3 + \cdot CH_2COCH_3 \quad (7.242)$$

2-Propanol is not formed at low pH where e_{aq}^- is scavenged by H^+, but the yields of hydrogen and dione are increased compared with neutral solutions. Observed H_2O_2 yields are always lower than $G_{H_2O_2}$, and since added H_2O_2 decreases the yield of 2-propanol and increases that of dione in neutral solutions, it is suggested that H_2O_2 reacts in part with the alcohol precursor, i.e.,

$$H_2O_2 + CH_3\dot{C}OHCH_3 \rightarrow CH_3COCH_3 + H_2O + OH \quad (7.243)$$

Carbohydrates

The radiation chemistry of the carbohydrates illustrates the greater complexity observed when polyfunctional rather than simple monofunctional compounds are irradiated. Though the polysaccharides are of greatest interest in radiation biology, the radiolysis of solutions of simple mono-saccharides is probably better understood and at the same time serves to illustrate most of the reactions expected with the polymeric carbohydrates with the exception of hydrolytic cleavage of the glycoside linkages.

D-Glucose is representative of the simpler sugars and reacts with the primary radicals with the following rate constants: $+e_{aq}^-$, $\sim 3 \times 10^5 \, M^{-1}$ sec^{-1}; $+H$, $4 \times 10^7 \, M^{-1} sec^{-1}$; $+OH$, $1.9 \times 10^9 \, M^{-1} sec^{-1}$. The low rate of reaction with the hydrated electron is consistent with the fact that the sugar is predominantly in the cyclic hemiacetal form rather than the aldehyde form, since e_{aq}^- reacts rapidly with carbonyl groups but not with alcohol or ether groups (HCHO, which is more highly hydrated than CH_3CHO in solution, reacts more slowly with e_{aq}^-). In solutions containing dissolved air or oxygen, e_{aq}^- will form O_2^-. Hydrogen atoms will also be scavenged by oxygen present in solution, though the extent of the scavenging will be dependent on the relative concentrations of glucose and oxygen. The reaction of both H and OH with glucose is expected to be similar to their reaction with alcohols, namely hydrogen abstraction, predominantly from C—H groups. Products formed by the irradiation of glucose solutions have been identified by means of paper chromatography and the use of radioactive tracers (224 to 229); some typical yields are listed in Table 7.14. As the irradiation progresses, the initial products are themselves attacked and converted to secondary products, as shown in the reaction scheme below (228, 229). In the absence of oxygen an acidic polymer is also formed as a secondary product, presumably by a sequence of reactions in which radicals combine to form larger molecules and these in turn are attacked by H and OH to give radicals of increased size which continue the process. When oxygen is present the radicals react with this rather than by combination and the polymer is not formed, glucuronic acid appearing in its place, suggesting that the polymer is formed as a result of hydrogen abstraction from carbon-6.

Degradation Pattern for D-*Glucose Irradiated in Oxygen-Free Solution*

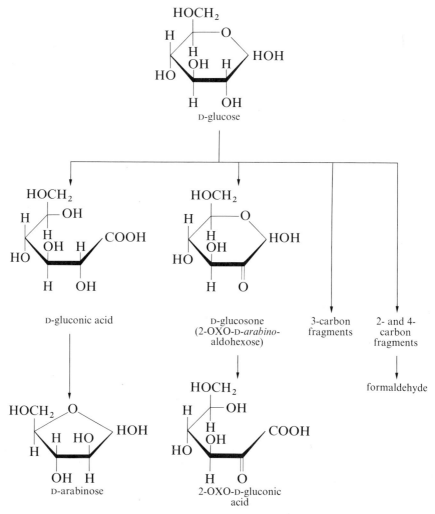

Two and three carbon products identified include glyoxal (OHC—CHO), glycollic aldehyde (OHC—CH$_2$OH), dihydroxyacetone

$$HOCH_2—CO—CH_2OH$$

and D-glyceraldehyde (OHC—CHOH—CH$_2$OH). The range of products identified shows that radical attack is not limited to a single point in the molecule but that, with the possible exception of carbon-5, it must occur in varying degree at each carbon atom present. The products D-gluconic acid, D-glucosone, and D-glucuronic acid, for example, are clearly the result

T A B L E 7.14   *Radiolysis of Aqueous* D-*Glucose Solutions*[a]

| Product | Initial G value | |
	Oxygen-Free Solution	Oxygen-Saturated Solution
G(− D-glucose)	3.5	3.5
Two-carbon fragments	0.85	0.8
Three-carbon fragments	0.8	—
D-Erythrose	—	0.25
D-Gluconic acid	0.35	0.4 to 0.5
D-Glucuronic acid	0	0.9
D-Glucosone	0.4	0

[a] Ref. 228; γ-irradiation, neutral 5×10^{-2} M.

of attack at carbons 1, 2, and 6, respectively, while the three carbon fragments must be formed following radical attack at carbon-3 or carbon-4. In the absence of oxygen, the organic radicals produced appear to disproportionate, to give products such as gluconic acid and glucosone, or to combine to give polymeric materials, while the addition of oxygen results in the eventual formation of oxidized products such as the sugar acids. In aerated solutions the overall yield of acid products, expressed in terms of G(monobasic acid), is 1.2 to 1.3, compared with a yield of 0.4 for deaerated solutions. When glucose solutions are employed as chemical dosimeters, the absorbed dose is estimated from an empirical relationship based on the change in optical activity of the solution and no knowledge of the precise chemical changes taking place is needed.

Oxygenated D-mannitol solutions have also been suggested for use in dosimetry (230). In this case the change is simpler, and the initial products are D-mannose (G 2.25) and H_2O_2 (G 3.0) when 5×10^{-3} M solutions of D-mannitol are irradiated with fast electrons, γ-rays, and 190 kV x-rays. The yield of mannose approximates to G_{OH} and that of H_2O_2 to $G_{H_2O_2} + G_{e_{aq}^-} + G_H$ so that the mechanism can be represented by

$$H \text{ (or } e_{aq}^-) + O_2 \rightarrow HO_2 \text{ (or } O_2^-) \tag{7.204}$$

$$\begin{array}{ccc}
CH_2OH & & \cdot CHOH \\
| & & | \\
(CHOH)_4 & + OH \rightarrow & (CHOH)_4 + H_2O \\
| & & | \\
CH_2OH & & CH_2OH
\end{array} \tag{7.244}$$

D-mannitol

$$\begin{array}{ccc}
\cdot CHOH & & CHO \\
| & & | \\
(CHOH)_4 & + HO_2 \text{ (or } O_2^-) \rightarrow & (CHOH)_4 + H_2O_2 \text{ (or } HO_2^-) + O_2 \\
| & & | \\
CH_2OH & & CH_2OH
\end{array} \tag{7.245}$$

D-mannose

with radical attack at one or other of the primary alcohol groups rather than on the secondary groups; in the case of D-mannitol oxidation of either terminal group yields D-mannose. D-Mannose is also formed in high yield by the action of Fenton's reagent (a mixture of ferrous ion and hydrogen peroxide which is believed to generate hydroxyl radicals) upon D-mannitol (231), although the reaction may involve complex formation rather than simple oxidation by OH.

ascorbic acid   dehydroascorbic acid

Ascorbic acid (AH_2) is readily oxidized to dehydroascorbic acid (A) by irradiation of its aqueous solutions. In acidified aerated solutions the consumption of ascorbic acid and oxygen, $G(-\text{ascorbic acid}) = 7.8$, $G(-O_2) = 7.49$ (232) are higher than G_{OH} and suggest that ascorbic acid is also being attacked by HO_2, the suggested mechanism being,

$$AH_2 + OH \rightarrow AH\cdot + H_2O \tag{7.246}$$

$$AH_2 + HO_2 \rightarrow AH\cdot + H_2O_2 \tag{7.247}$$

$$AH\cdot + O_2 \rightarrow AHO_2\cdot \tag{7.248}$$

$$2AHO_2\cdot \rightarrow 2A + H_2O_2 + O_2 \tag{7.249}$$

Examples of hydrogen abstraction by the perhydroxyl radical are rare and, as in this instance, are attributed to the presence of particularly labile hydrogen atoms in the solute.

Organic Halogen Compounds

Aliphatic compounds containing chlorine, bromine, or iodine are among the substances most sensitive to attack by hydrated electrons, which bring about release of the halogen as the halide ion. Hydrogen atoms may react in the same way, but generally the reaction is slower and is in competition with other hydrogen atom reactions so that, in practice, organic halides can be used to distinguish between e_{aq}^- and H.

Chloroacetic acid was used by several groups in early experiments to establish the nature and properties of the reducing species in irradiated water. Jortner and Rabani (101) showed that hydrogen atoms generated in an electrodeless discharge and introduced into a neutral chloroacetate solution did not release as much chloride as equivalent amounts of the reducing species produced by irradiating the neutral solution, although the yield of chloride increased if the hydrogen atoms were passed into strongly basic solutions (233, 234). In parallel experiments Hayon and Allen (100) showed that the yield of chloride from irradiated solutions was essentially constant in the range of pH from 3 to 8, though it fell below pH 3, where the reduction in chloride yield was accompanied by a corresponding increase in the yield of hydrogen (Fig. 7.6). These observations helped identify the reducing species in irradiated acid solutions as the hydrogen atom and the species in neutral solutions as the hydrated electron, indicating also that in very basic solutions the hydrogen atom might be converted to the hydrated electron. The processes occurring are

$$e_{aq}^- + ClCH_2COOH \text{ (or } ClCH_2COO^-) \rightarrow Cl^- + \cdot CH_2COOH$$
$$\text{(or } \cdot CH_2COO^-) \qquad (7.250)$$

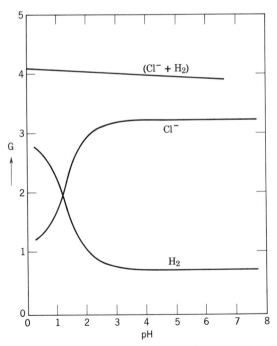

FIGURE 7.6 *Effect of pH on the yields of products from γ-irradiated* $ClCH_2COOH$ *solutions* (0.1 M, oxygen-free. Data from ref. 100).

$$H + ClCH_2COOH \text{ (or } ClCH_2COO^-) \rightarrow Cl^- + H^+ + \cdot CH_2COOH$$
$$\text{(or } \cdot CH_2COO^-) \tag{7.251}$$

$$\rightarrow H_2 + Cl\dot{C}HCOOH$$
$$\text{(or } Cl\dot{C}HCOO^-) \tag{7.252}$$

with, in acid solutions,

$$e_{aq}^- + H^+ \rightarrow H \tag{7.42}$$

and, in very basic solutions (pH about 13 or above),

$$H + OH^- \rightarrow e_{aq}^- + H_2O \tag{7.46}$$

The rate constants for reaction 7.250 are at least several hundred times greater than those for reactions 7.251 and 7.252, and the rate constants for reaction 7.252 are several times higher than those for reaction 7.251 (Table 7.7). In acid solutions containing 0.1 M $ClCH_2COOH$, reaction 7.42 can compete successfully with reaction 7.250 and the yield of chloride falls with decreasing pH while the hydrogen atoms formed contribute to the increasing yield of molecular hydrogen (Fig. 7.6). Hydroxyl radicals also abstract hydrogen from chloroacetic acid

$$OH + ClCH_2COOH \rightarrow H_2O + Cl\dot{C}HCOOH \tag{7.253}$$

but the main interest in this system lies with reactions 7.250 to 7.252, and the complete reaction mechanism has not been established.

A system that has been proposed for use as a chemical dosimeter is a saturated aqueous solution of chloroform (0.07 M $CHCl_3$) containing dissolved air ($\sim 2.5 \times 10^{-4}$ M O_2) which liberates HCl upon γ-irradiation with a G value of 27.5 (235, 236). Other products are CO_2 (G 8.4), H_2O_2 (G 2.04), and formic and oxalic acids ($G \sim 1$). With this solution the initial reactions are

$$e_{aq}^- + CHCl_3 \rightarrow Cl^- + \cdot CHCl_2 \qquad k_{254} = 3.0 \times 10^{10} \ M^{-1} \ sec^{-1} \tag{7.254}$$

$$H + O_2 \rightarrow HO_2 \qquad k_{57} = 1.9 \times 10^{10} \ M^{-1} \ sec^{-1} \tag{7.57}$$

$$H + CHCl_3 \rightarrow H^+ + Cl^- + \cdot CHCl_2 \qquad k_{255} = 1.2 \times 10^7 \ M^{-1} \ sec^{-1} \tag{7.255}$$

$$OH + CHCl_3 \rightarrow H_2O + \cdot CCl_3 \qquad k_{256} = 1.4 \times 10^7 \ M^{-1} \ sec^{-1} \tag{7.256}$$

with about 80% of the hydrogen atoms reacting with oxygen. The yield of chloride is an order of magnitude greater than $G_{e_{aq}^-}$ but a chain reaction is unlikely, since the yields are independent of dose rate over several orders of magnitude. It is more probable that the organic radicals formed in reactions 7.254 to 7.256 lose all their chlorine by hydrolysis, possibly after addition of oxygen since compounds containing the —O—C—Cl grouping are known to hydrolyze rapidly, and that hydrolysis of one or other of $\cdot CHCl_2$ and $\cdot CCl_3$ produces a radical that is able to react with an additional molecule of $CHCl_3$. A possible sequence of reactions might be:

$$\cdot CCl_3 \xrightarrow{H_2O} 3HCl + \cdot COOH \qquad (7.257)$$

$$\cdot COOH + O_2 \rightarrow CO_2 + HO_2 \qquad (7.180)$$

$$\cdot CHCl_2 \xrightarrow{H_2O} 2HCl + H\dot{C}O \qquad (7.258)$$

$$H\dot{C}O + O_2 \rightarrow \cdot O_2CHO \rightarrow CO_2 + OH \qquad (7.181)$$

$$\cdot O_2CHO + HO_2 \rightarrow HCOOH + H_2O_2 + O_2 \qquad (7.233)$$

$$2HO_2 \rightarrow H_2O_2 + O_2 \qquad (7.101)$$

$$2\cdot COO^- \rightarrow {}^-OOC—COO^- \qquad (7.191)$$

However other reactions are also possible and the complete mechanism has not been determined; a complicating factor is the change in pH, from near neutral to acid, during the course of the radiolysis. At pH below 3 the product yields fall as e_{aq}^- is increasingly scavenged by H^+, while at pH above 11 an increase in yields is observed, tentatively attributed to the conversion of $\cdot OH$ to O^- and reaction of the latter with $CHCl_3$ in a manner similar to e_{aq}^- (236).

$$O^- + CHCl_3 \rightarrow OCl^- + \cdot CHCl_2 \qquad (7.259)$$

Product yields are lower in the absence of oxygen (235, 236): for γ-radiation, $G(Cl^-) = 6.4$, $G(CO_2) = 0.4$, $G(H_2O_2) = 1.22$, $G(\text{monobasic acid}) = 7.2$ at near-neutral pH. The lower yields of acid products and carbon dioxide are attributed to slower hydrolysis of the intermediate organic radicals ($\cdot CHCl_2$ and $\cdot CCl_3$) in the absence of oxygen, and suggest that the species hydrolyzed in aerated solution are $\cdot O_2CHCl_2$ and $\cdot O_2CCl_3$. Chloride yields from oxygen-free solutions also increase at high pH in a manner consistent with a reaction between O^- and $CHCl_3$.

A second system of interest in dosimetry, since it offers the possibility of a very sensitive chemical dosimeter is an aqueous, air-saturated, solution of chloral hydrate containing 10^{-2} to 1 M $CCl_3CH(OH)_2$ (237–239). These solutions give acid products (mainly HCl) with G values ranging from 10 to several hundred depending upon the conditions, but the yields are dependent on such factors as temperature, chloral hydrate concentration $[G \propto (\text{chloral})^{0.5}]$, and dose rate $[G \propto \sim(\text{dose rate})^{-0.3}]$. The high yields and concentration and dose-rate dependencies are indicative of a chain reaction initiated by radical attack upon the chloral hydrate, i.e.,

$$e_{aq}^{-} + CCl_3CH(OH)_2 \rightarrow Cl^{-} + \cdot CCl_2CH(OH)_2 \qquad (7.260)$$

$$OH \text{ (or H)} + CCl_3CH(OH)_2 \rightarrow H_2O \text{ (or } H_2) + CCl_3\dot{C}(OH)_2 \qquad (7.261)$$

The steps in the chain reaction are not known although it probably resembles chain reactions between polyhalogen compounds and alcohols and aldehydes in organic solution (p. 405). In these, alternate steps involve abstraction of halogen from the halogen compound and of α-hydrogen from the alcohol or aldehyde, the chain being conditional upon the presence of relatively weakly bound halogen in the halide and of a labile hydrogen in the oxygen compound. In the case of the chloral hydrate chain reaction, the propagation steps probably include the formation and hydrolysis of an organic peroxy radical as the chain is dependent on the presence of oxygen; chloride yields in the absence of oxygen are of the order of 10 to 20 and can be accounted for by nonchain mechanisms. High yields of halide ion suggestive of chain reactions are obtained when bromal hydrate, $CBr_3CH(OH)_2$, and 2,2,2-tribromoethanol, CBr_3CH_2OH, are irradiated in aqueous solution in the presence of oxygen, but the yields of fluoride ion from solutions of $CF_3CH(OH)_2$ are low (240, 241).

Aromatic halogen compounds are also dehalogenated by reaction with hydrated electrons, although the overall mechanism is rendered more complex by the possibility of radical addition to the aromatic system. Irradiation of oxygen-free p-bromophenol solutions, for example, leads to debromination by

$$e_{aq}^{-} + HO\!\!-\!\!\bigcirc\!\!-\!\!Br \rightarrow Br^{-} + HO\!\!-\!\!\bigcirc \;\cdot$$

$$k_{262} = 1.2 \times 10^{10} \; M^{-1} \, sec^{-1} \qquad (7.262)$$

but addition of the hydroxyphenyl radical to the solute leads to two bromodihydroxybiphenyls and, via secondary reactions, to several substituted terphenyls (242, 243). All the chlorine present in the pesticide DDT $[(p\text{-}ClC_6H_4)_2CHCCl_3]$ can be released as Cl^{-} by irradiation, $10^{-5} \, M$ DDT requiring an absorbed dose of about $1.5 \times 10^{19} \; eV \; g^{-1} (2.4 \times 10^5 \, rad)$ when coated on silica gel and irradiated as a suspension in air-saturated water (244). Dechlorination is more efficient if the DDT can be irradiated as a solution in 2-propanol, since a chain reaction is possible under these circumstances (245).

Organic Nitrogen Compounds

Organic nitrogen compounds have received considerable attention from radiation chemists because of their relation to living systems and radiation biology. Since it is impossible to cover them adequately in a brief review, the more important functional groups containing nitrogen are treated here by describing the radiation chemistry of several relatively simple compounds. Many more detailed reviews are available (e.g., 246–248).

Hydrated electrons react at diffusion-controlled rates with aliphatic nitro compounds, e.g.,

$$e_{aq}^- + C(NO_2)_4 \rightarrow C(NO_2)_3^- + NO_2 \quad k_{77} = 5.0 \times 10^{10} \, M^{-1} \, sec^{-1} \quad (7.77)$$

The product from this reaction is a stable and highly colored ion that has been used to estimate the concentration of hydrated electrons in an irradiated solution and hence the extinction coefficient of e_{aq}^- (133). Tetranitromethane is also reduced by hydrogen atoms

$$H + C(NO_2)_4 \rightarrow C(NO_2)_3^- + H^+ + NO_2$$
$$k_{263} = 2.6 \times 10^9 \, M^{-1} \, sec^{-1} \quad (7.263)$$

and by HO_2 and O_2^- (the latter reacting over 10^3 times faster than the un-ionized radical)

$$O_2^- + C(NO_2)_4 \rightarrow C(NO_2)_3^- + NO_2 + O_2$$
$$k_{264} = 2 \times 10^9 \, M^{-1} \, sec^{-1} \quad (7.264)$$

but does not react with hydroxyl radicals which, in the absence of a hydroxyl radical scavenger, oxidize the nitroform ion

$$OH + C(NO_2)_3^- \rightarrow OH^- + C(NO_2)_3 \quad k_{265} \sim 3 \times 10^9 \, M^{-1} \, sec^{-1} \quad (7.265)$$

The nitroform produced may react with NO_2 to reform tetranitromethane, or with NO_2^- to give the nitroform ion and NO_2.

The cyanide or nitrile group, $-C{\equiv}N$, is somewhat less reactive than the aliphatic nitro group and forms addition products with radicals in a manner typical of unsaturated compounds. The hydroxyl radical, for example, adds to the cyanide ion to give, following reaction with water and rearrangement, the formamide radical,

$$OH + (C{\equiv}N)^- \rightarrow (HO-\dot{C}{=}N)^- \xrightarrow{+H_2O}$$
$$HO-\dot{C}{=}NH + OH^- \;(\rightarrow \cdot CONH_2) \quad (7.175)$$

Similar reactions occur when the primary radicals react with acetonitrile,

$$e_{aq}^- + CH_3C{\equiv}N \xrightarrow{H_2O} CH_3CH{=}\dot{N} \text{ (or } CH_3\dot{C}{=}NH) + OH^-$$
$$k_{266} = 2.5 \times 10^7 \; M^{-1} \sec^{-1} \quad (7.266)$$

$$H + CH_3C{\equiv}N \rightarrow CH_3CH{=}\dot{N} \text{ (or } CH_3\dot{C}{=}NH)$$
$$k_{267} = 3.5 \times 10^6 \; M^{-1} \sec^{-1} \quad (7.267)$$

$$OH + CH_3C{\equiv}N \rightarrow CH_3C(OH){=}\dot{N} \text{ (or } CH_3\dot{C}{=}NOH)$$
$$k_{268} = 5.5 \times 10^6 \; M^{-1} \sec^{-1} \quad (7.268)$$

The major products formed when oxygen-free 0.1 M CH_3CN solutions are irradiated with γ-radiation (pH 6) are acetaldehyde (G 1.50) and ammonia (G 1.00) (249), which are the products expected if the radicals formed in reactions 7.266 and 7.267 disproportionate to give an imine that subsequently hydrolyzes,

$$2CH_3CH{=}\dot{N} \text{ (or } CH_3\dot{C}{=}NH) \rightarrow CH_3CN + CH_3CH{=}NH \quad (7.269)$$

$$CH_3CH{=}NH + H_2O \rightarrow CH_3CHO + NH_3 \quad (7.270)$$

However, other processes are also occurring, since acetaldehyde and ammonia are formed, in reduced yield $[G(CH_3CHO) 0.86, G(NH_3) 0.36]$, when oxygen-saturated solutions are irradiated. The concentration of oxygen in the latter solutions should be sufficient to scavenge hydrogen atoms and most of the hydrated electrons.

Aliphatic amines react with the hydrated electron at the amino group, though the reaction is relatively slow with the free amines and only two or three times faster when these are protonated (250),

$$e_{aq}^- + CH_3CH_2NH_2 \xrightarrow{H_2O} CH_3CH_2\cdot + NH_3 + OH^-$$
$$k_{271} = 1 \times 10^6 \; M^{-1} \sec^{-1} \quad (7.271)$$

$$e_{aq}^- + CH_3CH_2NH_3^+ \rightarrow H + CH_3CH_2NH_2$$
$$k_{272} = 2.5 \times 10^6 \; M^{-1} \sec^{-1} \quad (7.272)$$

Hydroxyl radicals and hydrogen atoms abstract hydrogen, but the site of attack differs for the unprotonated and protonated forms, being predominantly from the amino group and α-position for the free amine but mainly from sites remote from the $-NH_3^+$ group with the protonated amine (250),

$$OH + CH_3CH_2NH_2$$
$$\rightarrow H_2O + CH_3CH_2\dot{N}H \Big\} \qquad (7.273)$$
$$\rightarrow H_2O + CH_3\dot{C}HNH_2 \Big\} \quad k = 6.2 \times 10^9 \; M^{-1} \sec^{-1} \quad (7.274)$$

$$OH + CH_3CH_2NH_3^+ \rightarrow H_2O + \cdot CH_2CH_2NH_3^+$$
$$k_{275} = 5 \times 10^8 \ M^{-1} \ sec^{-1} \quad (7.275)$$

In the absence of oxygen the radicals formed decay by second-order processes (i.e., radical combination or disproportionation) but when oxygen is present follow first-order decay kinetics (presumably addition of oxygen) followed by second-order decay of the new transients produced. Products formed when oxygenated ethylamine solutions are irradiated are (251) acetaldehyde ($G = 3.2$ at pH 11 to 12, but only 0.77 at pH 2 to 3), an oxime ($CH_3CH{=}NOH$, $G \sim 0.6$), nitroethane ($G \sim 0.6$), and hydrogen peroxide; the yields of oxime and nitroethane decrease at pH below 10 and become zero at pH 7 or below, while the yield of hydrogen peroxide increases from about 1.5 at pH 7 to about 4 at pH 1. Reactions consistent with these observations are, at high pH, addition of oxygen to the radicals formed in reactions 7.271, 7.273, and 7.274, followed by

$$CH_3CH_2O_2\cdot + HO_2 \rightarrow CH_3CHO + H_2O + O_2 \quad (7.276)$$

$$2\ CH_3CH_2NHO_2\cdot \rightarrow CH_3CH_2NO_2 + CH_3CH{=}NOH + H_2O \quad (7.277)$$

The pK for $CH_3CH_2NH_3^+$ is 10.5 so that at pH below about 10.5 the protonated form predominates and reactions 7.272 and 7.275 displace reactions 7.271, 7.273 and 7.274, with a consequent reduction in the yields of acetaldehyde, oxime, and nitroparaffin.

Radiolysis of aromatic amines lead to radical attack on the aromatic system rather than on the amino group (252).

In contrast to the simpler aliphatic nitrogen compounds, the α-amino acids, in common with many other bio-organic compounds, have been extensively studied by radiation chemists (246, 247, 253). Reaction generally centers on the amino acid grouping and will be illustrated by describing the behavior of glycine, NH_2CH_2COOH, although attack may also occur at other sites with the more complex amino acids. Acids containing thiol groups are particularly susceptible to attack at the sulfur grouping and are described separately in the next section. In neutral and near-neutral solutions the predominant form of glycine is the zwitterion, $NH_3^+CH_2COO^-$, which reacts with the primary radicals from water in the following manner:

$$e_{aq}^- + NH_3^+CH_2COO^- \rightarrow NH_3 + \cdot CH_2COO^- \text{ (major reaction)} \quad (7.278)$$
$$\rightarrow H + NH_2CH_2COO^- \quad (7.279)$$
$$k = 7 \times 10^6 \ M^{-1} \ sec^{-1}$$

$$H + NH_3^+CH_2COO^- \rightarrow H_2 + NH_3^+\dot{C}HCOO^-$$
$$k_{280} = 8 \times 10^4 \ M^{-1} \ sec^{-1} \quad (7.280)$$

$$OH + NH_3^+CH_2COO^- \rightarrow H_2O + NH_3^+\dot{C}HCOO^-$$
$$k_{281} = 9 \times 10^6 \ M^{-1} \ sec^{-1} \quad (7.281)$$

When oxygen is absent, these reactions are followed (253) by,

$$NH_3^+\dot{C}HCOO^- + \cdot CH_2COO^- \rightarrow NH_2^+{=}CHCOO^- + CH_3COO^- \quad (7.282)$$

$$2\,NH_3^+\dot{C}HCOO^- \rightarrow NH_2^+{=}CHCOO^- + NH_3^+CH_2COO^- \quad (7.283)$$

$$NH_2^+{=}CHCOO^- + H_2O \rightarrow NH_4^+ + OHCCOO^- \quad (7.284)$$

giving, with 1 M glycine solutions (Table 7.15), hydrogen, ammonia, acetic acid, and glyoxyllic acid as major products and aminosuccinic acid (aspartic acid) and diaminoscuccinic acid, formed by combination reactions of $NH_3^+\dot{C}HCOO^-$ and $\cdot CH_2COO^-$, as minor products. Small quantities of carbon dioxide and formaldehyde are also formed, possibly by hydrolysis of part of the intermediate iminoacetic acid,

$$NH_2^+{=}CHCOO^- + H_2O \rightarrow NH_3 + HCHO + CO_2 \quad (7.285)$$

The very low yield of hydrogen peroxide suggests that it is consumed by reaction with an intermediate radical, e.g.,

$$H_2O_2 + NH_3^+\dot{C}HCOO^- \rightarrow NH_2^+{=}CHCOO^- + H_2O + OH \quad (7.286)$$

When 1 M glycine solutions are irradiated with x-rays, an appreciable part (7.5% under the conditions applicable to Table 7.15) (254) of the energy absorbed is absorbed directly by the amino acid molecules and will lead to "direct" action, i.e., chemical effects that are not produced indirectly as a result of energy absorption by water. The products of such direct action will presumably be the same as those formed when glycine is irradiated in the solid state. After dissolution of the solid in water, these are (257) ammonia (G 4.8), acetic acid (G 2.3), and glyoxyllic acid (G 2.5) with smaller quantities (G 0.2) of hydrogen, carbon dioxide, and methylamine. Garrison (246) has suggested that the products are the result of ionic processes in the solid, represented by

$$NH_3^+CH_2COO^- \rightsquigarrow NH_3^+\dot{C}HCOO^- + H^+ + e^- \quad (7.287)$$

$$e^- + NH_3^+CH_2COO^- \rightarrow NH_3 + \cdot CH_2COO^- \quad (7.288)$$

which are followed by reactions 7.282 to 7.284 when the irradiated solid is dissolved in water. However, since the radicals are the same as those formed indirectly in reactions 7.278 to 7.281, no distinctly different products would be expected to result from any direct action that occurs.

Introduction of oxygen into the aqueous glycine system blocks reactions 7.278 to 7.280 by scavenging e_{aq}^- and H, but is without effect on the hydroxyl

T A B L E 7.15   *Radiolysis of Aqueous Glycine Solutions*[a]

Product	G(Product) Oxygen-Saturated Solutions (ref. 255)	G(Product) Oxygen-Free Solutions (ref. 254)
H_2		2.02
H_2O_2	3.6	<0.01
NH_3	4.3	3.97
CO_2		0.90
HCHO	1.1	0.53
HCOOH		0.085
CH_3NH_2	0.16	0.19
CH_3COOH		1.20
OHCCOOH	3.4	2.10
$HOOCCH_2CH(NH_2)COOH$		0.25^b
$HOOCCH(NH_2)CH(NH_2)COOH$		0.08^b

[a] 1 M unbuffered solutions, pH 6, irradiated with x-rays.
[b] Ref. 256.

radical reaction (Eq. 7.281). However, addition of oxygen to the radical produced in the latter reaction results in breakdown to ammonia and glyoxyllic acid,

$$NH_3^+\dot{C}HCOO^- + O_2 \xrightarrow{H_2O} NH_4^+ + OHCCOO^- + HO_2 \quad (7.289)$$

so that in the presence of oxygen, $G(NH_3) \simeq G(OHCCOO^-) \simeq G_{OH}$. This relationship is not obeyed exactly and is not applicable to more complex amino acids where the number of sites susceptible to hydroxyl radical attack is greater.

The reactions described above apply particularly to glycine solutions irradiated at near-neutral pH, which is the region of greatest interest in radiation biology. In strongly acid and strongly basic solutions the product yields and mechanism may be quite different (258), since both the ionic character of the solute and the nature of the attacking radicals can change. The rate of hydrated electron (Table 7.7) and hydroxyl radical (Table 7.10) attack is significantly different if the +ve ion present in acid solutions, $NH_3^+CH_2COOH$, or the −ve ion found in basic solutions, $NH_2CH_2COO^-$, is substituted for the zwitterion, and the site of attack may also change as with the simple amines. Ammonia yields from amino acids containing aromatic and heterocyclic rings are low, radical reaction occurring mainly with the cyclic system of such acids as phenylalanine, tyroscine, histidine, and tryptophan.

Simple amides (259, 260) have not been extensively investigated and attention has been focused instead on the peptides in which each amide (or peptide) group serves to link two amino acids, modeling the structure of the larger protein molecules. The zwitterion form of the simplest peptide, glycylglycine, reacts with e_{aq}^- and OH in a similar manner to glycine, the hydroxyl radical abstracting hydrogen from the carbon atom adjacent to the amide nitrogen rather than from the carbon next to the terminal NH_3^+ group (40, 261),

$$e_{aq}^- + NH_3^+CH_2CONHCH_2COO^- \rightarrow$$
$$NH_3 + \cdot CH_2CONHCH_2COO^- \quad (7.290)$$

$$\cdot OH + NH_3^+CH_2CONHCH_2COO^- \rightarrow$$
$$H_2O + NH_3^+CH_2CONH\dot{C}HCOO^- \quad (7.291)$$

Terminal amino groups do not make up a very significant part of protein molecules and more representative, simple, model compounds for proteins are *N*-acyl derivatives of the amino acids such as *N*-acetylglycine and *N*-acetylalanine.

Gamma radiolysis of oxygenated 0.05 *M* *N*-acetylalanine (262) gives H_2O_2 (*G* 2.2), NH_3 (*G* 2.9), CO_2 (*G* 2.0), and CH_3COOH (*G* 3.0) as the main products with smaller yields of CH_3CHO (*G* ~0.2), $CH_3COCOOH$ (*G* ~0.2), and unidentified hydroperoxides (*G* 0.5). The reaction of e_{aq}^- and H with *N*-acetylalanine is sufficiently slow that these species are scavenged by oxygen in oxygenated solutions. Hydroxyl radicals abstract hydrogen from the carbon next to the amide nitrogen, as with glycylglycine,

$$OH + CH_3CONHCH(CH_3)COO^- \rightarrow$$
$$H_2O + CH_3CONH\dot{C}(CH_3)COO^- \quad (7.292)$$

The radical formed by hydrogen abstraction adds oxygen to give a peroxy radical ($RO_2\cdot$), and this loses oxygen in a bimolecular reaction giving the corresponding alkoxy radical,

$$2\,RO_2\cdot \rightarrow 2\,RO\cdot + O_2 \quad (7.293)$$

The latter again reacts with oxygen and breaks down with loss of $HO_2\cdot$ (or O_2^-),

$$RO\cdot + O_2 \rightarrow HO_2\cdot + products \quad (7.294)$$

With *N*-acetylalanine (253, 262) the addition of oxygen to, and subsequent breakdown of, the alkoxy radical follows the sequence,

$$CH_3CONH\underset{\underset{O\cdot}{|}}{C}(CH_3)COO^- + O_2 \rightarrow CH_3CONHCOCH_3 + CO_2 + O_2^-$$

$$(7.295)$$

$$CH_3CONHCOCH_3 + H_2O \rightarrow CH_3CONH_2 + CH_3COOH \quad (7.296)$$

$$CH_3CONH_2 + H_2O \rightarrow CH_3COOH + NH_3 \quad (7.297)$$

and accounts for the formation of the main products. The nature and yields of radiolysis products from polyalanine are similar to those from N-acetyl-alanine, and the sequence of reactions shown is believed to follow hydroxyl radical attack on other polypeptides and proteins.

Oxygen will normally be present in living systems and will tend to scavenge electrons. If oxygen and other oxidizing solutes are absent, the electron adds to the peptide bond (259),

$$e_{aq}^- + CH_3CONHCH(CH_3)COO^- + H_2O \rightarrow$$
$$CH_3\dot{C}(OH)NHCH(CH_3)COO^- + OH^- \quad (7.298)$$

but the reaction leads to little permanent change, apparently because reaction with the radical formed by OH attack reforms the starting material,

$$CH_3\dot{C}(OH)NHCH(CH_3)COO^- + CH_3CONH\dot{C}(CH_3)COO^-$$
$$2 \, CH_3CONHCH(CH_3)COO^- \quad (7.299)$$

However, the back reaction can be blocked by solutes containing a labile hydrogen, e.g., cysteine and other thiols (RSH), by virtue of the hydrogen atom transfer reaction,

$$CH_3\dot{C}(OH)NHCH(CH_3)COO^- + RSH \rightarrow$$
$$CH_3CH(OH)NHCH(CH_3)COO^- + RS\cdot \quad (7.300)$$

which is followed by breakdown of the —CH(OH)—NH— group present in the product to give an aldehyde and an amine. Thus γ-irradiation of oxygen-free, molar (pH 7) solutions of acetamide and N-ethylacetamide gives acetaldehyde with G values of 2.4 and 2.8, respectively, in the presence of $4 \times 10^{-4} \, M$ cysteine, but G values of less than 0.1 in the absence of cysteine.

When N-acetylalanine and polyalanine are irradiated in the solid state the main products are ammonia (largely in the form of amide), acetaldehyde, propionic acid, pyruvic acid, and lactic acid, formed as a result of both ionization and excitation. Excitation, possibly by subexcitation electrons, also appears to play a part when concentrated aqueous solutions are irradiated, since yields of amide from 2 M N-acetylalanine solutions are not affected by the addition of radical scavengers but are reduced when compounds such as benzoic acid and benzaldehyde are added. Protection by the aromatic solute seems to involve energy transfer to the solute from an excited triplet state of N-acetylalanine as only solutes with triplet excited levels below that of the acetyl compound are effective (253, 263). The stoichiometry of the excited state reaction is

$$CH_3CONHCH(CH_3)COO^-{}^* + CH_3CONHCH(CH_3)COO^- \rightarrow$$
$$CH_3CONH\dot{C}(CH_3)COO^- + CH_3\dot{C}HCOO^- + CH_3CONH_2 \quad (7.301)$$

The heterocyclic nitrogen compounds that have received most attention are the pyrimidine bases, uracil, thymine, and cytosine, which are of interest in connection with radiation damage studies on nucleic acids and related compounds. The bases can exist in two tautomeric forms (carbonyl or amide, and hydroxy), but are shown here as having the amide structure, which is the more important with the 2 and 4-hydroxy pyrimidines.

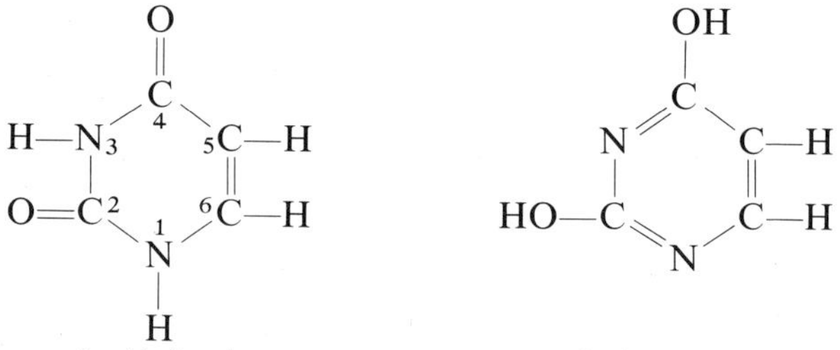

(amide form)   (hydroxy form)

uracil

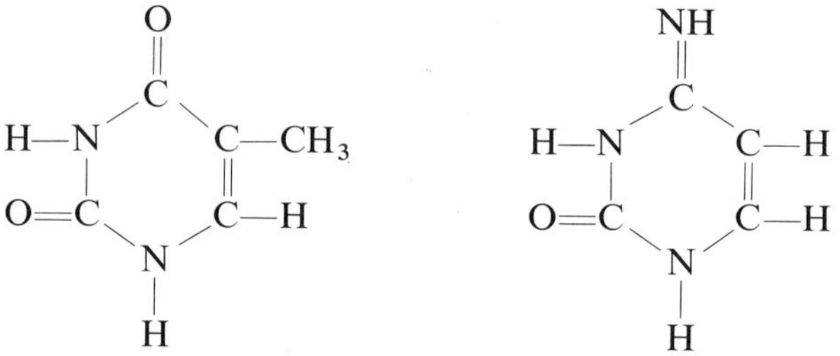

thymine   cytosine

Hydroxyl radicals and hydrogen atoms add to the pyrimidine ring at the 5,6-double bond giving, with uracil (264, 265),

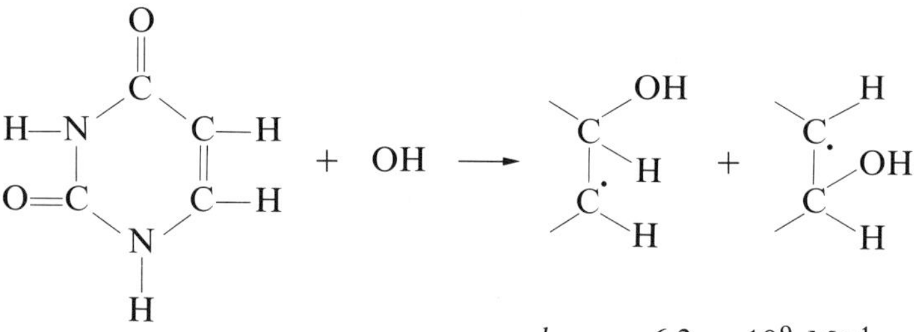

$$k_{302} = 6.3 \times 10^9 \; M^{-1} \; sec^{-1} \quad (7.302)$$

$$k_{303} = 3 \times 10^9 \ M^{-1} \ \text{sec}^{-1} \quad (7.303)$$

Hydrated electrons add to the 4-carbonyl group of uracil giving an adduct which exists in equilibrium with a protonated form ($pK = 7$),

$$k_{304} = 1.1 \times 10^{10} \ M^{-1} \ \text{sec}^{-1} \quad (7.304)$$

The hydroxyl-radical adducts of uracil can also ionize, and the reactions of the adducts are pH dependent.

Hunt and his colleagues (265) have shown that dimeric molecules, formed by combination of the radicals produced in reactions 7.302 to 7.304, are the main products when aqueous uracil solutions are irradiated in the absence of oxygen; smaller amounts of monomeric products are also formed. The latter include uracil glycol, uracil hydrate, and dihydrouracil, in which carbons 5 and 6 have the structures —CHOH—CHOH— (*cis* and *trans*), —CHOH—CH$_2$—, and —CH$_2$—CH$_2$—, respectively. The pH dependence of the product yields is illustrated in Table 7.16 and results from the different behavior of the neutral and ionized forms of the intermediate organic radicals. Monomeric products can be formed by disproportionation of the radicals, but at the higher pH other reactions must occur to regenerate uracil because $G(-\text{uracil})$ is considerably less than $0.5G(\text{radical})$; $G(\text{radical})$ is about 6. The

TABLE 7.16 *Radiolysis of Aqueous Uracil Solutions*[a]

	Nitrogen-Saturated		Air-Saturated	
	pH 5	pH 8.3	pH 5	pH 8.5
$G(-\text{uracil})$	3.3	0.8	2.6	2.4
$G(\text{uracil} \rightarrow \text{dimer})$	2.4	0.4	0	0
$G(\text{monomeric products})$	0.9	0.4		

[a] Ref. 265; 10^{-4} to 10^{-3} M.

explanation is probably a reaction, or reactions, analogous to reaction 7.299 in which the products of e_{aq}^- and OH reaction with the solute react together and reform two molecules of solute. In the case of uracil the extent of this back reaction must depend on the ionized state of the species involved.

When dilute uracil solutions are irradiated in the presence of oxygen, e_{aq}^- and H are scavenged to form O_2^- and HO_2, which do not react with uracil, and the hydroxyl radical adducts add oxygen as a prelude to further reaction; $G(-\text{uracil}) = G_{OH}$ (266, 267). A variety of products have been isolated, including about 20% of urea and related compounds which result from ring opening; however, for the most part the ring remains intact and the more important products are the uracil 5,6-glycols (*cis* and *trans*), isobarbituric acid, and dialuric acid. Several reaction pathways leading to these products can be written, but the following illustrates the types of reaction believed to occur following formation of the hydroxyl radical adduct (Eq. 7.302):

uracil hydroperoxides

cis- and *trans*-uracil glycols   isobarbituric acid   dialuric acid

Isobarbituric acid could also be formed by elimination of water from the *cis*-glycol and dialuric acid by elimination of hydrogen from the glycols.

The radiolysis of thymine is similar to that of uracil (268, 269) though in basic solution OH (or O^-) abstracts hydrogen from the 5-methyl group in competition with OH addition to the 5,6-double bond. The addition reaction in neutral solution has been used as a standard for measuring relative rate constants for hydroxyl radical reactions by the competition method; the extent of reaction with the thymine is readily followed, since addition to the 5,6-double bond destroys the chomophore responsible for the strong absorption of thymine at 264 nm.

Organic Sulfur Compounds

The thiol group is particularly susceptible to attack by free radicals and is the focus of the reaction when cysteine, $HSCH_2CH(NH_3^+)COO^-$ (represented by CySH), is irradiated in near-neutral aqueous solutions. This is in marked contrast to glycine, alanine, and the other simple amino acids where attack centers on the amino group and the adjacent carbon atom. The primary radicals from water react rapidly with the cysteine zwitterion as follows:

$$e_{aq}^- + CySH \rightarrow Cy\cdot + SH^- \; (\overset{H^+}{\rightarrow} H_2S)$$
$$k_{305} = 8.2 \times 10^9 \; M^{-1} \, sec^{-1} \quad (7.305)$$

$$H + CySH \rightarrow H_2 + CyS. \quad k_{306} = 1.0 \times 10^9 \; M^{-1} \, sec^{-1} \quad (7.306)$$

$$\rightarrow H_2S + Cy\cdot \quad k_{307} = 1.2 \times 10^8 \; M^{-1} \, sec^{-1} \quad (7.307)$$

$$OH + CySH \rightarrow H_2O + CyS\cdot \quad k_{308} \sim 9 \times 10^9 \; M^{-1} \, sec^{-1} \quad (7.308)$$

The S—H bond is weaker than most covalent bonds in organic compounds and the thiols also suffer hydrogen abstraction by O_2^- and organic radicals such as Cy·,

$$O_2^- + CySH \rightarrow HO_2^- + CyS\cdot \quad (7.309)$$

$$Cy\cdot + CySH \rightarrow CyH + CyS\cdot \quad (7.310)$$

When oxygen is absent, the CyS· radicals dimerize to cystine (CySSCy)

$$2 \, CyS\cdot \rightarrow CySSCy \quad (7.311)$$

so that the main radiolysis products are H_2, H_2S, alanine (CyH), and cystine (Table 7.17) (270–272). Hydrogen peroxide slowly oxidizes cysteine to cystine in neutral solutions, giving rise to a postirradiation effect in which additional quantities of cystine are produced for several hours after irradiation by

$$H_2O_2 + 2CySH \rightarrow H_2O + CySSCy \quad (7.312)$$

The reaction is much slower in acid solutions so that $G(H_2O_2) \, (= G_{H_2O_2})$ is given in Table 7.17 for pH 0 to 1 but not for pH 5 to 6, where reaction 7.312 has been allowed to go to completion. In neutral, oxygen-free, solutions the mechanism consists of reactions 7.305 to 7.312 and,

$$G(CyH) = G(H_2S) \quad (7.313)$$

$$G(-CySH) = 2G(CySSCy) + G(CyH) \quad (7.314)$$

TABLE 7.17   *Radiolysis of Aqueous Cysteine Solutions*[a]

	G(Product)		
Product	pH ~ 0[b] (ref. 270)	pH 1[b] (ref. 271)	pH 5 to 6 (ref. 272)
$G(-\text{cysteine})$ (CySH)	7.8		9.3
Cystine (CySSCy)	3.3	3.0, 3.8	3.4
Alanine (CyH)	0.87		2.6
H_2	3.35	3.05	1.1
H_2S	0.78	1.2	2.5
H_2O_2	0.68	0.68	

[a] ^{60}Co γ-rays; oxygen-free 10^{-2} M solutions.
[b] pH adjusted with $HClO_4$.

within the limits of experimental error. The yields of cystine, alanine, H_2S, and H_2 are related to the primary yields by

$$G(\text{CySSCy}) = G_{H_2O_2} + 0.5(G_{e_{aq}^-} + G_H + G_{OH}) \qquad (7.315)$$

$$G(\text{CyH}) = G(H_2S) = G_{e_{aq}^-} + G_H \left(\frac{k_{307}}{k_{306} + k_{307}} \right) \qquad (7.316)$$

$$G(H_2) = G_{H_2} + G_H \left(\frac{k_{306}}{k_{306} + k_{307}} \right) \qquad (7.317)$$

Substituting the primary yields given in Table 7.4 for neutral solutions in Eq. 7.315 to 7.317 gives $G(\text{CySSCy}) = 3.63$, $G(\text{CyH}) = G(H_2S) = 2.69$, and $G(H_2) = 0.94$, which are within the experimental limits for the values given in Table 7.17.

In acid oxygen-free solutions the yields of alanine and hydrogen sulfide are smaller, but this is balanced by a corresponding increase in the yield of hydrogen. With 1 M $HClO_4$ and 10^{-2} M cysteine, all but 0.4% of the hydrated electrons react

$$e_{aq}^- + H^+ \rightarrow H \qquad k_{42} = 2.35 \times 10^{10} \ M^{-1} \ \text{sec}^{-1} \qquad (7.42)$$

and the hydrogen atoms produced react mainly by reaction 7.306, so that more hydrogen but less hydrogen sulfide and Cy· radicals are formed and, because the latter are the precursors of alanine, less alanine. Substitution of primary yields for a solution of pH 0.46 from Table 7.4 in Eqs. 7.315 to 7.317, omitting $G_{H_2O_2}$ in Eq. 7.315, since hydrogen peroxide does not react in acid solutions, gives values for the product yields that are close to the experimental values; better agreement should be possible by substituting primary yields for a solution of pH ~0.

Hydrogen sulfide yields also decline at pH above 8 (273). In this region the thiol group (pK 8.4) is ionized, and the point of electron attack probably shifts to the amino group (ammonia is not an important product below pH 7),

$$e_{aq}^- + {}^-SCH_2CH(NH_3^+)COO^- \rightarrow NH_3 + {}^-SCH_2\dot{C}HCOO^- \quad (7.318)$$

(The rate of electron reaction with H_2S is also reduced, by a factor of over 10^4, when hydrogen sulfide is converted to the anion, HS^-.) The reaction of hydrogen atoms with cysteine has been investigated directly by passing a stream of hydrogen diluted with helium through an electric discharge and bubbling the emergent gases, which contain atomic hydrogen, through cysteine solutions (274). A major product in both acid and neutral solutions is cystine, consistent with hydrogen abstraction from the SH group (Eq. 7.306), but the yield of cystine falls at higher pH and reaches zero at pH 12, while the yield of hydrogen sulfide increases with increasing pH. These observations suggest that in basic solutions the predominant reaction of hydrogen atoms is

$$H + CyS^- \rightarrow HS^- + Cy\cdot \quad (7.319)$$

Cystine is also the main product when cysteine solutions are irradiated in the presence of oxygen, but the yields are considerably higher (up to $G = 70$; ref. 275) and are indicative of a chain reaction. Other products isolated in lower yield than cystine are, in order of decreasing importance (276), H_2O_2, serine (CyOH), cysteine sulfinic acid (CySO$_2$H), alanine (CyH), and lanthionine (CySCy); hydrogen sulfide is also formed. Steps in the chain reaction (276–278) probably include reactions 7.305 to 7.309 and

$$CyS\cdot + O_2 \rightarrow CySO_2\cdot \quad (7.320)$$

$$CySO_2\cdot + CySH \rightarrow CySO_2H + CyS\cdot \quad (7.321)$$

$$CySO_2H + CySH \rightarrow CySSCy + H_2O_2 \quad (7.322)$$

$$CySO_2H + 3CySH \rightarrow 2CySSCy + 2H_2O \quad (7.323)$$

$$Cy\cdot + O_2 \rightarrow CyO_2\cdot \quad (7.324)$$

$$CyO_2\cdot + CySH \rightarrow CySCy + HO_2 \quad (7.325)$$

$$CyO_2\cdot + 3CySH \rightarrow 3CyS\cdot + CyOH + H_2O \quad (7.326)$$

with the chain propagated chiefly by reactions 7.320 and 7.321. However, the complete mechanism is undoubtedly more complex than this; product yields are dependent on dose rate, pH, and cysteine concentration.

γ-Irradiation of solid (oxygen-free) cysteine gives cystine (G 5.0), H_2 (G 3.1), NH_3 (G 1.8), and H_2S (G 1.5) as the main products (253, 279). The greater proportion of ammonia in the solid state irradiation is attributed to competing reactions following electron capture, one of which results in deamination

$$e^- + HSCH_2CH(NH_3^+)COO^- \rightarrow NH_3 + HSCH_2\dot{C}HCOO^- \qquad (7.327)$$

$$\overset{H^+}{\rightarrow} H_2 + \cdot SCH_2CH(NH_3^+)COO^- \qquad (7.328)$$

$$\overset{H^+}{\rightarrow} H_2S + \cdot CH_2CH(NH_3^+)COO^- \qquad (7.322)$$

and to the contribution of solvation energy in aqueous solution acting in favour of reactions such as 7.305 in which ionic products are formed.

Compounds such as cysteine which are very readily attacked by radicals can often act as protective agents toward other solutes by reacting preferentially with radicals formed by irradiation of the substrate. Cysteine and glutathione, for example, protect the enzyme catalase against radiation-induced deactivation in aqueous solution (280), and many other examples have been reported (e.g., 281, 282). In some cases molecules such as cysteine (CySH), which contain weakly bound hydrogen atoms, can act as repair agents by donating hydrogen to a radical to replace a hydrogen atom lost as a result of radical attack; e.g., cysteine regenerates the organic solute (RH) in the following sequence,

$$OH + RH \rightarrow H_2O + R\cdot \qquad (7.323)$$

$$R\cdot + CySH \rightarrow RH + CyS\cdot \qquad (7.324)$$

Both radical scavenging and repair serve to protect the second solute against radiation damage at the expense of the cysteine, which is therefore referred to as a "sacrificial" protective agent.

The disulfide linkage is also very sensitive toward radical attack as the rate constants for the following reactions of the cystine zwitterion (pH ~ 6), $[SCH_2CH(NH_3^+)COO^-]_2$ ($= CySSCy$), show

$$e_{aq}^- + CySSCy \rightarrow (CySSCy)^- \overset{H_2O}{\rightarrow} CySH + CyS\cdot + OH^-$$
$$k_{325} = 1.3 \times 10^{10} \ M^{-1} \ sec^{-1} \qquad (7.325)$$

$$H + CySSCy \rightarrow CySH + CyS\cdot$$
$$k_{326} > 1.5 \times 10^9 \ M^{-1} \ sec^{-1} \qquad (7.326)$$

$$OH + CySSCy \rightarrow CySOH + CyS\cdot$$
$$k_{327} = 4.8 \times 10^9 \ M^{-1} \ sec^{-1} \ (pH\ 2) \quad (7.327)$$

In oxygen-free solutions the products from these reactions react to give cysteine (CySH), the sulfinic acid (CySO$_2$H), and a trisulfide (CySSSCy) as the main products (283), though other products are formed in lower yield; the yields are dose-rate dependent at low dose rates. Steps that have been suggested as contributing to the formation of the major products are

$$CySOH + CySH \rightarrow CySSCy + H_2O \qquad (7.328)$$

$$CySOH + H_2O_2 \rightarrow CySO_2H + H_2O \qquad (7.329)$$

$$2CySOH \rightarrow CySO_2H + CySH \qquad (7.330)$$

$$CyS\cdot + CySSCy \rightarrow CySSSCy + Cy\cdot \qquad (7.331)$$

$$Cy \rightarrow products \qquad (7.332)$$

When oxygen is present, O_2^- and HO_2 may be formed in competition with reactions 7.325 and 7.326, depending on the relative concentrations of oxygen and cystine, and oxygen will add to the intermediate organic radicals CyS· and Cy· giving $CySO_2\cdot$ and $CyO_2\cdot$. Under these conditions, the yield of cysteine is considerably reduced and those of the sulfinic acid and trisulfide reduced to a smaller extent, while cysteic acid ($CySO_3H$) becomes the main product (283). The latter may be formed by

$$CySOH + O_2^- \rightarrow CySO_3^- + H \qquad (7.333)$$

$$2\,CySO_2\cdot \rightarrow CySO_2SO_2Cy \xrightarrow{H_2O} CySO_2H + CySO_3H \quad (7.334)$$

Pulse radiolysis studies on the thiols (RSH) and disulfides (RSSR) benefit from the formation of $(RSSR)^-$, which exhibits a strong absorption in the visible region at 400 to 450 nm (247, 284). The ion is formed by addition of an electron to a disulfide and also via the radical RS·, formed by hydrogen abstraction from the corresponding thiol, and the following equilibrium,

$$RS\cdot + RS^- \rightleftharpoons (RSSR)^- \qquad (7.335)$$

Aromatic and Unsaturated Compounds

Addition to the unsaturated system is the characteristic reaction of the primary species from water with aromatic and unsaturated compounds. With benzene, this is represented by

$$k_{336} = 1.3 \times 10^7 \, M^{-1} \, sec^{-1} \quad (7.336)$$

$$+ \text{ H } \rightarrow \quad \overset{\text{H} \quad \text{H}}{\bigodot} \qquad\qquad k_{80} = 5.3 \times 10^8 \ M^{-1} \sec^{-1} \quad (7.80)$$

$$+ \text{ OH } \rightarrow \quad \overset{\text{H} \quad \text{OH}}{\bigodot} \qquad\qquad k_{92} = 5.3 \times 10^9 \ M^{-1} \sec^{-1} \quad (7.92)$$

where the symbols for the products indicate cyclohexadienyl systems with an unpaired electron that is not localized on a particular carbon atom. Electron spin resonance measurements (142) assign a positive electron density to the *ortho-* and *para*-carbon atoms, and the products would be expected to behave as a mixture of radicals with an unpaired electron at one or other of these positions. The dienyl radicals have a reasonably strong absorption at 310 to 315 nm and have been extensively investigated by pulse radiolysis (e.g., 285 to 289).

The radicals produced in reactions 7.336, 7.80, and 7.92 combine and disproportionate to give a complex mixture of products that include phenol, cyclohexadienes, hydroxycyclohexadienes, biphenyl, and reduced and hydroxylated biphenyls (e.g., 290, 291). The products are difficult to separate and to estimate separately so that G values reported for oxygen-free benzene solutions tend to be contradictory, e.g., values for G(phenol) range from 0.2 to 0.5 and for G(biphenyl) from 0.4 to 1.4 for neutral solutions. Products may be formed in postirradiation processes and be pH and dose-rate dependent, further complicating interpretation of the data. However, typical of the reactions believed to occur are

$$2 \ \overset{\text{H} \quad \text{OH}}{\bigodot} \rightarrow \overset{\text{OH}}{\bigodot} + \bigodot + \text{H}_2\text{O} \qquad (7.337)$$

$$\rightarrow \bigodot\!-\!\bigodot + 2\text{H}_2\text{O} \qquad (7.338)$$

which give phenol and biphenyl from hydroxycyclohexadienyl radicals.

Corresponding reactions of cyclohexadienyl radicals formed in reactions 7.336 and 7.80 give cyclohexadienes and reduced biphenyls, while hydroxy-cyclohexadienes and hydroxybiphenyls can be formed by the reaction of hydroxycyclohexadienyl and cyclohexadienyl radicals together. Total product yields appear to be lower than expected if all radicals ($G \sim 6$) disproportionate or dimerize, suggesting that some of the organic radicals regenerate benzene, for example, by

$$\text{(cyclohexadienyl)} + \text{(hydroxycyclohexadienyl, H\ OH)} \longrightarrow 2 \text{ (benzene)} + OH^- \tag{7.339}$$

In air-saturated benzene solutions about 97% of the hydrated electrons and 50% of the hydrogen atoms are scavenged by oxygen, leading to significant ($G \sim 2.8$) yields of hydrogen peroxide; H_2O_2 yields in deaerated solution do not exceed $G_{H_2O_2}$. Other products formed in the presence of oxygen include phenol, or phenol-like products, ($G \sim 1.5$ to 2.0), dialdehyde (α and β-hydroxymucondialdehyde) ($G \sim 0.7$ to 1.5), and small yields of dimeric products ($G \sim 0.2$) (e.g., 292, 293). Suggested reactions include

$$\text{(cyclohexadienyl H\ H)} + O_2 \longrightarrow \text{(H\ H\ H, O_2\cdot)} \longrightarrow \text{(benzene)} + HO_2 \tag{7.340}$$

$$\text{(hydroxycyclohexadienyl H\ OH)} + O_2 \longrightarrow \text{(H\ OH\ H, O_2\cdot)} \longrightarrow \text{(phenol, OH)} + HO_2 \tag{7.341}$$

$$\longrightarrow OHC\text{—}CH{=}C(OH)\text{—}CH{=}CH\text{—}CHO \tag{7.342}$$
β-hydroxymucondialdehyde

The latter reaction involves several steps, and the complete radiolysis mechanism of benzene is obviously complex.

At higher temperatures (about 200°C) the yield of phenol from oxygenated benzene solutions is increased to 10 to 15 times the yield at room temperature (294, 295) as the result of a short chain reaction.

Radiolysis of aqueous toluene solutions offers the possibility of attack either on the aromatic ring or on the methyl side chain. Pulse radiolysis (285, 287) shows that radical attack is predominantly on the ring system, producing adducts similar to those observed in the pulse radiolysis of benzene solutions. Initial steps in the radiolysis are therefore represented (296) as,

$$H + C_6H_5CH_3 \rightarrow \cdot C_6H_6CH_3 \qquad (7.343)$$

$$OH + C_6H_5CH_3 \rightarrow HO\dot{C}_6H_5CH_3 \qquad (7.344)$$

(each product represents several isomeric structures). The yield of hydrogen is found to be the molecular yield, even in acid solutions, showing that hydrogen atoms add to toluene and do not abstract hydrogen from it. Some of the products (Table 7.18), such as the cresols and certain of the dimeric products, can be accounted for by disproportionation or combination of the products from reactions 7.343 and 7.344, accompanied if necessary by loss of water. However, other products, such as 1,2-diphenylethane, are more readily accounted for if benzyl radicals, $C_6H_5CH_2\cdot$, are formed. Since the yield of diphenylethane is increased if the solution is saturated with nitrous oxide, converting e_{aq}^- to OH, it appears that the hydroxyl radical adduct may lose water to give a benzyl radical and, as diphenylethane yields are higher in acid solutions, that the reaction is catalyzed by acid,

$$HO\dot{C}_6H_5CH_3 \xrightarrow{H^+} C_6H_5CH_2\cdot + H_2O \qquad (7.345)$$

T A B L E 7.18   *Radiolysis of Aqueous Toluene Solutions*[a]

| | G(Product) | | | |
| | Argon-Saturated Solutions | | Air-Saturated Solutions | |
Product	Neutral	pH 3	Neutral	pH 3
H_2	0.48	0.49	0.44	0.48
H_2O_2	0	0.55	2.3	3.0
Acid (e.g., C_6H_5COOH)	0	0	1.1	1.7
C_6H_5CHO	0	0	0	0.33
Cresols ($HOC_6H_4CH_3$)	0.45	0.14	0.94	0.52
$C_6H_5CH_2CH_2C_6H_5$	0.02	0.49	0	0
Other nonreduced dimers	0.37	0.32	0	0
Reduced dimers	0.05	0.57	0	0

[a] Ref. 296; ^{60}Co γ-ray.

Yields of hydroxylated products (cresols) are lower from toluene than from benzene, particularly in the presence of oxygen, due to the formation of diphenylmethane and side-chain oxidation products. Hydrated electrons add to the aromatic ring of toluene more slowly than H and OH and in neutral, acid, and aerated solutions react preferentially with H_2O_2, H^+, and O_2, respectively, accounting for the absence of H_2O_2 in neutral oxygen-free solution and for the higher yield of reduced dimers in the acid solution.

Other aromatic compounds undergo radical attack on the aromatic ring system in a similar manner to benzene and toluene. Phenol, in the absence of oxygen (297), is oxidized to 1,2- and 1,4-dihydroxybenzene and, when oxygen is available, to *o*-quinone. The substituted dihydroxybenzenes, or hydroquinones, are also readily oxidized, and it has been possible to correlate their reactions with HO_2 with their standard electrode potentials (298). There are obvious similarities between the radiolysis of benzene and the aromatic pyrimidine system of uracil.

Addition reactions also predominate when compounds containing isolated multiple bonds are irradiated in aqueous solution. Acrylamide has been investigated by a number of groups (299) and reacts rapidly with the primary radicals,

$$e_{aq}^- + CH_2{=}CHCONH_2 \xrightarrow{H_2O} CH_3\dot{C}HCONH_2 + OH^-$$
$$k_{346} = 1.8 \times 10^{10} \ M^{-1} \ sec^{-1} \quad (7.346)$$

$$H + CH_2{=}CHCONH_2 \rightarrow CH_3\dot{C}HCONH_2$$
$$k_{347} = 1.8 \times 10^{10} \ M^{-1} \ sec^{-1} \quad (7.347)$$

$$OH + CH_2{=}CHCONH_2 \rightarrow HOCH_2\dot{C}HCONH_2$$
$$k_{348} = 4.5 \times 10^9 \ M^{-1} \ sec^{-1} \quad (7.348)$$

The organic radicals (which may have structures isomeric with those shown) subsequently dimerize, disproportionate, or react with the solute forming polymeric materials, the course of the reaction depending on the radiolysis conditions. Several other compounds containing C=C or C≡C bonds are listed in Table 7.19. Addition can also occur to C≡N (p. 329) and C=O (p. 320) bonds, but not to COOH; with organic compounds, hydrogen abstraction by H and OH often competes with the addition reaction.

Other Solutes

Table 7.19, which is very far from complete, lists data for a number of other compounds that have been irradiated in aqueous solution. Only major products are given, and for detailed information the reader is referred to the references and to the current literature; the *G* values listed generally refer to x- or γ-irradiation.

To the present, very little work has been carried out using nonaqueous inorganic solvents. After water, which is by far the most frequently used solvent, alcohols and other organic compounds have been employed most often.

TABLE 7.19 Radiolysis of Aqueous Solutions

Solute	Conditions	Products (G Values Given in Parentheses)	Ref.
Inorganic			
NH_3	$+O_2$	Formation NO_2^- via $NH_3 + OH \rightarrow NH_2 + H_2O$; NH_4^+ unreactive.	300
N_2H_4	O_2-free	(pH 6); NH_3 (5.5), H_2 (2.9), N_2 (2.7)	301
	$+O_2$	(pH <8); H_2, N_2, H_2O_2.	302
H_2S	O_2-free	$G(S) = 0.5(G_H + G_{OH})$; $G(H_2) = G_{H_2} + G_H$.	278
	$+O_2$	Sulfur formed by a short chain reaction; $G(S)$ 8 to 23 under conditions used; SO_4^{2-} formed in lower yield.	303
H_2O_2		Decomposed to $O_2 + H_2O$ by chain reaction; $G(O_2) \propto [H_2O_2]^{0.5}$ (dose rate)$^{-0.5}$.	304
CO	O_2-free	(pH ~0.7); H_2 (0.95), H_2O_2 (0), CO_2 (2.6), HCHO (0.5), HCOOH (0.4), OHCCHO (0.3). In basic solutions HCOOH is formed by a short chain reaction (~44).	202
CO_2	$+O_2$	(pH 6.9); H_2 (0.51), H_2O_2 (3.28), CO_2 (2.17), $-CO$ (2.16), $-O_2$ (2.51).	203
		Primary products formic acid and aldehydes, with oxalic acid and glycol as secondary products; yields depend on pH and total dose.	201
Arsenite, (H_3AsO_3)	$+O_2$	(pH 1.4); arsenate, H_3AsO_4 (3.2), H_2O_2 (4.2). As(III) oxidized by OH and HO_2.	305
BH_4^-	O_2-free	$G(H_2) = 6.8$ (immediate) $+ 6.3$ (released during several hours). $OH + BH_4^- \rightarrow OH^- + BH_4$ converts OH to reducing species.	306
$Cr_2O_7^{2-}$	O_2-free or $+O_2$	(pH 0.46); $G(-Cr_2O_7^{2-}) = 0.41$. Mechanism similar to that for reduction Ce^{4+}, dichromate reduced to Cr(III). Oxygen has little effect on stoichiometry.	307
NO_2^-	O_2-free	NO_2^- oxidized to NO_3^- by H_2O_2; H, OH and e_{aq}^- react with NO_2^- but reaction products reform NO_2^-.	308
NO_3^-	O_2-free	H_2O_2 (0.74), NO_2^- (0.3). NO_2^- and O_2 formed by direct action at high (>1 M) concentrations.	309 310

Phosphite (H_3PO_3)	$+O_2$	Phosphate formed by short chain reaction with G values up to about 30.	311
Borates, OH^-, CO_3^{2-}, ClO_4^-, PO_4^{3-}, SO_4^{2-}		Little permanent change when irradiated as dilute solutions.	
Organic			
Methane	O_2-free	H_2, H_2O_2, CH_3OH, C_2H_6.	312
Ethylene	O_2-free	(pH 1.2) H_2O_2 (0.4), aldehyde ($CH_3CHO + CH_3CH_2CH_2CHO$) (0.24), polymer.	313
	$+O_2$	pH 1.2; 1:1 ethylene-oxygen mixture) H_2O_2 (2.4), aldehyde ($HCHO + CH_3CHO$) (2.4), $HOCH_2CHO$ (2.4), RO_2H (0.4).	
Propene ($CH_3CH{=}CH_2$)	$+O_2$	(pH 1.2) $CH_3CH(O_2H)CH_2OH$ (2.6), H_2O_2 (1.9), HCHO (about 0.4), CH_3CHO (about 0.4), HCOOH (about 0.4), CH_3CH_2CHO (about 0.2).	314 315
Acetylene	$+O_2$	(pH 1.2; 1:1 acetylene-oxygen mixture) H_2O_2 (2.6), OHCCHO (14).	316
Glycollic acid ($HOCH_2COOH$)	O_2-free	CO_2 (0.5), HCOOH (0.5), glyoxylic acid (1.5), tartaric acid (2.1).	207
	$+O_2$	CO_2 (1.9), HCOOH (1.6), glyoxylic acid (2.75), tartaric acid (0.04).	317
Glyoxylic acid (OHCCOOH)	O_2-free	Oxalic acid, dimer, trimer.	318
	$+O_2$	Oxalic acid.	
Lactic acid ($CH_3CHOHCOOH$)	$+O_2$	(pH 1.2) H_2O_2 (4.2), $CH_3COCOOH$ (3.5), some CH_3CHO.	319
D-Fructose	$+O_2$	H_2O_2 (2.0), D-glucosone, glycollaldehyde, D-glyceraldehyde, glycerosone, etc.	320
Sucrose	$+O_2$	H_2O_2 (2.0), D-glucose, D-fructose, D-glucosone, D-gluconic acid, etc.	321
Diethylamine	$+O_2$	(a and b) (pH 2.8) H_2O_2 (2.58), CH_3CHO (about 0.77), ethylamine (about 0.8), oxime (0). (a)	322
	$+O_2$	(b) (pH 11) H_2O_2 (3.16), CH_3CHO (3.16), oxime (0.36). (b)	323
Acrylonitrile ($CH_2{=}CHCN$)	O_2-free	Polyacrylonitrile, repeating unit —$(CH_2$—$CH)$— CN	324

TABLE 7.19 *Radiolysis of Aqueous Solutions* (*Cont'd*)

Solute	Conditions	Products (G Values Given in Parentheses)	Ref.
Phenylalanine ($C_6H_5CH_2CH(NH_2)COOH$)	$+O_2$	Phenylpyruvic acid, products formed by hydroxylation of the phenyl ring, polymer.	325
Tryptophan	O_2-free	(0.02 M, pH 1); H_2 (0.37), H_2O_2 (0.63), NH_3 (0.51), CO_2 ($\sim$0), $G(-$tryptophan$) = 0.7$. Radicals add to indole ring.	326
Choline chloride ($[(CH_3)_3NCH_2CH_2OH]^+Cl^-$)		$G(-$choline chloride$) = 2$ to 4.5.	327
Thiourea (H_2NCSNH_2)	$+O_2$	Sulfur (G up to 10^4 at low dose rates).	328
Polyethylene oxide ($-CH_2-CH_2-)_n$	O_2-free	Attack by H and OH (e_{aq}^- unreactive) leads to increased viscosity (cross-linking).	329
Chlorobenzene	$+O_2$	o, m, and p-Chlorophenol.	330
Benzyl alcohol ($C_6H_5CH_2OH$)	$+O_2$	H_2O_2 (3), C_6H_5CHO (1.6), phenols (1.2) (roughly equal attack on side chain and ring).	331
Benzoic acid (C_6H_5COOH)	$+O_2$	(pH 7 to 8, 5×10^{-3} M); H_2 (0.8), H_2O_2 (2.6), CO_2 (0.74), o- (0.52), m- (0.27), and p- (0.22) hydroxybenzoic acids, dihydroxybenzoic acids (0.03), dialdehyde (0.55). CO_2 is also formed by direct action at high benzoic acid concentrations.	332

REFERENCES

Authoritative reviews of the radiation chemistry of aqueous solutions have been published by A. O. Allen, *The Radiation Chemistry of Water and Aqueous Solutions*, Van Nostrand, Princeton, New Jersey, 1961, by E. J. Hart and R. L. Platzman in *Mechanisms in Radiobiology* (eds. M. Errera and A. Forssberg), Academic, New York, Vol. 1, 1961, by M. Lefort in *Actions Chimiques et Biologiques des Radiations* (ed. M. Haissinsky), Masson et Cie, Paris, Vol. 1, 1955, by M. Anbar in *Fundamental Processes in Radiation Chemistry* (ed. P. Ausloos), Interscience, New York, 1968, by J. K. Thomas in *Advances in Radiation Chemistry* (eds. M. Burton and J. L. Magee), Wiley-Interscience, New York, Vol. 1, 1969, and by I. G. Draganić and Z. D. Draganić, *The Radiation Chemistry of Water*, Academic, New York, 1971.

1. A. Debierne, *Ann. Phys. (Paris)*, **2**, 97 (1914).
2. O. Risse, *Strahlentherapie*, **34**, 578 (1929).
3. J. Weiss, *Nature*, **153**, 748 (1944).
4. A. O. Allen, *J. Phys. Colloid Chem.*, **52**, 479 (1948).
5. M. Burton, *J. Phys. Colloid Chem.*, **51**, 611 (1947); *Ann. Rev. Phys. Chem.*, **1**, 113 (1950).
6. H. Fricke, *Cold Spring Harbor Symp.*, **2**, 241 (1934); **3**, 55 (1935).
7. R. L. Platzman in *Basic Mechanisms in Radiobiology, National Research Council Publication 305*, Washington, D.C., 1953.
8. D. E. Lea in *Actions of Radiations on Living Cells*, Cambridge University Press, Cambridge, 1946.
9. L. H. Gray, *J. Chim. Phys.*, **48**, 172 (1951).
10. A. H. Samuel and J. L. Magee, *J. Chem. Phys.*, **21**, 1080 (1953).
11. E. J. Hart, *J. Chem. Educ.*, **36**, 266 (1959).
12. A. O. Allen, *The Radiation Chemistry of Water and Aqueous Solutions*, Van Nostrand, Princeton, N.J., 1961.
13. A. O. Allen in *Actions Chimiques et Biologiques des Radiations* (ed. M. Haissinsky), Masson et Cie, Paris, Vol. 5, 1961, p. 11.
14. I. G. Draganic and Z. D. Draganić, *The Radiation Chemistry of Water*, Academic, New York, 1971.
15. J. W. T. Spinks and R. J. Woods, *An Introduction to Radiation Chemistry*, Wiley, New York, 1964.
16. R. S. Dixon, *Radiat. Res. Rev.*, **2**, 237 (1970).
17. D. Lewis and W. H. Hamill, *J. Chem. Phys.*, **51**, 456 (1969); L. M. Hunter, D. Lewis, and W. H. Hamill, *ibid.*, **52**, 1733 (1970).
18. S. Trajmar, W. Williams, and A. Kuppermann, *J. Chem. Phys.*, **54**, 2274 (1971); **58**, 2521 (1973).
19. K. Watanabe and M. Zelikoff, *J. Opt. Soc. Am.*, **43**, 753 (1953).
20. A. Johannin-Gilles and B. Vodar, *J. Phys. Radium*, **15**, 223 (1954).
21. M. Cottin, J. Masanet and C. Vermeil, *J. Chim. Phys.*, **63**, 959 (1966).
22. M. Cottin, *J. Chim. Phys.*, **56**, 1024 (1959).
23. K. Watanabe, *J. Chem. Phys.*, **26**, 542 (1957).
24. J. R. McNesby, I. Tanaka, and H. Okabe, *J. Chem. Phys.*, **36**, 605 (1962).
25. M. M. Mann, A. Hustrulid, and J. T. Tate, *Phys. Rev.*, **58**, 340 (1940).

26. P. S. Rudolph and C. E. Melton, *J. Chem. Phys.*, **45**, 2227 (1966).
27. K. J. Laidler, *J. Chem. Phys.*, **22**, 1740 (1954).
28. A. M. Peers and M. Cottin, *J. Chim. Phys.*, **63**, 1346 (1966).
29. C. E. Melton, *J. Phys. Chem.*, **74**, 582 (1970).
30. E. E. Muschlitz and T. L. Bailey, *J. Phys. Chem.*, **60**, 681 (1956).
31. P. F. Knewstubb and A. W. Tickner, *J. Chem. Phys.*, **38**, 464 (1963).
32. P. Kebarle, S. K. Searles, A. Zolla, J. Scarborough, and M. Arshadi, *J. Am. Chem. Soc.*, **89**, 6393 (1967).
33. J. L. Magee and K. Funabashi, *Radiat. Res.*, **10**, 622 (1959).
34. S. C. Lind, *Chemical Effects of Alpha Particles and Electrons*, A.C.S. Monograph 2 (1921), revised as A.C.S. Monograph 151 (1961).
35. A. R. Anderson in *Fundamental Processes in Radiation Chemistry* (ed. P. Ausloos), Interscience, New York, 1968, p. 298.
36. I. T. Myers in *Radiation Dosimetry* (eds. F. H. Attix and W. C. Roesch) 2nd ed., Academic, New York, Vol. 1, 1968, p. 317.
37. R. L. Platzman, *Radiat. Res.*, **17**, 419 (1962); *Vortex*, **23**, 372 (1962).
38. R. L. Platzman in *Radiation Research* (ed. G. Silini), North Holland Publishing Co., Amsterdam, 1967, p. 20.
39. I. Santar and J. Bednář, *Int. J. Radiat. Phys. Chem.*, **1**, 133 (1969).
40. E. J. Hart, *J. Chem. Educ.*, **34**, 586 (1957).
41. J. H. Baxendale and G. P. Gilbert, *Science*, **147**, 1571 (1965).
42. R. S. Dixon and M. G. Bailey, *Can. J. Chem.*, **46**, 1181 (1968).
43. R. F. Firestone, *J. Am. Chem. Soc.*, **79**, 5593 (1957).
44. N. E. Bibler and R. F. Firestone, *J. Phys. Chem.*, **76**, 621 (1972).
45. A. W. Boyd, C. Willis, and O. A. Miller, *Can. J. Chem.*, **51**, 4048, 4056 (1973).
46. A. R. Anderson, B. Knight, and J. A. Winter, *Nature*, **201**, 1026 (1964).
47. A. W. Boyd and O. A. Miller, *Can. J. Chem.*, **46**, 3773 (1968).
48. G. V. Buxton, *Radiat. Res. Rev.*, **1**, 209 (1968).
49. I. G. Draganić, M. T. Nenadović, and Z. D. Draganić, *J. Phys. Chem.*, **73**, 2564 (1969).
50. E. Collinson, F. S. Dainton, and J. Kroh, *Nature*, **187**, 475 (1960); *Proc. Roy. Soc.* (*London*), **A265**, 422 (1962).
51. A. Appleby and H. A. Schwarz, *J. Phys. Chem.*, **73**, 1937 (1969).
52. M. Lefort and X. Tarrago, *J. Phys. Chem.*, **63**, 833 (1959).
53. M. Lefort, *Ann. Rev. Phys. Chem.*, **9**, 123 (1958).
54. M. Imamura, M. Matsui, and T. Karasawa, *Bull. Chem. Soc. Jap.*, **43**, 2745 (1970).
55. A. Kuppermann in *Actions Chimiques et Biologiques des Radiations* (ed. M. Haissinsky), Masson et Cie, Paris, Vol. 5, 1961, p. 85, and in *Radiation Research* (ed. G. Silini), North Holland Publishing Corp., Amsterdam, 1967, p. 212.
56. M. Anbar and P. Neta, *Int. J. Appl. Radiat. Isotop.*, **18**, 493 (1967). See also M. Anbar, Farhataziz, and A. B. Ross, *Selected Specific Rates of Reactions of Transients from Water in Aqueous Solution. II. Hydrogen Atom*, NSRDS-NBS 51 (1975), U.S. Dept. Commerce-National Bureau Standards, Washington, D.C.
57. M. Anbar, M. Bambenek, and A. B. Ross, *Selected Specific Rates of Reactions of Transients from Water in Aqueous Solution. I. Hydrated Electron*, NSRDS-NBS 43 (1973), and A. B. Ross, Supplement to NSRDS-NBS 43 (1975), U.S. Dept. Commerce-National Bureau Standards, Washington, D.C.
58. L. M. Dorfman and G. E. Adams, *Reactivity of the Hydroxyl Radical in Aqueous Solutions*, NSRDS-NBS 46 (1973), U.S. Dept. Commerce-National Bureau Standards, Washington, D.C.
59. C. J. Hochanadel and J. A. Ghormley, *Radiat. Res.*, **16**, 653 (1962).

60. J. Rotblat and H. C. Sutton, *Proc. Roy. Soc. (London), Ser. A255*, 490 (1960).
61. A. R. Anderson and E. J. Hart, *J. Phys. Chem.*, **66**, 70 (1962).
62. J. K. Thomas and E. J. Hart, *Radiat. Res.*, **17**, 408 (1962).
63. J. K. Thomas, *Int. J. Appl. Radiat. Isotop.*, **16**, 451 (1966).
64. A. Mozumder and J. L. Magee, *J. Chem. Phys.*, **45**, 3332 (1966); *Radiat. Res.*, **28**, 203 (1966).
65. A. Mozumder in *Advances in Radiation Chemistry* (eds. M. Burton and J. L. Magee), Wiley-Interscience, New York, Vol. 1, 1969, p. 1.
66. A. Mozumder and J. L. Magee, *Radiat. Res.*, **28**, 215 (1966).
67. A. Mozumder, A. Chatterjee, and J. L. Magee, *Adv. Chem. Ser.*, **81**, 27 (1968).
68. D. A. Armstrong, E. Collinson, and F. S. Dainton, *Trans. Faraday Soc.*, **55**, 1375 (1959).
69. H. A. Schwarz, *J. Chem. Phys.*, **73**, 1928 (1969).
70. J. H. Baxendale, *Radiat. Res.*, **17**, 312 (1962).
71. U. Sokolov and G. Stein, *J. Chem. Phys.*, **44**, 2189, 3329 (1966).
72. N. Getoff and G. O. Schenck, *Photochem. Photobiol.*, **8**, 167 (1968).
73. M. Anbar in *Fundamental Processes in Radiation Chemistry* (ed. P. Ausloos), Interscience, New York, 1968, p. 651.
74. J. W. Boyle, J. A. Ghormley, C. J. Hochanadel, and J. F. Riley, *J. Phys. Chem.*, **73**, 2886 (1969).
75. J. W. Hunt, R. K. Wolff, M. J. Bronskill, C. D. Jonah, E. J. Hart, and M. S. Matheson, *J. Chem. Phys.*, **77**, 425 (1973).
76. C. D. Jonah, E. J. Hart, and M. S. Matheson, *J. Phys. Chem.*, **77**, 1838 (1973).
77. R. K. Wolff, M. J. Bronskill, J. E. Aldrich, and J. W. Hunt, *J. Phys. Chem.*, **77**, 1350 (1973).
78. W. H. Hamill, *J. Phys. Chem.*, **73**, 1341 (1969).
79. T. Sawai and W. H. Hamill, *J. Chem. Phys.*, **52**, 3843 (1970); *J. Phys. Chem.*, **74**, 3914 (1970).
80. J. E. Aldrich, M. J. Bronskill, R. K. Wolff, and J. W. Hunt, *J. Chem. Phys.*, **55**, 530 (1971).
81. H. Ogura and W. H. Hamill, *J. Phys. Chem.*, **77**, 2952 (1973).
82. D. R. Smith and W. H. Stevens, *Nature*, **200**, 66 (1963).
83. M. Anbar and J. K. Thomas, *J. Phys. Chem.*, **68**, 3829 (1964).
84. W. G. Brown and E. J. Hart, *Radiat. Res.*, **51**, 249 (1972).
85. H. A. Dewhurst, A. H. Samuel, and J. L. Magee, *Radiat. Res.*, **1**, 62, (1954).
86. F. S. Dainton, *Radiat. Res. Suppl. 1*, 1 (1959).
87. J. L. Magee, *Discuss. Faraday Soc.*, **36**, 233 (1963).
88. R. L. Platzman in *Radiation Biology and Medicine* (ed. W. D. Claus), Addison-Wesley, Reading, Mass., 1958.
89. C. J. Hochanadel in *Comparative Effects of Radiation* (eds. M. Burton, J. S. Kirby-Smith, and J. L. Magee), Wiley, New York, 1960, Chap. 8.
90. J. A. Ghormley and A. C. Stewart, *J. Am. Chem. Soc.*, **78**, 2934 (1956).
91. L. H. Piette, R. C. Rempel, H. E. Weaver, and J. M. Flournoy, *J. Chem. Phys.*, **30**, 1623 (1959).
92. J. Kroh, B. C. Green, and J. W. T. Spinks, *Can. J. Chem.*, **40**, 413 (1962).
93. R. Livingston, H. Zeldes, and E. H. Taylor, *Discuss. Faraday Soc.*, **19**, 166 (1955).
94. D. Schulte-Frohlinde and K. Eiben, *Z. Naturforsch.*, **17a**, 445 (1962).
95. T. Henriksen, *Radiat. Res.*, **23**, 63 (1964).
96. J. A. Ghormley, *J. Chem. Phys.*, **24**, 1111 (1956).
97. N. F. Barr and A. O. Allen, *J. Phys. Chem.*, **63**, 928 (1959).
98. A. R. Anderson and E. J. Hart, *J. Phys. Chem.*, **65**, 804 (1961).
99. G. Czapski, J. Jortner, and G. Stein, *J. Phys. Chem.*, **65**, 956, 960, 964 (1961).

100. E. Hayon and A. O. Allen, *J. Phys. Chem.*, **65**, 2181 (1961).
101. J. Jortner and J. Rabani, *J. Am. Chem. Soc.*, **83**, 4868 (1961); *J. Phys. Chem.*, **66**, 2078, 2081 (1962).
102. J. T. Allan and G. Scholes, *Nature*, **187**, 218 (1960).
103. G. Czapski and H. A. Schwarz, *J. Phys. Chem.*, **66**, 471 (1962).
104. E. Collinson, F. S. Dainton, D. R. Smith, and S. Tazuké, *Proc. Chem. Soc.*, 140 (1962).
105. A. Hummel and A. O. Allen, *Radiat. Res.*, **17**, 302 (1962).
106. E. J. Hart and J. W. Boag, *J. Am. Chem. Soc.*, **84**, 4090 (1962); J. W. Boag and E. J. Hart, *Nature*, **197**, 45 (1963).
107. *Solvated Electron* (ed. R. F. Gould), *Adv. Chem. Ser.*, **50** (1965).
108. E. J. Hart in *Actions Chimiques et Biologiques des Radiations* (ed. M. Haissinsky), Masson et Cie, Paris, Vol. 10, 1966, p. 1.
109. D. C. Walker, *Quart. Rev. (London)*, **21**, 79 (1967); *Can. J. Chem.*, **45**, 807 (1967); *Adv. Chem. Ser.*, **81**, 49 (1968).
110. J. K. Thomas in *Radiation Research* (ed. G. Silini), North Holland Publishing Co., Amsterdam, 1967, p. 179; *Radiat. Res. Rev.*, **1**, 183 (1968).
111. M. Anbar in *Advances in Physical Organic Chemistry* (ed. V. Gold), Academic, New York, Vol. 7, 1969, p. 115.
112. J. K. Thomas in *Advances in Radiation Chemistry* (eds. M. Burton and J. L. Magee), Wiley-Interscience, New York, Vol. 1, 1969, p. 103.
113. E. J. Hart and M. Anbar, *The Hydrated Electron*, Wiley-Interscience, New York, 1970.
114. P. Pagsberg, H. Christensen, J. Rabani, G. Nilsson, J. Fenger, and S. O. Nielsen, *J. Phys. Chem.*, **73**, 1029 (1969).
115. B. H. J. Bielski and J. M. Gebicki in *Advances in Radiation Chemistry* (eds. M. Burton and J. L. Magee), Wiley-Interscience, New York, Vol. 2, 1970, p. 177.
116. A. Ekstrom, *Radiat. Res. Rev.*, **2**, 381 (1969).
117. B. C. Webster and G. Howat, *Radiat. Res. Rev.*, **4**, 259 (1972).
118. E. A. Shaede and D. C. Walker, *Chem. Soc. Special Publ. 22*, 277 (1967); D. C. Walker, *Can. J. Chem.*, **44**, 2226 (1966).
119. J. Jortner, M. Ottolenghi, and G. Stein, *J. Phys. Chem.*, **66**, 2029, 2037, 2042 (1962); **67**, 1271 (1963); *J. Am. Chem. Soc.*, **85**, 2712 (1963).
120. L. I. Grossweiner, G. W. Swenson, and E. F. Zwicker, *Science*, **141**, 805, 1042 (1963).
121. A. J. Birch and H. Smith, *Quart. Rev. (London)*, **12**, 17 (1958).
122. H. Smith in *Chemistry in Non-Aqueous Ionizing Solvents* (eds. G. Jander, H. Spandau, and C. C. Addison), Interscience, New York, Vol. 1, part 2, 1963.
123. A range of syntheses are described in the volumes of *Organic Syntheses* (eds. R. Adams, J. B. Conant, H. T. Clarke and O. Kamm), Wiley, New York, Vol. 1, 1921.
124. J. H. Baxendale, *Radiat. Res. Suppl. 4*, 139 (1964).
125. *Handbook of Chemistry and Physics* (ed. R. C. Weast), 54th ed., CRC Press, Cleveland, Ohio, 1973.
126. P. Debye, *Trans. Electrochem. Soc.*, **82**, 265 (1942).
127. P. Neta, R. W. Fessenden, and R. H. Schuler, *J. Phys. Chem.*, **75**, 1654 (1971).
128. O. I. Mićić and V. Marković, *Int. J. Radiat. Phys. Chem.*, **4**, 43 (1972).
129. W. M. Latimer, *The Oxidation States of the Elements and Their Potentials in Aqueous Solution*, 2nd ed., Prentice Hall, Englewood Cliffs, N.J., 1952.
130. J. H. Baxendale, *Radiat. Res. Suppl. 4*, 114 (1964).
131. E. J. Hart, S. Gordon and E. M. Fielden, *J. Phys. Chem.*, **70**, 150 (1966).
132. S. O. Nielsen, *Nature (London)*, *Phys. Sci.*, **240**, 21 (1972).
133. J. Rabani, W. A. Mulac, and M. S. Matheson, *J. Phys. Chem.*, **69**, 53 (1965).
134. G. Navon and G. Stein, *Israel J. Chem.*, **2**, 151 (1964); *J. Phys. Chem.*, **69**, 1384 (1965).
135. M. S. Matheson and J. Rabani, *J. Phys. Chem.*, **69**, 1324 (1965).

136. J. Rabani and S. O. Nielsen, *J. Phys. Chem.*, **73**, 3736 (1969).
137. D. Behar, G. Czapski, J. Rabani, L. M. Dorfman, and H. A. Schwarz, *J. Phys. Chem.*, **74**, 3209 (1970).
138. G. Czapski, *Ann. Rev. Phys. Chem.*, **22**, 171 (1971).
139. G. V. Buxton, *Trans. Faraday Soc.*, **66**, 1656 (1970).
140. J. L. Weeks and M. S. Matheson, *J. Am. Chem. Soc.*, **78**, 1273 (1956).
141. H. J. H. Fenton, *J. Chem. Soc.*, **65**, 899 (1894).
142. K. Eiben and R. W. Fessenden, *J. Phys. Chem.*, **75**, 1186 (1971).
143. G. Czapski in *Radiation Chemistry of Aqueous Systems* (ed. G. Stein) Interscience, New York, 1968, p. 211.
144. J. H. Baxendale and J. A. Wilson, *Trans. Faraday Soc.*, **53**, 344 (1957).
145. G. Czapski, H. Levanon, and A. Samuni, *Israel J. Chem.*, **7**, 375 (1969).
146. M. Anbar, R. A. Munoz, and P. Rona, *J. Phys. Chem.*, **67**, 2708 (1963).
147. K. F. Nakken and A. Pihl, *Radiat. Res.*, **26**, 519 (1965).
148. R. C. Das, *Radiat. Res. Rev.*, **3**, 121 (1971).
149. J. P. Keene, *Radiat. Res.*, **22**, 14 (1964).
150. G. G. Jayson, B. J. Parsons, and A. J. Swallow, *Int. J. Radiat. Phys. Chem.*, **3**, 345 (1971); *J. C. S. Faraday Trans. I*, **68**, 2053 (1972); **69**, 236 (1973).
151. F. S. Dainton and H. C. Sutton, *Trans. Faraday Soc.*, **49**, 1011 (1953).
152. K. Coatsworth, E. Collinson, and F. S. Dainton, *Trans. Faraday Soc.*, **56**, 1008 (1960).
153. E. M. Fielden and N. W. Holm in *Manual on Radiation Dosimetry* (eds. N. W. Holm and R. J. Berry) Dekker, New York, 1970, p. 261.
154. H. A. Dewhurst, *J. Chem. Phys.*, **19**, 1329 (1951); *Trans. Faraday Soc.*, **48**, 905 (1952).
155. E. J. Hart, *J. Am. Chem. Soc.*, **74**, 4174 (1952).
156. J. Weiss, *Nature*, **165**, 728 (1950).
157. T. Rigg, G. Stein, and J. Weiss, *Proc. Roy. Soc. (London), Ser. A211*, 375 (1952).
158. N. Uri, *Chem. Rev.*, **50**, 375 (1952).
159. T. W. Davis, S. Gordon, and E. J. Hart, *J. Am. Chem. Soc.*, **80**, 4487 (1958).
160. G. Czapski, J. Jortner, and G. Stein, *J. Phys. Chem.*, **65**, 960 (1961).
161. H. A. Schwarz, *J. Phys. Chem.*, **67**, 2827 (1963).
162. D. Katakis and A. O. Allen, *J. Phys. Chem.*, **68**, 657 (1964).
163. J. W. Boyle, *Radiat. Res.*, **17**, 427, 450 (1962).
164. F. S. Dainton, A. R. Gibbs, and D. Smithies, *J. Chem. Soc. A*, 33 (1967).
165. R. W. Matthews, H. A. Mahlman, and T. J. Sworski, *J. Phys. Chem.*, **76**, 1265 (1972).
166. B. Lesigne, C. Ferradini, and J. Pucheault, *J. Phys. Chem.*, **76**, 3676 (1972); **77**, 2156 (1973).
167. H. A. Schwarz, *J. Am. Chem. Soc.*, **79**, 534 (1957).
168. E. J. Hart, *Radiat. Res.*, **2**, 33 (1955).
169. E. J. Hart and P. D. Walsh, *Radiat. Res.*, **1**, 342 (1954).
170. E. J. Hart, W. J. Ramler, and S. R. Rocklin, *Radiat. Res.*, **4**, 378 (1956).
171. P. Y. Feng, A. Brynjolfsson, J. W. Halliday, and R. D. Jarrett, *J. Phys. Chem.*, **74**, 1221 (1970).
172. R. W. Matthews, *Radiat. Res.*, **55**, 242 (1973).
173. R. W. Matthews, *Int. J. Appl. Radiat. Isotop.*, **22**, 199 (1971).
174. T. J. Sworski, *Radiat. Res.*, **4**, 483 (1956).
175. T. J. Sworski, *J. Am. Chem. Soc.*, **78**, 1768 (1956); *Radiat. Res.*, **6**, 645 (1957).
176. E. J. Hart and R. L. Platzman in *Mechanisms in Radiobiology* (eds. M. Errera and A. Forssberg), Academic, New York, 1961, Vol. 1, p. 93.
177. M. Haissinsky in *Actions Chimiques et Biologiques des Radiations* (ed. M. Haissinsky), Masson et Cie, Paris, Vol. 11, 1967, p. 133.
178. D. Zehavi and J. Rabani, *J. Phys. Chem.*, **76**, 3703 (1972); **78**, 1368 (1974).

179. G. G. Jayson, B. J. Parsons, and A. J. Swallow, *J.C.S. Faraday Trans. I*, **69**, 1597 (1973).
180. S. Khorana and W. H. Hamill, *J. Phys. Chem.*, **75**, 3081 (1971).
181. M. M. Fisher and W. H. Hamill, *J. Phys. Chem.*, **77**, 171 (1973).
182. E. R. Johnson and A. O. Allen, *J. Am. Chem. Soc.*, **74**, 4147 (1952).
183. C. J. Hochanadel, *J. Phys. Chem.*, **56**, 587 (1952).
184. D. Behar, P. L. T. Bevan, and G. Scholes, *J. Phys. Chem.*, **76**, 1537 (1972).
185. M. V. Vladimirova and I. A. Kulikov, *Khim. Vys. Energ.*, **5**, 432 (1970).
186. D. Behar and R. W. Fessenden, *J. Phys. Chem.*, **76**, 3945 (1972); D. Behar, *ibid.*, **78**, 2660 (1974).
187. E. J. Hart, *J. Am. Chem. Soc.*, **73**, 68 (1951).
188. E. J. Hart, *J. Am. Chem. Soc.*, **76**, 4198 (1954).
189. E. J. Hart, *J. Am. Chem. Soc.*, **76**, 4312 (1954).
190. E. J. Hart, *Radiat. Res.*, **1**, 53 (1954).
191. A. R. Anderson and E. J. Hart, *Radiat. Res.*, **14**, 689 (1961).
192. H. Fricke, E. J. Hart and H. P. Smith, *J. Chem. Phys.*, **6**, 229 (1938).
193. E. J. Hart, *J. Am. Chem. Soc.*, **73**, 68 (1951).
194. E. J. Hart, *J. Phys. Chem.*, **56**, 594 (1952).
195. P. Riesz and B. E. Burr, *Radiat. Res.*, **16**, 661 (1962).
196. A. Husain and E. J. Hart, *J. Am. Chem. Soc.*, **87**, 1180 (1965).
197. W. M. Garrison, W. Bennett, and M. Jayko, *J. Chem. Phys.*, **24**, 631 (1956).
198. W. M. Garrison, W. Bennett, and S. Cole, *Radiat. Res.*, **9**, 647 (1958).
199. D. Smithies and E. J. Hart, *J. Am. Chem. Soc.*, **82**, 4775 (1960).
200. G. E. Adams and E. J. Hart, *J. Am. Chem. Soc.*, **84**, 3994 (1962).
201. N. Getoff, G. Scholes, and J. Weiss, *Tetrahedron Lett.*, **18**, 17 (1960); N. Getoff, *Int. J. Appl. Radiat. Isotop.*, **13**, 205 (1962).
202. Y. Raef and A. J. Swallow, *Trans. Faraday Soc.*, **59**, 1631 (1963).
203. T. Balkas, F. S. Dainton, J. K. Dishman, and D. Smithies, *Trans. Faraday Soc.*, **62**, 81 (1966).
204. W. M. Garrison, W. Bennett, S. Cole, H. R. Haymond, and B. M. Weeks, *J. Am. Chem. Soc.*, **77**, 2720 (1955).
205. Lj. Josimović and I. Draganić, *Int. J. Radiat. Phys. Chem.*, **5**, 505 (1973).
206. W. M. Garrison, H. R. Haymond, D. C. Morrison, B. M. Weeks, and J. Gile-Melchert, *J. Am. Chem. Soc.*, **75**, 2459 (1953)
207. E. Hayon and J. Weiss, *J. Chem. Soc.*, 5091 (1960).
208. W. M. Garrison, H. R. Haymond, W. Bennett, and S. Cole, *J. Chem. Phys.*, **25**, 1282 (1956); *Radiat. Res.*, **10**, 273 (1959).
209. I. G. Draganić and O. Gal, *Radiat. Res. Rev.*, **3**, 167 (1971).
210. V. Marković, K. Sehested, and E. Bjergbakke, *Int. J. Radiat. Phys. Chem.*, **5**, 15 (1973).
211. P. Riesz and B. E. Burr, *Radiat. Res.*, **16**, 668 (1962).
212. K.-D. Asmus, H. Möckel, and A. Henglein, *J. Phys. Chem.*, **77**, 1218 (1973).
213. G. G. Jayson, G. Scholes, and J. Weiss, *J. Chem. Soc.*, 1358 (1957).
214. J. H. Baxendale and G. Hughes, *Z. Phys. Chem.*, **14**, 306 (1958); J. H. Baxendale and R. S. Dixon, *ibid.*, **43**, 161 (1964).
215. M. Simic, P. Neta, and E. Hayon, *J. Phys. Chem.*, **73**, 3794 (1969).
216. J. H. Baxendale and R. S. Dixon, *Z. Phys. Chem.*, **43**, 161 (1964).
217. J. Rabani, D. Klug-Roth, and A. Henglein, *J. Phys. Chem.*, **78**, 2089 (1974).
218. H. A. Schwarz, J. M. Caffrey, and G. Scholes, *J. Am. Chem. Soc.*, **81**, 1801 (1959).
219. A. Hummel and A. O. Allen, *Radiat. Res.*, **17**, 302 (1962).
220. J. T. Allan, E. M. Hayon, and J. Weiss, *J. Chem. Soc.*, 3913 (1959).
221. Y. Raef and A. J. Swallow, *J. Phys. Chem.*, **70**, 4072 (1966).

222. J. T. Allan, *J. Phys. Chem.*, **68**, 2714 (1964).
223. P. Riesz, *Radiat. Res. Suppl.* **4**, 152 (1964); *J. Phys. Chem.*, **69**, 1366 (1965).
224. P. M. Grant and R. B. Ward, *J. Chem. Soc.*, 2871 (1959).
225. S. A. Barker, P. M. Grant, M. Stacey, and R. B. Ward, *J. Chem. Soc.*, 2648 (1959).
226. G. O. Phillips, G. L. Mattok, and G. J. Moody, *Proc. 2nd Int. Conf. Peaceful Uses Atomic Energy*, United Nations, Geneva, **29**, 92 (1958); *J. Chem. Soc.*, 3522 (1958).
227. G. O. Phillips and G. J. Moody, *Int. J. Appl. Radiat. Isotop.*, **6**, 78 (1959).
228. G. O. Phillips, *Radiat. Res.*, **18**, 446 (1963).
229. G. O. Phillips, *Radiat. Res. Rev.*, **3**, 335 (1972).
230. G. O. Phillips, *J. Chem. Soc.*, 297 (1963).
231. H. J. H. Fenton and H. Jackson, *J. Chem. Soc.*, **75**, 1 (1899).
232. N. F. Barr and C. G. King, *J. Am. Chem. Soc.*, **78**, 303 (1956).
233. J. Jortner and J. Rabani, *J. Am. Chem. Soc.*, **83**, 4868 (1961).
234. J. Jortner and J. Rabani, *J. Phys. Chem.*, **66**, 2081 (1962).
235. J. Teply, *Collect. Czech. Chem. Commun.*, **25**, 24 (1960).
236. B. J. Rezansoff, K. J. McCallum, and R. J. Woods, *Can. J. Chem.*, **48**, 271 (1970).
237. H. L. Andrews and P. A. Shore, *J. Chem. Phys.*, **18**, 1165 (1950).
238. G. R. Freeman, A. B. Van Cleave, and J. W. T. Spinks, *Can. J. Chem.*, **31**, 1164 (1953); **32**, 322 (1954).
239. W. S. Moos, *Int. J. Appl. Radiat. Isotop.*, **23**, 538 (1972); W. S. Moos, L. H. Lanzl and H. Brunner, *ibid.*, **24**, 645 (1973).
240. R. J. Woods and J. W. T. Spinks, *Can. J. Chem.*, **38**, 77 (1960).
241. V. G. Sorensen, V. M. Bhale, K. J. McCallum, and R. J. Woods, *Can. J. Chem.*, **48**, 2542 (1970).
242. M. Namiki, T. Komiya, S. Kawakishi, and H. Aoki, *JCSD Chem. Comm.*, 311 (1970).
243. K. Bhatia and R. H. Schuler, *J. Phys. Chem.*, **77**, 1356 (1973).
244. R. J. Woods and S. Akhtar, *Agric. Food Chem.*, **22**, 1132 (1974).
245. R. Evans, E. Nesyto, C. Radlowski, and W. V. Sherman, *J. Phys. Chem.*, **75**, 2762 (1971).
246. W. M. Garrison, *Radiat. Res. Suppl.* **4**, 158 (1964); in *Current Topics in Radiation Research* (eds. M. Ebert and A. Howard), Wiley, New York, Vol. 4, 1968, p. 43.
247. G. E. Adams in *Radiation Chemistry of Aqueous Systems* (ed. G. Stein), Interscience, New York, 1968, p. 241, and in *Advances in Radiation Chemistry* (eds. M. Burton and J. L. Magee), Wiley-Interscience, New York, Vol. 3, 1972, p. 125.
248. G. Scholes in *Radiation Chemistry of Aqueous Systems* (ed. G. Stein), Interscience, New York, 1968, p. 259.
249. I. Draganić, Z. Draganić, Lj. Petković, and A. Nikolić, *J. Am. Chem. Soc.*, **95**, 7193 (1973).
250. N. Getoff and F. Schwörer, *Int. J. Radiat. Phys. Chem.*, **5**, 101 (1973).
251. G. G. Jayson, G. Scholes, and J. J. Weiss, *J. Chem. Soc.*, 2594 (1955).
252. H. Christensen, *Int. J. Radiat. Phys. Chem.*, **4**, 311 (1972).
253. W. M. Garrison, *Radiat. Res. Rev.*, **3**, 305 (1972).
254. C. R. Maxwell, D. C. Peterson, and N. E. Sharpless, *Radiat. Res.*, **1**, 530 (1954).
255. C. R. Maxwell, D. C. Peterson, and W. C. White, *Radiat. Res.*, **2**, 431 (1955).
256. B. M. Weeks and W. M. Garrison, *Radiat. Res.*, **9**, 291 (1958).
257. G. Meshitsuka, K. Shindo, A. Minegishi, H. Suguro, and Y. Shinozaki, *Bull. Chem. Soc. Jap.*, **37**, 928 (1964).
258. G. Stein and J. Weiss, *J. Chem. Soc.*, 3256 (1949).
259. J. Holian and W. M. Garrison, *J. Phys. Chem.*, **72**, 4721 (1968); *Nature*, **221**, 57 (1969).
260. E. Hayon, T. Ibata, N. N. Lichtin, and M. Simic, *J. Am. Chem. Soc.*, **92**, 3898 (1970); **93**, 5388 (1971).
261. R. L. S. Willix and W. M. Garrison, *Radiat. Res.*, **32**, 452 (1967).

262. W. M. Garrison, M. Kland-English, H. A. Sokol, and M. E. Jayko, *J. Phys. Chem.*, **74**, 4506 (1970).
263. W. M. Garrison, M. E. Jayko, M. A. J. Rodgers, H. A. Sokol, and W. Bennett-Corniea, *Adv. Chem. Ser.*, **81**, 384 (1968).
264. P. C. Shragge and J. W. Hunt, *Radiat. Res.*, **60**, 233 (1974).
265. P. C. Shragge, A. J. Varghese, J. W. Hunt, and C. L. Greenstock, *Radiat. Res.*, **60**, 250 (1974).
266. R. Ducolomb, J. Cadet, and R. Teoule, *CR*, **273D**, 2647 (1971).
267. G. A. Infante, P. Jirathana, E. J. Fendler, and J. H. Fendler, *JCS Faraday Trans. I*, **70**, 1162 (1974).
268. R. Teoule and J. Cadet, *Bull. Soc. Chim. Fr.*, 927 (1970).
269. G. A. Infante, P. Jirathana, J. H. Fendler, and E. J. Fendler, *JCS Faraday Trans. I*, **69**, 1586 (1973).
270. V. G. Wilkening, M. Lal, M. Arends, and D. A. Armstrong, *Can. J. Chem.*, **45**, 1209 (1967); M. Lal, *Radiat. Effects*, **2**, 225 (1970).
271. A. Al-Thannon, R. M. Peterson, and C. N. Trumbore, *J. Phys. Chem.*, **72**, 2395 (1968).
272. V. G. Wilkening, M. Lal, M. Arends, and D. A. Armstrong, *J. Phys. Chem.*, **72**, 185 (1968).
273. W. M. Dale and J. V. Davies, *Biochem. J.*, **48**, 129 (1951).
274. F. E. Littman, E. M. Carr, and A. P. Brady, *Radiat. Res.*, **7**, 107 (1957).
275. A. J. Swallow, *J. Chem. Soc.*, 1334 (1952).
276. M. Lal, *Radiat. Effects*, **22**, 49, 237 (1974).
277. A. A. Al-Thannon, J. P. Barton, J. E. Packer, R. J. Sims, C. N. Trumbore, and R. V. Winchester, *Int. J. Appl. Radiat. Isotop.*, **6**, 233 (1974).
278. J. E. Packer, *J. Chem. Soc.*, 2320 (1963).
279. D. B. Peterson, J. Holian, and W. M. Garrison, *J. Phys. Chem.*, **73**, 1568 (1969).
280. W. M. Dale and C. Russell, *Biochem., J.*, **62**, 50 (1956).
281. G. M. Gaucher, B. L. Mainman, G. P. Thompson, and D. A. Armstrong, *Radiat. Res.*, **46**, 457 (197).
282. G. Nucifora, B. Smaller, R. Remko, and E. C. Avery, *Radiat. Res.*, **49**, 96 (1972).
283. J. W. Purdie, *J. Am. Chem. Soc.*, **89**, 226 (1967).
284. G. E. Adams, G. S. McNaughton, and B. D. Michael in *The Chemistry of Ionization and Excitation* (eds. G. R. A. Johnson and G. Scholes), Taylor and Francis, London, 1967, p. 281.
285. L. M. Dorfman, R. E. Bühler, and I. A. Taub, *J. Chem. Phys.*, **36**, 549, 3051 (1962).
286. L. M. Dorfman, I. A. Taub, and D. A. Harter, *J. Chem. Phys.*, **41**, 2954 (1964).
287. M. C. Sauer and B. Ward, *J. Phys. Chem.*, **71**, 3971 (1967).
288. M. H. Studier and E. J. Hart, *J. Am. Chem. Soc.*, **91**, 4068 (1969).
289. D. G. Marketos, A. Marketou-Mantaka, and G. Stein, *J. Phys. Chem.*, **75**, 3886 (1971); **78**, 1987 (1974).
290. K. Bhatia, *Radiat. Res.*, **59**, 537 (1974).
291. M. K. Eberhardt, *J. Phys. Chem.*, **78**, 1795 (1974).
292. I. Loeff and G. Stein, *J. Chem. Soc.*, 2623 (1963).
293. T. K. K. Srinivasan, I. Balakrishnan, and M. P. Reddy, *J. Phys. Chem.*, **73**, 2071 (1969); I. Balakrishnan and M. P. Reddy, *ibid.*, **74**, 850 (1970).
294. M. A. Proskurnin and Y. M. Kolotyrkin, *Proc. 2nd Int. Conf. Peaceful Uses Atomic Energy*, United Nations, Geneva, **29**, 52 (1958).
295. M. A. Proskurnin, E. V. Barelko, and L. I. Kartasheva, *Dokl. Akad. Nauk SSSR.*, **121**, 671 (1958).
296. H. C. Christensen and R. Gustafsson, *Acta Chem. Scand.*, **26**, 937 (1972).

297. G. Stein and J. Weiss, *J. Chem. Soc.*, 3265 (1951).
298. B. H. J. Przybielski-Bielski and R. R. Becker, *J. Am. Chem. Soc.*, **82**, 2164 (1960).
299. K. W. Chambers, E. Collinson, and F. S. Dainton, *Trans. Faraday Soc.*, **66**, 142 (1970).
300. T. Rigg, G. Scholes, and J. Weiss, *J. Chem. Soc.*, 3034 (1952).
301. J. Belloni and M. Haissinsky, *Int. J. Radiat. Phys. Chem.*, **1**, 519 (1969).
302. M. Lefort and M. Haissinsky, *J. Chim. Phys.*, **53**, 527 (1956).
303. W. Karmann, G. Meissner and A. Henglein, *Z. Naturforsch.*, **22b**, 273 (1967).
304. K. M. Bansal and G. R. Freeman, *Radiat. Res. Rev.*, **3**, 209 (1971).
305. M. Daniels and J. Weiss, *J. Chem. Soc.*, 2467 (1958).
306. J. H. Baxendale, A. Breccia, and M. D. Ward, *Int. J. Radiat. Phys. Chem.*, **2**, 167 (1970).
307. A. R. Anderson and Farhataziz, *Trans. Faraday Soc.*, **59**, 1299 (1963).
308. H. A. Schwarz and A. O. Allen, *J. Am. Chem. Soc.*, **77**, 1324 (1955).
309. M. Daniels and E. E. Wigg, *J. Phys. Chem.*, **73**, 3703 (1969); M. Daniels, *ibid.*, 3710.
310. M. Faraggi, D. Zehavi and M. Anbar, *Trans. Faraday Soc.*, **67**, 701 (1971).
311. M. Cottin, *J. Chim. Phys.*, **53**, 917 (1956).
312. G. C. Stevens, R. M. Clarke, and E. J. Hart, *J. Phys. Chem.*, **76**, 3863 (1972).
313. P. G. Clay, G. R. A. Johnson, and J. Weiss, *J. Chem. Soc.*, 2175 (1958).
314. P. G. Clay, J. Weiss, and J. Whiston, *Proc. Chem. Soc.*, 125 (1959).
315. G. Scholes and J. Weiss, *Nature*, **185**, 305 (1960).
316. P. G. Clay, G. R. A. Johnson, and J. Weiss, *J. Phys. Chem.*, **63**, 862 (1959).
317. P. M. Grant and R. B. Ward, *J. Chem. Soc.*, 2654, 2659 (1959).
318. S. A. Barker, P. M. Grant, M. Stacey, and R. B. Ward, *Nature*, **183**, 376 (1959).
319. G. R. A. Johnson, G. Scholes, and J. Weiss, *J. Chem. Soc.*, 3091 (1953).
320. G. O. Phillips and G. J. Moody, *J. Chem. Soc.*, 754 (1960).
321. G. O. Phillips and G. J. Moody, *J. Chem. Soc.*, 762 (1960).
322. M. E. Jayko and W. M. Garrison, *J. Chem. Phys.*, **25**, 1084 (1956).
323. G. G. Jayson, G. Scholes, and J. Weiss, *J. Chem. Soc.*, 2594 (1955).
324. F. S. Dainton, *J. Phys. Colloid Chem.*, **52**, 490 (1948).
325. O. H. Wheeler and R. Montalvo, *Radiat. Res.*, **40**, 1 (1969).
326. R. C. Armstrong and A. J. Swallow, *Radiat. Res.*, **40**, 563 (1969).
327. R. M. Lemmon, P. K. Gordon, M. A. Parsons, and F. Mazzetti, *J. Am. Chem. Soc.*, **80**, 2730 (1958).
328. W. M. Dale and J. V. Davies, *Radiat. Res.*, **7**, 35 (1957).
329. P. A. King and J. A. Ward, *J. Polym. Sci.*, *A-1.*, **8**, 253 (1970).
330. G. R. A. Johnson, G. Stein, and J. Weiss, *J. Chem. Soc.*, 3275 (1951).
331. N. Bach, *Proc. Int. Conf. Peaceful Uses Atomic Energy*, United Nations, New York, **7**, 538 (1956).
332. I. Loeff and A. J. Swallow, *J. Phys. Chem.*, **68**, 2470 (1964).

CHAPTER 8

Organic Compounds

The radiation chemistry of a large number of organic compounds has been studied, but, in the limited space available, it is only possible to describe a few of them in this chapter to illustrate the types of behavior observed. Several gaseous organic compounds are described in Chapter 6 and some organic reactions of possible industrial importance in Chapter 10; the radiolysis of organic compounds dissolved in water is included in the previous chapter.

Radiolysis mechanisms for organic compounds can, in most cases, be represented by schemes made up of the following types of reaction (RH_2 represents an organic compound containing two hydrogen atoms joined to carbon):

$$RH_2 \rightsquigarrow \boxed{\begin{array}{c} \longrightarrow RH_2^{**} \longrightarrow \\ \\ \end{array}} \longrightarrow [RH_2^*, RH_2^+, e^-] + e^-$$

$$(8.1)$$

$[RH_2^+] \rightarrow [RH^+ + H]$	(ion dissociation)	(8.2)
$\rightarrow [R^+ + H_2]$	(ion dissociation)	(8.3)
$[RH_2^+ + RH_2] \rightarrow [RH_3^+ + RH\cdot]$	(ion-molecule reactions)	(8.4)
$[RH_2^+ + e^-] \rightarrow [RH_2*]$	(geminate recombination ions)	(8.5)
$e^- + n\,RH_2 \rightarrow e_{solv}^-$	(solvation charged species)	(8.6)
$[RH_2*] \rightarrow [RH_2]$	(energy dissipation without reaction) (8.7)	
$\rightarrow [RH\cdot + H]$	(dissociation to radicals)	(8.8)
$\rightarrow [R + H_2]$	(dissociation to molecular products) (8.9)	
$[RH\cdot + H] \rightarrow [RH_2]$	(recombination caged radicals) (8.10)	
$\rightarrow RH\cdot + H$	(diffusion from spur)	(8.11)
$RH_3^+ + e_{solv}^- \rightarrow RH_2 + H$	(neutralization)	(8.12)
$\rightarrow RH\cdot + H_2$	(neutralization)	(8.13)
$H + RH_2 \rightarrow H_2 + RH\cdot$	(hydrogen abstraction)	(8.14)
$2\,RH\cdot \rightarrow RH \quad RH$	(radical combination)	(8.15)
$\rightarrow RH_2 + R$	(radical disproportionation)	(8.16)

Reaction is initiated by the formation of excited molecules, which may be singlet, triplet, or superexcited (RH_2**) states, and ions along the tracks of ionizing particles (electrons in the case of x-, γ-, and fast electron irradiation). The ions and excited molecules may be formed singly, but a large proportion (depending on the LET of the radiation) are grouped together in spurs, blobs, and short tracks as described earlier (pp. 59 and 261); the products from reaction 8.1 are enclosed in brackets as a reminder that they will largely be grouped in this manner. A small proportion of the secondary electrons (δ-rays) are ejected with sufficient energy to create tracks of their own and inevitably become widely separated from their parent ions; these are shown as isolated electrons in Eq. 8.1. The time scale of events that follow the initial act of energy deposition is similar to that given in Chapter 7 for water (Table 7.5) and is not repeated here. Reactions 8.2 to 8.10 are very fast and follow within about 10^{-11} sec of the initial event; they are enclosed in brackets to show that they are over before the spur[1] has time to expand

---

[1] The term spur as used in this chapter should be taken to include spurs, blobs, and short tracks.

and dissipate by diffusion. The reactions shown include the dissociation of excited ions (Eqs. 8.2 and 8.3) and molecules (Eqs. 8.8 and 8.9), ion-molecule reactions (represented by Eq. 8.4), energy loss from excited species without dissociation (Eq. 8.7), slowing of secondary electrons to thermal energies followed by geminate recombination with a positive ion (Eq. 8.5) or solvation (Eq. 8.6) (solvation may precede geminate recombination), and reaction between two radicals formed close together in the same spur (Eq. 8.10) (disproportionation is possible in addition to the combination reaction shown). Not all ion and radical pairs will necessarily undergo geminate reaction, and those that do not will escape into the bulk of the liquid as the spurs expand by diffusion (Eq. 8.11). Diffusion takes considerably longer than the processes described to this point, and reactions in the bulk of the medium take place at times of the order of 10^{-8} sec or longer after the initial energy deposition in the system. Reactions 8.10 and 8.14 to 8.16 are typical of radical reactions in organic systems, although other types of reaction are possible (see Chapter 4; ions and excited molecules may also react in other ways than those shown). Radicals, ions, and excited molecules may also react with substances (deliberately added or impurities) present in small amounts. Radicals, for example, generally react very rapidly with oxygen, which, unless carefully removed, is always present.

The reaction scheme shown includes alternatives at various points. Ions, for example, may dissociate, take part in ion-molecule reactions, undergo geminate recombination with an ion of opposite charge, or become solvated. Excited molecules may dissociate or dissipate their energy without reaction. Radicals may reform the original material, or form dimers or unsaturated compounds. The actual mechanism and the proportions of reactive species following alternative paths depend on the irradiation conditions and the nature of the material being irradiated. Geminate recombination of ions and radicals, for example, is more favored in liquid and solid systems than in gases. High dose-rate and high LET can affect radiolysis yields by increasing the probability of interradical reactions in the bulk of the medium and in the particle tracks respectively, although only if the interradical reactions are in competition with alternative reactions that form different products (1). Solvation of ions, and diffusion of the solvated species into the bulk of the liquid, are particularly important with polar liquids (alcohols and amines are examples), whereas geminate recombination of ions within the spurs is the more probable process with nonpolar liquids such as the hydrocarbons. With some materials, notably the aromatic compounds, the energy of excited species may be dissipated particularly efficiently without dissociation, and radiolysis yields are low. At the other extreme, chain reactions leading to very high yields of radiolysis products are sometimes possible if the material contains relatively weak covalent bonds (several examples are described which involve rupture of weak C—Cl bonds).

Two species that are not included explicitly in the reaction scheme given above, although they contribute to the radiolysis reactions of organic compounds, are "hot" radicals and triplet excited states. Hot radicals are radicals formed in reactions such as 8.8 with energies appreciably above thermal energy and are a consequence of the formation of very highly excited states in radiation-induced processes. Though the excess energy of the "hot" radical will be dissipated by a rather small number of collisions with other molecules, hot radicals can react more rapidly than thermal radicals, since the extra energy can supply any activation energy necessary, and they may take part in reactions that would be endothermic if only thermal energies were available. Triplet and singlet excited states are formed together in the initial act of energy deposition (Eq. 8.1) and by ion recombination (Eq. 8.5), while triplet states alone are formed if excitation is brought about by subexcitation electrons (secondary electrons that have slowed to energies below that of the lowest singlet excited state). Triplet states are characterized by longer lifetimes than singlet excited states, and may persist long enough to diffuse from the spurs into the bulk of the liquid. The longer lifetime is most significant if the system contains a low concentration of solute able to react with the relatively long-lived species, but present in too low a concentration to compete with fast reactions occurring in the spurs.

In organic media, as in aqueous solutions, it is often possible to distinguish between "radical" and "molecular" products.[2] Radical products are those that are formed via scavengeable free radicals and that are eliminated by the presence of an efficient radical scavenger. The term may also be applied to the scavengeable free radicals themselves. Molecular products are those not affected by scavenger, and hence formed by unimolecular or other nonradical processes or by radical reactions which are inaccessible to scavenger (e.g., radical reactions occurring within a spur). In many radiolysis reactions a product may be formed partly by a radical and partly by a molecular mechanism. For instance, a particular radical-molecule reaction may give both types of product depending on whether the individual radicals react as hot radicals (molecular product) or are reduced to thermal energies before reacting (radical product). Radical-radical reactions can give products that are classed as molecular if the reaction occurs within a spur or solvent cage, and as radical products if the radicals diffuse into the bulk of the medium and react there. Molecular products may also be formed by processes that involve ions or excited states rather than free radicals, for example, reactions 8.3 and 8.9. Scavenger concentrations of the order of 10^{-3} M (1 mM) are generally sufficient to scavenge thermal radicals and other species which have escaped from the spurs. Higher scavenger concentrations (~ 1 M, or higher)

---

[2] Radical and molecular yields of product may be distinguished from overall product yields by the use of subscripts, as for aqueous solutions. For example, the radical and molecular yields of hydrogen would be G_H and G_{H_2}, respectively.

may scavenge radicals and other reactive species that would normally react within the spurs (Eqs. 8.2–8.10), increasing the apparent yield of "radical" products (the effect of low pH on the radical yields from water is an example of this; cf. Table 7.4). High concentrations are necessary in the latter case to increase the rate of the scavenging reaction sufficiently to compete with the very fast intraspur reactions.

Unsaturated compounds are effective radical scavengers and are often formed in organic systems upon irradiation, so that prolonged irradiation leads to self-scavenging and product yields that vary with the absorbed dose. To avoid this complication, radiolytic yields given in this chapter are, as far as possible, initial yields or yields extrapolated to zero absorbed dose. Radiolysis products may be potential (positive) ion scavengers if their ionization potential is lower than that of the solvent (RH_2) so that the reaction

$$RH_2^+ + S_{ion} \rightarrow RH_2 + S_{ion}^+ \tag{8.17}$$

is exothermic, or potential scavengers of excited solvent molecules if they have a lower excited state than the solvent, allowing

$$RH_2^* + S_{exc} \rightarrow RH_2 + S_{exc}^* \tag{8.18}$$

However these reactions require relatively high scavenger concentrations and are less likely to occur than radical scavenging by the radiolysis products.

Photochemical and mass spectrometric data are helpful in elucidating radiolysis mechanisms. Information derived from photochemistry generally applies to gaseous systems and must be applied with caution to liquids, where there is a greater possibility of excited states being quenched before any chemical change occurs and of radicals formed by dissociation of an excited molecule being caged and recombining. Mass spectrometric data can also be applied with more confidence to gaseous systems, which approach more closely to the conditions in the mass spectrometer.

It might be remarked that since ionizing radiation produces free radicals in all organic materials, any reaction in a mixture of compounds that normally proceeds by way of radical intermediates can be expected to occur under the influence of radiation. This opens up a wide field of radiation chemistry which is barely touched upon in this chapter, where we have confined our attention for the most part to pure compounds and dilute solutions. However, the preparative possibilities arising from the irradiation of organic mixtures are being investigated (e.g., 2, 3 and Chapter 10).

HYDROCARBONS

The radiolysis of saturated aliphatic hydrocarbons (alkanes) (4–11) offers an opportunity to correlate radiolysis products with the length of the carbon chain in the absence of reactive chemical groups. Upon irradiation both carbon–carbon and carbon–hydrogen bonds are broken, and it is found, making a rather rough generalization, that all carbon–carbon and all carbon–hydrogen bonds are equally likely to rupture. Furthermore, carbon–carbon and carbon–hydrogen bond scission are of comparable importance. This tendency toward random splitting of the chemical bonds present in the hydrocarbons can be illustrated by reference to the radiolysis products and to the radicals detected by scavenger experiments. Methane, for example, becomes less important a product as the proportion of terminal methyl groups present falls (i.e., as the chain length increases in straight-chain alkanes), whereas the yield of hydrogen shows much less variation, consistent with the relatively constant proportion of carbon–hydrogen bonds in these hydrocarbons (Table 8.1; refs. 15, 16).

Closer examination of the yields of products from irradiated alkanes shows that the hydrogen and methane yields are only approximately proportional to the proportions of C—H and C—CH$_3$ bonds in the compounds irradiated, and that bond scission is by no means completely random. It appears, for example, that tertiary carbon–carbon bonds are more readily

TABLE 8.1   *Yields of Hydrogen and Methane from Saturated Hydrocarbons*

Hydrocarbon	Structure	$G(H_2)$	$G(CH_4)$
Vapor phase (α-irradiation; ref. 12)			
Propane	$CH_3CH_2CH_3$	8.2	0.37
Butane	$CH_3(CH_2)_2CH_3$	9.0	1.2
Pentane	$CH_3(CH_2)_3CH_3$	7.3	0.81
Hexane	$CH_3(CH_2)_4CH_3$	5.6	0.78
Isobutane	$(CH_3)_2CHCH_3$	7.4	2.7
Neopentane	$(CH_3)_4C$	2.0	2.0
Liquid phase (800 kvp electron irradiation; refs. 13, 14)			
Cyclohexane[a]	$(CH_2)_6$	5.6	∼0.05
Pentane	$CH_3(CH_2)_3CH_3$	4.2	0.4
Hexane	$CH_3(CH_2)_4CH_3$	5.0	0.15
Heptane	$CH_3(CH_2)_5CH_3$	4.7	0.09
Octane	$CH_3(CH_2)_6CH_3$	4.8	0.08
Nonane	$CH_3(CH_2)_7CH_3$	5.0	0.07
Decane	$CH_3(CH_2)_8CH_3$	5.2	0.06
Dodecane	$CH_3(CH_2)_{10}CH_3$	4.9	0.05
Hexadecane	$CH_3(CH_2)_{14}CH_3$	4.8	0.04
2-Methylpentane	$(CH_3)_2CHCH_2CH_2CH_3$	4.0	0.5
2,2-Dimethylbutane	$(CH_3)_3CCH_2CH_3$	2.0	1.2

[a] See Table 8.3.

broken than secondary or primary carbon–carbon bonds since neopentane and 2,2-dimethylbutane, which both contain tertiary C—CH$_3$ bonds, give a particularly high proportion of methane. This is a fairly general observation, and normally tertiary carbon–carbon bonds ($\geqslant$C—R) break in preference to secondary ($>$CH—R) and secondary in preference to primary (—CH$_2$—R), in line with the lower bond dissociation energies in the series primary > secondary > tertiary. In the present examples R is —CH$_2$— or —CH$_3$, but the bond dissociation energies change in the same manner, decreasing in the series primary, secondary, tertiary, when R is some other group or atom, e.g., a hydrogen or halogen atom. Preferential (but not exclusive) breaking of tertiary and secondary bonds is also inferred from the mass spectra of hydrocarbons and, in radiolysis, is confirmed when the radical intermediates are identified by means of scavenger experiments. As would be expected, branched-chain alkanes give higher yields of radiolysis products resulting from carbon–carbon bond fission than do straight-chain alkanes.

Electron paramagnetic resonance data for frozen alkanes irradiated at low temperatures are also consistent with this pattern; straight-chain alkanes give spectra characteristic of the loss of a hydrogen atom, with little evidence of carbon–carbon bond breaking, while branched-chain alkanes give more complex spectra suggestive of carbon–carbon bond fission (17).

Radiolysis of the alkanes produces other products beside hydrogen and methane. These include low molecular weight saturated and unsaturated hydrocarbons with the same or fewer carbon atoms than the parent alkane, intermediate molecular-weight alkanes with a number of carbon atoms intermediate between those of the parent and its dimer, and dimeric products with twice as many carbon atoms as the parent compound; $G(-$alkane$)$ is generally between 6 and 10. With straight-chain alkanes, products with fewer carbon atoms than the parent tend to become less important as the chain length is increased—the higher alkanes giving mainly products with more carbon atoms than the parent. The effect of radiation on these compounds is therefore to increase the average molecular weight of the hydrocarbon, and this is shown by changes in the physical properties of the material. The most obvious change is the formation of an insoluble gel in liquid hydrocarbons and an increase in the melting point of solid hydrocarbons when these are exposed to rather large amounts of radiation. This is illustrated particularly well by the cross-linking of polymers such as polyethylene. The unsaturated products formed from alkanes are efficient radical scavengers and, as their concentration builds up, markedly influence the later stages of the radiolysis.

The main differences between the radiolysis of saturated and unsaturated hydrocarbons can be seen from Table 8.2, where radiolysis yields from similar saturated and unsaturated compounds are compared. It is clear that

T A B L E 8.2   *Comparison of the Radiolysis of Saturated, Unsaturated, and Aromatic Hydrocarbons*

Compound (n Carbon Atoms)	H$_2$	G(Product)[a]			Ref.
		Low Molecular Weight Products (C$_1$ to C$_{n-1}$)	Intermediate Molecular Weight Products (C$_n$ to C$_{2n-1}$)	High Molecular Weight Products (C$_{2n}$ and above)	
Gaseous hydrocarbons					
Ethane	6.8	0.61	0.59	1.64	Table 6.2
Ethylene	1.28	0.12	2.07	~13[b]	Table 6.3
Acetylene	0	0	0	~70[b]	Table 6.3
Liquid hydrocarbons					
Hexane	5.25	1.71	3.32	0.99	Table 8.4
1-Hexene	0.8	0.12	0.11	7	18
Cyclohexane	5.6	~0.2	4.2	2.1	Table 8.3
Cyclohexene	1.28	~0	~1.2	8.9	19
Benzene	0.039	0.02	0.03	0.94	Table 8.7

[a] Electron or γ-irradiation.
[b] G(– Parent compound).

the yield of hydrogen is lower from the unsaturated compounds while the yield of high molecular weight products is greater. The yield of low and intermediate molecular weight products which, with the exception of those with the same number of carbon atoms as the parent compound, are formed as a consequence of C—C bond fission, is generally greatest with the saturated compounds. These results are largely explicable on the assumption that unsaturated compounds are efficient radical scavengers, so that any radicals formed react with the substrate rather than with other radicals. Hydrogen is formed to a smaller extent from unsaturated compounds because hydrogen atoms can both add to the compound and abstract hydrogen from it,

$$\text{H} + \text{—CH}_2\text{CH}\text{=}\text{CH—} \rightarrow \text{—CH}_2\text{CH}_2\text{—}\dot{\text{C}}\text{H—} \quad (\text{or —CH}_2\dot{\text{C}}\text{H—CH}_2\text{—})$$
$$(8.19)$$

$$\rightarrow \text{H}_2 + \text{—}\dot{\text{C}}\text{H—CH}\text{=}\text{CH—} \quad (8.20)$$

and only the latter reaction gives molecular hydrogen; hydrogen atoms abstract hydrogen from saturated hydrocarbons. Radical scavenging will not affect products formed by molecular processes such as ion-dissociation, ion-molecule reactions, or the dissociation of excited molecules into molecular products, and the hydrogen formed from unsaturated compounds (Table 8.2) probably results in part from reactions of this sort. Radicals formed in reactions 8.19 and 8.20, and in similar reactions, often dimerize or add to another molecule of the unsaturated compound, leading to the dimeric and polymeric products observed. The lower yields in the low-intermediate range of products suggests that C—C bond scission makes a smaller contribution to the radiolysis of unsaturated compounds than to the radiolysis of saturated compounds. Aromatic compounds are generally more stable toward radiation than are the corresponding aliphatic compounds. Benzene, for example, is markedly more stable than cyclohexane. The stability is associated with the presence in the aromatic ring system of electrons in π orbitals, which reduces the probability that excited and ionized aromatic molecules will dissociate and favors alternative modes of energy dissipation which do not result in dissociation of the molecule. In the mass spectrometer, for instance, aromatic ions show little tendency to break down into smaller fragments by fission of an aromatic ring. Such dissociation is even less likely under normal radiolysis conditions, where ions can lose their extra energy rapidly by collisional processes. It should be emphasized that the stability of aromatic compounds is not the result of immunity to radical attack; free radicals react with these compounds quite readily. The stabilizing influence of an aromatic ring system extends to alkyl groups and substituted alkyl groups in the same molecule and even, under favorable circumstances, to aliphatic compounds which are merely mixed with the aromatic compound.

Cyclohexane

Cyclohexane has been studied extensively in both the vapor and liquid phase (e.g., 4–7, 10, 11, 20); representative values for the yields of products upon γ-irradiation are given in Table 8.3. The vapor phase yields are dependent on cyclohexane pressure and dose rate. Pressure effects are attributed to ion-neutralization at the vessel walls at low pressures (below about 100 mm Hg) (21) and to collisional deactivation of excited species or caging at high pressures (27). Competition between ion neutralization at the walls (favored by low dose rates) and neutralization in the gas phase (favored at high dose rates) is believed to be responsible for the dose-rate dependency (28). The hydrogen yield from liquid cyclohexane is independent of dose rate (29), LET (30), and temperature (31), and it has been suggested that determination of this product be used as a means of chemical dosimetry (13, 32), although the hydrogen yield falls as unsaturated radiolysis products build up in the liquid (e.g., 33). Doses of 3×10^{20} eV g^{-1} (4.8 Mrad) and 0.1 to 3×10^{20} eV g^{-1} (0.2 to 4.8 Mrad) were used in determining the G values given in Table 8.3 for vapor and liquid samples, respectively, corresponding to conversion of about 0.01 to 0.25% of the cyclohexane to radiolysis products; the G values listed for the liquid systems were obtained by extrapolating the experimental yields to zero dose.

T A B L E 8.3 *γ-Radiolysis of Cyclohexane*

	G(Product)		
Product	Vapor (40°C, >100 mm Hg; ref. 21)	Liquid (refs. 22 to 26)	Liquid + 2 × 10^{-3} M Oxygen (ref. 26)
H_2	6.6	5.6	
CH_4	0.22	~0.05	
C_2H_2	0.21	0.01	
C_2H_4	0.98	0.1	
C_2H_6	0.18	0.03	
C_3H_6	0.14		
C_3H_8	0.24		
Butenes	0.05		
Butanes	0.25		
1-Hexene		0.5	0.26
Hexane		0.1	<0.01
Methylcyclopentane	0.10	0.3	~0.03
Cyclohexene	0.50	3.2	1.49
C_7 to C_{11}	0.87	~0.1	
Bicyclohexyl	0.55	1.9	0.29
Other C_{12}		0.2	~0.04
Cyclohexanol			3.17
Cyclohexanone			2.63
Peroxides			0.61

The more important products formed when pure liquid cyclohexane is irradiated are hydrogen, cyclohexane, and bicyclohexyl, and the main features of the radiolysis can be accounted for by a relatively simple sequence of reactions since all C—H and all C—C bonds are equivalent:

$$C_6H_{12} \rightsquigarrow C_6H_{12}^+ + e^- \qquad (G \geqslant 4.2) \qquad (8.21)$$

$$\rightsquigarrow C_6H_{12}^* \qquad (G \geqslant 2.2) \qquad (8.22)$$

$$C_6H_{12}^+ + e^- \rightarrow C_6H_{12}^* \qquad (8.23)$$

$$C_6H_{12}^* \rightarrow \cdot C_6H_{11} + H \qquad (G = 4.8) \qquad (8.24)$$

$$\rightarrow C_6H_{10} + H_2 \qquad (G \sim 0.3) \qquad (8.25)$$

$$\rightarrow C_6H_{12} \qquad (8.26)$$

$$[\cdot C_6H_{11} + H] \rightarrow C_6H_{10} + H_2 \qquad (G \sim 0.8) \qquad (8.27)$$

$$H + C_6H_{12} \rightarrow H_2 + \cdot C_6H_{11} \qquad (G = 4) \qquad (8.28)$$

$$k_{28} = 5 \times 10^6 \, M^{-1} \, sec^{-1}$$

$$2 \cdot C_6H_{11} \rightarrow C_6H_{12} + C_6H_{10} \Big\}$$
$$\phantom{2 \cdot C_6H_{11} } \rightarrow C_{12}H_{22} }$$

$$2k = 2.5 \times 10^9 \ M^{-1} \ sec^{-1} \quad (G = 2.1) \quad (8.29)$$

$$\frac{k_{29}}{k_{30}} = 1.1 \quad\quad\quad\quad\quad (G = 1.9) \quad (8.30)$$

Cyclohexane is a nonpolar liquid and has a relatively low dielectric constant (2.02 at 20°C compared with 80.37 for water) so that almost all the electrons formed in reaction 8.21 are expected to be attracted back to their parent ion to undergo geminate recombination (Eq. 8.23); the parent ion is unreactive toward cyclohexane and does not take part in any ion-molecule reactions in pure cyclohexane (34). Excited molecules formed directly (Eq. 8.22) or by geminate recombination dissociate to cyclohexyl radicals and hydrogen atoms (Eq. 8.24) or cyclohexene and molecular hydrogen (Eq. 8.25), with the former predominating, or are quenched without detectable chemical reaction (Eq. 8.26). A proportion of the radical pairs formed in reaction 8.24 react together (e.g., Eq. 8.27) before diffusion separates them, but most hydrogen atoms abstract hydrogen from cyclohexane forming additional cyclohexyl radicals (Eq. 8.28), which disproportionate to cyclohexane and cyclohexene (Eq. 8.29) or combine to give bicyclohexyl (Eq. 8.30). For cyclohexyl radicals, disproportionation and combination are approximately equally probable. At high doses, polymeric products (C_{18}, C_{24}, etc.) are formed, apparently by reaction of cyclohexyl radicals with cyclohexene formed earlier in the irradiation (35). Reactions 8.21 to 8.27 are rapid and precede spur expansion, while the radical reactions 8.28 to 8.30 are slower and occur in part in the expanding spurs and in part after the radicals that escape the spurs have become homogeneously distributed through the liquid. Reactions 8.21 to 8.27 might all have been shown within brackets to signify that they are spur reactions although, for clarity and convenience, this has not been done. However the reactants in reaction 8.27, which could conceivably occur either within the spurs or in the bulk of the liquid, are enclosed in brackets to emphasize that this is regarded as a spur reaction.

The validity of a reaction mechanism such as that given above can be tested in various ways. Proposed reactions involving ions, excited states, and radicals must, obviously, be consistent with the known properties of these species as determined by mass spectrometric, photochemical, and classical organic techniques. While the mechanism should enable qualitative and, if possible, quantitative predictions to be made of the effect of changing dose rate, LET, and temperature and of the effect of a range of ion and radical scavengers upon radiolysis yields, more direct checks can be made if the transient species formed can be identified by esr or pulse radiolysis techniques. The latter can be particularly valuable, since it is readily applied to liquid systems and may provide kinetic data for some of the reactions taking place. Examples of the use of some of these sources of information, as applied to cyclohexane, are given below.

Important ions in the mass spectrum of cyclohexane are the parent ion, $C_6H_{12}^+$ (14.6%), $C_4H_8^+$ (20.7%), $C_4H_7^+$ (7.1%), and $C_3H_5^+$ (14.0%) (11), while a number of ion-molecule reactions of the fragment ions have been identified (36–38). Fragmentation of the parent ion and reactions of the fragment ions probably account in part for the greater yield of low molecular weight products when cyclohexane is irradiated in the vapor phase (Table 8.3). However, fragmentation in the mass spectrometer is reduced as the pressure is increased (27), increasing the probability that excited ions will be deactivated by collisional processes, and dissociation of excited ions must make a minor contribution to the radiolysis of liquid cyclohexane.

The yield of ions produced by irradiation can be measured directly in the case of gases (for cyclohexane vapor, $W = 22.7$ eV ion-pair^{-1}), but indirect methods must be used to estimate ion yields in irradiated liquids. The free ion yield in irradiated cyclohexane, i.e., the yield of ions that do not undergo geminate recombination, has been estimated from the radiation-induced conductance of the liquid, which leads to a value of G(free ion pairs) = 0.10 (40); geminate recombination is too fast for the total ion yield to be measured in this way. Independent estimates of the free ion yield have been made by pulse radiolysis of cyclohexane containing low ($\sim 10^{-3} M$) concentrations of electron or ion scavengers; the low concentration ensures that only ions that have escaped from the spurs are collected. Estimates of the free ion-pair yield using biphenyl (which scavenges both cations and electrons) (41) and anthracene (42) and the galvinoxyl radical (43) (which scavenge electrons) are in the range of 0.1 to 0.2 G units, in excellent agreement with the value estimated from conductance measurements. Many substances react very rapidly with electrons and, if present in sufficiently high concentration, are able to scavenge electrons that would normally undergo geminate recombination in the spurs. Methyl bromide is one such substance and irradiation of cyclohexane solutions containing methyl bromide (10^{-4} to 0.5 M) and a small amount of radioactive iodine ($\sim 10^{-3} M$) has been used to estimate both the free ion and the total ion yield (44). Electrons are scavenged by the methyl bromide forming methyl radicals, which subsequently react with the iodine; the function of the iodine is to form a radioactive product that can be estimated using sensitive counting techniques.

$$e^- + CH_3Br \rightarrow \cdot CH_3 + Br^- \qquad (8.31)$$

$$\cdot CH_3 + {}^{131}I_2 \rightarrow CH_3{}^{131}I + {}^{131}I \qquad (8.32)$$

Kinetic analysis of the yield of methyl iodide as a function of the methyl bromide concentration gave a lower limiting yield of iodide of about 0.1 G units, corresponding to the capture of free electrons and in agreement with the other estimates of G(free ion pairs). The limiting yield of iodide at high bromide concentrations corresponds to capture of both free and geminate electrons and gave G(ion pairs) $= 3.9 \pm 0.1$, which is a little lower than the value estimated from the value of W for cyclohexane vapor ($G = 100/W = 4.41$).

Nitrous oxide is very widely used as an electron scavenger, reacting according to the equation,

$$e^- + N_2O \rightarrow N_2 + O^- \tag{8.33}$$

When cyclohexane containing nitrous oxide is irradiated, the yield of nitrogen increases with increasing N_2O concentration but eventually reaches a plateau value at about 0.5 M N_2O corresponding to $G_{e^-} = 5.3 \pm 0.3$ (45). If the latter value is correct and nitrogen is only formed *via* reaction 8.33, W for liquid cyclohexane is 19 ± 1 eV ion-pair^{-1} rather than the gas-phase value of 22.7 eV ion-pair^{-1}. Radiation-induced conductance measurements with liquid inert gases (46) have shown that ionization yields in liquids can be greater than those in the corresponding gases, and this possibility was also raised in the case of water (p. 264). In any event, comparison of the free ion and total ion yields in liquid cyclohexane show that more than 97% of the ion pairs produced by irradiation undergo geminate recombination (cf. 47). In pure cyclohexane, ion recombination leads to the same products whether it occurs by geminate recombination in the spurs or by the combination of free ions in the bulk of the liquid. However the lifetimes of the ions are quite different in the two cases, being of the order of 10^{-11} sec for ions which undergo geminate recombination and about 10^{-3} sec for the free ions (40, 48). In the presence of nitrous oxide, the ion neutralization reaction is probably (49–51),

$$C_6H_{12}^+ + O^- \rightarrow \cdot C_6H_{11} + OH \tag{8.34}$$

followed by

$$OH + C_6H_{12} \rightarrow H_2O + \cdot C_6H_{11} \tag{8.35}$$

since the reaction products include water and a smaller quantity of cyclohexanol (52), and addition of N_2O decreases the yield of hydrogen but increases the yields of cyclohexyl radicals, cyclohexene, and bicyclohexyl. Kinetic analysis of the hydrogen yield as a function of the nitrous oxide concentration (53) indicated that the yield of hydrogen not formed as a result of ion recombination was 2.2 G units in pure cyclohexane; since the total yield is 5.6 units, the yield of hydrogen resulting from the neutralization reaction 8.23 is therefore 3.4 G units.

Though electron scavenging is more common, positive ions can also be scavenged. Thus ethanol (54) and ammonia (55, 56) have been used to scavenge positive ions in irradiated cyclohexane. The scavenging reaction forms hydrogen and to distinguish this from hydrogen produced by non-ionic processes, the scavenger is labeled with deuterium. Scavenging is the result of proton transfer from the positive ion,

$$C_6H_{12}^+ + C_2H_5OD \rightarrow \cdot C_6H_{11} + C_2H_5ODH^+ \tag{8.36}$$

$$C_6H_{12}^+ + ND_3 \rightarrow \cdot C_6H_{11} + ND_3H^+ \tag{8.37}$$

and the hydrogen (HD) is released following neutralization of the product ion.

Information regarding the excited states formed when cyclohexane is irradiated is obtained, in part, from photochemical studies. Photolysis of cyclohexane vapour at 147 and 123.6 nm gives hydrogen, acetylene, cyclohexene, and 1,3-butadiene (39) according to

$$C_6H_{12}{}^* \rightarrow C_6H_{10} + H_2 \tag{8.25}$$

$$\rightarrow CH_2{=}CH{-}CH{=}CH_2 + CH_2{=}CH_2 + H_2 \tag{8.38}$$

Ring scission (Eq. 8.38) becomes less important as the pressure is increased and is believed to be negligible in the liquid phase. Photolysis of liquid cyclohexane in the presence of scavengers (57) has shown that elimination of molecular hydrogen (Eq. 8.25) is the major process ($\sim 86\%$) although a smaller yield of hydrogen atoms is formed (Eq. 8.24); at 147 nm, the quantum yield for the formation of hydrogen is about 0.5. The low yield of hydrogen atoms from the direct excitation of cyclohexane has prompted the suggestion that these may be formed by the dissociation of excited triplet molecules in the radiolysis of cyclohexane.

Yields of singlet excited molecules cannot be estimated directly by scavenger techniques (cf. 58), but methods are available for the estimation of triplet excited molecules based on transfer of electronic energy from the excited solvent to a solute, e.g.,

$$^3C_6H_{12}{}^* + S \rightarrow C_6H_{12} + {}^3S^* \tag{8.39}$$

The triplet energy level of the solute must be equal to, or less than, that of the solvent (i.e., the reaction should not be endothermic). Solutes may be aromatic compounds such as anthracene and naphthalene, which form triplet excited states that may be detected spectroscopically in pulse radiolysis experiments, or substituted alkenes in which triplet excitation is detected by the *cis→trans*, or *trans→cis*, isomerization produced (it should be noted, however, that charge transfer to the alkene can also cause isomerization) (59, 60). Triplet yields estimated by pulse radiolysis of solutions of anthracene

374 An Introduction to Radiation Chemistry

and naphthalene in cyclohexane range from 0.2 to 3.7 (61), depending on solute concentration and irradiation conditions.

An estimate of the number of excited molecules dissociating to cyclohexene and molecular hydrogen (Eq. 8.25) has been made by irradiating dilute solutions of C_6D_{12} in C_6H_{12} and measuring the D_2 content of the hydrogen produced (62). The reactions concerned in the production of deuterium are:

$$C_6D_{12}^* \rightarrow \cdot C_6D_{11} + D \qquad (8.40)$$

$$\rightarrow C_6D_{10} + D_2 \qquad (8.41)$$

$$D + C_6D_{12} \rightarrow \cdot C_6D_{11} + D_2 \qquad (8.42)$$

$$D + C_6H_{12} \rightarrow \cdot C_6H_{11} + HD \qquad (8.43)$$

and

$$\frac{G(D_2)}{F} = G(D_2)_{41} + G(D)_{40} \frac{k_{42}}{k_{43}} F \qquad (8.44)$$

when F, the ratio $[C_6D_{12}]/[C_6H_{12}]$, is much smaller than 1. The expression allows for competition between reactions 8.42 and 8.43 and for energy absorption by C_6D_{12} and C_6H_{12} in proportion to their mole fractions in the mixture. A plot of $G(D_2)/F$ against F was found to be linear, as required by the expression, at values of F up to about 0.6. Extrapolation of the linear plot to $F = 0$ gave $G(D_2)_{41} = 0.31$; it is usually assumed that $G(H_2)_{25}$ is of the same order of magnitude.

A number of compounds have been used as radical scavengers with the irradiated alkanes, among them iodine, hydrogen iodide, oxygen, and ethylene and other alkenes (63). However, quantitative estimates of radical yields, particularly hydrogen atom yields, are sometimes uncertain because a proportion of the radicals may be "hot" and because the radical scavenger may also react with electrons or positive ions. Iodine was one of the earliest radical scavengers used and, in cyclohexane, reacts with both hydrogen atoms and cyclohexyl radicals,

$$H + I_2 \rightarrow HI + I \qquad (8.45)$$

$$\cdot C_6H_{11} + I_2 \rightarrow C_6H_{11}I + I \qquad (8.46)$$

The iodine atoms formed can also add to radicals, or they may dimerize,

$$I + \cdot C_6H_{11} \rightarrow C_6H_{11}I \qquad (8.47)$$

$$2\,I \rightarrow I_2 \qquad (8.48)$$

but in either event, $G(radical) = 2G(-I_2)$. Concentrations of iodine of the order of 10^{-3} M, or less, are used to minimize electron capture,

$$e^- + I_2 \rightarrow I^- + I \qquad (8.49)$$

and the products are analyzed by gas chromatography or, using radioactive $^{131}I_2$, by radiochemical techniques. Hydrogen iodide reacts with radicals with the formation of iodine atoms and a stable molecule,

$$\cdot C_6H_{11} \text{ (or H)} + HI \rightarrow C_6H_{12} \text{ (or } H_2) + I \qquad (8.50)$$

and oxygen and alkenes by addition reactions,

$$\cdot C_6H_{11} + O_2 \rightarrow \cdot O_2C_6H_{11} \qquad (8.51)$$

$$\cdot C_6H_{11} + CH_2{=}CH_2 \rightarrow C_6H_{11}CH_2\dot{C}H_2 \qquad (8.52)$$

Addition of hydrogen atoms to cyclohexene is partly responsible for the decreasing yield of hydrogen as radiolysis products build up in irradiated cyclohexane (charge and energy transfer and positive ion scavenging can also occur in the presence of accumulated radiolysis products). Hydrogen atom scavenging by cyclohexene,

$$H + C_6H_{10} \rightarrow \cdot C_6H_{11} \qquad (8.53)$$

does not affect the yield of bicyclohexyl (22, 64), since the hydrogen atoms produce cyclohexyl radicals whether they react directly with cyclocyclo-hexane (Eq. 8.28) or are scavenged by cyclohexene. Yields of thermal hydrogen atoms estimated by the reduction in $G(H_2)$ in the presence of 10^{-3} to 10^{-1} M scavenger are generally about 2 (24, 63 to 66), though the total yield of hydrogen atoms is greater than this as hot hydrogen atoms react with cyclohexane too rapidly to be scavenged by the low-moderate concentrations of scavenger used. The yield of cyclohexyl radicals estimated using low concentrations of iodine is about 5.7, but this falls to 4.2 G units with 10^{-2} M I_2 due to secondary reactions (51).

Oxygen is a particularly important radical scavenger, since traces of air and oxygen will always be present in organic liquids unless special precautions are taken to remove them. Yields of liquid products from cyclohexane irradiated in the presence of oxygen are included in Table 8.3. The concentration of oxygen used (2×10^{-3} M) is too low to scavenge hydrogen atoms or electrons that react by geminate recombination, and the main effect is reaction with thermal alkyl radicals. Addition of oxygen to thermal cyclohexyl radicals (Eq. 8.51) prevents the disproportionation and combination reactions forming cyclohexene (Eq. 8.29) and bicyclohexyl (Eq. 8.30), and the yields of these products fall, while cyclohexanol and cyclohexanone are formed, largely by

$$2\cdot O_2C_6H_{11} \rightarrow C_6H_{12}O + C_6H_{10}O + O_2 \qquad (8.54)$$

A smaller yield of peroxide (mainly $C_6H_{11}O_2H$) is also produced. The difference between the yields of cyclohexene and bicyclohexyl in the presence and absence of oxygen gives the yield of cyclohexyl radicals that normally form these products; since each molecule of product requires two cyclohexyl radicals, the relationship is

$$G(\cdot C_6H_{11})$$
$$= 2\{[G(C_6H_{10}) - G(C_6H_{10})_{o_2}] + [G(C_{12}H_{22}) - G(C_{12}H_{22})_{o_2}]\} \quad (8.55)$$

and, substituting the appropriate G values from Table 8.3,

$$G(\cdot C_6H_{11}) = 2[(3.2 - 1.49) + (1.9 - 0.29)]$$
$$= 6.6$$

The data also give the ratio of disproportionation to combination for cyclohexyl radicals as

$$\frac{k_{29}}{k_{30}} = \frac{(3.2 - 1.49)}{(1.9 - 0.29)} = 1.1 \quad (8.56)$$

The same value has been obtained by experiments in which cyclohexyl radicals are generated by the mercury photosensitized photolysis of cyclohexane (67). Oxygen does not reduce the yield of bicyclohexyl to zero, or the yield of cyclohexene to the value (~ 0.3 G units) estimated for the unimolecular dissociation reaction 8.25, because the concentration is not sufficient, even with a saturated solution, to scavenge radicals which react within the spurs. Bicyclohexyl is believed to be formed by combination of cyclohexyl radicals in the spurs (Eq. 8.30), and the disproportionation to combination ratio for cyclohexyl radicals indicates that this will be accompanied by the formation of about 0.3 G units of cyclohexene by disproportionation of the radicals (Eq. 8.29). The total "molecular" yield of cyclohexene ($G = 1.49$, from the oxygen scavenger experiments) therefore includes ~ 0.3 G units attributable to unimolecular dissociation (Eq. 8.25) and 0.3 G units formed by disproportionation of cyclohexyl radicals in the spurs, leaving about 0.8 to 0.9 G units unaccounted for. In the reaction scheme given above, the additional cyclohexene is attributed to a spur reaction between hydrogen atoms and cyclohexyl radicals,

$$[H + \cdot C_6H_{11}] \rightarrow C_6H_{10} + H_2 \quad (8.27)$$

which probably competes with combination of the radicals to reform cyclohexane,

$$[H + \cdot C_6H_{11}] \rightarrow C_6H_{12} \quad (8.57)$$

The reactions represent geminate reaction between radicals formed by dissociation of an excited cyclohexane molecule.

The G values shown with reactions 8.21 to 8.30 at the head of this section are estimates based on product yields and the scavenger experiments described above. Minor products have been ignored, but most result from additional reactions of excited ions and molecules that bring about ring opening; the minor products represent about 12% of the total chemical change. About 60% of the hydrogen atoms formed are hot, or react within the spurs, and are not scavenged by low to moderate concentrations of scavenger, but only about 25% of the cyclohexyl radicals are not scavenged. Based on susceptibility to radical scavenging, about 1.7 G units of the cyclohexene yield and 1.6 G units of the bicyclohexyl yield are classed as as "radical" yields, while the remaining cyclohexene (1.5 G units) and bi-cyclohexyl (0.3 G units) are "molecular" yields.

The spur diffusion model (e.g., 68) used in describing the nonhomogeneous distribution of radicals in irradiated water can also be applied to organic liquids and, as in water, predicts a decrease in radical yields, and an increase in the yields of molecular products as the LET of the radiation is increased. Experiments with cyclohexane have shown that the molecular yield of bicyclohexyl does indeed increase from about 0.3 to about 0.8 when polonium α-particles are used in place of γ-radiation (69), while the yield of iodine-scavenged cyclohexyl radicals falls from 4 to about 2. Overall product yields show smaller changes, but at high LET reduced yields of cyclohexene and bicyclohexyl are observed and, at very high LET, an increased yield of molecular hydrogen (e.g., 1, 69–72). These observations can be accounted for in part by competition between radical–solvent,

$$H + C_6H_{12} \rightarrow H_2 + \cdot C_6H_{11} \tag{8.28}$$

and radical–radical reactions,

$$H + \cdot C_6H_{11} \rightarrow H_2 + C_6H_{10} \tag{8.27}$$

$$\rightarrow C_6H_{12} \tag{8.57}$$

$$2H \rightarrow H_2 \tag{8.58}$$

with the latter relatively more favored at high LET, but the increased hydrogen yield at very high LET suggests that additional, nonradical, processes may occur with high-LET radiation.

Other Saturated Hydrocarbons

Product yields for the radiolysis of hexane (73), a typical straight-chain saturated hydrocarbon are given in Table 8.4. The vapor yields are relative to the hydrogen yield, for which a value of 5.0 G units was assumed. Radiolysis yields tend to be higher in the vapor than in the liquid phase so that the values for hexane vapor may be low; however, the relative yields clearly illustrate the different product distributions observed in vapor and liquid phases. Product yields for the liquid phase given in column three of Table 8.4 are for an absorbed dose of 22 Mrad (1.4×10^{21} eV g^{-1}) (75), while those in

T A B L E 8.4   *Radiolysis of Hexane*

Product	Vapor[a] (800 kVp e^-, ~25°C, 100 mm Hg; ref. 74)	Liquid (γ-Radiation; ref. 75)	Liquid[b] (γ-Radiation; ref. 76)
		G(Product)	
H_2	5.0	(3.96)	5.25
CH_4	0.5	0.17	0.15
C_2H_2	0.3	0.012	
C_2H_4	1.1	0.12	0.14
C_2H_6	1.0	0.36	0.22
C_3H_6	0.3	0.10	0.18
C_3H_8	2.3	0.37	0.31
Butenes	0.06	0.088	0.13
Butanes	2.7	0.376	0.36
Pentenes		0.032	0.06
Pentanes	0.60	0.077	0.09
Hexenes	0.10	1.81	2.54
C_7	0.50	0.063	0.05
C_8	1.10	0.212	0.09
C_9	0.47	0.125	0.109
C_{10}	0.14	0.109	0.114
C_{11}	0.10	0.015	0.017
C_{12}	0.40	0.99	1.26

[a] Calculated relative to the hydrogen yield assuming $G(H_2) = 5.0$.
[b] Values extrapolated to zero dose.

column four were obtained by plotting yields at several doses and extrapolating back to zero dose to correct for secondary reactions (76). Comparison of columns three and four shows that this correction is most significant for the yields of hydrogen and unsaturated products; the unsaturated products act as radical, and possibly ion, scavengers as their concentration builds up in the irradiated liquid. Products from C_4 to C_{12} are mixtures of isomeric compounds, a total of 41 different fractions being listed in the original reference (75).

The main difference between the radiolysis of hexane and cyclohexane is the larger number of products from the straight chain hydrocarbon. This is due in part to the fact that all C—C and all C—H bonds are not equivalent in hexane, so that loss of a hydrogen atom may give any one of three isomeric radicals from hexane but only a single radical from cyclohexane, and in part to a greater degree of C—C bond rupture with the straight chain compound. Thus there are a larger number of products in the C_1 to C_5 and C_7 to C_{11} range for hexane, and these products account for a greater fraction of the total decomposition for hexane than for cyclohexane (Table 8.5). Examination of the product yields from hexane shows that chain scission occurs predominantly at the inner C—C bonds as does alkyl radical addition; methyl addition, however, is predominantly to a terminal carbon atom so that heptane is the main C_7 product whereas branched chain isomers predominate from C_8 to C_{12} (75).

TABLE 8.5 *Product Distribution from γ-Irradiated Liquid Alkanes*

	G(Product)		
Product(s)	Cyclohexane	Hexane	2,3-Dimethylbutane[a]
H_2	5.6	5.25	3.77
C_1 to C_5	~0.2	1.71	7.12
C_6	4.2	1.81	0.35
C_7 to C_{11}	~0.1	1.51	1.09
C_{12}	2.1	0.99	0.22

[a] Ref. 77.

The radiolysis mechanism for hexane is complex, but the relatively larger yields of low and intermediate molecular weight products and the greater degree of C—C bond scission compared with cyclohexane can probably be attributed to the occurrence of ion-molecule reactions and to the dissociation of excited radicals. Futrell (78), for example, has suggested a number of feasible ion-molecule reactions involving fragment ions formed from hexane in the mass spectrometer. These are of the hydride-ion-transfer type,

$$R^+ + C_6H_{14} \rightarrow RH + C_6H_{13}^+ \tag{8.59}$$

and in several cases the neutral product retains sufficient energy to dissociate, e.g.,

$$C_2H_3^+ + C_6H_{14} \rightarrow C_2H_4^* + C_6H_{13}^+ \tag{8.60}$$

$$C_2H_4^* \rightarrow C_2H_2 + H_2 \tag{8.61}$$

The ion-molecule reactions would convert most of the primary ions to $C_6H_{13}^+$, which is assumed to give hydrogen atoms upon neutralization,

$$C_6H_{13}^+ + e^- \rightarrow C_6H_{13}^* \rightarrow C_6H_{12} + H \tag{8.62}$$

The suggested reactions also give rise to ethyl, propyl, and hexyl radicals which, by combination and disproportionation, can form many of the products found experimentally. Using the mass spectral ion-abundance pattern to represent the initial distribution of ions, ion-molecule reactions as suggested above, and reasonable interradical reactions, Futrell was able to calculate G values for the radiolysis products from hexane vapor that were remarkably close to those found experimentally. In the liquid phase there is a greater probability that excited ions will dissipate their energy without fragmentation and smaller yields of low and intermediate molecular weight products would be predicted, as is indeed observed. Other types of ion-molecule reaction may also contribute, and Williams (79), basing his conclusions on the energetics of ion reactions in liquid hydrocarbons, has suggested that dimeric products may be formed directly by ion-molecule reactions such as

$$C_6H_{14}^+ + C_6H_{14} \rightarrow C_{12}H_{26}^+ + H_2 \qquad (8.63)$$

Photolysis of straight chain alkanes (80) causes detachment of both molecular hydrogen and hydrogen atoms, e.g.,

$$C_6H_{14}{}^* \rightarrow C_6H_{12} + H_2 \qquad (8.64)$$

$$\rightarrow \cdot C_6H_{13} + H \qquad (8.65)$$

with the former processes predominating at low photon energies and the latter at higher energies. Elimination of low molecular weight alkanes such as methane is also possible, as is the formation of excited radicals which subsequently dissociate to an alkene and a smaller radical (81), e.g.,

$$\cdot C_6H_{13}{}^* \rightarrow C_2H_4 + \cdot C_4H_9 \qquad (8.66)$$

Electron (e.g., N_2O) and radical (e.g., I_2, O_2, and NO) scavengers reduce the yields of both saturated (alkane) and unsaturated (alkene) products from hexane, with the electron scavengers being most effective in reducing the alkene yields and the radical scavengers most effective in reducing the alkane yields; radical scavengers largely eliminate the intermediate (C_7 to C_{11}) products (81, 82). The action of the scavengers suggests that unsaturated products are formed largely by reactions that follow from ion neutralization, while the alkanes and intermediate molecular weight products are formed by radical combination and disproportionation. Radical yields for hexane, estimated using scavengers, are (63): methyl (G 0.7), ethyl (G 0.3), 1-propyl (G 0.3), butyl (G 0.27), pentyl (G 0.04), and hexyl (G 4.1). The most abundant organic products from hexane are those formed by disproportionation and combination of the most abundant (hexyl) radical, namely hexenes and dodecane,

$$2\cdot C_6H_{13} \rightarrow C_6H_{14} + C_6H_{12} \qquad (8.67)$$

$$\rightarrow C_{12}H_{26} \qquad (8.68)$$

Hydrogen atoms and the smaller alkyl radicals are able to abstract hydrogen from hexane,

$$H + C_6H_{14} \rightarrow H_2 + \cdot C_6H_{13} \qquad (8.69)$$

$$\cdot CH_3 + C_6H_{14} \rightarrow CH_4 + \cdot C_6H_{13} \qquad (8.70)$$

and competition between such radical–solvent reactions and interradical reactions such as 8.58 leads to dose rate and LET effects similar to those observed with cyclohexane (1). Higher dose rates increase the proportion of C_7 to C_{11} products, since the smaller alkyl radicals tend to react with other radicals rather than abstract hydrogen from the solvent (83).

When solid hexane is irradiated at $-196°C$ (84), the main products are hydrogen (G 3.65), hexenes (G 2.68), and dimer (G 0.72), and there is less C—C bond fission than in the liquid, fragmentation of excited ions and molecules being effectively inhibited by caging and recombination of the fragmentation products. Kevan and Libby (84) conclude that ionic (e.g., Eq. 8.63) rather than radical reactions are occurring in the irradiated solid and find that the distribution of C_{12} products is similar to that found in the liquid in the presence of a radical scavenger, which differs from the product distribution from pure hexane.

Radiolysis of other straight chain hydrocarbons is similar to that of hexane, increasing chain length having little effect on the yields of products except that of dimer, which increases slowly with increasing chain length (85). Branched chain hydrocarbons give a greater proportion of products resulting from C—C bond scission, the bonds adjacent to the branches being particularly susceptible. Thus the main products from 2,2-dimethylbutane are (86), H_2 (G 2.50), CH_4 (G 1.76), C_2H_4 (G 1.87), C_2H_6 (G 1.25), isobutene (G 2.07), and isomeric methyl butenes (G 1.7), indicative of chain scission between the tertiary carbon and the attached alkyl groups, as shown below.

$$
\begin{array}{cc}
\mathrm{CH_3} & \mathrm{CH_3\ \ CH_3} \\
| & \ \ \ \ |\ \ \ \ | \\
\mathrm{CH_3\!-\!C\!-\!CH_2\!-\!CH_3} & \mathrm{CH_3\!-\!CH\!-\!CH\!-\!CH_3} \\
| & \\
\mathrm{CH_3} & \\
\text{2,2-dimethybutane} & \text{2,3-dimethylbutane}
\end{array}
$$

Major products from 2,3-dimethylbutane are (77), H_2 (G 3.77), CH_4 (G 1.01), C_3H_6 (G 2.28), and C_3H_8 (G 2.63) indicative of bond rupture between carbons 2 and 3 and, to a lesser extent, between the terminal methyl groups and the chain. Yields of C_6-alkene and dimer are 1.18 and 0.8 G units, respectively, from the 2,2-isomer and 0.35 and 0.22 G units, respectively, from the 2,3-isomer, in each case lower than the corresponding yield from the isomeric straight chain hydrocarbon, hexane.

Polymers

Irradiation of hydrocarbon polymers (e.g., polyethylene, polypropylene) produces C—C bond rupture (degradation) and C—C bond formation (cross-linking) as with the lower molecular weight hydrocarbons, and the main interest arises from the small amount of chemical change required to produce a marked change in the physical properties of the polymer. A polymer becomes, effectively, one large molecule when an average of one cross-link per molecule has been formed. If each polymer molecule is composed of several thousand monomer units, this is equivalent to less than 0.1% chemical change, and changes well below this are sufficient to cause changes in the viscosity of the polymer (or its solutions). Other physical changes consequent on cross-linking include a lower solubility in solvents and a higher melting point. Thus unirradiated polyethylene softens in the range 70 to 90°C and melts to a viscous liquid at about 115 to 125°C; after a dose of about 2 Mrad, the polymer can be taken to about 250°C without losing its shape, though above its usual melting point it becomes flexible and rubbery. Modification of the properties of polymers by irradiation is important commercially, and several examples are given in Chapter 10.

The facts that the melting point and the viscosity of solutions of polyethylene increase with irradiation is evidence that with this material cross-linking predominates over degradation. However cross-linking is accompanied by C—H bond rupture with the formation of hydrogen and vinylene (—CH=CH—) unsaturation with G values of about 3 and 1.5, respectively; G(cross-links) is of the order of 1.5. The chemical reactions occurring include the formation of radicals (which can be detected by esr spectroscopy) and probably ionic reactions. Thus the formation of cross-links in polyethylene can be accounted for by addition of radicals located on adjacent chains,

$$
\begin{array}{ccc}
-CH_2\dot{C}HCH_2- & & -CH_2CHCH_2- \\
& \longrightarrow & | \\
-CH_2\dot{C}HCH_2- & & -CH_2CHCH_2-
\end{array}
\qquad (8.71)
$$

or an ion-molecule reaction between adjacent chains (84),

$$
\begin{array}{ccc}
-CH_2\overset{+}{C}H_2CH_2- & & -CH_2CHCH_2- \\
& \longrightarrow & | \quad\quad + H_2 \\
-CH_2CH_2CH_2- & & -CH_2\overset{+}{C}HCH_2-
\end{array}
\qquad (8.72)
$$

The ion-molecule reaction requires only that two polymer chains be close together at the point where one becomes ionized, but the radical reaction (Eq. 8.71) requires two radical centers to be positioned opposite each other on adjacent chains. Although the latter situation appears inherently less probable, it might arise in several ways. For example, if neutralization of a positive ion formed a hot hydrogen atom which abstracted hydrogen from a neighboring polymer chain, or if two radicals were formed close together in a spur, or the radicals could migrate along the polymer chain until they encountered a radical on a neighboring chain. It is not inconceivable that both radical and ionic mechanisms contribute to the formation of cross-links.

Unsaturation probably arises from a disproportionation reaction,

$$
\begin{array}{ccc}
-CH_2\overset{\cdot}{C}HCH_2- & & -CH_2CH_2CH_2- \\
& \longrightarrow & \\
-CH_2\overset{\cdot}{C}HCH_2- & & -CH=CH-CH_2-
\end{array}
\qquad (8.73)
$$

which competes with reaction 8.71, and by unimolecular processes such as elimination of hydrogen from an excited molecule. Irradiation of polyethylene in the presence of oxygen, or exposure of the irradiated material to air or oxygen after irradiation, gives rise to carbonyl compounds (which can be detected by their infrared absorption) and water. The reaction is probably addition of oxygen to polymer radicals followed by disproportionation of the peroxy radicals produced, as in the case of cyclohexane (Eqs. 8.51 and 8.54).

While polyethylene and polypropylene primarily cross-link upon irradiation, degradation predominates when polyisobutylene (I)

$$
[-CH_2-\underset{\underset{CH_3}{|}}{\overset{\overset{CH_3}{|}}{C}}-]_n \qquad I
$$

is irradiated. The difference between the molecules is the presence of a tertiary carbon atom in isobutylene which, as with other branched chain hydrocarbons, leads to increased scission of the carbon chain upon irradiation. This behavior is not limited to the hydrocarbon polymers and in general polymers whose formula is $[-CH_2-CHR-]_n$ cross-link upon irradiation while those whose formula is $[-CH_2-C(CH_3)R-]_n$ degrade. Polymers that cross-link include polyethylene, polypropylene, poly(vinyl chloride), polyacrylonitrile, polyacrylates, polystyrene, and natural rubber. Polymers that degrade on irradiation include polyisobutylene, polytetrafluoroethylene, and the polymethacrylates. Radicals formed in the polymers that degrade undergo disproportionation rather than combination reactions; e.g., radicals formed by the rupture of the polymethylmethacrylate chain might react.

$$
-CH_2-\underset{\underset{CO_2CH_3}{|}}{\overset{\overset{CH_3}{|}}{C}}\cdot \quad + \quad \cdot CH_2-\underset{\underset{CO_2CH_3}{|}}{\overset{\overset{CH_3}{|}}{C}}- \quad \longrightarrow
$$

$$
-CH=\underset{\underset{CO_2CH_3}{|}}{\overset{\overset{CH_3}{|}}{C}} \quad + \quad CH_3-\underset{\underset{CO_2CH_3}{|}}{\overset{\overset{CH_3}{|}}{C}}- \qquad (8.74)
$$

Reaction is not limited to the main chain in the case of substituted polymers, degradation of the ester groups of polymethylmethacrylate giving gaseous products (CO_2, CH_4) in a similar manner to the radiolysis of other esters.

Cyclohexene

The major products from cyclohexene are listed in Table 8.6. Minor products detected in the γ-radiolysis and formed with G values of less than 0.1 (19) are methane, 1,3- and 1,4-cyclohexadiene, 1,5-hexadiene, and benzene. The yields are independent of total dose to 2×10^{20} eV g^{-1} and of dose rate from 2.2×10^{15} to 1.7×10^{21} eV g^{-1} min^{-1}, but are LET dependent.

TABLE 8.6   *Radiolysis of Liquid Cyclohexene*

	G(Product)		
Product	^{60}Co γ-Ray (ref. 19)	^{60}Co γ-Ray (ref. 87)	1.5-MeV α-Ray (ref. 87)
H_2	1.28	1.26	2.95
Cyclohexane	0.95	0.95	0.3
2,2'-Bicyclohexenyl (II)	1.94	1.8	0.4
3-Cyclohexylcyclohexene (III)	0.60	0.5	0.5
Bicyclohexyl (IV)	0.23	0.15	0.1
Unidentified dimer	0.22		
Polymer, including dimer (as C_6 units)	8.9	9.8	6.1

The more obvious differences between the radiolysis of cyclohexene and the saturated hydrocarbon cyclohexane are that product yields from cyclohexene do not vary with increasing absorbed dose and, with the exception of the dimer and polymer yields, are all smaller than the corresponding yields from cyclohexane. Absence of a dose dependency is attributed to the fact that cyclohexene is itself unsaturated so that formation of unsaturated products has little effect upon the reactions taking place. The different product distribution is accounted for if excited species (ions, molecules, or radicals) from the alkene suffer less fragmentation than the corresponding species from the alkane, and if alkene radicals take part in combination rather than disproportionation reactions. Radical yields in cyclohexene have been estimated using $^{14}CH_3$, generated by electron capture by $^{14}CH_3I$, as a radical scavenger (88). The radical present in largest amount (G 3.4) is the 3-cyclohexenyl radical, formed by loss of hydrogen from a carbon adjacent to the double bond, but a smaller yield (G 1.6) of cyclohexyl radicals is also formed. Dissociation of an excited cyclohexene molecule can give the allylic radical,

$$C_6H_{10}^* \rightarrow \cdot C_6H_9 + H \qquad (8.75)$$

while both radicals are formed when hydrogen atoms react with cyclohexene,

$$H + \langle \rangle \longrightarrow H_2 + \langle \rangle \cdot \qquad (8.76)$$

$$H + C_6H_{10} \longrightarrow \cdot C_6H_{11} \qquad (8.53)$$

Combination of the radicals gives the observed dimeric products:

$$\langle \rangle \cdot + \langle \rangle \cdot \longrightarrow \langle \rangle-\langle \rangle \qquad (8.77)$$

II

$$\langle \rangle \cdot + \langle \rangle \cdot \longrightarrow \langle \rangle-\langle \rangle \qquad (8.78)$$

III

$$\langle \rangle \cdot + \langle \rangle \cdot \longrightarrow \langle \rangle-\langle \rangle \qquad (8.30)$$

IV

Cyclohexane may be formed in part by disproportionation of cyclohexyl radicals,

$$2 \cdot C_6H_{11} \rightarrow C_6H_{12} + C_6H_{10} \qquad (8.29)$$

but the ratio of disproportionation to combination for this radical is known to be 1.1 so that the measured yield of bicyclohexyl (G 0.23) limits the yield of cyclohexane from this source to 0.25 G units. Disproportionation of the radicals taking part in reaction 8.78 was shown to be improbable by Wakeford and Freeman (19), who suggest that the remaining cyclohexane may be formed by ionic reactions such as,

$$C_6H_{10}^+ + C_6H_{10} \rightarrow \cdot C_6H_9 + C_6H_{11}^+ \qquad (8.79)$$

$$C_6H_{11}^+ + C_6H_{10} \rightarrow C_6H_{12} + C_6H_9^+ \qquad (8.80)$$

Polymer may result from successive ion-molecule reactions, since the yield of polymer is not dose-rate dependent, as it would be if the polymer were formed by successive addition of cyclohexene to a radical in competition with the interradical reactions 8.77, 8.78, and 8.30.

Increasing LET (Table 8.6) increases the yield of hydrogen but reduces the yields of dimeric and polymeric products. The effect may be due in part to competition between the radical reactions already given and

$$2 H \rightarrow H_2 \qquad (8.58)$$

$$H + \cdot C_6H_9 \rightarrow C_6H_{10} \qquad (8.81)$$

in the tracks of high-LET particles, but this cannot account for the increased yield of hydrogen entirely, and Burns and Winter (87) have suggested that competition between

$$C_6H_{10}^* \rightarrow \cdot C_6H_9 + H \tag{8.75}$$

and

$$2\,C_6H_{10}^* \rightarrow H_2 + \text{unsaturated products} \tag{8.82}$$

also occurs; reaction 8.82 would be favored by the higher concentration of excited molecules in the tracks.

Other Unsaturated Hydrocarbons

Relatively few alkenes have been irradiated, but the limited data available suggest that others behave in a similar manner to cyclohexene, giving lower yields of hydrogen (89) than the corresponding alkane but larger yields of dimeric and polymeric products; very little rupture of the carbon chain occurs. The most abundant radical produced is an allylic species, formed by loss of hydrogen from a carbon atom adjacent to the double bond, followed in importance by a radical in which a hydrogen atom has added to the double bond. With the terminal olefins, $R\!-\!CH\!=\!CH_2$, an ionic mechanism results in addition to the double bond and the formation of dimers (90, 91):

$$R\!-\!CH\!=\!CH_2 \,\leadsto\, (R\!-\!\dot{C}H\!-\!C^+H_2 \leftrightarrow R\!-\!C^+H\!-\!\dot{C}H_2) + e^- \tag{8.83}$$

$$R\!-\!\dot{C}H\!-\!C^+H_2 + R\!-\!CH\!=\!CH_2 \rightarrow$$
$$R\!-\!\dot{C}H\!-\!CH_2\!-\!CH_2\!-\!C^+H\!-\!R \tag{8.84}$$

and

$$R\!-\!C^+H\!-\!\dot{C}H_2 + R\!-\!CH\!=\!CH_2 \rightarrow R\!-\!\overset{\displaystyle |}{\underset{\displaystyle \cdot CH_2}{CH}}\!-\!CH_2\!-\!C^+H\!-\!R \tag{8.85}$$

Neutralization of the positive ions by an electron, or a negative ion, gives neutral molecules that rearrange to monoolefins. Ayscough and Evans (92) studied the esr spectra of solid olefins that had been irradiated at $-196°C$ and also concluded that straight-chain terminal olefins react mainly by ion-molecule reactions, although they found that most other olefins form an allylic radical by proton transfer (Eq. 8.79).

Isobutene forms a long-chain polymer when irradiated at temperatures below $-50°C$, apparently by an ionic mechanism, since the reaction is not inhibited by radical scavengers. Furthermore, polymerization is not induced by chemically produced radicals but can be initiated by Friedel-Crafts catalysts (e.g., $AlCl_3$ or $FeCl_3$ with traces of water) or by irradiation with ultraviolet light with an energy greater than the ionization potential of of isobutene (93). The suggested mechanism (94) is

$$C_4H_8 \rightsquigarrow C_4H_8^+ + e^- \tag{8.86}$$

$$C_4H_8^+ + C_4H_8 \rightarrow C_4H_9^+ + \cdot C_4H_7 \tag{8.87}$$

followed by the chain reaction

$$\begin{array}{cccc} & CH_3 & & CH_3 & & CH_3 & CH_3 \\ & | & & | & & | & | \\ CH_3-C^+ & + & CH_2{=}C & \longrightarrow & CH_3-C-CH_2-C^+ & \text{etc.} \\ & | & & | & & | & | \\ & CH_3 & & CH_3 & & CH_3 & CH_3 \end{array} \tag{8.88}$$

in which successive isobutene molecules add to the growing chain. Termination occurs when the positive ion is neutralized by an electron or negative ion.

Isomerization of *cis* (**V**) and *trans* (**VI**) 2-butene to a mixture of the two forms occurs *via* a triplet excited state (in which the molecule is free to rotate about the central C—C bond) when either isomer is irradiated in a hydrocarbon solvent (e.g., 95).

$$\begin{array}{cc} H_3C \diagdown \qquad \diagup CH_3 & H_3C \diagdown \qquad \diagup H \\ \qquad C{=}C & \qquad C{=}C \\ H \diagup \qquad \diagdown H & H \diagup \qquad \diagdown CH_3 \end{array}$$

V	**VI**

Since the process involves transfer of energy from a triplet excited state of the solvent to the 2-butene, the extent of isomerization can be used to estimate the number of triplet excited solvent molecules present. Isomerization does not occur when pure *cis* and *trans* 2-butene are irradiated at $-88°C$ (96), since the reactive species involved are 1-methylallyl radicals, which retain their configuration, and not excited triplet states of the molecule.

Aromatic Hydrocarbons

Aromatic compounds (97) are very much more resistant to radiation damage than the alkanes and alkenes (Tables 8.2 and 8.7). Their stability stems from the fact that the π-electrons of the aromatic system are shared by the entire system and excitation of a π-electron does not represent a concentration of excitational energy at a particular location in the molecule. Furthermore, triplet excited states formed by excitation of π-electrons are normally the lowest excited states of aromatic molecules and, in the liquid, are readily reached by internal conversion from higher triplet excited states and by internal conversion and intersystem crossing from higher singlet excited states. Thus a large fraction of the excitational energy received by an aromatic molecule is channeled to relatively low-energy triplet excited states which have a low probability of dissociation. The aromatic system can also act as an energy sink for energy absorbed by other parts of the molecule and yields of hydrogen and low molecular weight products from substituted benzenes are generally no greater than those from benzene itself (Table 8.7).

T A B L E 8.7   *Radiolysis of Aromatic Hydrocarbons*

Compound (M)	Phase	Type of Radiation	G(Product)						Ref.
			H_2	CH_4	C_2H_2	C_2H_4	C_2H_6	$-M$	
Benzene	Vapor	α	0.30	0.01	0.42	0.02	0.006	4.8	98
		γ	0.011		0.11	0.05	0		99
	Liquid	$^{10}B(n, \alpha)^7Li$	0.57		0.26			2.1	1, 98
		γ	0.039	0.019				0.94	1, 98
Toluene	Liquid	α	0.58					1.8	101
		γ	0.11	0.008	0.002	0.002		1.1	101
p-Xylene	Liquid	γ	0.21	0.014	0.003	0.0002	0.0001	1.1	102
Ethylbenzene	Liquid	γ	0.16	0.026	0.002	0.007	0.006	1.57	102, 103
Isopropyl benzene	Liquid	γ	0.18	0.09	0.004	0.002	0.004	1.8	104
Mesitylene	Liquid	e^-	0.24	0.02					105
Biphenyl	Liquid	γ	0.0067	0	0.0003		0	~ 0.007	106

The major product from benzene is a viscous yellow liquid containing higher molecular weight compounds which is generally referred to as "polymer." This has been shown (107) to contain C_{12} compounds (biphenyl, phenylcyclohexadiene, phenylcyclohexene, and nonaromatic bicyclic compounds), C_{18} compounds (hydrogenated terphenyls), and higher molecular weight material of similar composition; the average molecular weight of the mixture increases with increasing absorbed dose. Small quantities of 1,3- and 1,4-cyclohexadiene (G 0.025 to 0.11) are also formed (100, 108).

The three most abundant ions in the mass spectrum of benzene are the parent ion $C_6H_6^+$ (66%), $C_4H_4^+$ (13%), and $C_6H_5^+$ (9%) so that the preferred fragmentation reactions appear to be

$$C_6H_6^{+*} \rightarrow C_6H_5^+ + H \tag{8.89}$$

and

$$\rightarrow C_4H_4^+ + C_2H_2 \tag{8.90}$$

Aromatic compounds readily form negative ions when irradiated in solution (e.g., anthracene, naphthalene, and biphenyl are employed as electron scavengers in pulse radiolysis experiments) and will presumably do so when the pure material is irradiated, at least in the liquid phase. The most probable neutralization reaction in liquid benzene is therefore

$$C_6H_6^+ + C_6H_6^- \rightarrow C_6H_6^* + C_6H_6 \tag{8.91}$$

Negative ions release less energy than electrons in neutralization reactions, and the probability that the products of the reaction will dissociate is therefore less. In the case of benzene, most of the excited neutralization products appear to degrade to the lowest triplet excited state without dissociation. Other substances may be more efficient electron scavengers than benzene and $0.4\ M\ N_2O$, for example, gives $G(N_2)$ about 2.5 in benzene due to electron scavenging.

Benzene is not readily dissociated photochemically except by short wavelength light in the vapour phase, where fulvene (or other high-energy isomers), acetylene, methane, and polymeric material are among the products (109). Photolysis of benzene solutions of olefins results in the formation of adducts (e.g., 110); a similar reaction between excited benzene molecules and unsaturated radiolysis products may be partly responsible for the formation of polymer, and for the low yields of unsaturated products, in irradiated benzene. The triplet excited state of benzene has not been observed directly by pulse radiolysis, but several estimates of the yield of triplet excited benzene have been made by adding a solute with a lower triplet energy level (e.g., anthracene or naphthalene) and observing the yield of solute triplet formed by energy transfer. Yields estimated in this way range up to about 4 G units (61); triplet yields of 4 to 5 G units have been estimated from the *cis-trans* isomerization of 2-butene (111) and 2-octene (112) in irradiated benzene. γ-Radiolysis of benzene solutions of 3,5-cycloheptadienone, which gives CO and hexatriene by energy transfer from singlet excited benzene and an isomer by energy transfer from triplet excited benzene, has given G(singlet) = 1.45 and G(triplet) = 4.0 for benzene and a value of 0.58 (from photochemical experiments) for the intersystem crossing efficiency from singlet to triplet excited state (113). The maximum yield of triplet excited molecules, assuming all singlet excited molecules undergo intersystem crossing with the measured efficiency, is then 4.85. Radical yields are lower than either singlet or triplet yields, and, even allowing for the formation of molecular products by dissociation or reaction of the excited molecules, it is clear that a large proportion of the excited molecules must dissipate their energy without forming products.

Optical absorption spectra attributed to cyclohexadienyl radicals have been observed in the pulse radiolysis of benzene (114, 115); phenyl radicals were not observed, probably because they rapidly add to benzene. Radicals of the cyclohexadienyl type may be formed by the addition of hydrogen atoms or phenyl radicals to benzene,

$$H + \langle\!\!\!\bigcirc\!\!\!\rangle \longrightarrow \begin{array}{c} H \\ \diagdown \\ H \end{array}\!\!\!\Big\langle\!\!\!\begin{array}{c} \\ \bullet \end{array}\!\!\!\Big\rangle \qquad (8.92)$$

$$\cdot C_6H_5 + \text{(structure)} \longrightarrow \text{(structure)} \qquad (8.93)$$

Scavenger and other experiments give a total radical yield of about 0.8 cyclohexadienyl and phenylcyclohexadienyl radicals for benzene (5). Radical scavengers have little effect on the yield of hydrogen (116) but irradiation of mixtures of C_6H_6 and C_6D_6 gives a significant yield of HD (117), so that it appears that hydrogen is formed in a bimolecular reaction that does not involve hydrogen atoms. Burns (1, 118, 119) has suggested that this may be reaction of two excited molecules, which competes with collisional deactivation by the solvent, i.e.,

$$C_6H_6^* + C_6H_6^* \rightarrow H_2 + C_{12}H_{10} \qquad (8.94)$$

or

$$\rightarrow H_2 + \text{radicals} \qquad (8.95)$$

competing with

$$C_6H_6^* + C_6H_6 \rightarrow 2\,C_6H_6 \qquad (8.96)$$

Competition between reactions 8.94 and 8.95 and reaction 8.96 can also account for higher yields of hydrogen and $G(-\text{benzene})$ at high LET where collision between two excited molecules will be more probable. Disproportionation and combination of cyclohexadienyl and phenylcyclohexadienyl radicals gives a number of isomeric cyclohexadienes and partially hydrogenated bi, tri, and tetra-phenyls such as are found in the "polymer" fraction of the radiolysis products, e.g.,

$$2 \; \text{(structure)} \longrightarrow \text{(structure)} + \text{(structure)} \qquad (8.97)$$

$$\longrightarrow \text{(structure)} \qquad (8.98)$$

$$\text{(structure)} + \text{(structure)} \longrightarrow \text{(structure)} + \text{(structure)} \qquad (8.99)$$

$$\longrightarrow \text{(structure)} + \text{(structure)} \qquad (8.100)$$

$$\longrightarrow \text{(structure)} \qquad (8.101)$$

etc.

Acetylene yields are not affected by fairly high concentrations of iodine (116) and are believed to result from the dissociation of excited ions (Eq. 8.90) or molecules,

$$C_6H_6{}^* \rightarrow 3\,C_2H_2 \qquad (8.102)$$

The other low molecular weight products are probably formed in a similar manner.

Radiolysis of the alkyl benzenes is similar to the radiolysis of benzene, and product yields are low (Table 8.7). Fission of bonds in the alkyl groups increases as these groups become larger, and it appears that C—C bonds β to the ring (i.e., C_6H_5C—C) are split more frequently than other C—C bonds (105, 120). However, a large part of the energy absorbed by the side chain must be transferred to the aromatic system and dissipated, since product yields are substantially lower than would be expected if there was no interaction between side chain and ring system. The efficiency of such energy transfer is reduced as the distance between the site of energy absorption and the phenyl group is increased (121).

Biphenyl and the higher polyphenyls (terphenyl, quaterphenyl, etc.) and the polynuclear hydrocarbons (naphthalene, anthracene, etc.) are particularly resistant to radiation damage and are used, or have been suggested for use, in nuclear reactors as coolants and moderators. As coolants they have some advantages over water and liquid metals; e.g., they do not corrode metals and have low vapor pressures, so that expensive corrosion-resistant construction materials and high-pressure systems can be avoided. Furthermore, little radioactivity is induced in the hydrocarbon if it is pure, and it will not react dangerously with uranium if a fuel element should fail at high temperature. However, disadvantages include a certain amount of radiation decomposition, low heat conductivity, and inflammability. The radiation resistance of the polyphenyls decreases at temperatures above about 400°C, setting an upper limit to the working temperature. The same considerations apply when organic compounds are used to moderate (i.e., slow down) fast neutrons in a reactor, but here larger quantities of the organic material are exposed to intense radiation and the radiation-induced decomposition is of even more concern. The polyphenyls are preferred over the polynuclear hydrocarbons, since the latter tend to form solid, coke-like, pyrolysis products while the polymers formed by pyrolysis or radiolysis of the polyphenyls are resinous and remain in solution.

Mixtures of Hydrocarbons

If a mixture of compounds is irradiated, the fraction of the total absorbed dose absorbed by molecules of each component is proportional to the fraction by weight of that component and to a mean collisional mass stopping power for the component and the ionizing particles present; mass stopping powers are energy dependent so that the mean value used must take into account the spectrum of particle energies in the system. In practice, the appropriate mean mass stopping powers are not readily determined and an approximation is usually used, the most common being to assume that the energy is partitioned in proportion to the electron fraction (ε) of each component present (the electron fraction is given by the right-hand side of Eq. 3.43, p. 115, and is the fraction of the total number of electrons present contributed by that component). Then for a mixture of two components, A and B, which both form a product P upon irradiation,

$$G(P) = G(P)_A \varepsilon_A + G(P)_B \varepsilon_B \qquad (8.103)$$

where $G(P)_A$ and $G(P)_B$ are the yields of product for pure A and pure B respectively, and ε_A and ε_B are the electron fractions of A and B. This expression is often referred to as the "mixture law," although it is important to remember that it is an approximation rather than a law.

Quite apart from considerations of the proportions in which the absorbed energy is divided between the two components, the mixture law also assumes that the subsequent behavior of the excited and ionized species produced and their reaction products are not altered by the presence of the other component(s). That is, it is assumed that energy and charge transfer, electron scavenging by one component, new ion-molecule reactions, and new radical reactions do not distort the product yield in favour of one or other of the components. These conditions are difficult to meet except in the case of closely related compounds, and there are many more examples of deviations from the law than of systems in which it is obeyed (122). Examples of the latter include hydrocarbons and their deuterated analogues (e.g., benzene + benzene-d_6) (100, 123), mixtures of hydrocarbons such as cyclopentane and cyclohexane (the yields of cyclopentyl and cyclohexyl radicals, as determined by iodine scavenging, are a linear function of the electron fraction) (124), and benzene and toluene (125, 126), and mixtures of benzene with pyridine in which the yield of polymer is a linear function of the mixture composition (127).

The classic example of a system that deviates from the mixture law is a mixture of benzene and cyclohexane. Schoepfle and Fellows (128) were the first to report that the mixture gave a lower yield of gaseous products upon irradiation than would be expected from irradiation of the two components separately. This observation has been amply confirmed by subsequent work, a plot of hydrogen yield against composition for such mixtures having the form shown in Fig. 8.1 (e.g., 125). If the mixture law were obeyed, the plot would be a straight line (the broken line in the diagram) between the values

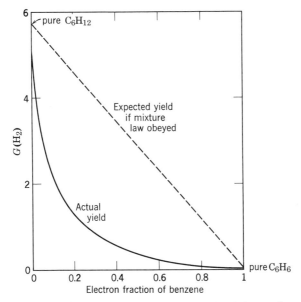

FIGURE 8.1 *Variation of* $G(H_2)$ *with the composition of cyclohexane-benzene mixtures* (*fast electron irradiation*).

of $G(H_2)$ for cyclohexane and benzene. Yields of other products from cyclohexane such as cyclohexene, cyclohexylhexene, and bicyclohexyl follow a similar curve to that for hydrogen (129, 130) while new products, such as phenylcyclohexane and dicyclohexadienyl, which are not formed from either pure component are also formed (129–131). When cyclohexane containing a few percent of C_6D_{12} is irradiated, the hydrogen formed contains both D_2 from unimolecular dissociation of C_6D_{12} and HD from bimolecular reactions in which the hydrogen atoms come from two different molecules (reactions 8.40 to 8.43). Addition of benzene to the mixture reduces the total hydrogen yield without changing the D_2-HD ratio, suggesting that the benzene reacts with a common precursor of both D_2 and D, since it is unlikely that benzene would reduce the extent of two independent reactions by exactly the same amount (62, 123). Precursors that might react with benzene are the cyclohexane ion, $C_6H_{12}^+$, and an excited molecule, $C_6H_{12}^*$, formed directly or by ion neutralization; both are precursors of molecular hydrogen and hydrogen atoms and of cyclohexene and cyclohexyl radicals, so that scavenging of either would bring about the observed effects. Scavenging of the ion

$$C_6H_{12}^+ + C_6H_6 \rightarrow C_6H_{12} + C_6H_6^+ \qquad (8.104)$$

is energetically feasible, since the ionization potential of cyclohexane (9.9 V) is greater than that of benzene (9.2 V), although relatively high concentrations of benzene might be required to compete with geminate ion recombination (Eq. 8.23). Energy transfer from excited cyclohexane,

$$C_6H_{12}{}^* + C_6H_6 \rightarrow C_6H_{12} + C_6H_6{}^* \qquad (8.105)$$

is also feasible energetically as benzene has lower lying excited states than cyclohexane. If the lifetime of $C_6H_{12}{}^*$ is longer than that of $C_6H_{12}^+$ (which is about 10^{-11} sec), lower concentrations of benzene might be effective in reaction 8.105 than in reaction 8.104; both charge and energy transfer are very rapid processes. However evidence in favor of both charge transfer and energy transfer has been published (e.g., 10, 57, 132–134), and it is possible that both mechanisms contribute. The decreased decomposition of the cyclohexane is accompanied by some increase in benzene decomposition; thus Burton and Patrick (135) found a relatively higher yield of D_2 from cyclohexane–deuterobenzene mixture and Manion and Burton (125) a higher yield of acetylene from cyclohexane–benzene mixture than from pure benzene. However, a large part of the energy transferred to the benzene is dissipated without producing any permanent chemical change. Earlier suggestions that the reduction in hydrogen yield might be due entirely to hydrogen atom scavenging by benzene (Eq. 8.92) have not been substantiated, although hydrogen atom and cyclohexyl radical scavenging by benzene undoubtedly occur and contribute to the effects observed (136, 137).

Benzene reduces the hydrogen yield from cyclohexane in the vapour phase, but experiments with mixtures containing C_6D_{12} have shown that in the vapor it does not inhibit the unimolecular process giving D_2 and C_6D_{10} although it does interfere with the bimolecular process forming HD, probably via an ionic mechanism (138). Results in solid mixtures of cyclohexane and benzene are similar to the liquid phase results if the solid is prepared by slow condensation of the premixed vapors on a cold surface so that a homogeneous mixture is obtained, but the solid system may obey the mixture law if the solid is prepared by cooling a liquid mixture as the components of the mixture can become separated under these conditions (139). Phenyl groups can also reduce radiation damage to cyclohexyl groups if they are present in the same molecule. This is more effective than simple mixing, since the yield of hydrogen from phenylcyclohexane (G 0.26) is lower than that (G 0.57) from a mixture of benzene and cyclohexane of the same molecular composition (140).

Energy and charge transfer effects are not restricted to mixtures of aliphatic and aromatic hydrocarbons. Hardwick (141), for example, found reduced hydrogen yields when hexane was mixed with compounds such as neohexane,

2,3-dimethylbutane, isopropyl alcohol, and diisopropyl ether. In each case
the added compound has a lower ionization potential than hexane, and it
was concluded that charge transfer was occurring,

$$C_6H_{14}^+ + S \rightarrow C_6H_{14} + S^+ \qquad (8.106)$$

Experiments by Dyne and Stone and their colleagues (134, 142–144) in
which binary mixtures of normal and deuterated hydrocarbons were irradiated led to the same conclusion, yields of D_2 and HD from the deuterated
compound being greater if it had the lower ionization potential but reduced
if the deuterated compound had the higher ionization potential of the two
hydrocarbons.

Energy transfer from solvent to (fluorescent) solute is employed in scintillation radiation detectors. Radiation energy absorbed by the solvent,
which is generally an aromatic hydrocarbon such as benzene, toluene, or
xylene, is transferred to the solute, generally a polynuclear or polyphenyl
aromatic compound (e.g., naphthalene, anthracene, biphenyl, terphenyl, or
their derivatives), which subsequently fluoresces:

$$C_7H_8 \rightsquigarrow C_7H_8^* \qquad (8.107)$$

$$C_7H_8^* + S \rightarrow C_7H_8 + S^* \qquad (8.108)$$

$$S^* \rightarrow S + h\nu \qquad (8.109)$$

Solid solutions (e.g., *p*-terphenyl in polystyrene) and single crystals of pure
phosphors (e.g., anthracene, stilbene, and *p*-terphenyl) can also be used.
Only a small part of the absorbed energy is emitted as light, and radiation
damage to the solvent is not significantly reduced (145).

Since addition of benzene to cyclohexane reduces the extent of radiation-
induced decomposition of the cyclohexane, the benzene is often said to
"protect" the cyclohexane against radiation damage. Protection in this
instance is largely the result of the physical processes of charge and energy
transfer and may be described as "physical" protection to distinguish it
from the radical scavenging and repair mechanisms that result in protection
in aqueous solutions (p. 342); the latter can be described as "chemical"
protection. Both physical and chemical protection can occur in the same
system but generally the physical mechanisms are applicable to mixtures
containing relatively large amounts of each component while chemical
protection is more applicable to dilute solutions in which the protective
agent is not present in high enough concentration to affect the physical
processes involving ions and excited molecules. Chemical protection in
another sense results if the added substance reacts chemically with the
material to mask a group that is particularly radiation sensitive. Thiol

groups, for example, are very susceptible to radiation damage but can be protected in the compounds cysteine and glutathione by reacting the amino acids with carbonyl compounds (e.g., glyoxal, pyruvic acid, and diacetyl) (146), forming a complex, e.g., for cysteine,

$$
RCHO + \begin{array}{c} HS-CH \\ | \\ H_2N-CH \\ | \\ COOH \end{array} \longrightarrow RCH \begin{array}{c} {}^{\diagup}S-CH \\ | \\ {}_{\diagdown}HN-CH \\ | \\ COOH \end{array} + H_2O \qquad (8.110)
$$

which is less sensitive to radiation than the free thiol group.

ORGANIC HALOGEN COMPOUNDS

Aliphatic halides, with the exception of the fluorides, are among the organic compounds most sensitive to radiation. In these compounds the carbon–halogen bonds are weaker than either carbon–carbon or carbon–hydrogen bonds, and the main effect of radiation is to break a carbon–halogen bond to give an organic free radical and a halogen atom. The high electron affinity of the halogens permits electron attachment reactions such as

$$
RX + e^- \rightarrow R\cdot + X^- \qquad (8.111)
$$

and

$$
\rightarrow R^+ + X^- + e^- \qquad (8.112)
$$

where X^- is a halogen ion, in addition to the more usual reactions listed at the beginning of this chapter. Fluorides form a class apart, since C—F bonds are generally stronger than the other bonds in the molecule, and irradiation leads more frequently to scission of C—C bonds.

The halogen atoms show a marked diminution in reactivity in the series $F > Cl \gg Br > I$. For example, chlorine atoms can abstract hydrogen from organic molecules, bromine atoms do so less readily, and iodine atoms not at all.[3] Thus the initial radiolysis products from chloro-compounds tend to include hydrogen chloride, whereas iodo-compounds tend to give iodine, and bromides give on occasion both hydrogen bromide and bromine.

---

[3] Although iodine atoms will not abstract hydrogen from an organic molecule, iodine may displace hydrogen under some circumstances. Willard and his colleagues (147) have shown that hot iodine atoms formed by the nuclear reaction $^{127}I(n, \gamma)^{128}I$ will displace hydrogen from methane,

$$
I + CH_4 \rightarrow CH_3I + H
$$

Iodine also displaces hydrogen when a mixture of iodine and methane is exposed to 184.9-nm light (148, 149),

$$
I_2{}^* + CH_4 \rightarrow CH_3I + HI
$$

Methyl Iodide

Irradiation of organic iodides liberates iodine, although a large fraction of the iodine formed may react with radicals to reform similar iodides. When the original iodide is reformed in this way, irradiation produces less chemical change than might be expected even though an appreciable number of molecules may have been dissociated in the course of the reaction. This is illustrated by the relatively low yields of radiolysis products from methyl (150) and ethyl (150, 151) iodides (Table 8.8), although irradiation of these iodides in the presence of radioactive iodine shows that for both compounds the exchange yield, G(radioactive iodide), is about 6 (152–154). The tracer experiments also show that most of the organic radicals formed when methyl and ethyl iodides are irradiated are methyl and ethyl radicals, respectively.

T A B L E 8.8 *γ-Radiolysis of Methyl and Ethyl Iodides*

	G(Product)		
Product	Methyl Iodide (Vapor, 25°C, 300 mm Hg; ref. 156)	Methyl Iodide (Liquid, ref. 150)	Ethyl Iodide (Liquid, ref. 150)
H_2	0.55	0.060	0.23
HI	0.13	~0.01	0.33
I_2	0.16	1.26	2.12
CH_4	2.90	0.77	0.01
C_2H_2	0.12		0.09
C_2H_4	0.065	0.081	2.20
C_2H_6	0.07	1.11	1.92
CH_2I_2	1.4		
C_4H_{10}			0.33

The effect of radiation on methyl iodide can be represented as causing fission of the C—I bond to give methyl radicals and iodine atoms,

$$CH_3I \rightsquigarrow \cdot CH_3 + I \tag{8.113}$$

which then combine to give the three major products, iodine, ethane, and methyl iodide. However, closer study of the radiolysis and related reactions and of the effect of added substances has shown that this is a considerable simplification (150).

The mass spectrum and ion-molecule reactions of methyl iodide have been studied by Hamill and his colleagues (150, 155). Major primary ions and their abundances relative to $CH_3I^+ = 100$ are, I^+ (53), CH_3^+ (28), CH_2I^+ (14), and CI^+ (5). The corresponding neutral fragments formed are $\cdot CH_3$, I, H, and H + H$_2$ respectively. Ion-molecule reactions proposed include

$$CH_3I^+ + CH_3I \rightarrow CH_3ICH_3^+ + I \qquad (8.114)$$

$$CH_3^+ + CH_3I \rightarrow CH_3ICH_3^+ \qquad (8.115)$$

$$CH_2I^+ + CH_3I \rightarrow I_2^+ + \cdot C_2H_5 \qquad (8.116)$$

$$I^+ + CH_3I \rightarrow I_2^+ + \cdot CH_3 \qquad (8.117)$$

Other mass spectrometric observations (157) have shown that the reaction

$$e^- + CH_3I \rightarrow \cdot CH_3 + I^- \qquad (8.118)$$

has a very low threshold energy and can be expected to occur in irradiated methyl iodide. The more important ion-neutralization reactions are therefore likely to be

$$CH_3ICH_3^+ + I^- \rightarrow 2\cdot CH_3 + 2I \qquad (8.119)$$
and

$$I_2^+ + I^- \rightarrow 3I \qquad (8.120)$$

The overall effect of these ionic processes is to produce methyl radicals and iodine atoms (which may also be formed by the dissociation of excited methyl iodide molecules) with smaller quantities of hydrogen atoms and ethyl radicals. Radical combination, both within the spurs and following radical diffusion into the bulk of the liquid, gives mainly methyl iodide, ethane, and iodine,

$$\cdot CH_3 + I \rightarrow CH_3I \qquad (8.121)$$

$$2\cdot CH_3 \rightarrow C_2H_6 \qquad (8.122)$$

$$2I \rightarrow I_2 \qquad (8.123)$$

Methyl iodide is also formed by reaction of methyl radicals with iodine as the radiolysis products accumulate in the liquid,

$$\cdot CH_3 + I_2 \rightarrow CH_3I + I \qquad (8.124)$$

Methane is formed by

$$[H + CH_3I] \rightarrow [HI + \cdot CH_3] \rightarrow CH_4 + I \qquad (8.125)$$

which occurs largely within the spurs and, probably, *via* hot methyl radicals (158),

$$\cdot CH_3^* + CH_3I \rightarrow CH_4 + \cdot CH_2I \qquad (8.126)$$

The latter reaction is believed to be the main source of methane in the vapor phase radiolysis of methyl iodide (156), where loss of excitational energy by collision is slower. Several reactions have been suggested as the source of ethylene in the liquid phase radiolysis (150, 159, 160),

$$\cdot C_2H_5 + I \rightarrow C_2H_4 + HI \tag{8.127}$$

$$2 \cdot CH_2I \rightarrow C_2H_4 + I_2 \tag{8.128}$$

$$CH_2 + \cdot CH_3 \rightarrow C_2H_4 + H \tag{8.129}$$

Methylene (CH_2) has been identified in the photolysis of methyl iodide (160) and is a possible precursor of acetylene,

$$2CH_2 \rightarrow C_2H_2 + H_2 \tag{8.130}$$

Hydrogen is formed by fragmentation of the excited parent ion and is not a major product in the liquid phase radiolysis.

Radiolysis of methyl iodide vapor is distinguished from the liquid phase radiolysis by the formation of considerably more methane and methylene iodide at the expense of the iodine and ethane yields. The higher yields of methane have been attributed to reaction of hot methyl radicals in the vapor (Eq. 8.126) (156), and since $\cdot CH_2I$ radicals are formed concurrently, this also accounts for the formation of CH_2I_2,

$$\cdot CH_2I + I_2 \rightarrow CH_2I_2 + I \tag{8.131}$$

Initial yields of ethane in the vapor are quite high, but rapidly fall to the value shown in Table 8.8 as the products (i.e., I_2) build up in the system. The relatively low yields of ethane and iodine are therefore attributed to scavenging of thermal radicals (Eqs. 8.124 and 8.131) by iodine.

Higher alkyl iodides behave in a similar manner to methyl iodide upon irradiation, though additional reactions leading to unsaturated products are possible. Hamill and his colleagues (150) have suggested that ethylene is formed by reaction 8.127 in irradiated ethyl iodide in competition with the addition reaction

$$\cdot C_2H_5 + I \rightarrow C_2H_5I \tag{8.132}$$

and that ethane is the product of

$$\cdot C_2H_5 + HI \rightarrow C_2H_6 + I \tag{8.133}$$

Ethylene may also be formed by unimolecular dissociation of excited ethyl iodide (151, 154, 161, 162),

$$C_2H_5I^* \rightarrow C_2H_4 + HI \tag{8.134}$$

Isomerization of Alkyl Halides

Irradiation of certain of the alkyl bromides and chlorides can bring about isomerization in which the location of the halogen atom is changed; the carbon skeleton of the molecule remaining unaltered. This is illustrated in Table 8.9 for the butyl chlorides (163, 164), irradiation of the normal chloride giving a high yield of the secondary isomer and irradiation of the iso compound a high yield of the tertiary isomer, while radiolysis of the secondary and tertiary compounds produces little isomerization. The reactions are catalyzed by HCl and the *G* values are such that a chain reaction must be occurring. Similar reactions have been observed with normal propyl chloride (164, 165) and normal (166) and iso (167) butyl bromides with, in the latter

T A B L E 8.9   *γ-Radiolysis of Liquid Butyl Chlorides (Ref. 163)*

Product	Normal $CH_3(CH_2)_3Cl$	Secondary $CH_3CH_2CHClCH_3$	Iso $(CH_3)_2CHCH_2Cl$	Tertiary $(CH_3)_3CCl$
H_2	1.4	1.0	0.9	0.6
HCl	1.0	1.1	—	0.2
Butane	6.9	5.6	0	0
Isobutane	0	0	6.9	3.7
n-Butyl chloride	—	0	0	0
s-Butyl chloride	60	—	0	0
Isobutyl chloride	0	0	—	1.6
t-Butyl chloride	0	0	24	—
Dichlorobutanes	1.5	1.0	2.2	1.5

cases, very much higher *G* values. The isomerizations are attributed to free radical chain reactions, since they are inhibited by radical scavengers such as iodine and can be initiated by exposure to ultraviolet light, and it is suggested that the propagation steps involve either an intramolecular shift of a halogen atom (164, 166), i.e.,

$$R_2\dot{C}CH_2X \rightleftharpoons R_2CX\dot{C}H_2 \qquad (8.135)$$

or equilibria involving the isomeric radicals, an alkene, and a halogen atom (166, 167), i.e.,

$$R_2\dot{C}CH_2X \rightleftharpoons R_2C{=}CH_2 + X \rightleftharpoons R_2CX\dot{C}H_2 \qquad (8.136)$$

Accepting the latter proposal for the purposes of illustration, the radiation-induced isomerization of isobutyl to *t*-butyl bromide can be attributed (167, 168) to

INITIATION

$$(CH_3)_2CHCH_2Br \rightsquigarrow (CH_3)_2CH\dot{C}H_2 + Br \qquad (8.137)$$

$$(CH_3)_2CH\dot{C}H_2 + (CH_3)_2CHCH_2Br \rightarrow (CH_3)_3CH + (CH_3)_2\dot{C}CH_2Br \qquad (8.138)$$

$$(CH_3)_2CH\dot{C}H_2 + HBr \rightarrow (CH_3)_3CH + Br \qquad (8.139)$$

PROPAGATION

$$Br + (CH_3)_2CHCH_2Br \rightleftharpoons HBr + (CH_3)_2\dot{C}CH_2Br \qquad (8.140)$$

$$(CH_3)_2\dot{C}CH_2Br \rightleftharpoons (CH_3)_2C{=}CH_2 + Br \rightleftharpoons (CH_3)_2CBr\dot{C}H_2 \qquad (8.141)$$

$$(CH_3)_2CBr\dot{C}H_2 + HBr \rightarrow (CH_3)_3CBr + Br \qquad (8.142)$$

TERMINATION

$$2 \text{ radicals} \rightarrow \text{products} \qquad (8.143)$$

The reaction produces very high initial yields of *t*-butyl bromide ($G \sim 18,000$) and low steady state concentrations of isobutene, $(CH_3)_2C{=}CH_2$, and HBr. Yields of isobutane, $(CH_3)_3CH$, which is formed in the initial steps, and 1,2-dibromo-2-methylpropane (a termination reaction product) increase continuously with increasing absorbed dose.

Carbon Tetrachloride

Carbon tetrachloride represents one of the simpler examples of radiolysis, since the radicals produced have a rather limited choice of reaction. Only two products are found when pure carbon tetrachloride is irradiated, chlorine and hexachloroethane, which can be accounted for by the following radical reactions:

$$CCl_4 \rightsquigarrow \cdot CCl_3 + Cl \qquad (8.144)$$

$$\cdot CCl_3 + Cl \rightarrow CCl_4 \qquad (8.145)$$

$$\cdot CCl_3 + Cl_2 \rightarrow CCl_4 + Cl \qquad (8.146)$$

$$2 \cdot CCl_3 \rightarrow C_2Cl_6 \qquad (8.147)$$

$$2Cl \rightarrow Cl_2 \qquad (8.148)$$

Once a small amount of chlorine has accumulated it will act as a scavenger for the trichloromethyl radicals which escape from the spurs, and the observed yields, $G(Cl_2) = G(C_2Cl_6) = 0.65$ to 0.75 for γ-radiation (169, 170), will be those for the "molecular" processes occurring. The latter include the dissociation of excited ions and molecules to give Cl_2 and reactions 8.147 and 8.148 in the spurs.

The primary dissociation reaction (Eq. 8.144) may take place directly through the dissociation of excited molecules or indirectly through ionic intermediates. The latter are likely to the ions observed in the mass spectrometer, namely (with their relative abundances (171), CCl_3^+ (100), Cl^+ (27), CCl^+ (22.5), and CCl_2^+ (22), and the parent ion, CCl_4^+, which has been reported in the radiolysis of liquid CCl_4 (172). Carbon tetrachloride is an efficient electron scavenger,

$$e^- + CCl_4 \rightarrow CCl_4^- \rightarrow \cdot CCl_3 + Cl^- \tag{8.149}$$

so that the main ion-neutralization reactions are expected to be

$$CCl_4^+ + Cl^- \rightarrow CCl_4 + Cl \tag{8.150}$$

and

$$CCl_3^+ + Cl^- \rightarrow \cdot CCl_3 + Cl \tag{8.151}$$

$$\rightarrow CCl_4 \tag{8.152}$$

Smaller amounts of molecular and radical products may be formed *via* ion fragmentation and neutralization of the fragment ions.

An estimate of the number of $\cdot CCl_3$ radicals that escape from the spurs can be made by adding radioactive chlorine and measuring the amount incorporated into the carbon tetrachloride by reaction 8.146. Using this technique, Schulte (173) found an exchange yield of 3.5 G units for γ-radiation, indicating a total radical yield ($\cdot CCl_3 + Cl$) of about 7. When bromine was used as a radical scavenger, Bibler (174) found the following products and G values, CCl_3Br 7.1, $BrCl$ 7, C_2Cl_6 0.48, C_2Cl_4 0.07, and CCl_2Br_2 0.12, which correspond with a radical yield roughly twice as great as that found by Schulte. Other radical scavengers (e.g., 174) have given values that range from about 2.5 to 20 for $G(\cdot CCl_3)$, although the higher values are probably the result of both radical and excited molecule scavenging (175). Tetrachloroethylene is only observed as a product in the presence of radical scavengers, since it reacts readily with radicals and with Cl_2; it must be formed by a "molecular" process, possibly *via* hot CCl_2 radicals,

$$CCl_2^* + CCl_4 \rightarrow C_2Cl_6^* \rightarrow C_2H_4 + Cl_2 \tag{8.153}$$

When carbon tetrachloride is irradiated in the presence of oxygen, carbonyl chloride and chlorine are formed with G values (for γ-radiation) of 4.3 (176). The mechanism probably includes

$$\cdot CCl_3 + O_2 \rightarrow \cdot O_2CCl_3 \rightarrow COCl_2 + \cdot OCl \tag{8.154}$$

$$\cdot OCl + CCl_4 \rightarrow \cdot CCl_3 + Cl_2O \tag{8.155}$$

In contrast with the chloroform-oxygen system, no peroxides have been detected in irradiated $CCl_4 + O_2$ mixtures, possibly because the absence of hydrogen precludes stabilization of the peroxy radical as CCl_3O_2H.

Chloroform

Products from the γ-radiolysis of liquid chloroform are listed in Table 8.10; chlorine is not formed. Yields are dependent on temperature and dose rate and are particularly susceptible to traces of oxygen and moisture, which

TABLE 8.10   γ-Radiolysis of Liquid
Chloroform

Product	G(Product)[a]	
	Ref. 177	Ref. 178
HCl	5.3[b]	
CH_2Cl_2	2.2	1.9
CCl_4	0.7	0.88
C_2Cl_4	0.076	0.09
$CHCl_2CHCl_2$	0.67	0.75
C_2HCl_5	1.48	1.6
C_2Cl_6	2.4	2.0

[a] 25 to 26°C, dose rate $\sim 5 \times 10^{19}$ eV $g^{-1} hr^{-1}$.
[b] Ref. 180.

induce chain reactions giving a large number of products (179). Ion (NH_3 and 1-butanol) scavengers reduce the yield of CCl_4 but have no effect on, or slightly increase, the yields of the remaining products; radical scavengers reduce the yields of C_2 products except C_2Cl_4, having no effect on the yield of C_2Cl_4 or of CCl_4 (180). Estimated radical yields (180) are Cl 5.4, $\cdot CHCl_2$ 5.6, CCl_2 0.4, $\cdot CCl_3$ 0.25, and H 0.2 G units.

The predominant ion in the mass spectrum of chloroform is $CHCl_2^+$ (181), and the main result of both ionization and excitation is the formation of Cl and $\cdot CHCl_2$; probable reactions include

$$CHCl_3 \rightsquigarrow CHCl_3^*, CHCl_3^+, e^- \tag{8.156}$$

$$e^- + CHCl_3 \rightarrow CHCl_3^- \rightarrow \cdot CHCl_2 + Cl^- \tag{8.157}$$

$$CHCl_3^+ \rightarrow CHCl_2^+ + Cl \tag{8.158}$$

$$CHCl_3^+ + Cl^- \rightarrow CHCl_3 + Cl \tag{8.159}$$

$$CHCl_2^+ + Cl^- \rightarrow \cdot CHCl_2 + Cl \tag{8.160}$$

$$CHCl_3^* \rightarrow \cdot CHCl_2 + Cl \tag{8.161}$$

Reaction of the radicals produced accounts for the major products:

$$Cl + CHCl_3 \rightarrow HCl + \cdot CCl_3 \qquad k_{162} = 6 \times 10^4\ M^{-1}\ sec^{-1} \tag{8.162}$$

$$\cdot CHCl_2 + CHCl_3 \rightarrow CH_2Cl_2 + \cdot CCl_3 \qquad k_{163} = 1.3\ M^{-1}\ sec^{-1} \tag{8.163}$$

$$\cdot CHCl_2 + \cdot CCl_3 \rightarrow C_2HCl_5 \qquad k_{164} = 7.5 \times 10^7\ M^{-1}\ sec^{-1} \tag{8.164}$$

$$2\cdot CHCl_2 \rightarrow CHCl_2CHCl_2 \qquad k_{165} = 5 \times 10^7\ M^{-1}\ sec^{-1} \tag{8.165}$$

$$2\cdot CCl_3 \rightarrow C_2Cl_6 \qquad k_{147} = 5 \times 10^7\ M^{-1}\ sec^{-1} \tag{8.147}$$

(Rate constants are from ref. 182.) Carbon tetrachloride is apparently formed by an ionic reaction, although this has not been identified, and tetrachloroethylene by a nonionic "molecular" reaction. Werner and Firestone (180) have suggested that the latter may be

$$CCl_2 + CHCl_3 \rightarrow C_2HCl_5{}^* \rightarrow C_2HCl_5 \qquad (8.166)$$

$$\rightarrow C_2Cl_4 + HCl \qquad (8.167)$$

with CCl_2 formed by the dissociation of an excited ion or molecule.

The effect of increasing dose rate is to increase the yields of C_2HCl_5 and $C_2H_2Cl_4$ at the expense of the CH_2Cl_2 and C_2Cl_6; the yields of CCl_4 and C_2Cl_4 are unchanged (177, 182). Dose rate effects generally arise from competition between radical reactions in the bulk of the medium and radical–radical reactions between radicals from different spurs; the latter are favored as the number of spurs is increased by increasing the dose rate, but require extremely high dose rates ($\sim 10^{23}$ eV g^{-1} hr^{-1}) before spur overlap becomes significant. In the case of chloroform, effects are observed at very much lower dose rates of the order of 10^{18} to 10^{20} eV g^{-1} hr^{-1}, and the mechanism is different, involving competition between relatively slow radical–solvent reactions (particularly reaction 8.163), which are favored by low dose rates, and much faster interradical reactions (8.164 and 8.165), which are favored by high dose rates. The yield of C_2Cl_6 falls as the dose rate is increased because fewer $\cdot CCl_3$ radicals are being formed by reaction 8.163.

In the presence of oxygen, chloroform takes part in a radiation-initiated chain reaction to give peroxide (probably CCl_3OOH) in relatively high yield; yields of HCl found after hydrolysis of the radiolysis products are often of the order of several hundred, though the yields are strongly influenced by impurities, temperature, and dose rate (183). Other products formed include Cl_2, HCl, HOCl, $COCl_2$, and C_2Cl_6. The chain propagating steps are probably

$$\cdot CCl_3 + O_2 \rightarrow \cdot O_2CCl_3 \qquad (8.168)$$

$$\cdot O_2CCl_3 + CHCl_3 \rightarrow CCl_3OOH + \cdot CCl_3 \qquad (8.169)$$

The hydroperoxide is unstable and may break down to carbonyl chloride and hypochlorous acid,

$$CCl_3OOH \rightarrow COCl_2 + HOCl \qquad (8.170)$$

When both oxygen and moisture are present the acid yields are rather lower (183), though two-phase chloroform-water systems saturated with air or oxygen have been employed as chemical dosimeters (184; see also Chapter 3). Most of the radiolysis products are hydrolyzed by water to give hydrochloric acid which can be detected or measured in various ways:

$$CCl_3OOH + H_2O \rightarrow 3HCl + CO_2 + \tfrac{1}{2}O_2 \qquad (8.171)$$

$$COCl_2 + H_2O \rightarrow 2HCl + CO_2 \qquad (8.172)$$

Aqueous, single-phase, solutions of chloroform do not decompose by a chain reaction and are briefly discussed in Chapter 7.

Mixtures Containing Polyhalogen Compounds

Carbon–halogen bonds in compounds such as $CHCl_3$, CCl_4, C_2Cl_6, and CBr_4 are weaker than those in most compounds in which a single halogen atom is attached to a carbon atom, and a variety of chain reactions are possible in which the weakened bonds are broken.

With saturated compounds, the chain reaction brings about an exchange of halogen from the polyhalogen compound with hydrogen from the saturated compound in a reaction represented by

$$RCCl_3 + RH \rightarrow RCHCl_2 + RCl \qquad (8.173)$$

Reaction is initiated by the formation of radicals upon exposure to radiation and the propagation steps of the chain reaction are of the form

$$R\cdot + RCCl_3 \rightarrow RCl + R\dot{C}Cl_2 \qquad (8.174)$$

$$R\dot{C}Cl_2 + RH \rightarrow RCHCl_2 + R\cdot \qquad (8.175)$$

In common with other radiation-induced chain reactions, the yield of products is dependent on such factors as reactant concentration, dose rate, temperature, and the purity of the reagents, since small quantities of impurities can often act as radical scavengers, terminating the chain. G values range from about 50 to several hundred or even thousand depending on the nature of the reactants and the irradiation conditions. The same chain reactions can be initiated by chemically produced radicals or by exposure to ultraviolet light and are not unique to radiation chemistry. Mixtures that have been irradiated include cyclohexane with either carbon tetrachloride (185, 186) or hexachloroethane (187), which form cyclohexyl chloride, and chloroform or pentachloroethane, carbon tetrachloride with alcohols (188, 189), which give chloroform and a carbonyl compound and HCl by breakdown of the intermediate chloroalcohol,

$$CCl_4 + R_2CHOH \rightarrow CHCl_3 + R_2C(Cl)OH (\rightarrow R_2CO + HCl) \quad (8.176)$$

and carbon tetrachloride and straight-chain aliphatic aldehydes (190), which form chloroform and the appropriate acid chloride,

$$CCl_4 + RCHO \rightarrow CHCl_3 + RCOCl \qquad (8.177)$$

Mixtures of carbon tetrachloride or chloroform with ethanol have been suggested for use as chemical dosimeters (191, 192), since irradiation produces relatively large yields of HCl, which is readily detected and measured. However, though sensitive to very small radiation doses, the systems suffer from several disadvantages including poor stability toward heat and light and high susceptibility to traces of impurity. Sherman and his colleagues (193) have pointed out that the organo-chlorine pesticide DDT (which contains a —CCl_3 group) can be efficiently dechlorinated by irradiation in alcoholic solution due to a chain reaction analogous to reaction 8.176, while Sawai, Shimokawa, and Shinozaki (194) have shown that polychlorinated biphenyls (PCBs) can be dechlorinated by irradiation in alkaline 2-propanol solution.

Polyhalogen compounds can add to compounds containing C=C or C≡C bonds (195), giving addition products

$$CCl_4 + R_2C{=}CR_2 \rightarrow R_2C(Cl){-}C(R_2)CCl_3 \qquad (8.178)$$

or initiating polymerization

$$CCl_4 + nR_2C{=}CR_2 \rightarrow Cl{-}(CR_2{-}CR_2)_n{-}CCl_3 \qquad (8.179)$$

Both reactions are radical chain reactions initiated by the formation of radicals in the irradiated material. Propagation steps for the addition and polymerization reactions are

$$\cdot CCl_3 + R_2C{=}CR_2 \rightarrow R_2\dot{C}{-}C(R_2)CCl_3 \qquad (8.180)$$

$$R_2\dot{C}{-}C(R_2)CCl_3 + CCl_4 \rightarrow R_2C(Cl){-}C(R_2)CCl_3 + \cdot CCl_3 \quad (8.181)$$

and

$$R_2\dot{C}{-}CR_2{-}(CR_2{-}CR_2)_n{-}CCl_3 + R_2C{=}CR_2 \rightarrow$$
$$R_2\dot{C}{-}CR_2{-}(CR_2{-}CR_2)_{n+1}{-}CCl_3 \qquad (8.182)$$

respectively, choice of the pathway followed depending on such factors as the strength of the carbon–halogen bond broken. Relatively weak carbon–halogen bonds, such as the C—Br bond in $BrCCl_3$, favor the addition reaction (196) while polymerization is favored if the carbon–halogen bonds are stronger (irradiation of $CHCl_3$–C_2H_4 mixtures, for example, tending to give products containing more than one ethylene molecule) (197).

Polyhalogen compounds, in common with other halogen compounds, are efficient electron scavengers and are frequently used for this purpose in attempting to establish radiolysis mechanisms. The quantities added as electron scavengers, however, are small ($\sim 10^{-3}$ M) whereas in the systems described above the halogen compound is a major component of the mixture.

Aromatic Halides

Aromatic halides have received less attention than the aliphatic halogen compounds although the yields of product from the halogen-substituted benzenes (Table 8.11) are comparable with those from the aliphatic halides,

T A B L E 8.11   *γ-Radiolysis of Liquid Chloro-, Bromo-, and Iodobenzene*

Product	G(Product)		
	Chlorobenzene (X=Cl; ref. 198)	Bromobenzene (X=Br; ref. 199)	Iodobenzene (X=I; ref. 200)
H_2	0.012	0.006^b	0.004
HX	1.4	2.3	0.008
X_2	0.006^b	0.2^b	2.0
C_6H_6	2.4	1.5^b	0.43
$C_6H_4X_2$	0.23	0.18	0.43
C_6H_5—C_6H_5	0.14	0.26	0.49
$C_6H_5C_6H_4X$	1.0^a	2.02	0.25
C_{24} (as C_6 units)	1.5		

^{a} Chlorobiphenyl and dichlorobiphenyl.
^{b} Ref. 200.

and appreciably greater than the product yields from benzene itself (Table 8.7). Radical yields estimated by radioactive bromine or iodine exchange, or by scavenger experiments, are in the region of 3 to 6 for the halides listed in Table 8.11 (200), compared with a radical yield of about 0.8 G units for benzene (5). It is apparent, therefore, that the presence of the aromatic ring does not stabilize the aromatic halides toward radiation decomposition to the extent that hydrocarbons are stabilized. This is due in large part to electron scavenging by the halide with the consequent formation of radicals, e.g.,

$$C_6H_5Cl + e^- \rightarrow \cdot C_6H_5 + Cl^- \qquad (8.182)$$

$$C_6H_5Cl^+ + Cl^- \rightarrow C_6H_5Cl + Cl \qquad (8.183)$$

The final products can largely be accounted for by the subsequent radical reactions,

$$Cl + C_6H_5Cl \rightarrow HCl + \cdot C_6H_4Cl \qquad (8.184)$$

$$\cdot C_6H_5 + C_6H_5Cl \rightarrow C_6H_6 + \cdot C_6H_4Cl \qquad (8.185)$$

$$\cdot C_6H_5 + \cdot C_6H_4Cl \rightarrow C_6H_5C_6H_4Cl \qquad (8.186)$$

etc.

Fluorides

The carbon–fluorine bond is strong compared with other covalent bonds in organic compounds and shows a greater tendency to remain intact in radiolysis reactions than other carbon–halogen bonds. Radiolysis of liquid trifluoroiodomethane (CF_3I), for example, gives CF_4 (G 0.37), C_2F_6(G 1.03), and I_2 (G 1.36) (201), and it is evident that C—I bond scission occurs to the virtual exclusion of C—F bond breaking. However, the preponderance of C—I bond scission is not unique to the fluorocompound, since the same is true for methyl iodide, and reflects rather the weakness of the C—I bonds; it does illustrate the preferential fission of other carbon–halogen bonds if fluorine and a second halogen are both present in the same molecule. Radicals produced by irradiation of liquid ethane and perfluoroethane (C_2F_6) have been investigated by electron spin resonance (202) and found to be predominantly $\cdot C_2H_5$ from ethane and $\cdot CF_3$ from the fluorocompound, demonstrating that C—H and C—C fission are preferred in the hydrocarbon and fluorocarbon respectively. However products identified in the γ-radiolysis of liquid C_2F_6 (203) indicate that the probability of C—F and C—C bond rupture are about equal (C—H scission predominates in the radiolysis of C_2H_6). With perfluorocyclohexane, C—F bond scission is the major process and radical yields estimated by iodine scavenging (204) are, G(perfluorocyclohexyl) 3.0, G(perfluorohexenyl) 0.8, G(perfluorohexyl) 0.1, and G(perfluorohexamethylene diradical) 0.1. The authors estimate a total radical yield of about 5.8 G units, which compares with a total radical yield for cyclohexane of about 8 G units indicating rather less dissociation with the perfluorocompound. Product yields for the γ-radiolysis of perfluorocyclohexane (204), G(perfluorobicyclohexyl) 0.95, G(perfluorocyclohexylhexene) 0.48, and G(perfluorocyclohexylhexane) 0.12, are lower than the corresponding yields from cyclohexane (Table 8.3). Perfluorocyclohexene is not formed since perfluoro radicals tend to combine rather than disproportionate (cyclohexene is predominantly a disproportionation product in the radiolysis of cyclohexane). Fluorocarbons, except CF_4 and C_2F_6, are electron scavengers in a similar manner to other halogen compounds (205), but dissociation of the negative ion formed is endothermic; Kennedy and Hanrahan (204) estimate that dissociative electron capture by perfluorocyclohexane requires an electron with an energy in excess of 1.8 eV,

$$C_6F_{12} + e^- \rightarrow C_6F_{12}^- \rightarrow \cdot C_6F_{11} + F^- \qquad (8.187)$$

Fluorine is not an important product from fluorocarbons because F atoms do not abstract fluorine to form F_2 (which has a low bond energy) and fluorine formed by combination of F atoms is readily attacked by radicals,

$$\cdot C_6F_{11} + F_2 \rightarrow C_6F_{12} + F \qquad (8.188)$$

One of the more frequently investigated fluorides is the polymer Teflon, perfluropolyethylene—$(CF_2)_n$—. Teflon is relatively resistant to radiation in the absence of oxygen but rapidly deteriorates when oxygen is present, becoming extremely brittle (206). A number of radiolysis products have been identified (207) including, in the absence of oxygen, CF_4, C_2F_6, C_3F_8, and higher fluorocarbons, and SiF_4 (from the glass container) and, when oxygen is present during the irradiation, CF_4 and C_2F_6 (in rather smaller amounts than are formed in the absence of oxygen), SiF_4, and a high proportion of COF_2. Golden (207) has suggested that these products are formed by random splitting of carbon–carbon bonds, and to a smaller extent carbon–fluorine bonds, to give radicals which in the absence of oxygen combine to give the observed products or else remain as relatively stable free radicals in the solid, i.e.,

$$-CF_2-CF_2-CF_2- \rightsquigarrow -CF_2-CF_2\cdot + \cdot CF_2- \qquad (8.189)$$

or, less often

$$-CF_2-CF_2-CF_2- \rightsquigarrow -CF_2-\dot{C}F-CF_2- + F \qquad (8.190)$$

Oxygen will add to the organic radicals to give peroxy radicals, which can react together to form alkoxy radicals,

$$-CF_2-CF_2\cdot + O_2 \rightarrow -CF_2-CF_2O_2\cdot \qquad (8.191)$$

$$2 -CF_2-CF_2O_2\cdot \rightarrow 2 -CF_2-CF_2O\cdot + O_2 \qquad (8.192)$$

Dissociation of the alkoxy radicals produces carbonyl fluoride and a new organic radical, smaller than the original radical by one $-CF_2-$ group,

$$-CF_2-CF_2O\cdot \rightarrow -CF_2\cdot + COF_2 \qquad (8.193)$$

A similar sequence of reactions can be envisaged for the secondary radicals and, by repetition of these reactions, an entire polymer molecule may be rapidly degraded once a free radical center has been formed in some part of it. This is analogous to the photochemical degradation of perfluorohalides in the presence of oxygen (208).

ALCOHOLS AND ETHERS

Alcohols are the organic analogues of water and, as polar compounds, their behavior upon irradiation is similar in many respects to that of water. The products obtained by irradiating pure alcohols (209, 210) include hydrogen, hydrocarbons, water, carbon monoxide, glycols, and aldehydes, or ketones. Primary alcohols are oxidized to aldehydes, secondary alcohols to a mixture of aldehydes and ketones, and tertiary alcohols to ketones alone.

In the absence of oxygen, glycols are invariably formed, and are predominantly α-glycols, $R_1R_2C(OH)$—$C(OH)R_1R_2$, consistent with the postulate that a common radiation-induced reaction of alcohols (R_1R_2CHOH) is loss of an α-hydrogen atom to give the radical $R_1R_2\dot{C}OH$. The yield of hydrogen is greatest with the straight-chain alcohols and decreases with increased branching, while hydrocarbon yields are greatest with branched-chain alcohols. As the length of the carbon chain is increased, the products tend to include smaller amounts of those characteristic of the alcohol group and more of those resulting from bond breaking in the hydrocarbon chain, suggesting that reaction is partly through random bond breaking in the molecule as a whole and partly through preferential reaction involving the functional group.

The radiolysis of the ethers (209) is rather similar to that of the alcohols though the predominant bond rupture in the primary act is C—O and C—C (adjacent to the ether oxygen) rather than O—H as with the alcohols. With both classes of compound, however, primary excitation and ionization involve the oxygen atom rather than the hydrocarbon chain(s).

Methanol

Representative values for the radiolysis yields from methanol are given in Table 8.12; alcohols are difficult to purify, and published yields tend to be far more scattered than those for the hydrocarbons and halides. The table

TABLE 8.12   *Radiolysis of Methanol*

Radiation	Solute	$G(H_2)$	$G(CO)$	$G(CH_4)$	$G(HCHO)$	$G(C_2H_6O_2)$	Ref.
Methanol vapor							
γ-Radiation	None	10.8	1.0	0.3	5.6	3.1	211, 212
Liquid Methanol							
$^{10}B(n, \alpha)^7Li$ recoil nuclei	0.1 M methyl borate	5.5	1.0	0.66	3.0	0.87	213
γ-Radiation	None	5.40[a]	0.13	0.7	2.0	3.65	214, 215
		5.0	0.06	0.43	2.2	3.2	213
	0.04 M iodine[b]	2.02	0.14	0.19	0.36	0.09	215
	>0.005 M H_2SO_4	6.05	0.12	0.60	2.55	3.8	214
	0.01 M CH_3ONa	5.50	0.11	0.87	2.05	3.9	214
	$Fe_2(SO_4)_3$ + 0.05 M H_2SO_4 + 3% water[c]	5.55	0.16	0.39	8.5		216
	1.5×10^{-3} M O_2[d]	1.9	0.09	0.18	8.7	0.1	213

[a] Unchanged by the addition of 3% water (214).
[b] $G(-I_2) = 1.78$.
[c] $G(Fe^{2+}) = 6.5$.
[d] Other products formed include CO_2 (G 0.1), HCOOH (G ~1.5), and peroxide (G 3.1).

illustrates two of the features in which the radiolysis of methanol resembles that of water, namely the pronounced effect of LET and the effect of pH. It also demonstrates, through the effect of radical scavengers, that "radical" and "molecular" processes can be distinguished with methanol as with water.

Radiolysis of methanol vapor gives H_2, CO, HCHO, and glycol as the major products (Table 8.12). Electron scavengers (CH_3Br, CCl_4) reduce $G(H_2)$ to about seven, and unsaturated compounds (C_2H_4, C_3H_6, or benzene) $G(H_2)$ to about 2 G units. Unsaturated compounds also reduce the yield of formaldehyde (to 3.7 G units) and eliminate glycol as a product. The scavenger experiments demonstrate that hydrogen is formed by more than one process and that these probably include ion neutralization, radical reactions, and other "molecular" processes, while glycol is probably formed entirely, and formaldehyde in part, by radical reactions.

Photolysis of methanol vapor (217, 218) shows that dissociation of excited methanol occurs mainly by

$$CH_3OH^* \rightarrow HCHO + H_2 \quad (G\ 1.6) \qquad (8.194)$$

$$\rightarrow CH_3O\cdot + H \quad (G\ 1.1) \qquad (8.195)$$

and that reaction 8.195 accounts for 70 to 80% of the reaction from 123.6 to 185 nm. The more important ions formed in the mass spectrometer with their relative abundances and the neutral fragments formed (219) are CH_3OH^+ (68), CH_2OH^+ (100, H), and CHO^+ (52, H_2 + H). Mass spectrometry also provides evidence that the ions can take part in a number of ion-molecule reactions (38, 220):

$$CH_3OH^+ + CH_3OH \rightarrow$$

$$CH_3OH_2^+ + \cdot CH_2OH \quad (and\ CH_3O\cdot) \quad (G\ 1.21) \quad (8.196)$$

$$CH_2OH^+ + CH_3OH \rightarrow CH_3OH_2^+ + HCHO \qquad (G\ 1.78) \quad (8.197)$$

$$CHO^+ + CH_3OH \rightarrow CH_3OH_2^+ + CO \qquad (G\ 0.63) \quad (8.198)$$

$$\rightarrow CH_2OH^+ + HCHO \qquad (8.199)$$

$$\rightarrow CH_3^+ + CO + H_2O \qquad (G\ 0.3) \quad (8.200)$$

$$CH_3^+ + CH_3OH \rightarrow CH_2OH^+ + CH_4 \qquad (G\ 0.3) \quad (8.201)$$

and that clustering occurs about the stable $CH_3OH_2^+$ ion (221),

$$CH_3OH_2^+ + (n - 1)CH_3OH \rightarrow H^+(CH_3OH)_n \quad (G\ 3.92) \quad (8.202)$$

Ion neutralization produces hydrogen atoms

$$H^+(CH_3OH)_n + e^- \rightarrow H + nCH_3OH \quad (G\ 3.92) \qquad (8.203)$$

which abstract hydrogen from methanol

$$H + CH_3OH \rightarrow H_2 + \cdot CH_2OH \quad (G\ 7.73) \qquad (8.204)$$

Other radical reactions, established independently and expected to occur in the radiolysis of methanol vapor, are

$$CH_3O\cdot + CH_3OH \rightarrow CH_3OH + \cdot CH_2OH \quad (G\ 1.1) \qquad (8.205)$$

$$2\cdot CH_2OH \rightarrow CH_3OH + HCHO \Big\} \qquad (8.206)$$
$$CH_3O\cdot + \cdot CH_2OH \rightarrow CH_3OH + HCHO \Big\} \quad (G\ 1.92) \qquad (8.207)$$

$$2\cdot CH_2OH \rightarrow HOCH_2CH_2OH \quad (G\ 3.1) \qquad (8.208)$$

A mechanism made up of reaction 8.209 and reactions 8.194 to 8.208 accounts qualitatively for the products formed by irradiation of methanol vapor.

$$CH_3OH \rightsquigarrow CH_3OH^*, CH_3OH^+, e^- \qquad (8.209)$$

The numbers shown in parentheses to the right of Eqs. 8.194 to 8.208 are estimated G values for the reactions based on the following assumptions: (1) that G(positive ions) is given by $100/W$, i.e., by $100/25.5$ eV ion-pair^{-1} = 3.92, and that G values for the individual positive ions are in proportion to the ion abundances observed in the mass spectrometer, (2) that CO and CH_4 are only formed *via* the ion-molecule reactions shown, (3) that molecular formaldehyde (G 3.7) is only formed *via* reactions 8.194 and 8.197, and (4) that the yield of $\cdot CH_2OH$ radicals is that required to form the radical formaldehyde (G 1.9) and all glycol. Estimated yields are compared with experimental yields in Table 8.13; the agreement is satisfactory but could undoubtedly be improved by minor adjustments in the G values for the intermediate steps. The estimated values meet several simple tests giving, for example, the correct ratios of radical to molecular yields for hydrogen and formaldehyde, and forming CO and CH_4 by molecular mechanisms and glycol by a radical reaction. The observed reduction of $G(H_2)$ by 3.5 to 4 G units in the presence of electron scavengers is explained if the scavengers replace reaction 8.203 by one that does not produce hydrogen atoms, reducing $G(H_2)$ by an amount equal to G(positive ion), or 2.92 G units. The suggested G values appear to violate the ratio of reactions 8.194 to 8.195 observed photochemically, but it could be argued that the population of excited states produced by irradiation would be richer in triplet states than that produced photochemically and that triplet excited states may dissociate to molecular rather than radical products. However, it must be emphasized that the fact that a set of assumptions and equations can be manipulated to give the correct G values is not evidence that the mechanism is necessarily correct, but merely that it may reasonably be taken as a working hypothesis in moving toward a complete mechanism. Further steps toward establishing the mechanism include attempts to study each step independently to determine rate constants and the effect of such variables as pressure and temperature, direct observation of intermediates by pulse radiolysis or electron spin resonance (in the vapor phase where possible, but otherwise in a condensed phase), and a study of the effect of as wide a range of ion, excited state, and radical scavengers as possible (e.g., studies on the radiolysis of methanol vapor reported in refs. 211, 212, 220, 222, and 223). It is often instructive to consider radiolysis mechanisms proposed for related compounds; in the case of methanol, for example, the radiolysis of water vapor is informative with regard to reactions that may be occurring and to experiments that might contribute useful information.

T A B L E 8.13 *Estimated and Experimental G Values*
for Methanol Vapor

Product	Experimental G value	Estimated G value[a]
H_2 (total)	10.8	10.26
(molecular)	~2	2.53
(radical)	8 to 9	7.73
CO	1.0	0.93
CH_4	0.3	0.3
HCHO (total)	5.6	5.60
(molecular)	3.7	3.68
(radical)	1.9	1.92
$C_2H_6O_2$	3.1	3.1

[a] Based on a mechanism made up of reactions 8.194 to 8.209 and the assumptions described in the text.

Products formed when liquid methanol is irradiated are the same as those formed in the vapor although most product yields and $G(-CH_3OH)$ are lower in the liquid; yields of methane and glycol may be higher in the liquid phase but depend on the LET of the radiation. Similar effects are observed when the radiolysis of liquid water and water vapor are compared, differences between the two phases being attributed to rapid solvation of charged species, more rapid dissipation of excitation energy by collisional processes, and slower diffusion in the liquid, with the latter giving rise to caging and LET effects in the liquid that are largely absent in the vapor phase radiolysis.

Solvation of the positive ion and electron in liquid methanol would be expected to delay recombination of a significant proportion of the ion-pairs formed, rendering this part of the electron yield susceptible to scavenging by moderate concentrations of scavenger and to observation by pulse radiolysis. Pulse radiolysis studies have indeed confirmed the presence of the solvated electron (λ_{max} 630 nm, $\varepsilon = 1.7 \times 10^4\ M^{-1}\ cm^{-1}$) (224) in irradiated methanol with a G value ranging from 3.3 at 150 psec (225) to 2.0 at times of the order of several microseconds (226, 227); some of the electrons responsible for the higher G value will subsequently undergo geminate recombination within the spurs, but the lower G value corresponds to "free" electrons that have escaped from the spurs. Solvation is slower in alcohols than in water and a change in the solvated electron absorption spectrum in methanol between 30 and 150 psec after the pulse (225) has been attributed to the conversion of "damp" electrons (i.e., electrons in shallow traps) to normal solvated electrons; similar spectral changes have been observed in alcohols cooled to temperatures near their melting points over a longer (nanosecond) time period (228). Nitrous oxide scavenges electrons in methanol forming nitrogen,

$$e^-_{solv} + N_2O \rightarrow N_2 + O^- \qquad (8.210)$$

At low concentrations of N_2O (e.g., $10^{-4}\ M$), the yield of nitrogen corresponds with the yield of "free" solvated electrons observed by pulse radiolysis at times of the order of microseconds, but at higher concentrations $G(N_2)$ increases to about 3 due to scavenging of electrons that would otherwise have undergone geminate recombination (229). Solvated electrons are also scavenged by hydrogen ions (i.e., $CH_3OH_2^+$) in the spurs and in acidified methanol,

$$e^-_{solv} + CH_3OH_2^+ \rightarrow CH_3OH + H$$
$$k_{211} = 5.2 \times 10^{10}\ M^{-1}\ sec^{-1}\ (230) \quad (8.211)$$

The increased yields of hydrogen, formaldehyde, and glycol in the presence of acid (Table 8.12) are probably due to competition between reaction (8.211) and spur reactions that normally reform methanol, e.g.,

$$e^-_{solv} + CH_3O\cdot \rightarrow CH_3O^- \qquad (8.212)$$

Product yields are not significantly altered by the presence of base. Solvated electrons that escape from the spurs will probably react with the solvent

$$e^-_{solv} + CH_3OH \rightarrow CH_3O^- + H$$
$$k_{213} = 1.4 \times 10^4\ M^{-1}\ sec^{-1}\ (231) \quad (8.213)$$

before encountering a positive ion. However, stoichiometrically, the result is the same, since the subsequent neutralization reaction produces only methanol

$$CH_3OH_2^+ + CH_3O^- \rightarrow 2CH_3OH \qquad (8.214)$$

Collisional deactivation of excited ions and molecules in the liquid phase is expected to reduce the yield of fragmentation products compared with the vapor phase radiolysis. Yields of CO and molecular HCHO are lower in the liquid phase, but the yield of CH_4 is higher. Methane may be formed by nonionic processes in the liquid that do not occur in the vapor, e.g., by

$$e_{solv}^- + CH_3OH \rightarrow \cdot CH_3 + OH_{solv}^- \qquad (8.215)$$

$$\cdot CH_3 + CH_3OH \rightarrow CH_4 + \cdot CH_2OH \qquad (8.216)$$

accounting for the higher yield. Scavenger studies have confirmed the presence of methyl radicals in irradiated liquid methanol and a reduction in $G(CH_4)$ in the presence of radical scavengers.

Radical and molecular yields estimated for γ-irradiated liquid methanol using radical scavengers (213) are given in Table 8.14; the tables corresponds

TABLE 8.14   *Radical and Molecular Product Yields for Liquid Methanol (213)*

	G(Product)	
Product	^{60}Co γ-Radiation	$^{10}B(n, \alpha)^7Li$ Recoil Irradiation
e_{solv}^- [a]	2.0	0.2
H [b]	1.1	1.2
OH	0.2	0
$\cdot CH_3$	0.2	0.1
$\cdot CH_2OH$	2.7	0
Σ(radicals)	6.2	1.5
H_2	1.9	2.4
CO	0.08	0.8
CH_4	0.2	0.6
HCHO	2.2	2.1
$C_2H_6O_2$	0	0.4

[a] From ref. 232.
[b] G_H from ref. 213 less G_e from ref. 232.

to Table 7.4 for water. Scavengers employed include ferric sulfate (Table 8.12) (214), which almost completely suppresses the formation of glycol but increases the yield of formaldehyde by an equivalent amount by scavenging $\cdot CH_2OH$ radicals,

$$Fe^{3+} + \cdot CH_2OH \rightarrow Fe^{2+} + HCHO + H^+ \qquad (8.217)$$

The yield of ferrous ion (G 6.5) provides a measure of the total radical yield, since H, $\cdot CH_3$, and $CH_3O\cdot$ all react with methanol to form $\cdot CH_2OH$;

hydrogen atoms, and solvated electrons may react directly with ferric ion or after reaction with the solvent. Isotopic studies using deuterium labeled methanol (215), esr investigation using a spin-trapping technique (233), and chemical trapping of the radicals to give identifiable products (234) show that both $CH_3O\cdot$ and $\cdot CH_2OH$ are formed in the initial reactions with the former predominating. However, methoxy radicals react rapidly with methanol (Eq. 8.205) to give $\cdot CH_2OH$ so that only the latter is included in Table 8.14. Yields of OH and $\cdot CH_3$ in Table 8.14 and of CH_4 in liquid systems (Table 8.12) suggest that C—O bond cleavage is more important in the liquid phase than in the vapor; this might be accomplished by reaction 8.215, as suggested in the previous paragraph, although the corresponding reaction in water is slow and this is not certain.

Increasing LET has the effect of decreasing the yields of scavengeable radicals (Table 8.14) and increasing the yields of "molecular" products (Tables 8.12 and 8.14); $G(-CH_3OH)$ decreases with increasing LET. These effects can be accounted for in a similar manner to LET effects in liquid water, namely an increase in intraspur reactions at high LET at the expense of radical escape from the spurs. In the case of methanol, the nature of the radicals involved may also be different, since alkoxy radicals ($CH_3O\cdot$) predominate during the brief spur lifetime but are converted to $\cdot CH_2OH$ over the longer time span required for spur expansion and diffusion. Interradical reactions occurring in the spurs probably include, therefore,

$$2\ CH_3O\cdot \rightarrow CH_3OH + HCHO \qquad (8.218)$$

$$H + CH_3O\cdot \rightarrow H_2 + HCHO \qquad (8.219)$$

$$CH_3O\cdot + \cdot CH_2OH \rightarrow CH_3OH + HCHO \qquad (8.207)$$

which have the effect of increasing the yield of formaldehyde at the expense of the glycol yield, and reactions such as 8.212 which reform methanol, reducing $G(-CH_3OH)$. The increased yield of CO at high LET may be due to secondary radical reactions such as

$$CH_3O\cdot + HCHO \rightarrow CH_3OH + \cdot CHO \qquad (8.208)$$

$$\cdot CHO + CH_3O\cdot \rightarrow CH_3OH + CO \qquad (8.209)$$

or to increased reaction between excited species.

In the presence of oxygen, glycol is not formed and the major product from methanol is formaldehyde (213); hydrogen, formic acid, and peroxide are also formed. Formaldehyde is also the major product when aqueous solutions of methanol containing oxygen are irradiated, and the mechanism is probably similar,

$$\cdot CH_2OH + O_2 \rightarrow \cdot O_2CH_2OH \qquad (8.210)$$

$$2 \cdot O_2CH_2OH \rightarrow 2\,HCHO + H_2O_2 + O_2 \qquad (8.211)$$

$$\cdot O_2CH_2OH + HO_2 \rightarrow HCHO + H_2O_2 + O_2 \qquad (8.212)$$

The perhydroxyl radical, HO_2, and its anion, O_2^-, are formed in both water and methanol by oxygen scavenging of hydrogen atoms and solvated electrons, respectively.

The radiolysis of solid alcohols has been reviewed by Kevan (235) and, briefly, by Freeman (210), who has emphasized that product yields depend strongly on both the purity of the alcohol and the structure of the solid phase. The radiolysis mechanism is similar to that for liquid alcohols with the restriction that diffusion is very limited so that radicals tend to react with neighboring molecules or radicals, or become trapped until the solid is melted. Trapped electrons can be freed from their traps by exposure to light and a considerable number of experiments have been concerned with such photo-bleaching.

Other Alcohols

The radiolysis of ethanol is similar in its main features to the radiolysis of methanol, though a rather larger number of minor products are produced (Table 8.15); G values often depend on irradiation conditions such as

T A B L E 8.15   *Radiolysis of Ethanol*

	G(Product)		
Product	Vapor (2 MeV e^- Irradiation 25°C, ~45 mm Hg; ref. 236)	Liquid (γ-Irradiation, 25°C; ref. 237)	Solid[b] (γ-Irradiation, −196°C; ref. 238)
H_2	10.8	5.0[a]	4.4
H_2O		0.5	
CO	1.2	0.06	0.3
CH_4	0.9	0.6	0.3
HCHO	0.9	0.13	
C_2H_2	0.30	<0.02	0.01
C_2H_4	1.6	0.14	0.20
C_2H_6	0.65	0.24	0.3
CH_3CHO	4.2	2.8[a]	2.8
$C_2H_5OC_2H_5$	0.07		
$(CH_3CHOH)_2$	1.2	2.2[a]	0.85
$CH_3CH(OH)CH_2OH$	0.16	0.13	
Other C_4 glycols		0.06	

[a] Extrapolated to zero dose.
[b] Products determined after melting the sample.

temperature, total dose, dose rate, and LET; a critical review of the effect of these and other variables on radiolysis yields from ethanol has been given by Freeman (239). Major products in all phases are hydrogen, acetaldehyde, and 2,3-butanediol, which are accounted for if the primary processes lead to loss of an α-hydrogen atom,

$$CH_3CH_2OH \rightsquigarrow CH_3\dot{C}HOH + H \qquad (8.213)$$

and the radicals formed react,

$$H + CH_3CH_2OH \rightarrow H_2 + CH_3\dot{C}HOH \qquad (8.214)$$

$$2 CH_3\dot{C}HOH \rightarrow CH_3CH_2OH + CH_3CHO \qquad (8.215)$$

$$\rightarrow CH_3CH(OH)CH(OH)CH_3 \qquad (8.216)$$

Reactions 8.214 to 8.216 also occur when aqueous solutions of ethanol are irradiated. The assumption that α-hydrogen atoms are lost in preference to β-hydrogen atoms, i.e., that the intermediate organic radicals are $CH_3\dot{C}HOH$ rather than $\cdot CH_2CH_2OH$, is supported by the nature of the products formed (2,3- rather than 1,4-glycols), by esr investigation of the radicals present in frozen irradiated ethanol (240), and by isotopic tracer experiments with deuterium labeled ethanol (241). As with methanol, the radicals formed initially include alkoxy radicals (233) which take part in interradical reactions or abstract hydrogen from the solvent,

$$CH_3CH_2O\cdot + CH_3CH_2OH \rightarrow CH_3CH_2OH + CH_3\dot{C}HOH \qquad (8.217)$$

Loss of an α-hydrogen atom in preference to hydrogen further removed from the hydroxyl group is characteristic of radical attack on alcohols, both pure and in aqueous solution. It is not, however, exclusive, and some C—H bond scission occurs at other locations (the "other C_4 glycol" group of products in Table 8.15 includes 1,3- and 1,4-butanediol which result from loss of a β-hydrogen).

In common with other compounds, irradiation of ethanol in the vapor phase gives a larger yield of products than irradiation in the liquid or solid phase. The proportion of low molecular weight products is also larger with the vapor, consistent with more extensive fragmentation of excited ions and molecules. Product yields from ethanol vapor are dependent on the irradiation conditions (210) so that G values given in Table 8.15 are restricted to the conditions specified; at temperatures above 200°C, chain reactions occur and most yields increase (210, 242). Yields of products that require two large radicals to come together (e.g., glycol) are lower for solid ethanol than for the liquid, but other yields are essentially unchanged. Photobleaching of the irradiated solid before melting increases the G values for H_2, CH_3CHO, and glycol to 4.9, 3.8, and 1.0, respectively. A possible mechanism is return of an electron to a cage containing a positive ion and an alcohol radical,

$$C_2H_5OH^+ + C_2H_5OH \rightarrow [C_2H_5OH_2^+ + \cdot C_2H_5OH] \qquad (8.218)$$

$$e^- + [C_2H_5OH_2^+ + \cdot C_2H_5OH] \rightarrow [C_2H_5O\cdot + H_2 + \cdot C_2H_5OH] \qquad (8.219)$$

$$[C_2H_5O\cdot + H_2 + \cdot C_2H_5OH] \rightarrow C_2H_5OH + H_2 + CH_3CHO \qquad (8.220)$$

Table 8.16 lists the major products from the simpler alcohols upon γ-irradiation and estimates of the extent to which C—H, C—C, and C—O bond cleavage contribute to the formation of the products. It should be noted that whereas C—C and C—O cleavage are caused by fragmentation of excited ions and molecules, C—H cleavage results in large part from the subsequent radical reactions. The data indicate that C—C scission becomes more important as the proportion of C—C bonds in the molecule increases and is likely to be particularly significant if the molecule contains a branched carbon chain as in *t*-butanol; C—C bonds adjacent to the hydroxyl group appear particularly susceptible, since ethane is a major product from 1-propanol and propane from 1-butanol. The proportion of carbonyl product to glycol increases as the chain length increases, apparently due to an increase in the ratio of disproportionation to combination with the larger radicals. However the high yields of acetone and methane from *t*-butanol can be attributed to β-scission of the alkoxy radical,

$$\underset{|}{\overset{R}{\underset{|}{\overset{\cdot\cdot}{C}}}}\!\!-\!\!O\cdot \rightarrow \underset{|}{-\overset{|}{C}}\!\!=\!\!O + R\cdot \qquad (8.221)$$

Formation of ethane from 1-propanol and propane from 1-butanol can be explained in the same manner.

Little has been published concerning the radiolysis of aromatic alcohols and phenols. Pulse radiolysis studies of benzyl alcohol (247) have suggested that the primary steps are

$$C_6H_5CH_2OH \rightsquigarrow C_6H_5CH_2OH^+ + e^- \qquad (8.222)$$

$$C_6H_5CH_2OH^+ + C_6H_5CH_2OH \rightarrow C_6H_5CH_2OH_2^+ + C_6H_5CH_2O\cdot \qquad (8.223)$$

$$e^- + C_6H_5CH_2OH \rightarrow C_6H_5CH_2\cdot + OH^- \qquad (8.224)$$

with a G value for each radical of 2.1; singlet (G 0.7) and triplet (G 1.1) excited states are also formed. If the alkoxy radical subsequently abstracts an α-hydrogen atom from another solvent molecule,

$$C_6H_5CH_2O\cdot + C_6H_5CH_2OH \rightarrow C_6H_5CH_2OH + C_6H_5\dot{C}HOH \qquad (8.225)$$

radical combination and disproportionation reactions are able to account for the observed products (248): benzaldehyde (G 1.32), benzylphenylcarbinol

TABLE 8.16 *Major Products from the γ-Radiolysis of Liquid Alcohols*

Alcohol	$G(H_2)$	$G(>C=O)^a$	G(glycol)	Other Products and Their G Values	Relative Frequency of Bond Breaking $(\%)^b$			Ref.
					C—H	C—C	C—O	
CH$_3$OH	5.4	2.0	3.65	CH$_4$ 0.7	88	0	12	Table 8.12
CH$_3$CH$_2$OH	5.0	2.8	2.2	CH$_4$ 0.6, C$_2$H$_6$ 0.24	84	7	9	Table 8.15
CH$_3$CH$_2$CH$_2$OH	4.4	2.9	1.5	C$_2$H$_6$ 2.0, HCHO 1.9	62	32	6	243
CH$_3$CH(OH)CH$_3$	3.7	1.5	0.2	CH$_4$ 1.53, CH$_3$CHO 0.3	69	27	4	244
CH$_3$(CH$_2$)$_2$CH$_2$OH	4.55	3.15	1.55	C$_3$H$_6$ 1.87, HCHO 1.90	65	30	5	245
(CH$_3$)$_3$COH	1.0	—	~0	CH$_4$ 3.56, C$_2$H$_6$ 0.69, CH$_3$COCH$_3$ 2.55, (CH$_3$)$_2$CHCHO 0.63	~11	~77	~12	246

^{a} R$_2$CHOH → R$_2$CO.
^{b} Based on product G values.

$(C_6H_5CH_2CHOHC_6H_5, G\ 0.68)$, DL$(G\ 0.22)$ and *meso* $(G\ 0.20)$, hydrobenzoin $(C_6H_5CHOHCHOHC_6H_5)$, and dibenzyl $(C_6H_5CH_2CH_2C_6H_5,\ G\ 0.29)$. Aromatic radicals generally combine rather than disproportionate, and the benzaldehyde found may have resulted from traces of oxygen in the system. Benzaldehyde and peroxides $(G \sim 50)$ (249) are formed *via* a chain reaction when benzyl alcohol is irradiated in the presence of oxygen. The key step in propagating the chain reaction is hydrogen abstraction by an intermediate peroxy radical, which is facilitated by the presence of a phenyl group adjacent to the alcohol grouping:

$$C_6H_5\dot{C}HOH + O_2 \rightarrow C_6H_5CH(O_2{\cdot})OH \qquad (8.226)$$

$$C_6H_5CH(O_2{\cdot})OH + C_6H_5CH_2OH \rightarrow$$
$$C_6H_5CH(O_2H)OH + C_6H_5\dot{C}HOH \quad (8.227)$$

$$C_6H_5CH(O_2H)OH \rightarrow C_6H_5CHO + H_2O_2 \qquad (8.228)$$

Ethers

Ethers (209) are decomposed by radiation to a somewhat greater extent than alcohols, in contrast to their stability toward many chemical reagents. Products formed by γ-irradiation of diethyl ether are listed in Table 8.17. In the liquid phase, products are formed primarily by C—O bond cleavage and, to a lesser extent, by C—H and C—C bond rupture; hydrogen is lost

TABLE 8.17 *γ-Radiolysis of Diethyl Ether (Ref. 250)*

Product	G(Product) Vapor (25°C, 0.5 atm)	Liquid
H_2	6.4	3.4
CH_4	1.3	0.4
HCHO	0.9	0.0
C_2H_4	2.1	0.6
C_2H_6	0.44	1.3
CH_3CHO	1.4	0.2
CH_3CH_2OH	0.65	2.1
$CH_2{=}CHOC_2H_5$	0.4	1.1
$C_2H_5CH(CH_3)OC_2H_5$	0.83	0.74
$CH_3CH(CH_3)OC_2H_5$	1.0	0.07
$(CH_3CHOC_2H_5)$	2.4	2.6
$(-\text{Ether})^a$	11.2	11.3

^{a} Based on the amount of carbon in the products.

mainly from a carbon atom adjacent to the ether oxygen (250):

$$(C_2H_5)_2O \rightsquigarrow (C_2H_5)_2O^*, (C_2H_5)_2O^+, e^- \qquad (8.229)$$

$$(C_2H_5)_2O^+ + (C_2H_5)_2O \rightarrow (C_2H_5)_2OH^+ + CH_3\dot{C}HOC_2H_5 \quad (8.230)$$

$$(C_2H_5)_2OH^+ + e^- \rightarrow (C_2H_5)_2O + H \qquad (8.231)$$

$$(C_2H_5)_2O^* \rightarrow C_2H_5O\cdot + \cdot C_2H_5 \qquad (8.232)$$

$$H + (C_2H_5)_2O \rightarrow H_2 + CH_3\dot{C}HOC_2H_5 \qquad (8.233)$$

$$\cdot C_2H_5 + (C_2H_5)_2O \rightarrow C_2H_6 + CH_3\dot{C}HOC_2H_5 \qquad (8.234)$$

$$C_2H_5O\cdot + (C_2H_5)_2O \rightarrow C_2H_5OH + CH_3\dot{C}HOC_2H_5 \qquad (8.235)$$

$$C_2H_5O\cdot + \cdot C_2H_5 \rightarrow C_2H_5OH + C_2H_4 \qquad (8.236)$$

$$\cdot C_2H_5 + CH_3\dot{C}HOC_2H_5 \rightarrow sec\text{-}C_4H_9OC_2H_5 \qquad (8.237)$$

$$\rightarrow C_2H_6 + CH_2{=}CHOC_2H_5 \qquad (8.238)$$

$$2\,CH_3\dot{C}HOC_2H_5 \rightarrow (CH_3CHOC_2H_5)_2 \qquad (8.239)$$

The free ion yield for diethyl ether (G 0.3) is considerably lower than that for ethyl alcohol ($G \sim 1.8$), and most ions undergo geminate recombination in the spurs. Radical scavengers reduce the yields of all liquid-phase products, confirming that they produced primarily by radical reactions. In the vapor phase, the product distribution is quite different with, as is usual, a greater proportion of smaller molecules. Ng and Freeman (250) attribute the differences to the dissociation of excited radicals in the vapor,

$$C_2H_5O^* \rightarrow CH_3CHO + H \qquad (8.240)$$

$$\rightarrow \cdot CH_3 + HCHO \qquad (8.241)$$

$$C_2H_5{}^* \rightarrow C_2H_4 + H \qquad (8.242)$$

to caging in the liquid which reduces the yield of methyl radicals by

$$(C_2H_5)_2O^* \rightarrow [\cdot CH_3 + \cdot CH_2OC_2H_5] \rightarrow (C_2H_5)_2O \qquad (8.243)$$

and to differences in the ratio of disproportionation to combination for radicals in the two phases.

Irradiation of diethyl ether in the presence of oxygen produces relatively large amounts of carbonyl compounds ($G \sim 30$) and peroxides ($G \sim 15$) [reported by N. A. Bach (251)], and it was suggested that these products arise in a chain reaction for which the propagation steps are

$$CH_3\dot{C}HOC_2H_5 + O_2 \rightarrow CH_3CH(O_2\cdot)OC_2H_5 \qquad (8.244)$$

$$CH_3CH(O_2 \cdot)OC_2H_5 + C_2H_5OC_2H_5 \rightarrow$$
$$CH_3CH(O_2H)OC_2H_5 + CH_3\dot{C}HOC_2H_5 \quad (8.245)$$

The hydroperoxide subsequently decays to give carbonyl products.

A survey of the hydrocarbon products from six other aliphatic ethers (252) showed that the hydrocarbons formed in greatest yield resulted from the rupture of the alkyl–oxygen bond, in contrast with the alcohols where the final products are mainly the result of C—H and C—C bond scission. However, the ethers resemble the alcohols in giving the highest yields of hydrogen from ethers with straight-chain alkyl groups and the highest hydrocarbon yields from ethers with branched-chain groups; for γ-irradiated dibutyl and di-isopropyl ether, for example, $G(H_2)$ are 2.9 and 2.4, respectively, and $G(CH_4)$ are 0.06 and 1.7, respectively (253).

A limited number of studies have been carried out on the cyclic ethers dioxan, tetrahydrofuran, and methyltetrahydrofuran. The more important products from dioxan (254), with their G values, are H_2 2.1, CO 0.3, C_2H_4 0.56, OHCCHO 0.28, CH_3CHO 0.15, and polymer 3.65. From tetrahydrofuran (255) the major products are H_2 2.64, C_2H_4 0.52, cyclopropane 0.78, C_3H_6 0.17, butanol 0.32, and polymer 5.0. With both ethers, irradiation causes ring opening at the ether linkage and the reactive biradicals formed take part in a series of radical reactions; the polymeric products result from secondary reactions involving unsaturated radiolysis products.

CARBONYL COMPOUNDS, ACIDS, AND ESTERS

As with the alcohols and ethers, radiolytic decomposition of these compounds centers on the functional group and the adjacent carbon atoms. Rupture of a C—O bond and loss of α-hydrogen occur with the ketones, decarboxylation with the acids, and elimination of CO and CO_2 with the esters; hydrocarbons are significant products in each case.

Carbonyl Compounds

The photochemical and thermal dissociation of acetone have been extensively studied and provide essential information on free radical reactions of the ketone. Acetone has also received considerable attention from radiation chemists (e.g., 256–262) and a partial list of radiolysis products is given in Table 8.18; G values are dose-rate and LET dependent.

Primary steps in the photochemical dissociation (109, 263) are

$$CH_3COCH_3 + h\nu \rightarrow CH_3\dot{C}O + \cdot CH_3 \quad (8.246)$$
$$\rightarrow H + \cdot CH_2COCH_3 \ (\text{or } H_2 + CHCOCH_3) \quad (8.247)$$

TABLE 8.18   γ-Radiolysis of Acetone (256)

Product	G(Product)[a]	
	Liquid	Liquid + Iodine
H_2	0.50	0.60
CO	0.72	0.72
CH_4	1.74	0.41
C_2H_6	0.24	0.30
CH_3CHO	0.09	0
$CH_3COC_2H_5$	0.18	0.01
$CH_3CH(OH)CH_3$	0.09	0.01
CH_3COOH	0.31[b]	
$(CH_3CO)_2$	0.28	0.05
$(CH_3COCH_2)_2$	0.27	
$CH_3COCH_2COCH_3$	0.11[b]	

[a] 22°C, 5×10^{16} eV ml^{-1} min^{-1}. Iodine concentration 3.6×10^{-3} M for gaseous products and 0.1 M for liquid products.
[b] 28°C, 2.4×10^{17} eV ml^{-1} min^{-1}.

In the vapor phase, and particularly at short wavelengths, the acetyl radical may be formed with sufficient energy to dissociate.

$$CH_3CO^* \rightarrow \cdot CH_3 + CO \tag{8.248}$$

However quantum yields are lower in liquid acetone, and Pieck and Steacie (264) have suggested that this is due to both deactivation of excited molecules and radicals and to caging of radical products, leading to back reactions that reform acetone.

The dominant ions in the mass spectrum of acetone are the parent ion, $CH_3COCH_3^+$, and CH_3CO^+, formed by loss of a methyl radical. Both ions enter into ion-molecule reactions (260),

$$CH_3COCH_3^+ + CH_3COCH_3 \rightarrow (CH_3)_2COH^+ + \cdot CH_2COCH_3 \tag{8.249}$$
$$\rightarrow CH_3COCH_3COCH_3^+ + \cdot CH_3 \tag{8.250}$$
$$CH_3CO^+ + CH_3COCH_3 \rightarrow CH_3COCH_3COCH_3^+ \tag{8.251}$$
$$CH_3COCH_3COCH_3^+ + CH_3COCH_3 \rightarrow$$
$$CH_2CO + H^+(CH_3COCH_3)_2 \tag{8.252}$$

The cation from the last reaction adds further acetone molecules to become a clustered or solvated proton in vapor and liquid respectively. Liquid acetone rapidly scavenges thermal electrons,

$$CH_3COCH_3 + e^- \rightarrow CH_3COCH_3^- \tag{8.253}$$

and the neutralization reaction for both geminate and free electrons is

$$H^+(CH_3COCH_3)_n + CH_3COCH_3^- \rightarrow$$
$$(CH_3)_2\dot{C}OH + nCH_3COCH_3 \quad (8.254)$$

Estimated ion and excited state yields, based on pulse radiolysis studies (257–259), are G(geminate ion) 2.3, G(free ion) 0.75, G(singlet) 0.34, and G(triplet) 1.0. The excited states are believed to be formed by direct excitation rather than by geminate recombination of ions (260).

Dissociation and reaction of excited molecules and ions as described above produces CO and possibly H_2 as molecular products and the radicals H, $\cdot CH_3$, $CH_3\dot{C}O$, $\cdot CH_2COCH_3$, and $(CH_3)_2\dot{C}OH$. Ketene, CH_2CO, is also formed. Combination and disproportionation of the radicals and hydrogen abstraction by H and $\cdot CH_3$ give most of the observed products, e.g.,

$$H + CH_3COCH_3 \rightarrow H_2 + \cdot CH_2COCH_3$$
$$k_{255} = 1.9 \times 10^6 \ M^{-1} \ \text{sec}^{-1} \quad (8.255)$$
$$\cdot CH_3 + CH_3COCH_3 \rightarrow CH_4 + \cdot CH_2COCH_3$$
$$k_{256} = 370 \ M^{-1} \ \text{sec}^{-1} \quad (8.256)$$
$$CH_3\dot{C}O + CH_3COCH_3 \rightarrow CH_3CHO + \cdot CH_2COCH_3 \quad (8.257)$$
$$2\cdot CH_3 \rightarrow C_2H_6 \quad (8.122)$$
$$2 \ CH_3\dot{C}O \rightarrow (CH_3CO)_2 \quad (8.258)$$
$$2\cdot CH_2COCH_3 \rightarrow (CH_3COCH_2)_2 \quad (8.259)$$
$$CH_3\dot{C}O + \cdot CH_2COCH_3 \rightarrow CH_3COCH_2COCH_3 \quad (8.260)$$
$$2 \ (CH_3)_2\dot{C}OH \rightarrow CH_3COCH_3 + CH_3CH(OH)CH_3 \quad (8.261)$$
$$\cdot CH_3 + (CH_3)_2\dot{C}OH \rightarrow CH_4 + CH_3COCH_3 \quad (8.262)$$

In keeping with these reactions, radical scavengers such as iodine reduce most product yields except those of H_2, CO, and C_2H_6. Carbon monoxide is formed by the dissociation of excited acetyl radicals (Eq. 8.248) and would not be affected by radical scavengers, but thermal hydrogen atoms and methyl radicals would be scavenged by the concentrations of iodine used by Barker (Table 8.18) and, indeed, the yield of methane is reduced. Hydrogen, therefore, is either formed *via* hot hydrogen atoms (which react too rapidly to be scavenged by moderate concentrations of iodine), or reaction 8.247 which produces molecular hydrogen rather than hydrogen atoms. Ethane is apparently immune to moderate concentrations of scavenger because it is formed in the spur or cage reaction,

$$CH_3COCH_3^* \rightarrow [CH_3CO^* + \cdot CH_3] \rightarrow$$
$$[\cdot CH_3 + CO + \cdot CH_3] \rightarrow [C_2H_6 + CO] \quad (8.263)$$

Consistent with this belief is the observation that irradiation of solid mixtures of CH_3COCH_3 and CD_3COCD_3 at $-196°C$ gives ethane consisting mainly of C_2H_6 and C_2D_6 (265); in the liquid phase, where diffusion within spurs is possible, the ethane formed also contains CH_3CD_3. Methane is formed mainly by thermal methyl radicals, but in part ($\sim 25\%$) by hot methyl radicals which are not scavenged. Acetic acid may be formed by the addition of water, present as an impurity or a product of radiolysis, to ketene

$$H_2O + CH_2{=}C{=}O \rightarrow CH_3COOH \qquad (8.264)$$

but this cannot be regarded as established. Other acetates, for example isopropyl acetate ($G \sim 0.15$; 262), have been identified as radiolysis products and might be formed by the addition of alcohols to ketene,

$$CH_3CH(OH)CH_3 + CH_2CO \rightarrow CH_3COOCH(CH_3)_2 \qquad (8.265)$$

The yield of isopropanol does in fact decrease with increasing absorbed dose, suggesting that it is being consumed in a secondary reaction as the irradiation progresses.

Barker (256) has shown that the yields of gaseous products and of methyl-ethyl ketone are dependent on both dose rate and LET, the relative yields of ethane and methylethyl ketone being increased by increasing dose rate or LET and the yield of methane decreased. These effects can be explained by competition between reaction 8.256, the relatively slow scavenging of methyl radicals by the solvent to form methane, and interradical reactions of $\cdot CH_3$ such as 8.122 and 8.266,

$$\cdot CH_3 + \cdot CH_3COCH_3 \rightarrow C_2H_5COCH_3 \qquad (8.266)$$

High dose rates increase the probability of the interradical reactions by increasing the steady-state radical concentration, but any increase in these reactions occurs at the expense of the radical-solvent reaction (Eq. 8.256). High LET favors the interradical reactions by increasing the proportion of energy deposited in blobs and short tracks, where the initial concentration of radicals is high compared with the concentration in isolated spurs. More recently, Matsui and Imamura (261) have used cyclotron-accelerated ^{12}C and ^{14}N ions, which represent the highest range of LET presently attainable, to irradiate acetone (Table 8.19) and conclude that additional processes must be postulated to account for the enhanced yields of H_2 and CO with these very high LET particles. Possibilities that they suggest are that highly (vibrationally or electronically) excited states formed in the high LET tracks may decompose differently to the lower excited states produced by lower-LET particles, or that thermal decomposition of radicals, represented by

$$C_nH_{2n+1} \rightarrow C_nH_{2n} + H \qquad (8.267)$$

T A B L E 8.19   *Effect of LET on the Gaseous Products from Acetone (261)*

Radiation	^{60}Co-γ	6.9-MeV He-ions	67-MeV C-ions	65.7-MeV N-ions
$-dE/dx$ (eV/Å)	0.02	13.1	39	55.3
$G(H_2)$	0.96	1.47	2.36	2.71
$G(CO)$	0.56	0.80	1.05	1.22
$G(CH_4)$	1.76	0.97	0.99	0.96
$G(C_2H_4)$	0.04	0.12	0.21	0.24
$G(C_2H_6)$	0.30	0.50	0.56	0.64

may occur in high temperature zones within the tracks (the thermal spike model).

Radiolysis of the higher ketones (209, 262, 266) is generally similar to that of acetone with predominantly C—CO bond rupture. Thus ethane rather than methane is the predominant hydrocarbon from diethyl ketone, for which the major products and their G values (262) are H_2 1.10, CO 1.11, CH_4 0.12, C_2H_4 0.48, C_2H_6 1.94, C_2H_5CHO 0.16, and butane 0.17. The yields suggest that a larger fraction of the acyl radicals dissociate

$$C_2H_5CO^* \rightarrow \cdot C_2H_5 + CO \qquad (8.268)$$

than in the case of acetone (Eq. 8.248) and that ethylene is formed by

$$C_2H_5COC_2H_5^* \rightarrow C_2H_5CHO + C_2H_4 \qquad (8.269)$$

The product distribution for methylethyl ketone (262) indicates that cleavage occurs predominantly between the carbonyl group and the larger of the two alkyl groups to give $CH_3\dot{C}O + \cdot C_2H_5$ rather than $\cdot CH_3 + C_2H_5\dot{C}O$.

Little has been published on the radiation chemistry of the aldehydes although it has been established that they are reactive and give predominantly polymeric products, even at low temperatures (209). Aldehydic products formed during the radiolysis of ketones and other organic compounds tend, like the alkenes, to take part in secondary radical-induced reactions so that G(aldehyde) falls with increasing dose.

Carboxylic Acids

Radiolysis yields for acetic acid, the carboxylic acid most thoroughly studied, are given in Table 8.20. Newton (267) also identified acetaldehyde, acetone, and methyl acetate as products but made no estimate of their yields; it seems probable that other higher-molecular weight products are also formed. Electron spin resonance studies on irradiated solid acetic acid (268, 270) showed that the predominant paramagnetic species trapped at $-196°C$ was the negative ion $CH_3\dot{C}OOH^-$ and that this decayed as the temperature was raised toward room temperature; the parent cation,

T A B L E 8.20   *Radiolysis of Acetic Acid and Potassium Acetate*

	G(Product)		
Product	Liquid CH_3COOH (α-Irradiation; ref. 267)	Liquid CH_3COOH (γ-Irradiation; ref. 268)	Solid CH_3COOK (γ-Irradiation; ref. 268)
H_2	0.52	0.53	0.71
H_2O	2.15		
CO	0.38	0.22	low
CO_2	4.04	5.36	0.17^b
CH_4	1.38	3.90	0.28
C_2H_6	0.85	0.48	0.056^b
CH_3COCH_3		1.0^a	

^{a} Ref. 269.
^{b} After dissolution in 0.05 M H_2SO_4.

$CH_3\dot{C}OOH^+$, was not seen. Based in part of the interpretation of the esr experiments, Ayscough and his colleagues (270) proposed a radiolysis mechanism for liquid acetic acid which includes the following steps:

$$CH_3COOH \rightsquigarrow CH_3COOH^*, CH_3COOH^+, e^- \quad (8.269)$$

$$CH_3COOH^+ + CH_3COOH \rightarrow CH_3C(OH)_2^+ + CH_3COO\cdot \quad (8.270)$$

$$CH_3COO\cdot \rightarrow \cdot CH_3 + CO_2 \quad (8.271)$$

$$CH_3COOH + e^- \rightarrow CH_3\dot{C}OOH^- \rightarrow$$
$$CH_3COO^- + H \quad (8.272)$$

$$CH_3C(OH)_2^+ + CH_3COO^- \rightarrow 2CH_3COOH \quad (8.273)$$

$$H + CH_3COOH \rightarrow H_2 + CH_3COO\cdot \quad (20\%) \quad (8.274)$$

$$\rightarrow CH_3\dot{C}(OH)_2 \rightarrow$$
$$CH_3\dot{C}O + H_2O \quad (80\%) \quad (8.275)$$

$$CH_3\dot{C}O \rightarrow \cdot CH_3 + CO \quad (8.248)$$

$$\cdot CH_3 + CH_3COOH \rightarrow CH_4 + \cdot CH_2COOH \quad (8.276)$$

$$\rightarrow CH_4 + CH_3COO\cdot \quad (8.277)$$

These are followed by combination and disproportionation of the radicals formed to give such products as ethane, acetaldehyde, acetone, and methyl acetate. The mechanism is able to account quantitatively for the products observed without including a contribution from the dissociation of excited molecules, although photolysis of acetic acid vapor (109) has been shown to give CO, CO_2, and CH_4 and the radicals H, OH, and $\cdot CH_3$. The mechanism is consistent with a decrease in the yields of CO_2 and CH_4 and an increase in C_2H_6 yield at higher LET, which will favor interradical reactions at the expense of radical dissociation and radical–solvent reactions.

Potassium acetate is more stable toward radiation damage than acetic acid and, upon irradiation at $-196°C$, was found to contain $\cdot CH_2COO^-$ and $\cdot CH_3$ as the main paramagnetic species rather than $CH_3\dot{C}OOH^-$ (268). The greater stability of potassium acetate compared with acetic acid is attributed to the ionic character of the solid which allows electrons to migrate through the solid until they encounter, and neutralize, a positive hole, without producing any chemical change. The structure of acetic acid (both liquid and solid), however, favors the ion-molecule reaction 8.270, since the molecules are strongly hydrogen bonded to each other, $RC(OH)O \cdots HOCOR$, and electron trapping by a solvent molecule (Eq. 8.272), both of which lead to chemical reaction.

Branched-chain acids give higher yields of carbon dioxide than the straight-chain isomer (271), particularly when the carboxyl group is attached at the point of branching. Thus isobutyric acid, $(CH_3)_2CHCOOH$, gives an exceptionally high yield of $CO_2(G\ 14.4)$, which is considerably greater than the yield from the straight-chain isomer, butyric acid $(G\ 5.0)$.

Oxalic acid and, to a lesser extent, succinic acid have generated considerable interest as chemical dosimeters, the absorbed dose being determined from the amount of degradation produced (272). For anhydrous oxalic acid, the initial yields are $G(-\text{oxalic acid}) = G(CO_2) = $ about 7, and the early steps in the radiolysis mechanism are probably similar to those described for acetic acid, namely an ion-molecule reaction between the parent positive ion and an adjacent, hydrogen-bonded, acid molecule, and electron capture followed by dissociation:

$$HOOC-COOH^+ \cdots HOOC-COOH \rightarrow$$
$$HOOC-C(OH)_2^+ + \cdot OOC-COOH \quad (8.278)$$

$$\cdot OOC-COOH \rightarrow CO_2 + \cdot COOH \quad (8.279)$$

$$HOOC-COOH + e^- \rightarrow$$
$$(HOOC-COOH)^- \rightarrow H\dot{C}OOH^- + CO_2 \quad (8.280)$$

The radicals shown have been identified in irradiated oxalic acid by esr measurements (272), but the subsequent reactions and radiolysis products, other than CO_2, are uncertain. The presence of water of crystallization affects the radiolysis mechanism, and oxalic acid dihydrate gives about half the yield of carbon dioxide of the anhydrous acid. The oxalates are more stable than oxalic acid itself, but the stability depends on the character of the bonds present, being greater the more ionic in character the metal-oxalate bonds; carbon dioxide or metal carbonate are normally products of the reaction.

Esters

Radiolysis products from methyl acetate, a fairly typical ester, are given in Table 8.21. Hummel (274) obtained similar results and also identified

T A B L E 8.21   *γ-Radiolysis of Methyl Acetate (273)*

Product	G(Product)			
	Vapor (27°C)	Vapor + Iodine (65°C)	Liquid (27°C)	Liquid + 0.01 M Iodine (27°C)
H_2	2.14	0.54	0.76	0.57
CH_4	1.76	0.05	2.03	0.50
C_2H_6	0.43	0.04	0.34	0.27
CO	2.68	2.00	1.64	1.51
CO_2	1.10	1.0	0.95	0.90
CH_3OCH_3	0.034	0	0.15	0.14

formaldehyde (G 0.42) and 16 liquid products, of which the more abundant were methanol (G 0.80) and acetic acid (G 0.50). Ausloos and Trumbore (273) were able to identify the methyl group from which products originate and to gain other information about the radiolysis process by irradiating the deuterium substituted ester, CH_3COOCD_3.

Although a complete radiolysis mechanism is not available for methyl acetate, the radiolysis results can be explained on the basis of known free-radical reactions and the assumption that the primary stages are

$$CH_3COOCH_3 \rightsquigarrow CH_3CO\cdot + \cdot OCH_3 \rightarrow \cdot CH_3 + CO + \cdot OCH_3 \quad (8.281)$$

$$\rightsquigarrow \cdot CH_3 + CH_3COO\cdot \text{ (or } \cdot COOCH_3) \rightarrow$$
$$2\cdot CH_3 + CO_2 \quad (8.282)$$

$$\rightsquigarrow H + \cdot CH_2COOCH_3 \text{ (or } CH_3COOCH_2\cdot) \quad (8.283)$$

The first two reactions will almost certainly take place with the intermediate formation of $CH_3CO\cdot$ and $CH_3COO\cdot$ (or $\cdot COOCH_3$), respectively. Both reactions have been observed in the photolysis and mercury photosensitized decomposition of methyl acetate, where the first (Eq. 8.281) predominates. However, in these photochemical reactions, the radical $CH_3CO\cdot$ is sufficiently stable to react to a large extent without dissociation, and is the precursor of such products as biacetyl, acetaldehyde, and acetone. These products are only formed in small amounts upon irradiation, suggesting that here the acetyl radical is formed with sufficient energy to dissociate almost immediately, as implied by Eq. 8.281.

The primary stages will be followed by a variety of radical reactions including, in all probability, the following:

$$H + CH_3COOCH_3 \rightarrow H_2 + \cdot CH_2COOCH_3$$
$$(\text{or } CH_3\overset{.}{C}OOCH_2 \cdot) \quad (8.284)$$

$$\cdot CH_3 + CH_3COOCH_3 \rightarrow CH_4 + \cdot CH_2COOCH_3$$
$$(\text{or } CH_3\overset{.}{C}OOCH_2 \cdot) \quad (8.285)$$

$$CH_3O \cdot + CH_3COOCH_3 \rightarrow CH_3OH + \cdot CH_2COOCH_3$$
$$(\text{or } CH_3\overset{.}{C}OOCH_2 \cdot) \quad (8.286)$$

$$2\cdot CH_3 \rightarrow C_2H_6 \quad (8.122)$$

$$CH_3O \cdot + \cdot CH_3 \rightarrow CH_4 + HCHO \quad (8.287)$$

$$\rightarrow CH_3OCH_3 \quad (8.288)$$

$$2\cdot CH_2COOCH_3 \rightarrow (CH_2COOCH_3)_2 \quad (8.289)$$

though this can only be a partial list.

By carrying out scavenger experiments in both liquid and vapor phases it is possible to distinguish between the formation of "molecular" products by radical reactions in a spur and by unimolecular dissociation of an excited molecule. In the radiolysis of methyl acetate, for example, ethane may be formed by one or both of the molecular processes:

$$[2\cdot CH_3] \rightarrow C_2H_6 \quad (8.122)$$

$$CH_3COOCH_3{}^* \rightarrow C_2H_6 + CO_2 \quad (8.290)$$

In the liquid phase the yield of ethane is little affected by added scavenger (Table 8.21), and it might be formed by either reaction. However, in the vapor phase, where spur formation and radical caging are not significant effects, the ethane yield is markedly reduced by scavenger, suggesting that ethane is formed by the radical reaction. In the liquid phase this reaction must occur in spurs, or within a solvent cage where a pair of methyl radicals are formed together (e.g., by reaction 8.282). The limited reduction in carbon dioxide yield in the presence of scavenger excludes the possibility that the scavenger quenches an excited state (which would otherwise dissociate, giving ethane and carbon dioxide, as in Eq. 8.290), or acts as an electron trap, interfering with an ion-neutralization process that leads to both ethane and carbon dioxide.

The scavenging experiments with methyl acetate also show that methane is formed almost entirely by radical reactions, though in the liquid these are not entirely accessible to scavenger. Hydrogen is apparently formed in part by unimolecular dissociation or, more probably, by hydrogen abstraction by hot hydrogen atoms, which are not scavenged by iodine. Dimethyl ether, CH_3OCH_3, may be formed in a similar manner to ethane by caging of the radicals formed in reaction 8.281.

NITROGEN COMPOUNDS

Few pure nitrogen compounds have been investigated in detail to the present time, most of the interest in these compounds being confined to their solutions in water or to their use as solvents in pulse radiolysis studies of the solvated electron. Radiolysis data for simple compounds which have been irradiated are summarized in Table 8.22. Yields are for the γ-irradiated liquid unless otherwise stated; minor products are not listed. Yields for nitrobenzene

TABLE 8.22   Radiolysis of Organic Nitrogen Compounds

Compound	Products and G Values	Ref.
CH_3NO_2	HCHO 2.0, CH_3ONO 1.1, cis-$(CH_3NO)_2$ 0.53.	275, 276
$C_6H_5NO_2$	N_2 0.16, C_6H_5NO 0.13, C_6H_5OH 0.088, dinitrobenzenes 0.012, nitrobiphenyls 0.16, dinitrobiphenyls 0.038.	277
CH_3CN	H_2 0.67, HCN 0.2, CH_4 0.65, CH_3CH_2CN 0.16, $(CH_3CN)_2$ 0.39, polymer 4.2.	278 to 280
CH_3NH_2	H_2 5.4, CH_4 0.18, $CH_3N_2CH_3$ 0.7, $(CH_2NH_2)_2$ 3.4.	281
	$G(e_{solv}^-)$ 2.2; 4.75 in basic soln.	282
	Vapor (25°C, 100 to 700 mm Hg): H_2 12.5, CH_4 0.4.	283
$(CH_3)_2NH$	Vapor: H_2 11.5, CH_4 2.0.	283
$(CH_3)_3N$	Vapor: H_2 9.5, CH_4 1.6.	283
$C_2H_5NH_2$	Vapor: H_2 9.5, CH_4 2.42.	283
$C_6H_5NH_2$	H_2 0.12, NH_3 0.25, C_6H_6 0.04, $(C_6H_5)_2NH$ 0.039, aminodiphenylamines 0.042, diaminobiphenyls 0.038.	284
$HCON(CH_3)_2$	H_2 0.14, CO 2.6, CH_4 0.93, $(CH_3)_2NH$ 2.6.	285
$CH_3CONHC_2H_5$	H_2 0.8, CO 0.4, CH_3CHO 1.1, $C_2H_5NH_2$ 1.2.	286
Pyrrole	H_2 0.20, polymer ~ 2.05	287
3-Pyrroline	H_2 2.34	287
Pyrollidine	H_2 6.35, polymer ~ 2.41	287
Pyrazole	H_2 0.04, N_2 0.12	287
Tetrazole	H_2 trace, N_2 0.96	287
Pyridine	H_2 0.025, polymer 3.1 to 3.8	287
Pyridazine	H_2 0.046, N_2 0.47, polymer 2.75	287, 288
Pyrimidine	H_2 0.030, polymer 3.5	287, 288

and aniline are low, and the aromatic system is clearly stabilizing the molecule compared with nitromethane and the aliphatic amines. Heterocyclic compounds that have aromatic character are also resistant toward radiation if the yield of hydrogen is taken as the criterion, although they may give appreciable yields of polymer. Benzene, for comparison, gives $G(H_2)$ 0.039 and G(polymer) 0.94. Heterocyclic compounds that have two nitrogen atoms adjacent to each other in the ring (e.g., pyrazole, tetrazole, and pyridazine) give significant amounts of nitrogen. Partially hydrogenated heterocyclic rings (e.g., 3-pyrroline and pyrrolidine) are not stabilized toward radiation damage and give hydrogen yields two orders of magnitude greater than the rings with aromatic character.

The nitrogen compound with the most intriguing radiation chemistry, however, is choline chloride. Crystalline choline chloride is abnormally sensitive to radiation, breaking down to trimethylamine hydrochloride and acetaldehyde with G values as high as 55,000 (289); the G values are dependent on dose rate and temperature and the type of radiation.

$$[(CH_3)_3N^+CH_2CH_2OH]Cl^- \rightsquigarrow (CH_3)_3NH^+Cl^- + CH_3CHO \qquad (8.291)$$

The decomposition is obviously a chain reaction, and it appears unique among chain decomposition reactions in being limited to the solid, crystalline, state; aqueous or alcoholic solutions of choline chloride decompose with G values less than 5 (290). A number of compounds chemically similar to choline chloride have been irradiated (290), but only choline bromide, $[(CH_3)_3N^+CH_2CH_2OH]Br^-$, approaches the sensitivity of the chloride, and this decomposes with yields about one third or one quarter those for choline chloride. The radiation decomposition of choline chloride increases with increasing temperature in the region 18 to 50°C, but the material is much more stable toward radiation at 150°C than it is at room temperature (291). It appears that the chain-decomposition is retarded by a phase transition which occurs in the crystals at about 75°C (292). Thus the high yields in the decomposition reaction are evidently closely related to the crystal structure in solid choline chloride.

Lemmon and Lindblom (290) have shown that substitution of deuterium for hydrogen in choline chloride only lowers the radiation sensitivity appreciably when hydrogen in the hydroxyl-substituted methylene group is replaced. Hydrogen abstraction from this group is therefore a likely step in the decomposition. Subsequently, on the basis of data from kinetic and esr studies, Lemmon and his colleagues (293) proposed a chain mechanism with the chain propagating step,

$$[:CH_2CH_2OH^*]^- + (CH_3)_2N^+CH_2CH_2OH \rightarrow$$
$$CH_3CHO + (CH_3)_3N^+H + [:CH_2CH_2OH^*]^- \qquad (8.292)$$

The chain carrier is an excited biradical derived from ethanol.

REFERENCES

The radiation chemistry of organic compounds is described in several chapters in *Fundamental Processes in Radiation Chemistry* (ed. P. Ausloos) Interscience, New York, 1968, and in more specialized reviews in *Actions Chimiques et Biologiques des Radiations* (ed. M. Haissinsky) Masson et Cie, Paris, Vol. 1, 1955 and subsequent volumes, and *Radiation Research Reviews,* Elsevier, Amsterdam, Vol. 1, no. 1, March 1968 and quarterly thereafter.

1. W. G. Burns and R. Barker in *Progress in Reaction Kinetics*, Pergamon, Oxford, Vol. 3, 1965, p. 303, and in *Aspects of Hydrocarbon Radiolysis* (eds. T. Gäumann and J. Hoigné), Academic, New York, 1968, p. 33.
2. J. G. Wilson, *Radiat. Res. Rev.*, **4**, 71 (1972).
3. B. Chutný and J. Kučera, *Radiat. Res. Rev.*, **5**, 1, 55, 93, 135 (1974).
4. *Radiolysis of Hydrocarbons* (ed. A. V. Topchiev, transl. R. A. Holroyd), Elsevier, New York, 1964.
5. T. J. Hardwick in *Actions Chimiques et Biologiques des Radiations* (ed. M. Haissinsky), Masson et Cie, Paris, Vol. 10, 1966, p. 125.
6. P. Ausloos and S. G. Lias in *Actions Chimiques et Biologiques des Radiations* (ed. M. Haissinsky), Masson et Cie, Paris, Vol. 11, 1967, p. 1.
7. A. O. Allen in *Current Topics in Radiation Research* (eds. M. Ebert and A. Howard), Wiley, New York, Vol. 4, 1968, p. 1.
8. G. R. Freeman, *Radiat. Res. Rev.*, **1**, 1 (1968).
9. *Aspects of Hydrocarbon Radiolysis* (eds. T. Gäumann and J. Hoigné), Academic, New York, 1968.
10. R. A. Holroyd in *Fundamental Processes in Radiation Chemistry* (ed. P. Ausloos), Interscience, New York, 1968, p. 413.
11. G. G. Meisels in *Fundamental Processes in Radiation Chemistry* (ed. P. Ausloos), Interscience, New York, 1968, p. 347.
12. R. A. Back and N. Miller, *Trans. Faraday Soc.*, **55**, 911 (1959).
13. H. A. Dewhurst, *J. Phys. Chem.*, **61**, 1466 (1957).
14. H. A. Dewhurst, *J. Am. Chem. Soc.*, **80**, 5607 (1958).
15. T. J. Hardwick, *J. Phys. Chem.*, **66**, 1611 (1962).
16. R. H. Schuler and R. R. Kuntz, *J. Phys. Chem.*, **67**, 1004 (1963).
17. M. S. Matheson, *Nucleonics*, **19**, 57 (October 1961).
18. M. S. Kharasch, P. C. Chang, and C. D. Wagner, *J. Org. Chem.*, **23**, 779 (1958).
19. B. R. Wakeford and G. R. Freeman, *J. Phys. Chem.*, **68**, 2635 (1964).
20. W. A. Cramer in *Aspects of Hydrocarbon Radiolysis* (eds. T. Gäumann and J. Hoigné), Academic, New York, 1968, p. 153.
21. H. Milton and T. Gäumann, *Helv. Chim. Acta.*, **57**, 721 (1974).
22. H. A. Dewhurst, *J. Phys. Chem.*, **23**, 813 (1959).
23. P. J. Dyne and J. A. Stone, *Can. J. Chem.*, **39**, 2381 (1961).
24. L. J. Forrestal and W. H. Hamill, *J. Am. Chem. Soc.*, **83**, 1535 (1961).
25. M. Burton in *Actions Chimiques et Biologiques des Radiations* (ed. M. Haissinsky), Masson et Cie, Paris, Vol. 3, 1958, p. 19.
26. S. K. Ho and G. R. Freeman, *J. Phys. Chem.*, **68**, 2189 (1964).
27. K. H. Jones, *J. Phys. Chem.*, **71**, 709 (1967).
28. T. W. Woodward and R. A. Back, *Can. J. Chem.*, **41**, 1463 (1963).
29. P. J. Dyne and J. W. Fletcher, *Can. J. Chem.*, **38**, 851 (1960).

30. R. H. Schuler and A. O. Allen, *J. Am. Chem. Soc.*, **77**, 507 (1955).
31. M. Hamashima, M. P. Reddy, and M. Burton, *J. Phys. Chem.*, **62**, 246 (1958).
32. J. Prévé and G. Gaudemaris, *CR*, **250**, 3470 (1960).
33. W. S. Guentner, T. J. Hardwick, and R. P. Nejak, *J. Chem. Phys.*, **30**, 601 (1959).
34. L. W. Sieck, S. K. Searles, R. E. Rebbert, and P. Ausloos, *J. Phys. Chem.*, **74**, 3829 (1970).
35. R. Barker and M. R. H. Hill, *Nature*, **194**, 277 (1962).
36. J. Milhaud and J. Durup, *CR*, **260**, 6363 (1965).
37. F. P. Abramson and J. H. Futrell, *J. Phys. Chem.*, **71**, 1233 (1967).
38. J. H. Futrell and T. O. Tiernan in *Fundamental Processes in Radiation Chemistry* (ed. P. Ausloos), Interscience, New York, 1968, p. 171.
39. R. D. Doepker and P. Ausloos, *J. Chem. Phys.*, **42**, 3746 (1965).
40. G. R. Freeman and J. M. Fayadh, *J. Chem. Phys.*, **43**, 86 (1965).
41. J. P. Keene, E. J. Land, and A. J. Swallow, *J. Am. Chem. Soc.*, **87**, 5284 (1965).
42. E. Hayon, *J. Chem. Phys.*, **53**, 2353 (1970).
43. T. Capellos and A. O. Allen, *J. Phys. Chem.*, **74**, 840 (1970).
44. J. M. Warman, K.-D. Asmus, and R. H. Schuler, *J. Phys. Chem.*, **73**, 931 (1969).
45. G. R. Freeman and T. E. M. Sambrook, *J. Phys. Chem.*, **78**, 102 (1974).
46. M. G. Robinson and G. R. Freeman, *Can. J. Chem.*, **51**, 641 (1973).
47. R. H. Schuler and P. P. Infelta, *J. Phys. Chem.*, **76**, 3821 (1972).
48. P. P. Infelta and S. J. Rzad, *J. Chem. Phys.*, **58**, 3775 (1973).
49. S. Sato, R. Yugeta, K. Shinsaka, and T. Terao, *Bull. Chem. Soc. Jap.*, **39**, 156 (1966).
50. N. H. Sagert and A. S. Blair, *Can. J. Chem.*, **45**, 1351 (1967).
51. K. M. Bansal and R. H. Shuler, *J. Phys. Chem.*, **74**, 3924 (1970).
52. Y. Hatano, K. Takeuchi, and S. Takao, *J. Phys. Chem.*, **77**, 586 (1973).
53. M. G. Robinson and G. R. Freeman, *J. Chem. Phys.*, **48**, 983 (1968); G. R. Freeman, *Quad. Ric. Sci.*, **67**, 9 (1970).
54. J. W. Buchanan and F. Williams, *J. Chem. Phys.*, **44**, 4377 (1966).
55. W. R. Busler, D. H. Martin, and F. Williams, *Discuss. Faraday Soc.*, **36**, 102 (1963); F. Williams, *J. Am. Chem. Soc.*, **86**, 3954 (1964).
56. K.-D. Asmus, *Int. J. Radiat. Phys. Chem.*, **3**, 419 (1971).
57. R. A. Holroyd, J. Y. Yang, and F. M. Servedio, *J. Chem. Phys.*, **46**, 4540 (1967).
58. F. Dainton, M. B. Ledger, R. May, and G. A. Salmon, *J. Phys. Chem.*, **77**, 45 (1973).
59. R. R. Hentz, D. B. Peterson, D. B. Srivastava, H. F. Barzynski, and M. Burton, *J. Phys. Chem.*, **70**, 2362 (1966).
60. J. W. F. van Ingen and W. A. Cramer, *J. Phys. Chem.*, **74**, 1134 (1970).
61. A. Singh, *Radiat. Res. Rev.*, **4**, 1 (1972).
62. P. J. Dyne and W. M. Jenkinson, *Can. J. Chem.*, **38**, 539 (1960); **39**, 2163 (1961).
63. R. A. Holroyd in *Aspects of Hydrocarbon Radiolysis* (eds. T. Gäumann and J. Hoigné), Academic, New York, 1968, p. 1.
64. W. A. Cramer and G. J. Piet, *Trans. Faraday Soc.*, **66**, 850 (1970).
65. C. E. Klots, Y. Raef, and R. H. Johnsen, *J. Phys. Chem.*, **68**, 2040 (1964).
66. J. W. F. van Ingen and W. A. Cramer, *Trans. Faraday Soc.*, **66**, 857 (1970).
67. W. A. Cramer, *J. Phys. Chem.*, **71**, 1112 (1967).
68. A. Kupperman in *Actions Chimiques et Biologiques des Radiations* (ed. M. Haissinsky), Masson et Cie, Paris, Vol. 5, 1961, p. 85, and in *Radiation Research* (ed. G. Silini), North Holland Publishing Co., Amsterdam, 1967, p. 212.
69. R. Blackburn, A. Charlesby, and J. F. Read, *Radiat. Res.*, **28**, 793 (1966); R. Blackburn and A. Charlesby, *Proc. Roy. Soc.* (*London*), **A293**, 51 (1966).
70. J. W. Falconer and M. Burton, *J. Phys. Chem.*, **67**, 1743 (1963).
71. A. W. Boyd and H. W. J. Connor, *Can. J. Chem.*, **42**, 1418 (1964).

72. W. G. Burns and C. R. V. Reed, *Trans. Faraday Soc.*, **66**, 2159 (1970); *J. Chem. Soc. Faraday Trans. I*, **68**, 67 (1972); W. G. Burns, M. J. Hopper, and C. R. V. Reed, *Trans. Faraday Soc.*, **66**, 2182 (1970).
73. T. Gäumann in *Aspects of Hydrocarbon Radiolysis* (eds. T. Gäumann and J. Hoigné), Academic, New York, 1968, p. 213.
74. H. A. Dewhurst, *J. Am. Chem. Soc.*, **83**, 1050 (1961).
75. E. Pancini, V. Santoro, and G. Spadaccini, *Int. J. Radiat. Phys. Chem.*, **2**, 147 (1970).
76. H. Widmer and T. Gäumann, *Helv. Chim. Acta.*, **46**, 944, 2766, 2780 (1963).
77. G. Castello, F. Grandi, and S. Munari, *Radiat. Res.*, **58**, 176 (1974).
78. J. H. Futrell, *J. Am. Chem. Soc.*, **81**, 5921 (1959).
79. T. Ff. Williams, *Trans. Faraday Soc.*, **57**, 755 (1961).
80. L. W. Sieck in *Fundamental Processes in Radiation Chemistry* (ed. P. Ausloos), Interscience, New York, 1968, p. 119.
81. E. N. Avdonina, *Radiat. Effects*, **20**, 245 (1973).
82. K. Shinsaka and S. Shida, *Bull. Chem. Soc. Jap.*, **43**, 3728 (1970).
83. H. A. Dewhurst and E. H. Winslow, *J. Chem. Phys.*, **26**, 969 (1957).
84. L. Kevan and W. F. Libby, *J. Chem. Phys.*, **39**, 1288 (1963); and in *Advances in Photochemistry* (eds. W. A. Noyes, G. S. Hammond, and J. N. Pitts), Interscience, New York, Vol. 2, 1964, p. 183.
85. T. Gäumann, S. Rappoport, and A. Ruf, *J. Phys. Chem.*, **76**, 3851 (1972); S. Rappoport and T. Gäumann, *Helv. Chim. Acta.*, **56**, 531 (1973).
86. G. Castello, F. Grandi, and S. Munari, *Radiat. Res.*, **45**, 399 (1971).
87. W. G. Burns and J. A. Winter, *Discuss. Faraday Soc.*, **36**, 124 (1963).
88. W. G. Burns, R. A. Holroyd, and G. W. Klein, *J. Phys. Chem.*, **70**, 910 (1967).
89. G. Cserép and G. Földiák, *Int. J. Radiat. Phys. Chem.*, **5**, 235 (1973).
90. P. C. Chang, N. C. Yang, and C. D. Wagner, *J. Am. Chem. Soc.*, **81**, 2060 (1959).
91. P. C. Kaufman, *J. Phys. Chem.*, **67**, 1671 (1963).
92. P. B. Ayscough and H. E. Evans, *Trans. Faraday Soc.*, **60**, 801 (1964).
93. C. Vermeil, F. Muller, M. Matheson, and S. Leach, *Bull. Soc. Chim. Belges.*, **71**, 837 (1962).
94. E. Collinson, F. S. Dainton, and H. A. Gillis, *J. Phys. Chem.*, **63**, 909 (1959).
95. R. B. Cundall and P. A. Griffiths, *Discuss. Faraday Soc.*, **36**, 111 (1963); *Trans. Faraday Soc.*, **61**, 1968 (1965).
96. R. A. Holroyd and G. W. Klein, *J. Phys. Chem.*, **69**, 194 (1965).
97. J. Hoigné in *Aspects of Hydrocarbon Radiolysis* (eds. T. Gäumann and J. Hoigné), Academic, New York, 1968, p. 61.
98. V. P. Henri, C. R. Maxwell, W. C. White, and D. C. Peterson, *J. Phys. Chem.*, **56**, 153 (1952).
99. J. P. Manion and M. Burton, *J. Phys. Chem.*, **56**, 560 (1952).
100. T. Gäumann, *Helv. Chim. Acta.*, **46**, 2873 (1963).
101. J. Hoigné, W. G. Burns, W. R. Marsh, and T. Gäumann, *Helv. Chim. Acta.*, **47**, 247 (1964).
102. D. Verdin, *J. Phys. Chem.*, **67**, 1263 (1963).
103. H. Hofer and H. Heusinger, *Z. Phys. Chem. (Frankfurt)*, **69**, 47 (1970).
104. R. R. Hentz, *J. Phys. Chem.*, **66**, 1622 (1962).
105. R. R. Hentz and M. Burton, *J. Am. Chem. Soc.*, **73**, 532 (1951).
106. K. L. Hall and F. A. Elder, *J. Chem. Phys.*, **31**, 1420 (1959).
107. S. Gordon, A. R. Van Dyken, and T. F. Doumani, *J. Phys. Chem.*, **62**, 20 (1958).
108. M. K. Eberhardt, *J. Phys. Chem.*, **67**, 2856 (1963).
109. J. G. Calvert and J. N. Pitts Jr., *Photochemistry*, Wiley, New York, 1966.

110. D. Bryce-Smith and A. Gilbert, *Chem. Comm.*, 643 (1966).
111. R. B. Cundall and P. A. Griffiths, *Trans. Faraday Soc.*, **61**, 1968 (1965); *J. Phys. Chem.*, **69**, 1866 (1965).
112. M. A. Golub and C. L. Stephens, *J. Phys. Chem.*, **70**, 3576 (1966); M. A. Golub, C. L. Stephens, and J. L. Brash, *J. Chem. Phys.*, **45**, 1503 (1966).
113. R. R. Hentz and L. M. Perkey, *J. Phys. Chem.*, **74**, 3047 (1970).
114. L. M. Dorfman, I. A. Taub, and R. E. Bühler, *J. Chem. Phys.*, **36**, 3051 (1962).
115. A. MacLachlan and R. L. McCarthy, *J. Am. Chem. Soc.*, **84**, 2519 (1962).
116. R. H. Schuler, *J. Phys. Chem.*, **60**, 381 (1956).
117. S. Gordon and M. Burton, *Discuss. Faraday Soc.*, **12**, 88 (1952).
118. W. G. Burns, *Trans. Faraday Soc.*, **58**, 961 (1962); W. G. Burns and C. R. V. Reed, *ibid.*, **59**, 101 (1963).
119. W. G. Burns and J. D. Jones, *Trans. Faraday Soc.*, **60**, 2022 (1964).
120. T. J. Sworski, R. R. Hentz, and M. Burton, *J. Am. Chem. Soc.*, **73**, 1998 (1951).
121. A. Zeman and H. Heusinger, *J. Phys. Chem.*, **70**, 3374 (1966); *Radiochim. Acta.*, **8**, 149 (1967).
122. J. Kroh and S. Karolczak, *Radiat. Res. Rev.*, **1**, 411 (1969).
123. P. J. Dyne and W. M. Jenkinson, *Can. J. Chem.*, **40**, 1746 (1962).
124. G. A. Muccini and R. H. Schuler, *J. Phys. Chem.*, **64**, 1436 (1960).
125. J. P. Manion and M. Burton, *J. Phys. Chem.*, **56**, 560 (1952).
126. J. Hoigné and T. Gäumann, *Helv. Chim. Acta.*, **47**, 590 (1964).
127. C. E. Klots and R. H. Johnsen, *J. Phys. Chem.*, **67**, 1615 (1963).
128. C. S. Schoepfle and C. H. Fellows, *Ind. Eng. Chem.*, **23**, 1396 (1931).
129. G. R. Freeman, *J. Chem. Phys.*, **33**, 71 (1960).
130. J. A. Stone and P. J. Dyne, *Radiat. Res.*, **17**, 353 (1962).
131. T. Gäumann, *Helv. Chim. Acta.*, **44**, 1337 (1961).
132. G. R. Freeman, *J. Chem. Phys.*, **33**, 957 (1960).
133. J. M. Ramaradhya and G. R. Freeman, *Can. J. Chem.*, **39**, 1769 (1961).
134. P. J. Dyne, *Can. J. Chem.*, **43**, 1080 (1965).
135. M. Burton and W. N. Patrick, *J. Phys. Chem.*, **58**, 421 (1954).
136. J. G. Burr and F. C. Goodspeed, *J. Chem. Phys.*, **40**, 1433 (1964).
137. J. Y. Yang, B. Scott, and J. G. Burr, *J. Phys. Chem.*, **68**, 2014 (1964).
138. J. Blachford and P. J. Dyne, *Can. J. Chem.*, **42**, 1165 (1964).
139. P. J. Dyne and J. Denhartog, *Nature*, **202**, 1105 (1964).
140. J. F. Merklin and S. Lipsky, *J. Phys. Chem.*, **68**, 3297 (1964).
141. T. J. Hardwick, *J. Phys. Chem.*, **66**, 2132 (1962).
142. P. J. Dyne and J. Denhartog, *Can. J. Chem.*, **40**, 1616 (1962).
143. P. J. Dyne, J. Denhartog, and D. R. Smith, *Discuss. Faraday Soc.*, **36**, 135 (1963).
144. J. A. Stone, A. R. Quirt, and O. A. Miller, *Can. J. Chem.*, **44**, 1175 (1966).
145. M. Burton and W. N. Patrick, *J. Chem. Phys.*, **22**, 1150 (1954).
146. F. E. Littman, E. M. Carr, and J. K. Clauss, *Science*, **125**, 737 (1957).
147. J. F. Hornig, G. Levey, and J. E. Willard, *J. Chem. Phys.*, **20**, 1556 (1952).
148. G. M. Harris and J. E. Willard, *J. Am. Chem. Soc.*, **76**, 4678 (1954).
149. T. A. Gover and J. E. Willard, *J. Am. Chem. Soc.*, **82**, 3816 (1960).
150. H. A. Gillis, R. R. Williams, and W. H. Hamill, *J. Am. Chem. Soc.*, **83**, 17 (1961).
151. R. N. Schindler, *Radiochim. Acta.*, **2**, 69 (1963).
152. R. H. Schuler and W. H. Hamill, *J. Am. Chem. Soc.*, **74**, 6171 (1952).
153. R. C. Petry and R. H. Schuler, *J. Am. Chem. Soc.*, **75**, 3796 (1953); **78**, 3954 (1956).
154. E. O. Hornig and J. E. Willard, *J. Am. Chem. Soc.*, **79**, 2429 (1957).
155. R. F. Pottie, R. Barker, and W. H. Hamill, *Radiat. Res.*, **10**, 664 (1959).

156. J. M. Donovan and R. J. Hanrahan, *Int. J. Radiat. Phys. Chem.*, **3**, 491 (1971).
157. V. H. Dibeler and R. M. Reese, *J. Res. Nat. Bur. Stand.*, **54**, 127 (1955).
158. J. W. Buchanan and R. J. Hanrahan, *Radiat. Res.*, **44**, 305 (1970).
159. G. J. Mains and D. Lewis, *J. Phys. Chem.*, **74**, 1694 (1970).
160. C. C. Chou, P. Angelberger, and F. S. Rowland, *J. Phys. Chem.*, **75**, 2536 (1971).
161. R. J. Hanrahan and J. E. Willard, *J. Am. Chem. Soc.*, **79**, 2434 (1957).
162. R. H. Luebbe and J. E. Willard, *J. Am. Chem. Soc.*, **81**, 761 (1959).
163. E. B. Dismukes and W. S. Wilcox, *Radiat. Res.*, **11**, 754 (1959).
164. M. Takehisa, G. Levey, and J. E. Willard, *J. Am. Chem. Soc.*, **88**, 5694 (1966).
165. R. H. Wiley, W. Miller, C. H. Jarboe, J. R. Harrell, and D. J. Parish, *Radiat. Res.*, **13**, 479 (1960).
166. D. H. Martin and F. Williams, *J. Am. Chem. Soc.*, **92**, 769 (1970).
167. D. K. Bakale and H. A. Gillis, *J. Phys. Chem.*, **74**, 2074 (1970).
168. K. M. Bansal and G. R. Freeman, *Radiat. Res.*, **3**, 209 (1971).
169. T. H. Chen, K. Y. Wong, and F. J. Johnston, *J. Phys. Chem.*, **64**, 1023 (1960).
170. F. P. Abramson, B. M. Buckhold, and R. F. Firestone, *J. Am. Chem. Soc.*, **84**, 2285 (1962).
171. R. B. Bernstein, G. P. Semeluk, and C. B. Arends, *Anal. Chem.*, **25**, 139 (1953).
172. R. Cooper and J. K. Thomas, *Adv. Chem. Ser. No. 82*, 351 (1968).
173. J. W. Schulte, *J. Am. Chem. Soc.*, **79**, 4643 (1957).
174. N. E. Bibler, *J. Phys. Chem.*, **75**, 24 (1971); **77**, 167 (1973).
175. S. Ciborowski, N. Colebourne, E. Collinson, and F. S. Dainton, *Trans. Faraday Soc.*, **57**, 1123 (1961).
176. Z. Spurný, *Int. J. Appl. Radiat. Isotop.*, **14**, 337 (1963).
177. F. P. Abramson and R. F. Firestone, *J. Phys. Chem.*, **70**, 3596 (1966).
178. J. N. Baxter and N. E. Bibler, *J. Chem. Phys.*, **53**, 3444 (1970).
179. M. Ottolenghi and G. Stein, *Radiat. Res.*, **14**, 281 (1961).
180. H. R. Werner and R. F. Firestone, *J. Phys. Chem.*, **69**, 840 (1965).
181. D. L. Hobrock and R. W. Kiser, *J. Phys. Chem.*, **68**, 575 (1964).
182. N. E. Bibler, *J. Phys. Chem.*, **75**, 2436 (1971).
183. J. W. Schulte, J. F. Suttle, and R. Wilhelm, *J. Am. Chem. Soc.*, **75**, 2222 (1953).
184. G. V. Taplin in *Radiation Dosimetry* (eds. G. J. Hine and G. L. Brownell), Academic, New York, 1956, p. 357.
185. A. Henglein, E. Heckel, Y. Ojima, and G. Meissner, *Ber. Bunsenges. Phys. Chem.*, **67**, 988 (1963).
186. J. A. Stone and P. J. Dyne, *Can. J. Chem.*, **42**, 669 (1964).
187. A. Horowitz and L. A. Rajbenback, *J. Phys. Chem.*, **74**, 678 (1970).
188. K. Hannerz, *Research*, **9**, S1 (1956).
189. C. Radlowski and W. V. Sherman, *J. Phys. Chem.*, **74**, 3043 (1970).
190. EL-A. I. Heiba and L. C. Anderson, *J. Am. Chem. Soc.*, **81**, 1117 (1959).
191. Z. Spurný, *Proc. 2nd Int. Conf. Peaceful Uses Atomic Energy*, United Nations, Geneva, **23**, 419 (1958).
192. G. L. Clark and P. E. Bierstedt, *Radiat. Res.*, **2**, 199, 295 (1955).
193. W. V. Sherman, R. Evans, E. Nesyto, and C. Radlowski, *Nature*, **232**, 118 (1971); *J. Phys. Chem.*, **75**, 2762 (1971).
194. T. Sawai, T. Shimokawa, and Y. Shinozaki, *Bull. Chem. Soc. Jap.*, **47**, 1889 (1974).
195. C. Walling, *Free Radicals in Solution*, Wiley, New York, 1957. Tables 6.3 to 6.6.
196. EL-A. I. Heiba and L. C. Anderson, *J. Am. Chem. Soc.*, **79**, 4940 (1957).
197. F. W. Mellows and M. Burton, *J. Phys. Chem.*, **66**, 2164 (1962).
198. J. Wendenburg and A. Henglein, *Z. Naturforschg.*, **17b**, 590 (1962).

199. R. E. Bühler and J. M. Bossy, *Int. J. Radiat. Phys. Chem.*, **6**, 95 (1974).
200. S. U. Choi and J. E. Willard, *J. Phys. Chem.*, **66**, 1041 (1962).
201. I. McAlpine and H. Sutcliffe, *J. Phys. Chem.*, **74**, 1422 (1970).
202. R. W. Fessenden and R. H. Schuler, *J. Chem. Phys.*, **43**, 2704 (1965).
203. A. Sokolowska and L. Kevan, *J. Phys. Chem.*, **71**, 2220 (1967).
204. G. A. Kennedy and R. J. Hanrahan, *J. Phys. Chem.*, **78**, 360 (1974).
205. L. A. Lajbenbach, *J. Am. Chem. Soc.*, **88**, 4275 (1966).
206. L. A. Wall and R. E. Florin, *J. Appl. Polymer Sci.*, **2**, 251 (1959).
207. J. H. Golden, *J. Polymer Sci.*, **45**, 534 (1960).
208. W. C. Francis and R. N. Haszeldine, *J. Chem. Soc.*, 2151 (1955).
209. J. Teplý, *Radiat. Res. Rev.*, **1**, 361 (1969).
210. G. R. Freeman in *Actions Chimiques et Biologiques des Radiations* (ed. M. Haissinsky), Masson et Cie, Paris, Vol. 14, 1970, p. 73.
211. J. H. Baxendale and R. D. Sedgwick, *Trans. Faraday Soc.*, **57**, 2157 (1961).
212. M. Meaburn and F. W. Mellows, *Trans. Faraday Soc.*, **61**, 1701 (1965).
213. S. U. Choi and N. N. Lichtin, *J. Am. Chem. Soc.*, **86**, 3948 (1964).
214. J. H. Baxendale and F. W. Mellows, *J. Am. Chem. Soc.*, **83**, 4720 (1961).
215. G. Meshitsuka and M. Burton, *Radiat. Res.*, **8**, 285 (1958); L. M. Theard and M. Burton, *J. Phys. Chem.*, **67**, 59 (1963).
216. G. E. Adams and J. H. Baxendale, *J. Am. Chem. Soc.*, **80**, 4215 (1958).
217. R. P. Porter and W. A. Noyes, *J. Am. Chem. Soc.*, **81**, 2307 (1959).
218. J. Hagege, S. Leach, and C. Vermeil, *J. Chim. Phys.*, **62**, 736 (1965).
219. L. Friedman, F. A. Long, and M. Wolfsberg, *J. Chem. Phys.*, **27**, 613 (1957).
220. P. Wilmenius and E. Lindholm, *Arkiv Fys.*, **21**, 97 (1962); *Arkiv Kemi*, **20**, 255 (1963).
221. E. P. Grimsrud and P. Kebarle, *J. Am. Chem. Soc.*, **95**, 7939 (1973).
222. E. Klosová, J. Teplý, and Z. Prášil, *Int. J. Radiat. Phys. Chem.*, **2**, 177 (1970).
223. T. Yamamoto, *Bull. Chem. Soc. Jap.*, **43**, 340 (1970).
224. A. Habersbergerová, I. Janovsky, and P. Kouřim, *Radiat. Res. Rev.*, **4**, 123 (1972).
225. K. Y. Lam and J. H. Hunt, *J. Phys. Chem.*, **78**, 2414 (1974).
226. F. Busi and M. D. Ward, *Int. J. Radiat. Phys. Chem.*, **5**, 521 (1973).
227. K. N. Jha, G. L. Bolton, and G. R. Freeman, *J. Chem. Phys.*, **76**, 3876 (1972).
228. J. H. Baxendale and P. Wardman, *J. Chem. Soc., Faraday Trans. I.*, **69**, 584 (1973).
229. H. Seki and M. Imamura, *Bull. Chem. Soc. Jap.*, **38**, 1229 (1965); *J. Phys. Chem.*, **71**, 870 (1967).
230. J. H. Baxendale, *Int. J. Radiat. Phys. Chem.*, **4**, 113 (1972).
231. E. Watson and S. Roy, *Selected Specific Rates of Reactions of the Solvated Electron in Alcohols*, NSRDS-NBS 42 (1972), U.S. Dept. Commerce-National Bureau Standards, Washington, D.C.
232. M. Imamura and H. Seki, *Bull. Chem. Soc. Jap.*, **40**, 1116 (1967); *J. Chem. Phys.*, **48**, 1866 (1968).
233. F. P. Sargent and E. M. Gardy, *Can. J. Chem.*, **52**, 3645 (1974).
234. A. Ekstrom and J. L. Garnett, *J. Phys. Chem.*, **70**, 324 (1966).
235. L. Kevan in *Actions Chimiques et Biologiques des Radiations* (ed. M. Haissinsky), Masson et Cie, Paris, Vol. 13, 1969, p. 57.
236. L. W. Sieck and R. H. Johnsen, *J. Phys. Chem.*, **69**, 1699 (1965).
237. J. J. J. Myron and G. R. Freeman, *Can. J. Chem.*, **43**, 381 (1965).
238. J. W. Fletcher and G. R. Freeman, *Can. J. Chem.*, **45**, 635 (1967).
239. G. R. Freeman, *Radiation Chemistry of Ethanol: A Review of Data on Yields, Reaction Rate Parameters, and Spectral Properties of Transients*, NSRDS-NBS 48 (1974), U.S. Dept. Commerce-National Bureau Standards, Washington, D.C.

240. B. Smaller and M. S. Matheson, *J. Chem. Phys.*, **28**, 1169 (1958).
241. J. G. Burr, *J. Am. Chem. Soc.*, **79**, 751 (1957); *J. Phys. Chem.*, **61**, 1477 (1957).
242. K. M. Bansal and G. R. Freeman, *J. Am. Chem. Soc.*, **92**, 4173 (1970); H. J. van der Linde and G. R. Freeman, *ibid.*, **92**, 4417 (1970).
243. R. A. Basson and H. J. van der Linde, *J. Chem. Soc.*, *A*, 1182 (1967).
244. R. H. Johnsen and D. A. Becker, *J. Phys. Chem.*, **67**, 831 (1963).
245. L. G. J. Ackerman, R. A. Basson, and H. J. van der Linde, *J. Chem. Soc. Faraday Trans.*, *I*, **68**, 1258 (1972).
246. D. Verdin, *Int. J. Radiat. Phys. Chem.*, **2**, 201 (1970).
247. A. Kira and J. K. Thomas, *J. Chem. Phys.*, **60**, 766 (1974).
248. G. A. Swan, P. S. Timmons, and D. Wright, *Proc. 2nd Int. Conf. Peaceful Uses Atomic Energy*, United Nations, Geneva, **29**, 115 (1958).
249. M. A. Proskurnin and E. V. Barelko, reported by N. Bach, *Proc. Int. Conf. Peaceful Uses Atomic Energy*, United Nations, New York, **7**, 538 (1956).
250. M. K. M. Ng and G. R. Freeman, *J. Am. Chem. Soc.*, **87**, 1635, 1639 (1965).
251. N. A. Bach, *Radiat. Res. Suppl. 1*, 190 (1959).
252. A. S. Newton, *J. Phys. Chem.*, **61**, 1490 (1957).
253. F. Kiss and J. Teplý, *Int. J. Radiat. Phys. Chem.*, **3**, 503 (1971).
254. Y. Llabador and J. P. Adloff, *J. Chim. Phys.*, **61**, 681 (1964).
255. Y. Llabador and J. P. Adloff, *J. Chim. Phys.*, **61**, 1467 (1964).
256. R. Barker, *Trans. Faraday Soc.*, **59**, 375 (1963).
257. M. A. J. Rodgers, *Trans. Faraday Soc.*, **67**, 1029 (1971).
258. M. A. J. Rodgers, *J. Chem. Soc. Faraday Trans.*, *I*, **68**, 1278 (1972).
259. A. J. Robinson and M. A. J. Rodgers, *J. Chem. Soc.*, *Faraday Trans. I*, **69**, 2036 (1973).
260. L. W. Sieck and P. Ausloos, *Radiat. Res.*, **52**, 47 (1972).
261. M. Matsui and M. Imamura, *Bull. Chem. Soc. Jap.*, **47**, 1113 (1974).
262. S. M. S. Akhtar, R. J. Woods, and J. A. E. Bardwell, *Int. J. Radiat. Phys. Chem.*, **7**, 603 (1975).
263. A. G. Leiga and H. A. Taylor, *J. Chem. Phys.*, **41**, 1247 (1964).
264. R. Pieck and E. W. R. Steacie, *Can. J. Chem.*, **33**, 1304 (1955).
265. P. Ausloos, *J. Am. Chem. Soc.*, **83**, 1056 (1961).
266. J. A. Slivinskas and J. E. Guillet, *J. Polymer Sci., Polymer Chem. Ed.*, **11**, 3043 (1973).
267. A. S. Newton, *J. Chem. Phys.*, **26**, 1764 (1957).
268. S. Lukáč, J. Teplý, and K. Vacek, *J. Chem. Soc., Faraday Trans. I*, **68**, 1377 (1972).
269. W. M. Garrison, W. Bennett, S. Cole, H. R. Haymond, and B. M. Weeks, *J. Am. Chem. Soc.*, **77**, 2720 (1955).
270. P. B. Ayscough, K. Mach, J. P. Oversby, and A. K. Roy, *Trans. Faraday Soc.*, **67**, 360 (1971).
271. R. H. Johnsen, *J. Phys. Chem.*, **63**, 2041 (1959).
272. I. G. Draganić and O. Gal, *Radiat. Res. Rev.*, **3**, 167 (1971).
273. P. Ausloos and C. N. Trumbore, *J. Am. Chem. Soc.*, **81**, 3866 (1959).
274. R. W. Hummel, *Trans. Faraday Soc.*, **56**, 234 (1960).
275. R. B. Cundall, A. W. Locke, and G. C. Street in *The Chemistry of Ionization and Excitation* (eds. G. R. A. Johnson and G. Scholes), Taylor and Francis, London, 1967, p. 131.
276. J. L. Cory and R. F. Firestone, *J. Phys. Chem.*, **74**, 1425 (1970).
277. J. A. Knight, *Radiat. Res.*, **52**, 17 (1972).
278. D. Bradley and J. Wilkinson, *J. Chem. Soc. A*, 531 (1967).
279. P. B. Ayscough, H. Drawe, and P. Kohler, *Radiat. Res.*, **33**, 263 (1968).
280. J. L. Baptista and H. D. Burrows, *J. Chem. Soc., Faraday Trans. I*, **70**, 2066 (1974).
281. D. Smithies and A. J. Whitworth, *J. Chem. Soc. A*, 1987 (1969).

282. W. A. Seddon, J. W. Fletcher, F. C. Sopchyshyn, and J. Jevcak, *Can. J. Chem.*, **52**, 3269 (1964).
283. M. A. Sami and D. Smithies, *J. Chem. Soc., Faraday Trans. I*, **70**, 51 (1974).
284. J. A. Knight, *Radiat. Res.*, **51**, 590 (1972).
285. N. Colebourne, E. Collinson, and F. S. Dainton, *Trans. Faraday Soc.*, **59**, 886 (1963).
286. H. A. Makada and W. M. Garrison, *Int. J. Radiat. Phys. Chem.*, **3**, 179 (1971).
287. S. Berk and H. Gisser, *Radiat. Res.*, **56**, 71 (1973).
288. F. Lahmani and N. Ivanoff, *Proc. 2nd Tihany Symp. Radiat. Chem.*, Akadémiai Kiadó, Budapest, 1967, p. 327.
289. R. O. Lindblom, R. M. Lemmon, and M. Calvin, *J. Am. Chem. Soc.*, **83**, 2484 (1961).
290. R. M. Lemmon and R. O. Lindblom, *Proc. 2nd Int. Conf. Peaceful Uses Atomic Energy*, United Nations, Geneva, **22**, 409 (1958).
291. I. Serlin, *Science*, **126**, 261 (1957).
292. R. L. Collin, *J. Am. Chem. Soc.*, **79**, 6086 (1957); P. Shanley and R. L. Collin, *Radiat. Res.*, **16**, 674 (1962).
293. Y. Tomkiewicz, R. Agarwal, and R. M. Lemmon, *J. Am. Chem. Soc.*, **95**, 3144 (1973).

CHAPTER 9

Effects of Radiation on Solids

Metals • Ionic Crystals • Glass • Graphite • Chemical Decomposition •
Reactions in Solids at Low Temperatures • Catalysts • References •

The advance of nuclear technology has prompted the study of the behavior of solid materials when exposed to nuclear radiations. For example, for a material to be useful within a nuclear reactor it must be able to withstand high intensities of nuclear radiations for long periods of time, and materials that suffer little damage upon irradiation are sought. In other fields the beneficial effects of radiation on solids are examined and effects that, if not beneficial in the usual sense of the word, may lead to increased knowledge of the solid state.

From the historical point of view, one of the earliest examples of the action of radiation on solids was the production of pleochroic haloes in mica by the radiation from inclusions of radioactive substances such as uranium or thorium (1, p. 269). A related effect is the production of metamict minerals in which the regular crystal structure of a mineral such as gadolinite has been disordered by nuclear radiations. Such minerals, on heating, often release the stored energy as heat or light (2, p. 6).

Early studies of the action of nuclear radiations on solids showed, among other effects, that colorless glass became colored by exposure to radiation, the coloration being discharged by the action of heat or light, that the rays from radium exert a destructive action on paper, linen and silk, that rubber becomes hard and stopcock grease is destroyed, that nitrogen iodide will explode when exposed to a sufficient intensity of α-rays, and that silver halides are decomposed. Quite obviously, a variety of effects may be observed, depending on the nature of the radiation and the solid. Much of the information is concerned with physical rather than chemical effects (3), and no attempt is made here at an exhaustive treatment. The aim is rather to touch on a field closely related to radiation chemistry for the sake of completeness.

All types of ionizing radiation are able to produce ionized and excited atoms in the solid, and in addition, heavy particles (protons, deuterons, α-particles, etc.) can cause an appreciable number of atoms to be displaced from their normal position. γ-Rays, x-rays, and electrons produce mainly excitation and ionization but, if the radiation is sufficiently energetic (greater than 0.5 MeV for an atom of atomic weight between 50 and 100), may also cause a small amount of atomic displacement (4). For γ-rays of sufficient energy one speaks of the displacement cross section $\sigma_{c,d}^{\gamma}$ where the Compton process is effective. For example, for γ-rays of 1.22-MeV energy, $\sigma_{c,d}^{\gamma}$ for $Z = 22$ is 0.114b, for $Z = 50$ it is 0.0161b (2, p. 28). The extent to which displacements occur in copper with several different types of radiation is shown in Table 9.1.

TABLE 9.1 *Displacements Produced by Various Radiations (4)*

Radiation	Energy (MeV)	Flux Density $(cm^{-2} sec^{-1})$	Time	Number of Displacements per Atom in Copper
Neutrons	2	10^{13}	1 month	10^{-1}
Deuterons	10	6×10^{12}	10 hr	10^{-2}
Electrons	2	6×10^{14}	10 hr	10^{-3}
γ-Rays	1.3	10^{11}	1 month	10^{-8}

Another useful concept is that of the "thermal spike," in which changes in the solid are attributed to brief, intense heating of minute volumes. For example, a copper atom of say 300 eV energy (resulting from megaelection volt deuteron bombardment) has a range of about 30 Å in copper. Dissipation of the atom's energy in a sphere of 30-Å radius would raise the material to the melting point (1083°C) in 5×10^{-12} sec. The temperature would fall about equally rapidly. Elaborations of the simple concept include electron spikes, displacement spikes, plasticity spikes, and fission fragment spikes. In contrast to the evidence for some of the spike theory elaborations, evidence for fission fragment spikes is relatively good (2, p. 44).

The nature of the changes produced in a solid depends on the type of material irradiated. In the treatment that follows materials are divided into metals, ionic crystals, glasses, and organic compounds.

Metals

Metals consist of a regular array of positive ions in a sea of electrons. Thus the ionization produced by radiation in a metal is without effect, since the positive hole left by ionization is rapidly refilled by an electron from the common pool of electrons (the conduction band). On the other hand, atoms displaced by heavy particle irradiation may not be able to return to their original position if they are displaced more than a few atoms distance, and the result will be a vacancy at the original position and an (interstitial) atom in a position not occupied by one in the original lattice. Physically, this may result in hardening of the metal and increased electrical resistance (e.g., 10^{17} 12-MeV deuterons cm^{-2} increase the resistance of pure copper by a factor of about 2), since a very regular, or perfect, lattice is needed for the lowest electrical resistance.

Semiconductors (e.g., germanium) are similar to metals but have a much smaller number of common electrons. The electrical resistance of semiconducting metals is very much more sensitive to radiation than that of typical metals because the radiation-induced defects may markedly alter the number of electrons available for conduction. Irradiation of one type of germanium with 10^{15} 9.6-MeV deuterons cm^{-2}, for example, increased the resistance by a factor of about 10^5 to 10^6. It has been suggested that the change in resistance of semiconductors on irradiation may provide a very useful means of dosimetry e.g., p. 109.

Appreciable changes in volume and density also accompany radiation-induced changes in metals. In copper, for example, a dose of 10^{17} 10-MeV deuterons cm^{-2} produces a fractional increase in length of 3.8×10^{-4}.

Radiation damage in uranium, and in fuel elements generally, has been exhaustively studied and is complicated by the production and retention of fission fragments, some of them gaseous, within the solid. Dimensional changes are an order of magnitude greater than with most other, nonfissionable, metals. For single uranium crystals, a lengthening takes place in the (010) direction and a contraction in the (100) direction (5, p. 29). Considerations of radiation damage are of great importance in the design of fuel elements and the excessively high amounts of damage occurring in uranium have led to the consideration of alternative materials such as uranium oxide, which apparently has a much greater resistance to radiation damage.

Ionic Crystals

When alkali halide crystals are irradiated, absorption bands in the visible and ultraviolet regions are developed. The color in the visible region varies with the nature of the crystal; e.g., lithium chloride gives a yellow color and caesium and potassium chlorides blue colors. The absorption bands respon-sible for these colors are called *F bands* and the defects in the crystal that give rise to them *F centers*. Other absorption bands are also produced as shown, for potassium chloride, below:

Name	Wavelength (Å)	Region
V bands	2200	Ultraviolet
	3550	
F band	5500	Yellow-green
R bands	6700	Red
	7200	
F' band	7400	Red-infrared
M band	8200	Infrared

Yet other absorption bands are produced when the crystal contains small amounts of impurities, calcium chloride, e.g., giving a series of *Z* bands and hydrides a series of *U* bands.

   The colors can be bleached by heating or by irradiation with light and are believed to be formed as follows. The crystal is made up of a regular three-dimensional array of positive and negative ions as shown in **I**. However, holes (vacancies) may exist in the structure where either

```
 − + − + − +        − + − +        − + − +
 + − + − + −        +   + −        + −   −
 + − + − + −        − + − +        − + − +
 − + − + − +        + − + −        + − + −
      I                 II             III
                  negative-ion     positive-ion
                    vacancy          vacancy
```

a negative ion **II** or a positive ion **III** is missing (similar holes are believed to be formed by irradiation). Irradiation causes electrons to be ejected from some of the atoms in the crystal lattice and, while most of these electrons will return to their parent atom or a similar atom which has lost an electron, some will be trapped and held in a negative-ion vacancy **IV**. This constitutes an *F* center. The electron can be detected

```
 − +   − +
 + e⁻ + −
 − +   − +
 + −   + −
      IV
   F center
```

by epr measurements, which show that it interacts quite strongly with the surrounding six (in three dimensions) positive ions. The electron can be made to move to some other negative-ion vacancy, in the direction of the anode, by applying an electric potential to the crystal. The opposite process to that giving an F center gives a V_1 center, which is associated with a positive-ion vacancy. This is best illustrated by reference to a particular crystal, and a positive-ion vacancy and a V_1 center in potassium chloride are illustrated in **V** and **VI** respectively. The

$$
\begin{array}{ccc}
K^+ & Cl^- & K^+ \\
Cl^- & & Cl^- \\
K^+ & Cl^- & K^+
\end{array}
$$

V
positive-ion
vacancy

$$
\begin{array}{ccc}
K^+ & Cl^- & K^+ \\
Cl^- & & Cl\cdot \\
K^+ & Cl^- & K^+
\end{array}
$$

VI
V_1 center

formation of a V_1 center follows the loss of an electron from one of the negative chlorine ions surrounding a positive-ion vacancy, giving a chlorine atom. The positive-ion vacancy-chlorine atom system is stable and electrically neutral; in practice the chlorine atom is in equilibrium with the five chloride ions surrounding the positive-ion vacancy, and the five electrons associated with these ions are shared between the six nuclei.

If the irradiated crystal is heated to a few hundred degrees Centigrade, the F and V_1 centers created by ionizing radiation can be removed, the electrons being released from their traps (F centers) to combine with the electron-deficient V_1 centers.

Color centers (F and V_1) can also be formed by electrolysis of an alkali halide crystal below its melting point, and F and V_1 centers separately by heating the crystal in the presence of the appropriate alkali metal vapor or halogen. Potassium vapor and chlorine, for example, produce F and V_1 centers respectively in potassium chloride crystals. Under these conditions the potassium (or chlorine) is deposited on the surface of the crystal and new crystal layers are built up as chloride ions diffuse to this surface. In doing so the chloride ions leave negative-ion vacancies in the crystal and these, or similar vacancies, in the interior of the crystal trap the electrons released by the potassium atoms in forming ions, giving F centers.

Centers other than F and V_1 can arise from the displacement of atoms and ions from their normal position in the crystal, in much the same way as the dislocation of atoms in metals. Impurity ions in the lattice may also form other types of absorption center. In irradiated potassium chloride, displaced chlorine atoms can combine with chloride ions to give the radical-ion Cl_2^-,

$$Cl\cdot + Cl^- \rightarrow Cl_2^- \tag{9.1}$$

which has been detected and identified by its absorption at 3650 Å.

Chemical changes can be observed when the irradiated crystals are dissolved in water. Potassium chloride, for example, gives a slightly alkaline oxidizing solution, presumably through reaction of the trapped electrons and holes (chlorine atoms) with the water, e.g.,

$$e^- + H_2O \rightarrow e^-_{aq} \tag{9.2}$$

$$Cl\cdot \text{ (crystal)} \rightarrow Cl\cdot \text{ (solution)} \tag{9.3}$$

followed by

$$2e^-_{aq} \rightarrow H_2 + 2OH^- \tag{9.4}$$

$$2Cl\cdot \rightarrow Cl_2 \quad \text{etc.} \tag{9.5}$$

Radiation also affects physical properties of the crystals, such as ionic conductivity, density, hardness, etc. (2, Chapter 8). Thus potassium chloride exposed to 6×10^6 R of cobalt-60 γ-radiation shows a decrease in ionic conductivity (σ) of an order of magnitude. By contrast, exposure to 3×10^{18} fast neutrons cm^{-2} increases the ionic conductivity by two orders of magnitude. In the case of the γ-irradiated material, σ can be brought near to its preirradiation value by annealing at 240°C. x-Rays produce a decrease in density in the alkali halides (the fractional decrease at saturation is about 7×10^{-5}), indicating strongly that lattice defects are produced. Changes in the x-ray diffraction pattern corresponding to the lattice expansion are also observed. Similar effects have been found with heavy particle irradiation. Proton and electron bombardment of potassium chloride crystals produce a marked increase in hardness, while lithium fluoride crystals show an increase in yield stress after exposure to neutrons.

Glass

The coloring of glass exposed to radiation probably resembles the coloring of ionic crystals, though the greater complexity of glass means that there are more possibilities for forming color centers. Thus the radiation-induced absorption in glass generally consists of overlapping absorption bands rather than discrete bands as are formed in ionic crystals.

Silver-activated phosphate glass has been mentioned earlier (p. 107) as a means of dosimetry. In this instance, silver ions are believed to be reduced to metallic silver, the silver atoms being responsible for the orange fluorescence observed when the irradiated glass is exposed to ultraviolet light. The process is analogous to the reduction of silver ions in a photographic emulsion on exposure to light.

Manganese in glass causes a purple color to develop on irradiation, which is most likely due to oxidation of manganous to manganic ions,

$$Mn^{2+} + h\nu \rightarrow Mn^{3+} + e^- \tag{9.6}$$

The electron is trapped in some other part of the system, possibly by a ferric ion,

$$Fe^{3+} + e^- \rightarrow Fe^{2+} \tag{9.7}$$

A similar reaction may be responsible for the protection against radiation-induced coloration that 1 to 2% of cerium oxide (CeO_2) affords glass, the ceric ions acting as efficient electron traps,

$$Ce^{4+} + e^- \rightarrow Ce^{3+} \tag{9.8}$$

Irradiated glass contains unpaired electron centers which give rise to epr spectra, and attempts have been made to correlate these with the optical absorption bands. Fast particle bombardment of silica glass produces two prominent ultraviolet absorption bands, the C and E bands. At the same time a narrow system of epr lines and a broad epr system are produced, which have been correlated with the C and E bands respectively. Furthermore, the C band has been identified as an electron localized at an oxygen ion vacancy and the E band interpreted as a hole trapped at an interstitial oxygen ion (2, p. 264). Electron paramagnetic resonance in crystals has been summarized by Shulman (6).

Graphite

Graphite has been of great interest technologically, starting with the early speculations of E. P. Wigner on possible radiation damage. Of particular engineering concern are such changes as a doubling in mechanical strength, a fiftyfold reduction in thermal conductivity, a linear expansion of 3%, and an accumulation of stored energy in excess of 500 calories g^{-1}, all of which have been observed with an exposure of about 2×10^{21} neutrons cm^{-2} at 30°C. The stored energy is released on heating, the temperature-release curve showing a marked peaking. A knowledge of this and related phenomena is obviously of first importance in graphite moderated reactor technology.

Chemical Decomposition

Covalent bonds present in solids may be broken upon irradiation and bring about chemical changes. This has already been mentioned in the case of organic polymers, and other organic solids as well as nitrates, chlorates, perchlorates, etc., decompose in this way.

Nitrates decompose giving oxygen and nitrite, and it is possible to demonstrate that the final products are formed in the crystal by magnetic-susceptibility measurements, which have the value expected if the oxygen formed is present as molecular oxygen, and by optical spectroscopy, which shows absorption at 3450 Å that can be attributed to nitrite ions (4). If the irradiated nitrate is dissolved in water, both of the products can be measured and the ratio of nitrite to oxygen is found to be 2:1, as expected for,

$$NO_3^- \rightarrow NO_2^- + \tfrac{1}{2}O_2 \tag{9.9}$$

$G(-NO_3^-)$ values from 0.01 to 3 are found for different nitrates (7). Absorption centers similar to F centers are also formed, but the situation is more complicated than in the case of the alkali halides because the centers may react with the products, NO_2^- and O_2. Thermal annealing causes some recombination of the radiation-produced fragments (8).

Potassium chlorate decomposes under the influence of reactor radiations (G is about 2) but potassium sulfate, lithium sulfate, potassium chromate, and calcium carbonate crystals are only colored in a reactor, without apparent decomposition (9, p. 383). Patrick and McCallum (10) have studied the radiation decomposition of a number of alkali chlorates and have found perchlorate, chlorite, hypochlorite, chloride, and oxygen as products. The G values for these species depend on the cation present, and are profoundly influenced by thermal annealing or ultraviolet irradiation of the solid following exposure to γ-rays. The mechanism of the changes produced in the solid by heat treatment appears to be complicated. When silver oxalate is preirradiated with γ-radiation the kinetics of its thermal decomposition at 128°C follows a cube root law rather than the exponential law found for unirradiated material (11).

A number of explosive salts can be detonated by ionizing radiations; e.g., barium azide using x-rays or electrons, sodium azide by electrons, and nitrogen azide by fission fragments and polonium α-particles (10, p. 383). Bowden and Singh (12) concluded from a study of the explosive decomposition of a number of azides on exposure to an electron beam that explosion results from the production of a "hot spot" of the order of 10^{-4} cm in diameter.

Radiation effects in organic solids are generally similar to those for the same compound in the liquid state when allowance is made for the restricted mobility of the active species in the solid. However, this is not invariably true. Choline chloride, for example, is about a hundred times as sensitive to γ-radiation in the solid state as in solution (cf. p. 433), apparently because the molecular alignment in the crystalline material is conducive to a chain decomposition.

Reactions in Solids at Low Temperatures

In the last few years many studies have been made of the effect of radiations on solids at low temperatures. Such studies have the advantage that at low temperatures the intermediates produced by the radiation will not ordinarily react with the substrate, and they may therefore be examined by such techniques as optical and emission spectroscopy, electron paramagnetic resonance, etc. The theoretical possibility of high energy reactions between radicals has added to the interest in the subject (13–15).

Hydrogen atoms and deuterium atoms are readily produced in frozen, acid or alkaline, H_2O and D_2O by irradiation with γ-radiation or tritium-β radiation (16, 17). $G(H)_\gamma$ for a 0.125 mole fraction perchloric acid solution is close to 2, but the yield in 16% hydrofluoric acid solution is much lower, and the yield in alkaline solutions containing lithium, sodium, or potassium hydroxide is lower still. Hayon (18) has discussed the results in terms of solvated electrons. The radiation chemistry of both liquid and frozen aqueous solutions of nitrous oxide and ferrous ion has been described by Dainton and Jones (19) for the temperature range -196 to $77°C$. It was found that whenever a transition from a crystalline or a fluid state to a glassy state occurred, $G(N_2)$ increased abruptly, though $G(H_2)$ and $G(O_2)$ were largely unaffected. It was suggested that electrons generated in the primary act migrate distances of about 50 Å in the glass, and invariably react with any solute molecule (N_2O or Fe^{2+}) they encounter during this migration. Polymer chains with repeating units such as $(-SO_3-)_n$, which are probably responsible for the glassy structure, may also furnish preferred paths for electron migration.

Collinson, Conlay, and Dainton (20) claim to have strong evidence for energy transfer in certain systems where there is considerable overlap between the absorption spectrum of the solute and that of the solvent. In a solution of ferric chloride in diphenyl ether just above the melting point, or in the supercooled liquid, $G(-FeCl_3)$ was low and independent of solute concentration and absorbed dose. In the solid phase, however, $G(-FeCl_3)$ became much larger and increased with decrease in temperature and decrease in solute concentration. In solvents with absorption bands which do not overlap those of the solute, ferric chloride, $G(-FeCl_3)$ is the same in the solid and in the liquid at temperatures near the melting point. An ordered structure of the donor (solvent) molecules is apparently necessary for this type of energy transfer (21).

The thermal behavior of trapped radicals seems to indicate that phase changes are of considerable importance. Gurman et al. (22) have reported that the epr spectrum of the radicals in frozen aqueous hydrogen peroxide irradiated at $-196°C$ decays rapidly between -125 and $-120°C$, but irradiation of the sample above $-115°C$ leads again to the production of radicals, which are stable up to $-53°C$. Thermograms of nonirradiated specimens showed an exothermic phase change at $-116°C$ and an endothermic change at $-53°C$ (23).

Numerous epr studies have been made of the thermal stability of radicals produced by the irradiation of organic solids at low temperatures. Thus the radicals in irradiated cyclohexane disappear between -113 and $-98°C$, and those from cyclohexyl iodide, cyclohexyl bromide, and cyclohexyl chloride between -158 to $-143°C$, -88 to $-68°C$, and -103 to $-53°C$, respectively. Activation energies for the radical disappearance, which follow second-order kinetics, are about 20 to 25 kcal mole^{-1} (24–26). Electron paramagnetic resonance spectra of a sample of n-hexadecene-1 irradiated at $-196°C$ and then warmed with temperature increments of $20°C$ showed a rapid fall in the number of trapped radicals during heating. Most of the radicals had disappeared by the time the temperature reached $-77°C$. It was suggested that the disappearance of a proportion of the radicals each time the temperature was raised in the region between -196 and $-77°C$ was due to the progressive release of a mobile species, probably electrons, from traps of various energies, since massive molecular motions were thought unlikely at these temperatures (27).

A few studies of conventional reactions carried out at low temperatures have appeared. The radiolysis of equimolar mixtures of hydrogen bromide and ethylene at liquid nitrogen and liquid oxygen temperatures, for example, leads to the production of virtually pure ethyl bromide (28, 29). Similar results were obtained just above the freezing point ($-165°C$) of the mixture. The reaction is a chain reaction with G of the order of 10^5, and it would appear that radiation produces reactive entities (radicals) in the frozen and liquid mixtures which subsequently initiate the reaction chain. The chain reaction must occur during melting of the frozen samples, and in support of this it was found that the reaction could be initiated in an unirradiated mixture by condensing it onto irradiated frozen material at liquid nitrogen temperature, and then melting the combined samples. When, for example, equal amounts of unirradiated and irradiated mixture were used, the amount of product formed in the combined sample, on melting, was 60% greater than that formed in the irradiated mixture alone. Electron paramagnetic resonance studies at liquid nitrogen temperature ($-196°C$) indicate that hydrogen atoms are present in irradiated hydrogen bromide and large amounts of ethyl radicals, with a small amount of hydrogen atoms, in irradiated ethylene. An irradiated mixture of hydrogen bromide and ethylene shows no evidence of ethyl radicals and only a weak spectrum similar to that obtained by irradiating ethyl bromide, indicating a large measure of reaction of the primary radiation-produced radicals even at $-196°C$.

The importance of phase transitions in low temperature reactions has been emphasized by Semenov (30).

The foregoing section indicates that the study of low temperature radiolysis reactions is still in its infancy. However, it would seem to be worth pursuing further since the resulting radical reactions might be expected to be of considerable theoretical and perhaps even practical interest. It is of particular interest in view of the importance of the direct experimental observation and measurement of the primary products and transient intermediates for our understanding of radiolytic processes (31).

Catalysts

In several cases radiation has been shown to affect the activity of solid catalysts (cf. 32), and it is very likely that this field of study will grow rapidly within the next few years.

Typical of the systems examined are hydrogen-deuterium exchange over silica, alumina, and other oxides; *ortho-para* hydrogen conversion over the oxides of iron, molybdenum, and other metals; oxidation of sulfur dioxide over vanadium pentoxide, etc. (4).

The effects observed range from zero to as much as a thousandfold change in activity and have been attributed to a variety of causes, such as an effect on catalyst poisons, the formation of holes, trapped electrons, or displaced atoms, and a change in adsorptive capacity. In relation to the latter, it is of interest that Charman and Dell (33) reported that nickel oxide and magsium oxide preirradiated with doses up to 6×10^{19} fast neutrons cm^{-2} showed a strong enhancement of oxygen adsorption and also an increase in hydrogen adsorption. They suggested that the irradiation effects may be explained in terms of the creation of excess metal and excess oxygen centers (cf. 34).

REFERENCES

1. G. Hevesy and F. A. Paneth, *Radioactivity*, Oxford University Press, 1938.
2. D. S. Billington and J. H. Crawford, *Radiation Damage in Solids*, Princeton University Press, 1961.
3. *Radiation Effects in Inorganic Solids, Faraday Soc. Discuss. No. 31*, 1961.
4. E. H. Taylor, *J. Chem. Educ.*, **36**, 396 (1959).
5. J. J. Harwood, H. H. Hausner, J. G. Morse, and W. G. Rauch, *Effects of Radiations on Materials*, Reinhold, New York, 1958.
6. R. G. Shulman, *Ann. Rev. Phys. Chem.*, **13**, 326 (1962).
7. E. R. Johnson and J. Forten, *Faraday Soc. Discuss. No. 31*, 1961, p. 238.
8. A. G. Maddock and S. R. Mohanty, *Faraday Soc. Discuss. No. 31*, 1961, p. 193.
9. M. Haissinsky, *La Chimie Nucléaire*, Masson et Cie., Paris, 1957.
10. P. F. Patrick and K. J. McCallum, *Nature*, **194**, 766 (1962).
11. R. M. Haynes and D. A. Young, *Faraday Soc. Discuss. No. 31*, 1961, p. 229.
12. F. P. Bowden and K. Singh, *Proc. Roy. Soc.* (*London*), Ser. *A*, **227**, 22 (1954).
13. M. S. Matheson, *Nucleonics*, **19**, 57 (October 1961).

14. A. M. Bass and H. P. Broida, *Formation and Trapping of Free Radicals*, Academic, New York, 1960.
15. G. J. Minkoff, *Frozen Free Radicals*, Interscience, New York, 1960).
16. J. Kroh, B. C. Green, and J. W. T. Spinks, *Can. J. Chem.*, **40**, 413 (1962).
17. R. Livingston and A. J. Weinberger, *J. Chem. Phys.*, **33**, 499 (1960).
18. E. Hayon, *Nature*, **194**, 737 (1962).
19. F. S. Dainton and F. T. Jones, *Radiat. Res.*, **17**, 388 (1962).
20. E. Collinson, J. J. Conlay, and F. S. Dainton, *Nature*, **194**, 1074 (1962).
21. P. J. Dyne, D. R. Smith, and J. A. Stone, *Ann. Rev. Phys. Chem.*, **14**, (1963).
22. G. B. Sergeev, V. S. Gurman, V. I. Papissova, and E. I. Yakovenko, *5th Int. Free Radical Symp. Uppsala*, 1961.
23. V. S. Gurman, E. I. Yakovenko, and V. I. Papissova, *Zh. Fiz. Khim.*, **34**, 1126 (1960).
24. R. Bensasson, K. Liebler, R. Marx, and H. Szwarc, *Bull. Ampére. 9ᵉ Année, Fasc. Special 1960*, 303.
25. K. Leibler and H. Szwarc, *J. Chim. Phys.*, 1109 (1960).
26. K. Leibler, *J. Chim. Phys.*, 1111 (1960).
27. P. B. Ayscough, A. P. McCann, C. Thomson, and D. C. Walker, *Trans. Faraday Soc.*, **57**, 1487 (1961).
28. D. A. Armstrong and J. W. T. Spinks, *Can. J. Chem.*, **37**, 1002 (1959).
29. F. W. Mitchell, B. C. Green, and J. W. T. Spinks, *Can. J. Chem.*, **38**, 689 (1960).
30. N. Semenov, *Plenary Lecture at IUPAC Conf. Montreal 1961; Pure and Applied Chem. XVIII International Congress, Montreal, 1961*, Butterworth, London, 1962.
31. M. S. Matheson, *Ann. Rev. Phys. Chem.*, **13**, 77 (1962).
32. E. H. Taylor, *Nucleonics*, **20**, 53 (January 1962).
33. H. B. Charman and R. M. Dell, *Trans. Faraday Soc.*, **59**, 470 (1963).
34. W. H. Cropper, *Science*, **137**, 955 (1962).

Additional Bibliography

A. Bishay (Ed.), *Interactions of Radiation with Solids*, Plenum, New York, 1967.

R. G. Di Martini and S. R. Huang, Radiation Effects in Inorganic Solids, in *Advances in Nuclear Science and Technology*, Academic, New York, Vol. 3, 1966.

E. H. Taylor, The Effect of Ionizing Radiation on Solid Catalysts, in *Advances in Catalysis*, Academic, New York, Vol. 18, 1968, p. 111.

M. W. Thompson, *Defects and Radiation Damage in Metals*, Cambridge University Press, Cambridge, 1969.

E. R. Johnson, *The Radiation-Induced Decomposition of Inorganic Molecular Ions*, Gordon and Breach, New York, 1970.

K. Kaindl and E. H. Graul, *Strahlenchemie*, A. Hüthig Verlag, Heidelberg, 1967, see Chap. 10 particularly.

L. A. Blumenfeld, V. V. Voevodskii, and A. G. Semenov, *Electron Paramagnetic Resonance in Chemistry*, Academy of Science, USSR, 1962.

CHAPTER 10

Chemical Radiation Synthesis:
Industrial Uses of Radiation

Radiation Cost • Sources of Radiation • Chemical Syntheses: Industrial •
Ethyl Bromide Synthesis • Synthesis of Gammexane (Gamma Benzene
Hexachloride) • Nitrogen Fixation • Some Other Chemical Syntheses:
Laboratory and Pilot Plant • Sulfochlorination of Hydrocarbons • Sul-
fonic Acid Production • Addition Reactions to Unsaturates • Hydroha-
logenation • Addition to H_2S to Olefins • Chain Reactions Involving
Saturated Molecules • Oxidation of Hydrocarbons • Perchlorinated Hy-
drocarbons • Telomerization • Synthesis of Organosilicon Compounds •
Fluorination • Carboxylation and Carbonylation • Synthesis of Nitrogen
Containing Organic Compounds • Organophosphorus Compounds • Or-
ganotin Compounds • Irradiation of Monomers and Polymers • Poly-
merization of Monomers • Cross-Linking of Polymers • Cross-Linked
Wire and Cable Insulation • Graft and Block Copolymerization • Com-
posites of Wood and Plastics • Battery Separators • Curing of Surface
Coatings • Radiation Sterilization • Food Irradiation • Control of In-
sects • Sprout Inhibition • Sterilization of Drugs and Pharmaceuticals •
Sterilization of Medical Supplies • Water Pollution Control • References •

The preceding chapters have indicated that radiation
may be used to bring about many different types of
chemical change, and that it should, in principle, be
possible to establish a radiation-chemical industry. While
the development of such an industry is largely a matter of
economics, it also depends to a large degree on the
acceptability of radiation methods by industry and the
presence of a sufficiently large number of suitably trained
industrial chemists and chemical engineers.

454

Radiation Cost

It can easily be shown that the number of pounds of product produced per kilowatt-hour of radiation energy absorbed is $G \times M \times 8.3 \times 10^{-4}$, where M is the molecular weight of the product. Nuclear energy power production is already in the neighborhood of 20 million kW (1974). If we suppose that 15% of this energy is available for industrial radiation processes, the potential output of chemicals from a radiation-chemical industry is 22×10^6 GM lb, or 11×10^3 GM tons year^{-1}. Putting M equal to 100, this gives 11×10^5 G tons year^{-1}. If G is 1, the calculated output is only a million tons per year, which is not very attractive. However, if G is 100, or better still 10^5, we get a possible production rising into the tens of millions of tons and one could hope to base several world industries on it, provided the energy is not too expensive and the process possesses some attractive competitive features. Thus one requires a chain reaction, to give a high G, or alternatively a process in which a small percentage chemical change brings about the desired change in the product. A good many reactions of these two types, leading to desirable products, are known so that the first criterion is met. The estimated radiation costs found in the literature vary all the way from a fraction of a cent to several cents per pound so that the second criterion is certainly met in a number of cases (1). The radiation cost per pound is easily calculated knowing the G value and the cost of the absorbed radiation (2). The relationship between absorbed dose, radiation yield, molecular weight, and radiation cost is shown graphically in the nomogram in Fig. 10.1. Scales A, B, and C allow one to read off the kilowatt-hours absorbed per pound; scales D and E then allow one to calculate the radiation cost. As an example, the lines drawn in Fig. 10.1 are for the formation of benzene hexachloride, for which the molecular weight is 291 and the G value 85,000. It is seen from scale C that this reaction requires an absorbed dose of 4.9×10^{-5} kW-hr lb^{-1} of benzene hexachloride formed. Moving from scale C to scales D and E and assuming the cost of radiation to be \$8.00 kW-hr^{-1} absorbed, we find the cost of chlorinating benzene to be 0.025¢ lb^{-1} or 50¢ ton^{-1} of product.

Radiation chemical processes possess several attractive features as compared to conventional processes.

(i) They may lead to the synthesis of materials which cannot easily be prepared otherwise, e.g., graft polymers such as nylon grafted to styrene.

(ii) They may modify a material in a desirable manner, e.g., polyethylene cross-linked by irradiation.

(iii) They may result in a purer product, e.g., where radiation replaces a catalyst, contamination resulting from incomplete removal of the catalyst is not a problem.

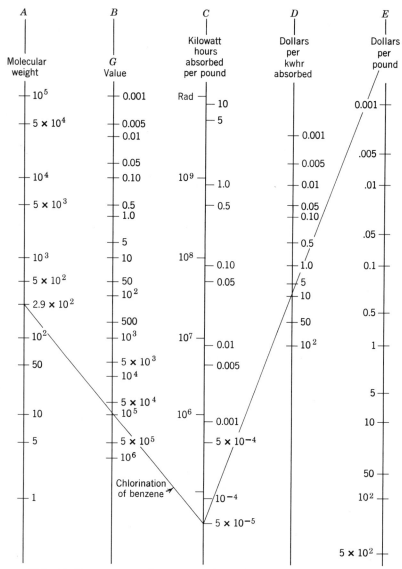

A	B	C	D	E
		Kilowatt hours absorbed per pound	Dollars per kwhr absorbed	Dollars per pound
Molecular weight	G Value			

FIGURE 10.1 *Radiation cost in relation to G value and molecular weight (ref. 2). Courtesy United States Atomic Energy Commission, Washington, D.C.*

(iv) The quality of a product produced using a solid catalyst often depends on the catalyst activity, which may vary with the method of forming the catalyst or because of catalyst poisoning. This variation is absent when a constant radiation intensity is used to trigger the reaction.

(v) Radiation-induced reactions can be carried out at relatively low temperatures, and this may lead to new products or the absence of competing side reactions.

(vi) The possible elimination of some stages of a synthesis.

(vii) Compared to reactions initiated by ultraviolet light, ionizing radiation offers the advantage of greater penetrating power so that more uniform reaction takes place in larger volumes of reactants, avoiding the buildup of product on the walls of the reaction vessel. The use of metal rather than glass vessels is also possible.

(viii) The penetrating power of ionizing radiation also makes it possible to irradiate materials in pressure vessels without the provision of special equipment.

(ix) The intensity of radiation can be easily controlled and can be continuous or pulsed. If the reaction becomes too vigorous, the radiation can be cut off instantaneously.

As an example of iii, ethyl bromide can be synthesized with 99.5% purity by irradiating a mixture of ethylene and hydrogen bromide at room temperature; the G value is about 10^5 (3). An example of vii is the formation of Gammexane (gamma benzene hexachloride) by a radiation-induced chain reaction with G about 10^5 (4); the overall reaction is

$$C_6H_6 + 3Cl_2 \rightarrow C_6H_6Cl_6 \qquad (10.1)$$

Radiation should have several advantages in comparison with the usual photochemical method using glass reaction vessels. The reaction liberates a considerable amount of heat, and there is also a rapid fall in light intensity as light passes through the mixture, leading to large local temperature variations. These are largely obviated using γ-rays. Furthermore, using metal tubes, transparent to γ-rays but not to light, greater safety and ease of cooling are realized (5, 6).

Industrial interest in radiation chemistry is indicated by the number of patents, estimated at 540 in the United States alone in the period 1960 to 1968 (1).

The degree of interest in radiation synthesis is shown by the estimate that the number of papers in this field increased from 6 in 1955 to 125 in 1968 but subsequently dropped off to 23 in 1973 (7). It is also indicated by the fact that the recent review paper by Chutny and Kučera (7) has 649 references and lists the following commercial syntheses (Part III, p. 167): U.S. ethyl bromide, 500 tons year^{-1}; USSR alkyl sulphochlorides, 5500 tons year^{-1}, tetrachloropentane, 400 tons year^{-1}, decachloropentanes, 50 tons year^{-1}, dichloroacetyl chloride, 50 tons year^{-1}, trichloroacetyl chloride, 50 tons year^{-1}.

Vereshchinskii (8) suggests that in spite of a quarter of a century of work in radiation chemical synthesis, this method has not yet become a generally practicable tool. The organic chemist does not consider the method to be a conventional synthetic technique and the radiation chemist prefers to devote his attention to the study of reaction mechanisms in relatively simple chemical systems.

Other reviews have been written by Danno (9), Wagner (10), Bayer et al. (11), and Wilson 11(a).

Sources of Radiation

Sources of radiation include fission products, artificial isotopes, reactors, linear accelerators, and so on. These have been discussed in Chapter 2.

CHEMICAL SYNTHESIS: INDUSTRIAL

As mentioned earlier, the actual interest of industry in radiation chemistry is difficult to estimate, and details of industrial chemical processes are, for the most part, lacking, the notable exception being the ethyl bromide synthesis. Information concerning a number of other reactions of potential interest has been published, however, and is briefly mentioned.

Ethyl Bromide Synthesis

The announcement by the Dow Chemical Company in 1962 that it had constructed a radiation facility with a capacity of a million pounds of ethyl bromide per year caused quite a stir. Shortly after the initial announcement, process details were published (12, 13, p. 205). In 1959 Armstrong and Spinks published a paper on the reaction of hydrogen bromide and ethylene under the influence of ^{60}Co γ-rays in the gas, liquid, and solid phases. The ion-pair yield was of the order of 10^5 (3, 14) and the product was 99.8% ethyl bromide. The publication of these two articles in Canada apparently stimulated an interest in other parts of the world (13, p. 207 and 9, p. 145) but in any event Dow Chemical shortly thereafter carried the process through to a successful conclusion using a two-phase process.

In preliminary experiments it was found that a number of halogenated solvents could be used as solvent, but that ethyl bromide gave particularly good results. Ethylene and hydrogen bromide were used in 1:1 stoichiometric amounts. It was desirable to use radiation produced ethyl bromide as the start up liquid, since chemically produced ethyl bromide gave a long induction period. (This is not unexpected in a chain reaction where yield can be drastically reduced by the presence of impurities). The G values for the two phase process were about 1/10 those for the gas phase reaction (see p. 240 for a discussion of the mechanism of the reaction).

In the Dow process, the reaction vessel has a capacity of 160 liters and is constructed of nickel (see Figs. 10.2 to 10.5). It is located underground in a steel lined, concrete encased pit. The Co source holder is located in a nickel well in the center of the reaction vessel. The ethyl bromide is continuously recirculated through the reaction vessel and a heat exchanger. Ethylene and hydrogen bromide are added to the recycle ethyl bromide and react as they pass through the irradiation zone. Excess ethyl bromide formed in the reaction overflows from a surge tank and proceeds to a product finishing step (13).

Calculations indicate that only 3% of the production costs is attributable to radiation facilities (9, p. 216).

The source strength employed is 1800Ci ^{60}Co. ^{60}Co has a half life of 5.3 years so about 20% replenishment will be required each year, costing ~$360 per annum. This very nominal catalyst cost is below the cost of other free-radical catalyst systems considered.

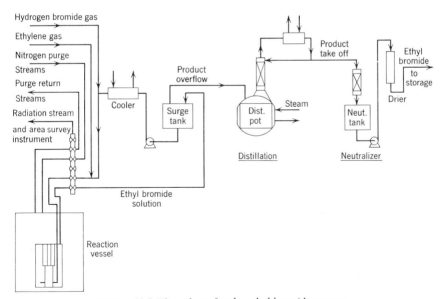

FIGURE 10.2 *Flow-sheet for the ethyl bromide process.*

OPERATING CONDITION

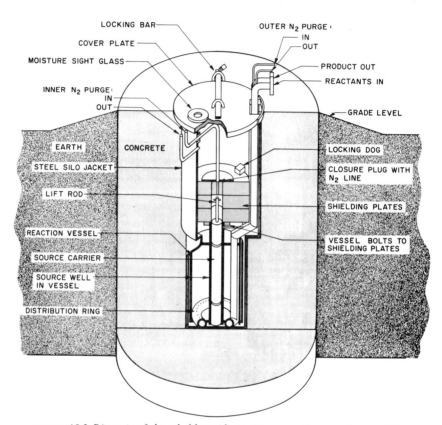

FIGURE 10.3 *Diagram of the ethyl bromide reaction vessel in operating condition.*

Synthesis of Gammexane (Gamma Benzene Hexachloride)

$$C_6H_6 + 3Cl_2 \rightarrow C_6H_6Cl_6 \qquad (10.1)$$

Gamma benzene hexachloride, which is one of the six stereoisomers formed by the addition of chlorine to benzene, is widely used as an insecticide, usually admixed with a large excess of the other isomers. The usual method of manufacture is by photochlorination using glass vessels. Pilot plant experiments have shown, however, that it can also be prepared with ionizing radiation to initiate the chlorination (4); G(benzene hexachloride) is about

FIGURE 10.4 *Shell for reaction vessel being lowered into position.*

10^5. By using ^{60}Co γ-rays the proportion of γ-isomer in the product is about 12%, which is very similar to the proportion produced photochemically. The reaction is highly exothermic and requires efficient heat exchange. Here γ-rays have the advantage that a metal reaction vessel can be used in place of a glass one, at once improving the mechanical safety and the overall rate of heat transfer. γ-Radiation also has the advantage of providing

FIGURE 10.5 *Shell for reaction vessel installed.*

a relatively uniform radiation intensity in the reactor in contrast with the high local intensities using light. The cost of the radiation-produced γ-isomer has been estimated as 1.51 and 1.60 dollars kg^{-1} using gross fission products and caesium-137 respectively as the source of radiation (15, cf. also 16), which is in line with the cost of commercial γ-benzene hexachloride produced photochemically (quoted as 1.90 to 3.00 kg^{-1}).

The radiation-induced chlorination of toluene is known to be intensity dependent, and it is likely that the chlorination of benzene is also. In this event, the arrangement of the radiation sources and the resultant radiation intensity distribution would be important factors in the overall economics. For example, it would be better to have a number of radiation sources arranged in such a way as to give a low intensity through a large volume rather than to give a high intensity in a small volume (cf. p. 189). Another, and perhaps not so immediately obvious factor, is the rate of flow and the rate of recirculation of the material through the reaction zone. For a chain reaction of the $I^{1/2}$ type, increasing the rate of flow of the material through the reaction zone will be expected, in general, to increase the effective G value. However, if the material is recirculated, a limiting G value will be reached, the optimum recirculation time being related to the mean lifetime of the free radicals involved in the chain reaction (cf. p. 191).

Nitrogen Fixation

Harteck and Dondes (17; 16, p. 234) have shown that nitrogen can be oxidized in a reactor core and that fission energy can thus be converted directly to chemical energy. The yield of nitrogen oxides is increased when the irradiation takes place in the presence of enriched U_3O_8 powder, and the authors found a maximum yield of NO_2 from a 4:1 nitrogen-oxygen mixture at a pressure of 10 atm and a temperature of 200°C. Under these optimum conditions the G value for NO_2 production was about 4 and for N_2O production about 2. It was calculated that the burnup of 1 mole of ^{235}U would "fix" 70 to 80 tons of nitrogen as NO_2.

SOME OTHER CHEMICAL SYNTHESES: LABORATORY AND PILOT PLANT

For chemical synthesis in a laboratory, with a ^{60}Co source giving an average dose rate of 1 Mrad hr^{-1} to an organic system of Vl, the numbers of grams product W, generated in t hr is $W = 0.00104\ GM_PIVt\rho$ (10). For usual conditions $\rho = 1$ g ml^{-1}, $I = 1$ Mrad hr^{-1}, $M_P \sim 100$, $G = 1$ to 5, $W \approx 0.1$ g hr^{-1}. This is, of course, not much use but if G is ≈ 1000 or more, $W = 100$ g hr^{-1} and the process becomes attractive.

A few examples of radiation chemical syntheses are given to illustrate the potential of the method. In some cases the process has been carried out on a pilot-plant scale.

Sulfochlorination of Hydrocarbons

Liquid cyclohexane reacts with sulfur dioxide and chlorine under γ-irradiation to give high yields of cyclohexanesulfonyl chloride and, to a smaller extent, chlorocyclohexane and cyclohexanedisulfonyl chloride (18). The reaction rate is proportional to the square root of the radiation intensity up to a dose rate of 40 rads min^{-1}, but becomes independent of the dose rate at dose rates above 400 rads min^{-1}; the G value is over 10^6. The radiation sources employed consisted of spent fuel elements, providing a γ-ray intensity of up to 2.5 Mrads hr^{-1}. A basic plant design was developed for continuous sulfochlorination, and it was concluded that the reaction might become attractive economically as cheap radiation sources became available.

Radiation-initiated sulfochlorination of hydrocarbons seems to occur generally. Thus the sulfochlorination of heptane has been reported (19), with G values up to 10^6, and similar high yields have been reported for the sulfochlorination of dodecane.

Sulfochlorination of synthine has been carried to the pilot plant stage in the USSR (8, p. 81).

Sulfonic Acid Production

Dodecane reacts with SO_2 and O_2 under the influence of γ-rays to give dodecylpersulfonic acid. The acid decomposes to various substances which may serve as intermediates in the manufacture of biodegradable detergents (7). Effects of dose rate, SO_2/O_2 ratio, pressure, and temperature were studied (see also 9, p. 220).

Addition Reactions to Unsaturates

A common type of reaction involves an alternating chain

$$A\cdot + U \rightarrow AU\cdot$$

$$AU\cdot + AB \rightarrow AUB + A\cdot$$

where U is the unsaturate; B is usually hydrogen or a halogen; and A· is a radical. An extensive table compiled by Wagner (10, p. 2) indicates an adequate chain length (> 100) for reactions such as the addition of aldehydes to isobutylene, chloroform to ethylene, CCl_4 to various olefins and allyl compounds, $BrCCl_3$ to various diolefins, $HSiR_3$ to various olefins, $H—PHCH_2CH_2CN$ to various olefins, $H—SH$ to α-olefins and acetylenes, HCl to C_2H_4, Cl_2 to benzene and toluene, and HBr to C_2H_4 and α-olefins. These can all be considered as potential commercial processes.

Hydrohalogenation

The ethyl bromide synthesis is referred to as the first industrial radiation chemical synthesis. The addition of HBr to various α-olefins leads in general to primary bromoalkanes, with G values of 10^3 in the gas phase and about 10^5 in the liquid phase.

Of interest is the production of an anesthetic 1,1,2-trifluoro-1-bromo-2-chloroethane from halogenated ethylene in 89% yield (8, p. 85).

Addition to H_2S *to Olefins*

In an early study, *n*-butyl mercaptan was found to react with pentene-1 to give the dialkyl sulphide with a *G* value of about 10^5 (20). H_2S reacts similarly with C_2H_4 to give ethyl mercaptan and diethyl sulphide (8, p. 90).

Chain Reactions Involving Saturated Molecules

Sometimes a pair of abstraction reactions results in a chain reaction

$$A \cdot + RH \rightarrow R \cdot + AH$$

$$R \cdot + AX \rightarrow A \cdot + RX$$

Examples of chains > 100 are Cl_2 + carboxylic acids and

$$Br_2 + Me_2CHCHMe_2$$

to give the dibromide (10, p. 218).

Oxidation of Hydrocarbons

The straight oxidation of hydrocarbons, initiated by radiation, is of no great industrial interest. However, the oxidation of halogenated hydrocarbons is of interest; e.g., oxidation of tetrachloroethylene using O_2 and radiation leads to tetrachloroethylene epoxide and trichloroacetyl chloride at 120°C. The production of trichloroacetic by this method would seem to be commercially attractive (8, p. 83).

Trichloroethylene can be similarly oxidized to dichloroacetyl chloride and this converted to methyl-dichloroacetate at a reasonable cost (8, p. 83).

Radiation oxidation of isoeugenol gives vanillin while isosafrol gives piperonal (8, p. 84).

Perchlorinated Hydrocarbons

Unsaturated perchlorohydrocarbons produce more complicated perchlorohydrocarbons on radiolysis; e.g., tetrachloroethylene gives a variety of products including hexachlorobutadiene.

Telomerization

Telomers are produced in high yields using, e.g., ethylene and chloroform or carbon tetrachloride (8, p. 86).

Higher alcohols can be similarly synthesized from aliphatic alcohols and ethylene. Liquid propanol-2 with ethylene gives tertiary amyl alcohol and higher alcohols (8, p. 87).

Methyl alkyl ketones have been synthesized from acetaldehyde plus ethylene (or propylene, etc.) and radiation (8, p. 88).

Synthesis of Organosilicon Compounds

A great variety of substituted chlorosilanes have been prepared using radiation; e.g., 2,3 bis(trifluoromethyl) phenyltrichlorosilane can be prepared by irradiation of a mixture of tetrachlorosilane and benzotrifluoride. G values in the range 15 to 300 are usual at temperatures in the 170 to 300°C range (8, p. 93).

Fluorination

Fluorocarbons have been used with radiation as fluorination agents, e.g., benzene plus carbon tetrafluoride yields, on irradiation, fluorobenzene and benzotrifluoride (8, p. 94).

Carboxylation and Carbonylation

Direct introduction of CO_2 into organic compounds takes place under the influence of radiation to give a variety of products with a low G value; e.g., *n*-pentane gives 2-ethylbutyric acid, 2-ethylvaleric, formic, capric, butyric, and other acids (8, p. 95).

Synthesis of Nitrogen Containing Organic Compounds

Irradiation of N_2 + hydrocarbons gives HCN with G values between 0.1 and 1.0. Similar results are obtained using hydrocarbons plus ammonia (8, p. 98).

Acetonitrile adds to olefins in the presence of radiation with G values about unity (8, p. 99).

Other syntheses based on nitrogen oxides or mixtures of NO and HCl have been reported—one might almost say that the range of possible reactions is limited mainly by the imagination of the experimenter!

Organophosphorus Compounds

Reactions of PCl_3 with a number of hydrocarbons have been reported by Henglein (21) and Vereshchinskii. Alkyl or cycloalkyl chlorophosphines are the main products (8, p. 104), some of which may be of interest industrially.

Using yellow phosphorus and alkyl chlorides, a variety of compounds such as PCl_3 and CCl_3PCl_2 result (using CCl_4); PCl_3 also reacts with olefins under the influence of radiation to produce 1:1 adducts. Relatively high G values up to several hundreds, occur.

Organotin Compounds

Synthesis of compounds of the R_2SnBr_2 type is based on the γ-induced reaction of the alkyl bromide with Sn. G values as high as 10^3 to 10^4 are obtained in the temperature range 20 to 95°C with good chemical yields (80 to 85%). Processes have been taken to the pilot-plant stage in the USSR and the economics seem to be attractive (8, p. 115).

Summarizing, it would appear that thus far there are very few radiation-initiated chemical syntheses in the production stage although a number have reached the stage of process investigation. However, the inevitable large increase in nuclear power generation by the end of the century cannot help but favor a much more rapid development in the future.

IRRADIATION OF MONOMERS AND POLYMERS

Radiation initiates the polymerization of monomers, graft, and block copolymerization, and also causes cross-linking and degradation in polymers. These reactions have been reviewed in Chapter 8 but are considered briefly here from the industrial standpoint.

Polymerization of Monomers

Many monomers have been polymerized by radiation, including such compounds as ethylene, vinyl chloride, methyl methacrylate, and styrene. However the properties of the polymers produced resemble those of the conventional products, and there is no particular cost advantage. Thus the probability of commercial development seems low and will probably wait the appearance of a radiation product possessing superior characteristics (2, 9).

Cross-Linking of Polymers

One of the earliest discoveries in the field of radiation chemistry to attract the attention of industry was the disclosure that thermoplastic polymers such as polyethylene become cross-linked upon irradiation. The vulcanized products exhibit new properties such as better resistance to heat and chemical action and improved mechanical strength (28). Commercial cross-linked products include nonmelting electrical insulating tape, extruded wire insulation, and special electronic parts. Radiation appears to be advantageous for cross-linking plastic film and possibly small diameter wire insulation (2, p. 152; 9, p. 186).

Radiovulcanization of rubber has also been accomplished, but for the moment the cost seems unfavorable compared with conventional methods.

Cross-Linked Wire and Cable Insulation

One of the most economically significant radiation processes involves electron irradiation cross-linked polyethylene, filled with pigments, flame retardation agents, etc. Irradiation produces an improved thermal stability, permitting greater current loads and higher operating temperatures (22).

Radiation cross-linking is also used in the formation of heat shrinkable films and cross-linked polyethylene foam (22).

Graft and Block Copolymerization

When polymers are irradiated, radicals are formed that can react with monomers to produce graft or block copolymers. These products often possess unusual properties having commercial possibilities. Suggested applications are production of flexible polymers with high melting points, production of oil-resistant rubbers, and the improvement of materials such as cotton.

Durable press articles have been produced by radiation-grafting of a monomer, N-methylol acrylamide, to a polyester/cotton blended fabric (22).

Favorable features of radiation-induced graft and block copolymerization include low dose requirements, close control of dose and dose rate, reaction in depth if necessary, and the possibility of continuous processing if desired. Cost estimates look favorable and there is undoubtedly a great deal of research activity in this field.

Radiation-induced grafting of certain vinyl monomers to cellulose materials has been described by Chapiro (23). The substrates are materials such as rayon, paper, wood pulp, cotton cloth, yarn, and jute; the monomers grafted on to these include styrene, methyl methacrylate, and other acrylic monomers. Cellulose grafting opens up new possibilities for dyed textiles, since cellulose materials are easily dyed and may then be grafted with a material such as polystyrene to give a lasting color. Furthermore, by grafting a hydrophobic material such as polystyrene to a hydrophilic material such as cotton cloth, the latter may be made water repellent (9, p. 185).

Mechanical properties such as tensile strength and elasticity may also be manipulated by grafting (24).

Composites of Wood and Plastics

In this process, a wood substrate is impregnated with liquid monomer which is then irradiated. Improved hardness and resistance to abrasion results, e.g., in parquet flooring (1, 25).

Battery Separators

Battery separators are being produced by what is essentially a graft polymer process in the amount of about 100,000 lb year^{-1} (1).

Curing of Surface Coatings

Paint formulations, consisting essentially of a prepolymer and a reactive vinyl monomer, are bonded to a surface in a matter of seconds using radiation. The product has improved adhesive properties (Ford Motor Co., 1970, refs. 1, 25a).

Concrete-polymer composite materials with improved properties have been produced by radiation polymerization techniques (25b).

RADIATION STERILIZATION

The ability of high-energy radiation to destroy microorganisms without appreciable temperature rise in the substrate offers an attractive means of sterilizing foods, drugs, and miscellaneous complex biological systems.

Since 1896, 1 year after Roentgen's discovery of x-rays, the bactericidal effects of radiation have been under study. Early patents covering both sterilization and deinfestation were granted and have long since expired. Early experimenters were troubled by two major problems, lack of an economic source of radiation, and insufficient information about the biological and biochemical effects of radiation. While these two problems have not been entirely solved, much progress has been made toward their eventual solution.

One of the most favorable markets for waste fission products appears to be the field of food sterilization. While this is not a direct application of radiation chemistry, the development of the large radiation sources that would be required for food sterilization is of first importance to a radiation chemical industry, and furthermore such facilities might initially be built with a multipurpose use. Brief mention of food irradiation, the control of infestation and sprouting, and the sterilization of drugs and pharmaceuticals is therefore appropriate. The doses required for these purposes are roughly as follows: to destroy the eggs of insects, 2000 to 5000 rads; to inhibit the sprouting of tubers, 5000 to 10,000 rads; to bring about the reproductive sterilization of adult insects in bulk stored grain, about 20,000 rads; and to bring about the radiosterilization of foods, about 2 to 5 Mrads.

A recent review article by Ley states that a total of 20 million Ci of cobalt-60 is currently in use in γ-radiation sterilization plants; 52 plants in 25 countries. National health authorities have approved a minimum dose of 2.5 Mrad for sterilization purposes, and there is an IAEA Recommended Code of Practice on Radiation Sterilization (25c).

Food Irradiation

Radiation-sterilization of food is not without its problems. Foods contain a few key components which, although present in micromole per gram concentrations, regulate the flavor and nutritional value. Experimental evidence indicates that some of these key components are very radiation-labile, and radiation-sterilization often brings about deleterious transformations in flavor, odor, and aesthetic quality. Proctor and Goldblith (26) have discussed the problems of radiation-sterilization. They concluded that it is possible to destroy all types of microorganism with radiation and were able to demonstrate that the species of organism is the prime factor in determining the magnitude of the sterilizing dose. They also point out that spoilage of certain foods occurs partly through enzymatic action so that it is also necessary to inactivate these enzymes. In the case of thermal sterilization the heat applied is sufficient to do this. However radiation-sterilization does not produce heating and some other means of inactivating the enzyme would have to be provided. High frequency electronic heating, to heat evenly the whole mass of the food, is suggested as a possibility. Here the exterior surface would not be overheated and only the minimum quantity of heat necessary for enzyme inactivation would be applied to the interior of the food.

Several means have been suggested for obviating the undesirable side reactions which lead to changes in color, texture, and flavor in irradiated food. These include irradiation of the food in the frozen state or in an inert atmosphere, or the addition of free-radical acceptors.

Although large doses are necessary to completely sterilize a food, much smaller doses have the useful property of extending the storage life of the food. Costwise, this application presents a more favorable picture than does the use of radiation as a sterilization agent and, even more important, a low radiation dose brings about almost no changes in the quality of most foods. Irradiation does appear a distinctly promising alternative to refrigeration for extending the storage life of food. Aebersold (26a), for example, has stated that radiation doses in the range 100,000 to 800,000 rads are sufficient to depress and delay microbial multiplication in fish, and hence to effect radio-pasteurization; about ten times this dose is required to kill all the organisms found in food. It was suggested that 1 ton of pasteurized product per hour would be produced by a cobalt-60 irradiation unit containing 300,000 Ci of the isotope.

A recent review by Josephson (26b) indicates the practicality of food preservation by ionizing radiation and states that experimental data on wholesomeness of irradiated foods obtained during the last 20 years has not demonstrated danger to the consumer. Fourteen countries have approved at least 1 of 17 irradiated foods but the process of getting approval by regulatory bodies such as the U.S. Food and Drug Administration in the United States is both time consuming and costly (see also 26c).

Control of Insects

Reproduction of the confused flour beetle, the insect most commonly found in flour, is halted by a dose of about 25,000 rads. Flour irradiation facilities have been proposed, based on spent fuel elements as radiation sources (27, p. 355). Irradiation of grain using a 15,000 curie cobalt-60 source, at a cost of approximately 1¢ bushel^{-1}, has also been proposed (27, p. 357).

Sprout Inhibition

Tests in several countries have indicated that a dose of about 10,000 rads effectively inhibits sprouting in potatoes (28, 29, 27, p. 291) while sprouting in onions is effectively controlled by a dose of 7000 to 10,000 rads (27, p. 296).

Rice (30, 31) has described a mobile irradiator containing 8500 curies of cobalt-60 designed to treat potatoes and similar crops. The unit gives a dose of 8000 rads to 6000 lb of material per hour, and it is thought that the cost might be brought down to about 0.2¢ lb^{-1}.

Sterilization of Drugs and Pharmaceuticals

Radiation sterilization of drugs and pharmaceuticals has already been carried out on a limited commercial scale. In comparison with food steriliza-tion this field presents an ideal case. One is usually dealing with a concen-trated, reasonably well-defined single substance, and since pharmaceuticals are usually not regularly consumed, no notice need be taken of chemical changes that may produce small amounts of cumulative poisons. The economic position is also favorable, since pharmaceuticals are generally products of high unit cost, and if irradiation sterilization should cost a trifle more, its advantages with respect to product standards could easily outweigh the difference.

Very little work has been published in this field but the following types of material have been successfully sterilized by irradiation:

(i) General products—vitamins, patent medicines, and household supplies. Experiment has shown that vitamins, when present as concentrates, do not lose much potency upon irradiation.
(ii) Chemotherapeutic agents—drugs such as aureomycin, penicillin, strep-tomycin, and many others have been irradiated without loss in potency.
(iii) Medical and surgical specialities—such items as sutures, dressings, bonebank bone, and surgical catgut may be radiation-sterilized (32).

Sterilization of Medical Supplies

This is an extremely promising field for the application of radiation, which has already been employed to sterilize such bulky items as cotton gauze, adhesive bandages, and surgical dressings. Sterilization of medical supplies by penetrating radiation such as γ-rays presents many advantages including:

(i) The radiation is extremely penetrating.
(ii) Complete sterilization is possible even in the most tightly packaged containers.
(iii) There are no limitations to the types of packaging that can be used.
(iv) A wider range of colored packaging materials may be used, enabling manufacturers to use color more effectively to promote sales appeal.
(v) There is no damage to heat-sensitive materials.
(vi) Continuous processing is possible.

It is of course possible that large doses of radiation might produce an undesired effect, such as embrittlement, in some wrappings.

Water Pollution Control

A dose of 10^5 rads is said to be effective in liquid waste irradiation pro-cesses (33).

REFERENCES

1. D. E. Harmer and D. S. Ballantine, *Chem. Eng.*, 98 (April 1971) and 91 (May 1971).
2. Radiation—A Tool for Industry, *ALI*, 52 (1959).
3. D. A. Armstrong and J. W. T. Spinks, *Can. J. Chem.*, **37**, 1210 (1959).
4. D. E. Harmer, L. C. Anderson, and J. J. Martin, *Nuclear Eng.*, Part I, *Chem. Eng. Progr.*, *Symp. Ser. 50*, no. 11 (1954) (AECU 2981).

5. H. Mohler, *Chemische Reaktionen Ionisierenden Strahlen*, Verlag H. R. Sauerlander, Aarau and Frankfort am Main, 1958, p. 241.
6. H. Drawe, *Angewandte Strahlenchemie*, Hüthig Verlag, Heidelberg, 1973.
7. B. Chutny and J. Kučera, *Radiat. Res. Rev.*, **5**, 1 (1974).
8. *Advances in Radiation Chemistry* (eds. M. Burton and L. G. McGee), No. 3, Wiley-Interscience, New York, 1972. Article by I. V. Vereshchinskii, p. 75.
9. *Actions Chimiques et Biologiques des Radiations* (ed. M. Haissinsky), Masson et Cie, Paris, 1967, 13th series. A Danno, Industrial applications.
10. *Advances in Radiation Chemistry* (eds. M. Burton and J. L. Magee), Vol. 1, 1969 (C. D. Wagner, Chemical Synthesis by Ionising Radiation, p. 199).
11. A. A. Bayer et al., Large-scale radiation-chemical plants, *Fourth U.N. Int. Conf. Peaceful Uses of the Atom*, No. P466, Geneva (1971).
11a. J. G. Wilson, *Radiat. Res. Rev.*, **4**, 71 (1972).
12. D. E. Harmer and J. S. Beale, *Nucleonics*, **20**, 18 (December 1962); **21**, 76 (October 1963).
13. D. E. Harmer, J. S. Beale, C. T. Pumpelly, and B. W. Wilkinson, *Industrial Uses of Large Radiation Sources*, Vol. 2, I.A.E.A., Vienna, 1963, p. 205.
14. D. A. Armstrong and J. W. T. Spinks, *Can. J. Chem.*, **37**, 1002 (1959).
15. L. C. Anderson, B. G. Bray, and J. J. Martin, *Proc. Int. Conf. Peaceful Uses Atomic Energy*, United Nations, New York, **15**, 235 (1956).
16. D. E. Harmer, *Engineering Research Institute*, University of Michigan, Progress Report No. 7, AECU 3077 (1954).
17. P. Harteck and S. Dondes, *Nucleonics*, **14**, 22 (July 1956).
18. A. Schneider, *ANL*, 5863 (1958).
19. A. Henglein and H. Url, *Z. Physik. Chem.* (*Leipzig*), **9**, 285, 516 (1956).
20. A. Fontijn and J. W. T. Spinks, *Can. J. Chem.*, **35**, 1384 (1957).
21. A. Henglein, *Int. J. Appl. Radiat. Isotop.*, **8**, 156 (1960).
22. A. S. Hoffman, *Appl. Indust. Radiat. Chem. Monomers and Polymers, 5th Int. Cong. Radiat. Res.*, Seattle (1974).
23. A. Chapiro, *Nucleonics*, **19**, 65 (October 1961).
24. W. H. Rapson, *Can. Nucl. Tech.*, no. 3, 20 (1962).
25. *Bull. Inf. Sci. Tech. No. 171*, 112 (June 1972). Commissariat a l'energie atomique, France.
25a. K. G. McLaren, *Roy. Austr. Chem. Inst. Proc.*, 340 (December 1970).
25b. K. G. McLaren and A. Samarin, U. of New South Wales, Conference on engineering materials, 137 (1974).
25c. F. J. Ley, Radiation Sterilization—An Industrial Process, *5th Int. Congr. Radiat. Res.*, Seattle (1974).
26. B. E. Proctor and S. A. Goldblith, *Proc. Int. Conf. Peaceful Uses Atomic Energy*, United Nations, New York, **15**, 245 (1956).
26a. P. C. Aebersold, speech to National Fisheries Institute, New Orleans (April 1962).
26b. E. S. Josephson et al., The Use of Ionizing Radiation for Preservation of Food and Feed Products, *5th Int. Congr. Radiat. Res.*, Seattle (1974).
26c. C. W. S. Gard, *Food Irradiation*, AECL Commercial Products (October 1974).
27. L. E. Brownell, *Radiation Uses in Industry and Science*, U.S. Atomic Energy Commission (1961).
28. A. H. Sparrow and E. Christensen, *Nucleonics*, **12**, 16 (August 1954).
29. G. Tripp, *Int. J. Appl. Radiat. Isotop.*, **6**, 174 (1959).
30. *Large Radiation Sources in Industry*, IAEA, Vienna, Vol. I, 1960, p. 180.
31. *Gamma Irradiation in Canada*, Atomic Energy of Canada Limited, Report no. AECL 1329, 67 (1961).
32. C. Artandi and W. Van Winkle, *Nucleonics*, **17**, 86 (March 1959).
33. K. L. Murphy, *Water and Pollution Control*, April 1974.

APPENDIX 1

Tables of Data

Values of general physical constants and energy conversion factors are from E. R. Cohen, "Fundamental Constants" in *The Encyclopedia of Physics* (ed. R. M. Besancon), Van Nostrand Reinhold, New York, 1974, p. 170, and were used in computing constants and other numerical quantities given in the text.

Physical Constants

Avogadro constant $\quad N_A = 6.0220 \times 10^{23}$ mole^{-1}
Electron
$\quad$ Charge $\quad e = 1.6022 \times 10^{-19}$ C
$\qquad\qquad\quad = 4.8032 \times 10^{-10}$ esu
$\quad$ Rest mass $\quad m_e = 9.1095 \times 10^{-31}$ kg
$\quad$ Energy equivalent of rest mass
$\qquad\qquad m_e c^2 = 511{,}003$ eV
Boltzmann constant
$\quad k = 1.3807 \times 10^{-23}$ J K^{-1}
$\qquad = 1.3805 \times 10^{-16}$ erg molecule^{-1} K^{-1}
Faraday constant $\quad F = 96{,}485$ C mole^{-1}
$\qquad\qquad\qquad = 23.053$ kcal equiv^{-1} volt^{-1}
Gas constant $\quad R = 8.314$ J^{-1} mole^{-1}
$\qquad\qquad\quad = 82.06$ cm^3 atm K^{-1} mole^{-1}
$\qquad\qquad\quad = 1.9872$ cal mole^{-1} K^{-1}
Planck constant $\quad h = 6.6262 \times 10^{-34}$ J sec
$\qquad\qquad\qquad = 4.1357 \times 10^{-21}$ MeV sec
Proton mass $\quad m_p = 1.6727 \times 10^{-27}$ kg
Neutron mass $\quad m_n = 1.6750 \times 10^{-27}$ kg
Velocity of light in a vacuum
$\qquad\qquad c = 2.9979 \times 10^8$ m sec^{-1}

Mathematical Constants

$\pi = 3.1416$ $\log_{10} e = 0.43429$ (log e)
$e = 2.7183$ $\log_e 10 = 2.3026$ (ln 10)
   $\log_e 2 = 0.69315$ (ln 2)

Units and Conversion Factors

Ampere $1\,A = 1\,C\,sec^{-1} = 6.2415 \times 10^{18}$ electrons sec^{-1}
Ångstrom $1\,\mathring{A} = 10^{-10}$ m
Atmosphere pressure (standard) $1\,atm = 101{,}325\,N\,m^{-2}$
Barn $1\,b = 10^{-24}\,cm^2$
Calorie (thermochemical) $1\,cal_{th} = 4.1840\,J$
Coulomb $1\,C = 2.9979 \times 10^9$ esu charge
Curie $1\,Ci = 3.7 \times 10^{10}$ disintegrations sec^{-1} (exactly)
Electron volt $1\,eV = 1.6022 \times 10^{-19}\,J$
   $= 8065.5\,cm^{-1}$ (wave numbers)
1 Electron volt per molecule $= 9.6484 \times 10^4\,J\,mole^{-1}$
   $= 2.3060\,cal_{th}\,mole^{-1}$
1 Electron volt per second $= 1.6022 \times 10^{-19}\,W$
1 Electron volt per gram $= 1.6022 \times 10^{-16}\,J\,kg^{-1}$
   $= 1.6022 \times 10^{-14}$ rad
1 Electron volt per $cm^3 = 1.6022 \times 10^{-16}/\rho\,J\,kg^{-1}$
   $= 1.6022 \times 10^{-14}/\rho$ rad
Erg $1\,erg = 10^{-7}\,J$
   $= 1\,cm^2\,g\,sec^{-2}$
Esu (charge) $1\,esu = 3.3356 \times 10^{-10}\,C$
1 Kilocalorie per mole $= 4.3365 \times 10^{-2}\,eV\,molecule^{-1}$
Rad $1\,rad = 0.01\,J\,kg^{-1}$
   $= 100\,erg\,g^{-1}$
   $= 6.2415 \times 10^{13}\,eV\,g^{-1}$
   $= 6.2415\rho \times 10^{13}\,eV\,cm^{-3}$
   $= 2.3901 \times 10^{-6}\,cal_{th}\,g^{-1}$
   $= 2.7777 \times 10^{-9}\,W\,hr\,g^{-1}$
   $= 10^{-5}\,W\,sec\,g^{-1}$
1 Roentgen x- or γ-radiation absorbed in air
   (W_{air}, 33.73 eV ion-pair^{-1}) $= 0.870$ rad
1 Roentgen x- or γ-radiation absorbed in water
   ($W_{air} = 33.73$ eV ion-pair^{-1}) $= 0.967$ rad
Watt $1\,W = 1\,J\,sec^{-1}$
   $= 100\,kg\,rad\,sec^{-1}$

G value (measure of radiation-induced yield equivalent to one molecule of the specified material formed or destroyed per 100 eV energy absorbed)

$$1 \; G \text{ unit} = 1.0364 \times 10^{-7} \text{ mole J}^{-1}$$

1 rad changes $1.0364G \times 10^{-9}$ mole material per kilogram (G is the G value for the change).

1 Mrad produces a temperature rise in water of 2.39°C.

Quantum energy $\times$ wavelength $\quad hv \times \lambda = 12.399$ keV Å
$$= 1239.9 \text{ eV nm}$$
$$= 1.2399 \times 10^{-9} \text{ MeV cm}$$

Radiation Chemistry Problems

These simple worked examples are intended to illustrate the use of the equations given in the earlier chapters.

Problem 1

It is necessary to reduce the radiation level several feet from a small cobalt-60 source from 10 R min^{-1} to 10 mR hr^{-1}. What is the minimum thickness of lead shielding (to the nearest centimeter) required to do this? Assume a half-thickness value of 1.06 cm for ^{60}Co γ-rays in lead.

The linear absorption coefficient (μ) can be calculated from the half-thickness value, since

$$\mu = \frac{0.693}{\text{half-thickness value}}$$

Thus for ^{60}Co γ-rays in lead,

$$\mu = \frac{0.693}{1.06}$$
$$= 0.654 \text{ cm}^{-1}$$

This value can be substituted in

$$I = I_0 e^{-\mu x} \qquad (2.3; 2.18)$$

to determine the thickness of lead needed to reduce the radiation intensity (or exposure rate) at a point to any given value. In the present example I_0 is 600 R hr^{-1} and

I is to be 0.01 R hr^{-1}; thus

$$0.01 = 600e^{-\mu x}$$

$$\log_{10}\left(\frac{0.01}{600}\right) = -\log_{10} e \times 0.654x$$

$$\bar{5}.2218 = -0.4343 \times 0.654x$$

$$-4.7782 = -0.284x$$

$$x = 16.8 \text{ cm}$$

i.e., *17 cm to the nearest centimeter.*

Problem 2

The exposure rate at a point near a cobalt-60 source was 235 R min^{-1} on October 15, 1975. Calculate the exposure rate at the same point on August 15, 1976 and on February 15, 1977. Take 5.27 years as the half-life of cobalt-60.

$$5.27 \text{ years} = 63.24 \text{ months}$$

The decay constant for a radioactive nuclide is given by

$$\lambda = \frac{0.693}{\text{half-life}}$$

and therefore, for ^{60}Co,

$$\lambda = \frac{0.693}{63.24}$$

$$= 0.01096 \text{ month}^{-1}$$

The intensity of the radiation emitted by a radioactive source, and the dose rate near it, fall at the same rate as the activity of the source falls due to radioactive decay, and

$$I_t = I_0 e^{-\lambda t} \qquad\qquad (2.6)$$

$$\log_{10}\left(\frac{I_t}{I_0}\right) = -\lambda t \log_{10} e$$

In this example t has the values 10 months and 16 months and λ is 0.01096 month^{-1}. The ratios of I_t/I_0 calculated by substituting these figures in the preceding equation are 0.896 and 0.839 for 10 months and 16 months, respectively, and the exposure rates at the point of interest after these periods of radioactive decay are therefore 235 × 0.896 and 235 × 0.839 R min^{-1}, i.e., *211 R min^{-1}* and *197 R min^{-1}*, respectively.

Problem 3

From the range-energy data for protons given below calculate the range in air of 6-MeV and 12-MeV tritium nuclei ($^3_1$H) and 6-MeV and 12-MeV α-particles.

Proton energy (MeV):   1   1.5   2   3   4   5   6   7   8   9
Range in air (cm):   2.3   4.3   7.2   14.1   23.1   33.9   46.7   61.2   77.3   95.3

This problem makes use of the equations:

$$\text{range of particle A} = \frac{m_A Z_B^2}{m_B Z_A^2} \times \text{range of particle B} \qquad (2.13)$$

when

$$\text{energy of particle A} = \frac{m_A}{m_B} \times \text{energy of particle B} \qquad (2.14)$$

For tritium nuclei, $m = 3$ and $Z = 1$, while for protons $m = 1$ and $Z = 1$. Thus the range of a tritium nucleus is equal to three times the range of a proton with one third the energy, i.e.,

range of a *6-MeV tritium nucleus* in air = 3 × range 2-MeV proton

$$= 21.6 \ cm$$

range of a *12-MeV tritium nucleus* = 3 × range 4-MeV proton

$$= 69.3 \ cm$$

For α-particles, $m = 4$ and $Z = 2$, and the range of an α-particle is equal to that of a proton with one quarter the energy. However, in the case of α-particles a constant, found by experiment to be 0.20 cm in air, must be added to the calculated range (see p. 36) and

range *6-MeV α-particle* in air = range 1.5-MeV proton + 0.20 cm

$$= 4.5 \ cm$$

range of a *12-MeV α-particle* = range 3-MeV proton + 0.20 cm

$$= 14.3 \ cm$$

Problem 4

The exposure rate 1.2 in. from a cobalt-60 source is 300 R min^{-1}. How thick must concrete shielding walls be to reduce the radiation level (i) at the operating face, 4 ft from the source, to less than 1 mR hr^{-1} and (ii) in an adjoining laboratory, 20 ft from the source, to less than 10 mR week^{-1}. Give the result to the nearest 3 in. and take the half-thickness value for ^{60}Co γ-rays in concrete as 2.0 in.

In this problem the combined effect of distance and shielding must be estimated.

(i) *To estimate the shielding required between source and operating face.* The reduced exposure rate at the operating face due to distance alone is found by applying the inverse square relationship; i.e., since the exposure rate 1.2 in. from the source is 300 R min^{-1}, the exposure rate 48 in. from the source will be $300(1.2/48)^2 = 0.1875$ R min^{-1} or 11.25 R hr^{-1}. The concrete shielding is therefore required to reduce the exposure rate at the operating face from 11.25 to 0.001 R hr^{-1}. The thickness of concrete necessary is found by substituting these values for I_0 and I respectively in the expression

$$I = I_0 e^{-\mu x} \qquad (2.3; 2.18)$$

taking the linear absorption coefficient (μ) as $0.693/2.0 = 0.3465$ in.$^{-1}$; the procedure is that used in Problem 1.[1] The thickness required is 26.9 in. or, to the nearest 3 in., *2 ft 3 in.*

(ii) *To estimate the shielding required between source and laboratory.* In this case distance reduces the exposure rate at the laboratory to $300(1.2/240)^2 = 0.0075$ R min^{-1} or 75.6 R week^{-1}. Proceeding as in part i, the thickness of concrete needed to reduce the exposure rate from 75.6 to 0.01 R week^{-1} is found to be 25.8 in. or, to the nearest three inches, *2 ft.* However, in shielding calculations it is usual to err on the side of safety and this result would normally be given to the next higher three inches, i.e., as *2 ft 3 in.*; this is implied by the question, which asks for the shielding necessary to reduce the exposure rate to *less* than 10 mR week^{-1}.

Problem 5

If the mass energy absorption coefficient for cobalt-60 γ-rays in water is 0.0296 cm^2 g^{-1}, what are the values of the corresponding linear, atomic, and electronic energy absorption coefficients?

The linear energy absorption coefficient (μ_{en}) is simply the mass coefficient multiplied by the density, which is 1 in the case of water so that

$$(\mu_{en})_{H_2O} = \left(\frac{\mu_{en}}{\rho}\right)_{H_2O} = 0.0296 \ cm^{-1}$$

[1] The calculation is simplified if the expression shown (Eqs. 2.3, 2.18) is put in log form and rearranged slightly, when

$$x = \frac{2\cdot303}{\mu}\log_{10}\left(\frac{I_0}{I}\right)$$

The atomic, energy absorption coefficient $(_a\mu_{en})$ is given by

$$_a\mu_{en} = \frac{A}{N_A}\left(\frac{\mu_{en}}{\rho}\right) \tag{2.35}$$

Thus for water,

$$(_a\mu_{en})_{H_2O} = \frac{18}{6.022 \times 10^{23}} \times 0.0296$$

$$= 8.84 \times 10^{-25}\ cm^2\ molecule^{-1}\ or\ 0.884\ b\ molecule^{-1}$$

The electronic energy absorption coefficient $(_e\mu_{en})$ is given by

$$_e\mu_{en} = \left(\frac{_a\mu_{en}}{Z}\right) \tag{2.34}$$

For a compound, Z is the sum of the atomic numbers of the constituent elements; i.e., 10 for water. Thus

$$(_e\mu_{en})_{H_2O} = 8.84 \times 10^{-26}\ cm^2\ electron^{-1}\ or\ 0.0884\ b\ electron^{-1}$$

Problem 6

Calculate the mass absorption coefficient for 1-MeV γ-rays for sodium iodide and for calcium metaphosphate, $Ca(PO_3)_2$, from data for the elements present. The atomic absorption coefficients (b atom^{-1}) for the elements are oxygen, 1.69; sodium, 2.32; phosphorus, 3.17; calcium, 4.22; iodine, 12.03 (for 1-MeV photons).

SODIUM IODIDE.   The atomic absorption coefficient for a compound is equal to the sum of the atomic absorption coefficients for the constituent atoms, i.e., for sodium iodide

$$(_a\mu)_{NaI} = (_a\mu)_{Na} + (_a\mu)_I$$
$$= 2.32 + 12.03$$
$$= 14.35\ b\ molecule^{-1}$$

While the mass absorption coefficient can be obtained from the atomic coefficient using the relationship,

$$\frac{\mu}{\rho} = \frac{N_A}{A}(_a\mu) \tag{2.35}$$

where A is the molecular weight of the compound. For sodium iodide,

$$\left(\frac{\mu}{\rho}\right)_{NaI} = 14.35 \times 10^{-24} \times \frac{6.022 \times 10^{23}}{150}$$
$$= 0.0576\ cm^2\ g^{-1}$$

CALCIUM METAPHOSPHATE. The atomic and mass absorption coefficients are calculated as shown for sodium iodide; thus

$$(_a\mu)_{\text{Ca phosphate}} = (_a\mu)_{\text{Ca}} + 2(_a\mu)_{\text{P}} + 6(_a\mu)_{\text{O}}$$
$$= 4.22 + 6.34 + 10.14$$
$$= 20.70 \text{ b molecule}^{-1}$$

$$\left(\frac{\mu}{\rho}\right)_{\text{Ca phosphate}} = 20.70 \times 10^{-24} \times \frac{6.022 \times 10^{23}}{198}$$
$$= 0.0630 \ cm^2 \ g^{-1}$$

Problem 7

Small samples of pure methanol and pure acetic acid are irradiated with cobalt-60 γ-rays at a position at which the exposure rate (measured by means of an ionization chamber) is 360 R min^{-1}. How much energy (in rads) will the samples absorb if they are irradiated for 10 hr?

METHANOL.   Organic compounds, apart from those containing appreciable amounts of heavy elements, interact with cobalt-60 γ-radiation by the Compton process almost exclusively. Under these circumstances the absorbed dose is related to the exposure by

$$D_M = 0.870 X_A \times \frac{(\overline{Z/A})_M}{(\overline{Z/A})_A} \tag{3.5}$$

where the subscripts A and M signify air and organic material, respectively.
In the present example X_A is the exposure accumulated during 10 hr, i.e., $360 \times 600 = 2.16 \times 10^5$ R, and $(\overline{Z/A})_A$ is 0.499 (Table 3.6). $(\overline{Z/A})_{\text{methanol}}$ is equal to the sum of the atomic numbers of the atoms present in methanol divided by the molecular weight; i.e., $18/32 = 0.562$. Therefore the absorbed dose in the methanol is

$$D_{\text{methanol}} = 0.870 \times 2.16 \times 10^5 \times \left(\frac{0.562}{0.499}\right)$$
$$= 2.12 \times 10^5 \ rads$$

ACETIC ACID. For acetic acid $(\overline{Z/A})_{\text{acetic acid}} = 32/60 = 0.533$ and

$$D_{\text{acetic acid}} = 0.870 \times 2.16 \times 10^5 \times (0.533/0.499)$$
$$= 2.01 \times 10^5 \ rads$$

Problem 8

The apparent exposure rate inside an aluminum block, measured with an air-filled cavity ionization chamber, is 79 R min^{-1}. The ionization chamber has walls of bakelite. What is the absorbed dose rate at this point in the aluminum if the chamber walls are (i) very thin, (ii) thick, compared with the range of the secondary electrons produced?

For cobalt-60 γ-rays; $(\mu_{en}/\rho)_{Al} = 0.0258$ cm^2 g^{-1}, $(\mu_{en}/\rho)_{Bakelite} = 0.0284$ cm^2 g^{-1}, $(s_m)_{air}^{Al} = 0.89$, and $(s_m)_{air}^{Bakelite} = 1.07$.

THIN-WALLED IONIZATION CHAMBER. For a thin-walled, air-filled, cavity ionization chamber the absorbed dose in the surrounding medium is related to the ionization produced in the cavity (Q) by

$$D_M = 0.870Q(s_m)_{air}^{medium} \tag{3.9}$$

Substituting numerical values for Q and $(s_m)_{air}^{Al}$, the absorbed dose rate at the point of interest in the aluminum will be

$$D_{Al} = 0.870 \times 79 \times 0.89$$
$$= 61.2 \, rads \, min^{-1}$$

THICK-WALLED IONIZATION CHAMBER. If the ionization chamber used has a relatively thick Bakelite wall the absorbed dose rate in the aluminum will be given by

$$D_{Al} = 0.870Q(s_m)_{air}^{Bakelite} \frac{(\mu_{en}/\rho)_{Al}}{(\mu_{en}/\rho)_{Bakelite}} \tag{3.10}$$

$$= 0.870 \times 79 \times 1.07 \times \left(\frac{0.0258}{0.0284}\right)$$

$$= 66.8 \, rads \, min^{-1}$$

Problem 9

A mixture composed of equal parts by weight of methanol and acetic acid is irradiated with cobalt-60 γ-rays. What part of the total energy absorbed is absorbed by the methanol?

Only Compton absorption need be considered here, and the fraction of the total energy absorbed that is absorbed by the methanol is

$$\frac{D_{methanol}}{D_{mixture}} = w_m \times \frac{(\overline{Z/A})_{methanol}}{(\overline{Z/A})_{mixture}} \tag{3.43}$$

where w_m is the fraction by weight of methanol.

Values of $(\overline{Z/A})$ for methanol and acetic acid are 0.562 and 0.533 respectively (Problem 7) while $(\overline{Z/A})$ for the mixture of methanol and acetic acid is the mean of these values, since each component contributes equally by weight, i.e., is 0.5475. Substituting these values in the equation above,

$$\frac{D_{methanol}}{D_{mixture}} = \frac{1}{2} \times \frac{0.562}{0.5475} = 0.513$$

i.e., *51.3%* of the energy absorbed is absorbed by the methanol.

It should be pointed out that this calculation assumes that the energy absorbed by a particular component is governed entirely by the extent to which it interacts with the primary radiation; secondary electrons are assumed to deposit their energy with the components in the same proportions as these interact with the primary radiation. The assumption appears to be reasonably valid for mixtures containing saturated, chemically similar, compounds, but in other cases its validity is uncertain.

Problem 10

Five hundred microcuries of tritium oxide is mixed with 250 mg of ethanol (the weight of the T_2O is negligible). Assuming that all the radiation emitted is absorbed by the ethanol, how much energy (rads) will be absorbed in (i) 1 day, (ii) 1 week, (iii) 1 month (30 days)?

The average energy of the β-particles emitted by tritium is 5.5 keV.

The absorbed dose rate for an internal irradiation is given by

$$T_M = 0.593 \times C \times \bar{E} \text{ rad sec}^{-1} \tag{3.20}$$

where C is the concentration of the radioisotope in millicuries per gram and $\bar{E}$ (MeV) is the mean energy released to ionizing particles per disintegration. In this problem $C = 0.5/0.25 = 2$ mCi g^{-1}, $\bar{E}$ is 5.5×10^{-3} MeV, and the absorbed dose rate is

$$T_M = 0.593 \times 2 \times 5.5 \times 10^{-3}$$
$$= 6.52 \times 10^{-3} \text{ rads sec}^{-1}$$

Thus the absorbed dose in 1 day (8.64×10^4 sec) is   *564 rads*

in 7 days (6.05×10^5 sec) is   *3950 rads*

in 30 days (2.60×10^6 sec) is   *16,900 rads*

Since the half-life of tritium is long (12.26 years), compared with the duration of the experiment, it is not necessary to correct here for radioactive decay.

Problem 11

A solution containing phosphorus-32 was counted on Monday and found to contain 1.50 mCi of ^{32}P. On Wednesday, 48 hr later, the solution was added to a solution of ferrous sulfate in sulfuric acid such that the mixed solution was 0.001 M with respect to $FeSO_4$ and 0.4 M with respect to H_2SO_4; the final volume of the mixture was 10 ml. The mixture was left in a dark cupboard at room temperature.

On Friday, 48 hr after mixing, the absorbance of the mixed solution was measured at 304 nm (24°C) and found to be 0.324, a blank solution similar in composition to the mixture had an absorbance of 0.003; both were measured using a 1-cm quartz cell. Calculate G(ferric) for phosphorus-32 β-rays.

How great would the absorbed dose be if the mixture were left until all the radioactive phosphorus had decayed?

The molar extinction coefficient for ferric ion at 304 nm, 24°C, is 2205 and density of 0.4 M sulfuric acid at 24°C is 1.024 g cm^{-3}. Phosphorus-32 has a half-life of 14.22 days and the average energy of the β-rays emitted by it is 0.70 MeV.

In this example allowance will be made for the decay of the radioactive material between counting and making up the experimental solution and also during the irradiation itself.

The activity present at the start of the irradiation, two days after counting, is calculated using the equation for radioactive decay,

$$C_t = C_0 e^{-\lambda t} \tag{2.4}$$

and putting $C_0 = 1.50$ mCi, $t = 2$ days, and $\lambda = 0.639/14.22 = 0.04873$ day^{-1}; thus $C_t = 1.361$ mCi.

The absorbed dose received during irradiation is given by

$$D_M = \frac{5.121 \times 10^4 \times \bar{E} \times C_0}{\lambda}(1 - e^{-\lambda t}) \text{ rads} \tag{3.25}$$

Here C_0 is the activity present at the start of the irradiation in units of milli-curies per gram of solution. The absorbed dose is found by substituting $\bar{E} = 0.70$ MeV, $C_0 = 1.361/10.24 = 0.1329$ mCi g^{-1}, $t = 2$ days, and $\lambda = 0.04873$ day^{-1}; when

$$D_M = \frac{5.121 \times 10^4 \times 0.70 \times 0.1329}{0.04873}(1 - e^{-0.09746})$$

$$= 9.776 \times 10^4(1 - 0.9071)$$

$$= 9079 \text{ rads}$$

(The absorbed dose can also be calculated by assuming that the mean dose rate during the irradiation is the dose rate at the midpoint of the experiment; 3 days after counting the radioactivity in the present instance. The dose rate at this time can be calculated as in Problem 10 after calculating the activity present at the midpoint. In this example the activity present three days after counting is 1.296 mCi, and this leads to a value of 9073 rads for the absorbed dose, in very good agreement with the value calculated using the integrated expression. However, when the duration of the experiment is longer, relative to the half-life of the isotope being used, the agreement will be poorer and the integrated equation should be used.)

For the ferrous sulfate solution the absorbed dose is given by

$$D_M = \frac{9.647 \times 10^8 \, \Delta A}{\Delta \varepsilon l p G(Fe^{3+})} \text{ rads} \tag{3.31}$$

Since the absorbed dose is known, G(ferric) can be found by substitution;

$$G(Fe^{3+}) = \frac{9.647 \times 10^8 (0.324 - 0.003)}{9079 \times 2205 \times 1 \times 1.024}$$

and

$$G(Fe^{3+}) = 15.1 \, (for \ phosphorus\text{-}32 \ \beta\text{-}rays)$$

If the sample were left until all the phosphorus had decayed, the total absorbed dose would be

$$D_M = \frac{5.121 \times 10^4 \times \bar{E} \times C_0}{\lambda} \tag{3.26}$$

$$= \frac{5.121 \times 10^4 \times 0.70 \times 0.1329}{0.04873}$$

$$= 9.78 \times 10^4 \ rads$$

Although equation 3.31 calls for $\Delta \varepsilon$ (i.e. the difference between the molar extinction coefficients for ferric and ferrous ion at 304 nm), the value for ferrous ion is small and can be neglected without introducing a significant error; for ferric and ferrous ion, $\varepsilon_{304} = 2205$ and $1 \ M^{-1} \ cm^{-1}$, respectively.

Problem 12

A sample of Fricke dosimeter solution was irradiated for 20 min with cobalt-60 γ-rays and then the absorbancies of the irradiated and a non-irradiated solution measured; they were found to be 0.254 and 0.002 respectively (304 nm, 24°C). A solution of ceric sulfate in 0.4 M sulfuric acid was irradiated in the same position for 100 min, causing the absorbance at 320 nm (the ceric absorption peak) to change from 0.695 to 0.214. Assuming that the same silica cells were used for all the optical density measurements, calculate $G(-\text{ceric ion})$.

Assume G(ferric) for cobalt-60 γ-rays to be 15.5 and the extinction coefficients for ferric ion at 304 nm (24°C) and ceric ion at 320 nm to be 2205 and 5610 liters mole^{-1} cm^{-1} respectively.

The absorbed dose in both ferrous and ceric solutions is given by

$$D_M = \frac{9.647 \times 10^8\, \Delta A}{\Delta \varepsilon l \rho G(Fe^{3+} \text{ or } Ce^{3+})} \tag{3.31}$$

Thus, for the ferrous solution,

$$D_{Fe} = \frac{9.647 \times 10^8(0.254 - 0.002)}{2205 \times l \times \rho \times 15.5} \text{ rads 20-min}^{-1}$$

and, for the ceric solution,

$$D_{Ce} = \frac{9.647 \times 10^8(0.695 - 0.214)}{5610 \times l \times \rho \times G(Ce^{3+})} \text{ rads 100-min}^{-1}$$

The absorbed dose will be the same for both solutions if they are irradiated for the same length of time, since they are both predominantly 0.4 M sulfuric acid, i.e., $5D_{Fe} = D_{Ce}$. Since l and ρ are the same for the two solutions G(cerous), which is equal to $G(-$ceric$)$, can be determined; thus

$$G(-Ce^{4+}) = G(Ce^{3+}) = 2.33 \text{ (for cobalt-60 } \gamma\text{-rays)}$$

Problem 13

A 5-ml sample of pure chloroform was irradiated with cobalt-60 γ-radiation for 10 min. Acid products were extracted by stirring with water and were titrated with dilute alkali; 15 μequiv (15 $\times$ 10^{-6} mole monobasic acid) of acid were found. The dose rate was determined with the Fricke dosimeter; 60 min irradiation causing the absorbance of the solution to change from 0.003 to 0.341 (304 nm, 24°C) when measured in a 1-cm cell. Calculate the G value for the production of monobasic acid from chloroform.

The density and $\overline{Z/A}$ of the Fricke dosimeter solution are 1.024 and 0.553 respectively; G(ferric) for cobalt-60 γ-radiation is 15.5. For ferric ion the molar extinction coefficient at 304 nm (24°C) is 2205. The density and $\overline{Z/A}$ of chloroform are 1.50 and 0.486 respectively.

The dose absorbed by the Fricke dosimeter solution is given by

$$D_D = 2.76 \times 10^4\, \Delta A \text{ rads} \tag{3.33}$$

when $\varepsilon = 2205$, $\rho = 1.024$, $l = 1$, and $G(Fe^{3+}) = 15.5$. Thus in the example above,

$$D_D = 2.76 \times 10^4 \times (0.341 - 0.003) \text{ rads 60-min}^{-1}$$
$$= 1555 \text{ rads 10-min}^{-1}$$

Energy absorption in both the dosimeter solution and in chloroform is predominantly by the Compton process with cobalt-60 γ-rays and the absorbed dose in the two media are related by

$$D_{\mathrm{CHCl_3}} = D_{\mathrm{dosimeter}} \times \frac{(\overline{Z/A})_{\mathrm{CHCl_3}}}{(\overline{Z/A})_{\mathrm{dosimeter}}} \tag{3.38}$$

$$= 1555 \times (0.486/0.553)$$

$$= 1366 \text{ rads } 10\text{-min}^{-1}$$

The number of molecules of acid formed

$$= 15 \times 10^{-6} \times 6.022 \times 10^{23} \text{ molecules 5-ml}^{-1}$$

or

$$= \frac{15 \times 6.022 \times 10^{17}}{5 \times 1.50} \text{ molecules g}^{-1}$$

$$= 1.205 \times 10^{18} \text{ molecules g}^{-1}$$

while

$$G(\mathrm{acid}) = \frac{\text{molecules of acid formed}}{100 \text{ eV absorbed}}$$

$$= \frac{\text{molecules of acid formed g}^{-1}}{\text{absorbed dose (rad)}} \times 1.602 \times 10^{-12}$$

$$= \frac{1.205 \times 10^{18} \times 1.602 \times 10^{-12}}{1366}$$

$$= \mathit{1415}$$

An alternative, though similar, method of calculating the G value is as follows. It is based on energy units of eV ml^{-1} rather than rads. The dose absorbed by the dosimeter solution is then

$$(D_v)_D = 1.76 \times 10^{18} \Delta A \text{ eV ml}^{-1} \tag{3.35}$$

$$= 1.76 \times 10^{18} \times 0.338 \text{ eV ml}^{-1} \text{ in 60 min}$$

$$= 4.957 \times 10^{17} \text{ eV 5-ml}^{-1} \text{ in 10 min}$$

This is converted to the dose absorbed by 5 ml of chloroform in a 10-min irradiation by

$$(D_v)_{\mathrm{CHCl_3}} = (D_v)_{\mathrm{dosimeter}} \times \frac{(\overline{Z/A})_{\mathrm{CHCl_3}}}{(\overline{Z/A})_{\mathrm{dosimeter}}} \times \frac{\rho_{\mathrm{CHCl_3}}}{\rho_{\mathrm{dosimeter}}} \tag{3.40}$$

$$= 4.957 \times 10^{17} \times \frac{0.486}{0.553} \times \frac{1.5}{1.024}$$

$$= 6.382 \times 10^{17} \text{ eV 5-ml}^{-1}$$

The number of molecules of acid formed is, as above,

$$= 15 \times 10^{-6} \times 6.022 \times 10^{23} \text{ molecules 5-ml}^{-1}$$

$$= 9.033 \times 10^{18} \text{ molecules 5-ml}^{-1}$$

and

$$G(\text{acid}) = \frac{\text{molecules of acid formed/5-ml}}{100 \text{ eV absorbed/5-ml}}$$

$$= \frac{9.033 \times 10^{18}}{6.382 \times 10^{15}}$$

$$= 1415$$

Problem 14

Pulse radiolysis of a nitrogen-purged 10^{-4} M perchloric acid solution gave a transient species whose absorption at 700 nm decayed as shown in Fig. 5.10 (page 198). The absorbance of the solution 0.1, 0.2, 0.3, 0.4, 0.6, 0.8, 1.0, 1.2 and 1.5 μsec after midpulse was 0.224, 0.174, 0.136, 0.106, 0.065, 0.040, 0.024, 0.014 and 0.007 respectively. Determine whether first order or second order kinetics best describe the decay and, if the data are sufficient, determine the rate constant for the reaction. Assuming the transient to be the hydrated electron, which reacts with H^+, estimate $k(e_{aq}^- + H^+)$.

To determine whether the transient decay obeys first or second order kinetics, log A and $1/A$ are plotted against time from midpulse. The necessary data are given in Table A.1 and are plotted in Fig. A.1.

TABLE A.1

Time From Midpulse (μsec)	Absorbance (A)	Log A	$1/A$
0.1	0.224	−0.65	4.5
0.2	0.174	−0.76	5.7
0.3	0.136	−0.87	7.4
0.4	0.106	−0.97	9.4
0.6	0.065	−1.19	15.4
0.8	0.040	−1.40	25.0
1.0	0.024	−1.62	41.7
1.2	0.014	−1.85	71.4
1.5	0.007	−2.17	143

The linear plot obtained when log A is plotted against time indicates first order kinetics (second order decay would give a linear plot when $1/A$ is plotted against time) and the first order rate constant is $-2.303 \times$ slope, i.e. 2.5×10^6 sec^{-1}. (Estimation of a second order rate constant requires additional information in the form of an extinction coefficient or the concentration of the transient at some point.)

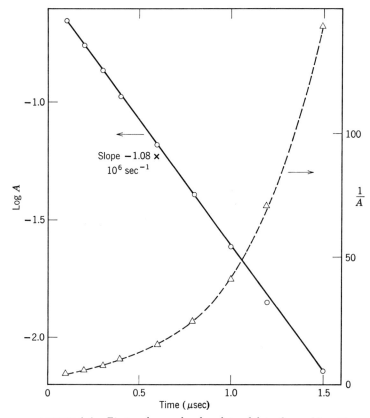

FIGURE A.1   *First and second order plots of data for problem 14.*

If the transient under observation is the hydrated electron, the decay reaction in the relatively acid solution will be

$$e_{aq}^- + H^+ \rightarrow H \tag{1}$$

The decay follows first order kinetics because the concentration of hydrogen ion is many times that of the hydrated electron and therefore changes little during the course of the reaction; the first order rate constant observed is equal to $k(e_{aq}^- + H^+)[H^+]$, and $k(e_{aq}^- + H^+)$ is $2.5 \times 10^6/10^{-4}$ i.e., $2.5 \times 10^{10}\ M^{-1}\ \text{sec}^{-1}$. The currently accepted value for this rate constant is $2.35 \times 10^{10}\ M^{-1}\ \text{sec}^{-1}$.

INDEX

Absorbed dose, calculation of, from absorbed dose in dosimeter, 110-115
definition, 67
effect on product yields, 364, 377
and *G* value, 92, 110
from ionization measurements (x- and γ-radiation), 76-83
measurement of, by calorimetry, 71
by cavity ionization, 80-83
by chemical dosimetry, 92
for charged-particle radiation, 84
for internal radioactive sources, 88-91
for neutrons, 87
partition in mixtures, 115
per unit exposure, 76-80
units, 69
Absorbed dose rate, definition, 68
units, 69
Absorption coefficients, 38-39, 45-49, 53-57, 70
atomic, 39, 47
electronic, 39, 46
energy, 53, 56
interconversion of, 39, 46
linear, 38, 47
mass, 39, 47. *See also* Mass energy absorption coefficient
scatter, 41, 53, 54. *See also* Scattering coefficients

symbols and units, 54
total, 18
true, 53
see also Attenuation coefficients
Absorption cross sections, 39. *See also* Absorption coefficients
Accelerators, 24-29
Acetaldehyde, aqueous, 320
Acetamide, aqueous, 335
Acetate, potassium, 428, 429
Acetic acid, 427-429
aqueous, 315
Acetone, 423-427
aqueous, 320
dose rate and LET effects, 426
photodissociation, 134, 150
Acetonitrile, aqueous, 330
Acetoxy radicals ($CH_3COO\cdot$), 151, 157, 430
N-Acetylalanine, aqueous, 334-335
solid, 335
Acetylene, 4, 237-240
aqueous, 349
benzene mixtures, 240
helium mixtures, 238
propane mixtures, 241
Acetyl radicals ($CH_3CO\cdot$), 150, 157, 425, 430
Acids, carboxylic, 427-430